T0260063

Ecological Mechanics

Ecological Mechanics

Principles of Life's Physical Interactions

Mark Denny

PRINCETON UNIVERSITY PRESS | PRINCETON AND OXFORD

To my mentors:

Floyd W. Denny, Jr.
Stephen A. Wainwright
John M. Gosline
Robert T. Paine

Contents

Part III Solid Mechanics

Part IV Ecological Mechanics

ONLINE RESOURCES

Additional supporting materials can be found at
http://press.princeton.edu/titles/10641.html

List of Illustrations

List of Tables

Acknowledgments

A small army of friends and colleagues provided ideas and critical comments during the long gestation of this text; I thank them all: Marissa Baskett, Andy Biewener, Lisandro Benedetti-Cecchi, Hong En Chew, John Crimaldi, Wes Dowd, Brian Gaylord, John Gosline, Chris Harley, Tom Hata, Brian Helmuth, Luke Hunt, Megan Jensen, Diana LaScala-Gruenewald, Luke Miller, Dianna Padilla, George Somero, Johan van de Koppel, Herb Waite, Dick Zimmerman, and the students of Biology 150/151. Jérôme Casas, Per Jonsson, and Art Woods read the entire manuscript, and their critical feedback helped immeasurably. Danna Staaf's editing of an early draft taught me textual discipline. Special thanks are due to Felicia King, whose tireless efforts in checking the math, questioning the ideas, and polishing the prose were invaluable. It has once again been a pleasure to work with the folks at Princeton University Press: Alison Kalett, who (rightly and thankfully) convinced me that shorter is better; Jenny Wolkowicki, who deftly put the pieces together; Linda Thompson, who vetted the prose; and Dimitri Karetnikov and his crew, who produced the figures. My passion for ecomechanics has been abetted by generous funding from the National Science Foundation and PISCO, a consortium funded by the David and Lucile Packard Foundation and the Gordon and Betty Moore Foundation.

And, as always, I am indebted to my wonderful wife, Susan Denny, for remaining upbeat while cohabiting with yet another book project.

Mark Denny
Hopkins Marine Station of Stanford University
Pacific Grove, California
December, 2014

Chapter 1

Ecological Mechanics
An Introduction

"The key to prediction and understanding lies in the elucidation of mechanisms underlying observed patterns." (Levin 1992)

mech-an-ism, *n.* **1.** an assembly of moving parts performing a complete functional motion, often being part of a large machine; linkage. **2.** the agency or means by which an effect is produced or a purpose is accomplished. (*Random House Dictionary of the English Language, Second Edition*)

This is a book about the mechanisms of biology: an exploration of how organisms function as individuals, how they are linked to each other and to the physical environment, and how these linkages affect the dynamics of populations and ecological communities. What sets this text apart from most other ecological investigations is its emphasis on biomechanics, the branch of physiology that uses the principles and perspectives of physics and engineering to study plants and animals. When properly applied, a biomechanical approach to ecology can provide valuable information unattainable by more traditional means, information we need to understand how organisms, populations, and communities function. My intent in this text is to open your eyes to this new way of thinking. A brief example will give you a feeling for how this approach works and where this book is headed.

1   MECHANICS AND ECOLOGY COMBINED
Coral Reefs

Foundation species (also known as ecosystem engineers) create or modify physical habitats, thereby influencing community structure and dynamics. Beavers are literal engineers—they build dams, which can have important impacts on stream ecology. Trees in rain forests and giant seaweeds in kelp forests are other obvious examples of foundation species. But the poster children of ecosystem engineering are corals.

Tropical coral reefs are perhaps the largest biological construction projects on Earth, hosting a diversity of life that is among the highest on the planet (Figure 1.1A). However, these reefs exist in delicate balance. Corals and macroalgae (seaweeds) compete for light and space, and a small shift in environment or community dynamics could potentially turn the world's coral reefs into algal reefs (Knowlton and Jackson 2001).

Recently, the application of principles drawn from ecology, engineering, and fluid dynamics has produced a major advance in our understanding of how coral reefs interact with their physical environment. Working at Lizard Island (part of the Great Barrier Reef in Australia), Josh Madin and his colleagues used a biomechanical approach to link ocean climate and chemistry to reef diversity and species distributions.

Much of the food available to corals comes from symbiotic microalgae in the coral polyps (not be be confused with the seaweeds that help to form reef structure). These microalgae need to absorb sunlight, which confines corals to shallow water where they are exposed to sufficient illumination, but also subjects them to rapid water velocities as waves pass overhead. The goal of Madin's research was threefold:

- first, to quantify the mechanisms by which this flow environment controls the relative abundance of coral species,
- then, to develop a model using those mechanisms to explain the observed spatial pattern of coral species on the reef,
- and finally, to use that model to predict how reef community dynamics will shift in response to climate change.

First, the mechanics. The ability of a coral colony to survive the hydrodynamic forces imposed by waves depends on several factors: the strength of both the colony's skeleton and the substratum, the size and shape of the colony, and the maximum water velocity encountered. Madin (2005) began his investigation of these factors by quantifying the mechanical attributes of the corals. He collected samples of coral skeleton and the substratum to which they were anchored, and by pulling on them with a materials testing machine he measured the strength of each. He then used beam theory (a branch of engineering) to calculate how much hydrodynamic force it would take to dislodge coral colonies of different sizes and shapes. He found that the strength of the substratum was the limiting factor, and that tablelike corals such as *Acropora hyacinthus*, which have relatively small bases, break under smaller loads than do blocky corals such as *Acropora palifera*.

The next step was to estimate the minimum water velocity necessary to impose these breaking forces. This was accomplished using a well-known result from fluid dynamics: force is proportional to the area of coral exposed to flow and to the square of water velocity.

At this point, all that Madin needed to predict which species would be living where on the reef was a record of the water velocities corals had experienced while growing to their present size. It would have been handy if someone had recorded velocities on the reef for the decades its took the corals to mature, but records of this sort are exceedingly rare, and none were available for Lizard Island. Instead, Madin et al. (2006) used their knowledge of physical oceanography to gather the required data.

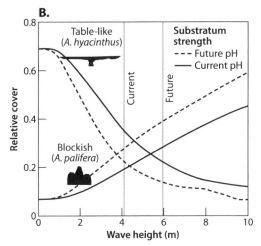

Figure 1.1 A. Coral reefs are home to an extraordinary diversity of life forms (www.noaa.gov/features/climate/images/fig1_reef_fullsize.jpg). **B.** Relative cover of species changes with wave height and ocean pH. Currently, tablelike *Acropora hyacinthus* is more abundant than blockish *A. palifora*, but a reduction in pH or an increase in wave height could reverse the order. Redrawn from J. S. Madin, M. J. O'Donnell, and S. R. Connolly. 2008. "Climate mediated changes to post-disturbance coral assemblages." *Biol. Lett.* 4:490–93, by permission of the Royal Society.

They first obtained a 37-year record of wind speeds near Lizard Island (unlike records of water velocity, wind-speed records are common) and used the theory of wave generation to translate the wind's history into a a 37-y record of wave heights. To check the accuracy of their results, they videotaped waves during known wind speeds and found that the theory was remarkably accurate.

They then used wave mechanics (a branch of fluid dynamics) to predict how wave height evolves as waves move onto the reef and what velocities would consequently be imposed on corals. The net result was a 37-y record—a hindcast—of the flow experienced by corals at every point in the habitat. Again, they checked their results with measurements made on the reef.

With these data in the bag, Madin and Connolly (2006) had the raw ingredients for a mathematical model of the life and death of corals. Knowing how water velocity varied through time at every point on the reef, they could predict how big each of the 1158 colonies they studied could grow without being broken. Their predicted size limits matched those observed on the reef, suggesting that their model of coral survival—from wind to waves to the mechanics of individual coral heads—was a valid description of reality. Unlike previous work, which had noted *correlations* between water motion and coral size (e.g., Done 1983), Madin and Connolly's model provided a mechanistic, quantitative *explanation* of the observed distribution.

Madin et al. (2008) then looked to the future and asked how climate change might affect the reef. In particular, they focused on two aspects of the environment that are likely to shift dramatically in the next few decades. First, climate models predict that the frequency and intensity of typhoons will increase at Lizard Island (Elsner et al. 2008; Young et al. 2011), leading to an increase in the forces imposed on corals. Second, due to the absorption of carbon dioxide by seawater, ocean acidity is predicted to intensify on the Great Barrier Reef (Kleypas et al. 1999, Hoegh-Guldberg et al. 2007), resulting in a decrease in the strength of the reef's carbonate substratum. Considering these effects in tandem, Madin et al. (2008) predicted that if, in a more acidic future ocean, a typhoon produces waves 6 m high (compared to the average maximum of 4.2 m over the 37 y of their hindcast), the fraction of space occupied by tablelike species will drastically decrease (from almost 40% to near 10%), while the abundance of blockish colonies will increase (Figure 1.1B). Note that the shift in species abundance is amplified by the interaction of waves and acidity: each factor on its own would have less drastic consequences.

This model and its predictions are an important advance in coral-reef ecology because they provide a robust basis for predicting the consequences of climate change—not just for corals, but for the whole reef community. For instance, the density of herbivorous fish on reefs is tied to the abundance of branching and tablelike corals, the fishes' preferred habitat (Luckhurst and Luckhurst 1978; Knowlton and Jackson 2001; Almany 2004). The predicted decrease in cover of these delicate corals at Lizard Island may, therefore, cause a decrease in herbivory, shifting the interaction between corals and macroalgae (Mumby 2006) and potentially pushing the reef to the point where seaweeds dominate.

Madin's studies convey the spirit of this text's approach. Without a thorough understanding of the mechanisms linking wind to waves to force and ultimately to breakage and climate change, it would be impossible to accurately model the dynamics of coral species' distribution. Similarly, without ecologists' understanding of fish behavior and natural history, it would be impossible to extrapolate from information regarding species distributions to predict the ensuing effects of climate change on community dynamics. Individually, biomechanical and ecological approaches are incomplete, but working together they provide valuable tools for predicting the future.

2   RESPONSE FUNCTIONS

As illustrated by this example, my strategy in this text is to explore *response functions*, the relationships between the conditions imposed on a system and how well the system performs. More specifically, response functions are mathematical descriptions of the cause-and-effect linkages between an input variable (such as wave height) and an output variable (such as hydrodynamic force). *Measured response functions* describe existing data without reference to underlying mechanism; *mechanistic response functions* use established principles of physics, physiology, and behavior to model reality. Mechanistic response functions have two important advantages: they allow for accurate extrapolation beyond measured data, and we can judge their validity by comparing their predictions to empirical observations.

An example again helps to explain what I mean. Inspired by Madin's work, a researcher might propose to study the dynamics of sea urchins foraging on seaweeds. At the study site, when hydrodynamic forces are small, urchins are free to roam, and their foraging can decimate the local seaweeds. By contrast, if forces are sufficient, urchins are dislodged or confined to their burrows, and seaweeds can flourish. As a first step toward investigating this process, our researcher might use a flow meter to measure the relationship between wave height and the ensuing hydrodynamic forces imposed on urchins (the dots in Figure 1.2A). Note that for practical reasons the researcher is able to make measurements only when wave conditions are relatively benign. This relationship is a measured (empirical) response function.

Phenomenological descriptions of this sort are valuable, but my desire in this text is to take the next step: to investigate not only what happens, but also *why* it happens. Knowing why can be important. For instance, if our researcher were naively to extrapolate from the empirical information in hand (Figure 1.2A), he or she would predict that waves of a height encountered in storms could impose forces on urchins large enough to disrupt their feeding. According to this prediction, any future increase in ocean storminess would increasingly restrict urchin foraging and favor seaweed dominance.

However, before relying on this extrapolation and its ecological implications, it would be wise for our researcher to investigate the fluid dynamics of waves to understand the mechanism behind his or her measurements. In this case, theory would reveal that, at the relatively low wave heights of the measured data, waves are intact as they pass over the urchins, and force should indeed increase in direct proportion to wave height. This match between theory and measurement serves as a check that the theory is correct. But theory also predicts a limit to this linear relationship—in the relatively shallow water of the field site, high waves become unstable and break before reaching the urchins, dissipating much of their energy and reducing their height (Helmuth and Denny 2003). This understanding of mechanism—the why of the process—allows the researcher to formulate a mechanistic response function, one that predicts that force will plateau before reaching levels that would bother urchins (Figure 1.2B). Because it is based on mechanics in addition to observations, this response function provides increased accuracy when extrapolating beyond the measured data. In this case, the researcher would predict that increasing storm

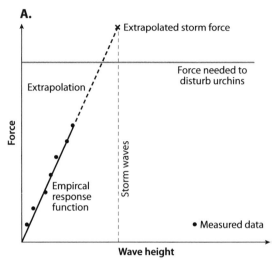

Figure 1.2 A. An empirical response function (solid line) fits the measured data (the dots), and a linear extrapolation suggests that storm waves will dislodge urchins. **B.** By contrast, a mechanistic response function also fits the data, but its more accurate extrapolation suggests that urchins can weather the storms.

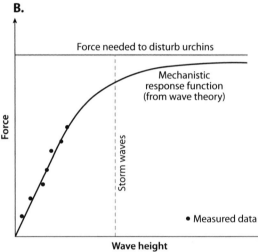

intensity would have negligible effect on urchin grazing. As a result, he or she should propose that, even if the ocean gets wavier, seaweed abundance won't increase.

Response functions (both measured and mechanistic) are commonplace in many fields of biology. Biomechanics (the field in which I was raised) revels in its ability to use Newtonian physics—an extraordinarily useful set of response functions—to explain how individual plants and animals work. Ecological physiology investigates the functional relationships between individual organisms and the physical environment. And in ecology, response functions are used to specify, for instance, the rate of prey capture as a function of prey density.

Despite the ubiquity of reponse functions, biologists are seldom taught how to recognize, classify, and use them. It doesn't help that different fields use different synonyms for what I call response functions. In physiology, for instance, measured response functions are commonly referred to as reaction norms. In ecology, predator-prey response functions are known as the functional response. In physics and engineering, some mechanistic response functions are so well established they are called laws. This text is intended, in part, to push through the barriers of jargon. By raising the awareness of response functions in general, I hope to impress on you how the basic concepts of response functions can unify different perspectives on biology, and to demonstrate that an understanding of mechanistic response functions can be used to great advantage in ecology.

3   TRANSPORT PHENOMENA

Many of the response functions we will encounter in this text involve transport phenomena, the movement of heat, mass, and momentum. Consider, for example, body temperature in the sea star, *Pisaster ochraceus. Pisaster* is the principle predator of mussels on the west coast of North America, acting as a keystone species (Paine 1966). As *Pisaster* consumes mussels (the dominant competitor for space on rocky shores), it opens up space for less capable competitors, thereby increasing species diversity. But, the rate at which *Pisaster* feeds depends on its body temperature. An increase of only 4°C (from 9.5°C to 13.5°C) doubles the rate at which *Pisaster* consumes mussels (Sanford 2002), so even a slight increase in temperature could have a substantial effect on mussel-bed community structure.

Engineering theory (which we will explore in Chapters 11 and 12) tells us that *Pisaster*'s body temperature depends on the rate at which heat is transported into the animal (from the absorption of sunlight, for instance) relative to the rate at which it is transported out (by evaporation, convection, or conduction). Thus, if we desire to predict *Pisaster*'s body temperature—and, thereby, its effect on community dynamics—we would be well advised to begin by studying the transport of heat.

Similar advice applies to the transport of mass. A plant's growth rate often depends on the rate at which it can acquire carbon dioxide and nutrients, forms of mass that must be transported to the organism. As we will see, transport of mass depends on the pattern of flow adjacent to a plant—laminar or turbulent—and that pattern depends on the speed of the medium—water or air—and the shape and size of the plant. Thus, if we want to predict the growth rate of a kelp or cactus, we need to understand how fluid flow governs the transport of mass.

The relevance of momentum transport is perhaps less immediately obvious, but we will find in Chapter 2 that the application of a force requires the transfer of momentum. When waves impose hydrodynamic forces on corals, for example, they do so because some of the moving fluid's momentum is, in effect, absorbed by the animals. Indeed, the transport of momentum is an underlying theme in all Newtonian mechanics.

Clearly it is valuable to understand the transport of heat, mass, and momentum, and each process is important as a separate phenomenon. But engineers have developed a theoretical perspective that can compound their utility. Because the movements of heat, mass, and momentum are all transport phenomena, knowledge about one can inform the others. For example, we will see in Chapter 11 that it is easy to measure the rate at which heat is transported out of a plant or animal by flowing fluid—in a nutshell, you heat the organism and measure how quickly it cools. By contrast, it can be much more difficult to directly measure the rate at which mass is transported to or from the same organism. For instance, the ability of a male moth to find a mate depends on the delivery of picograms of pheromone to its antennae. Directly measuring the delivery of such minuscule mass would be extremely difficult. But because the mechanism of mass transfer is similar to that of heat, information about the flow of heat from the antennae—which is relatively easy to obtain—can be used to estimate the rate of mass exchange. Similarly, measurements of the temperature of a leaf can be used to estimate the rate at which it takes up carbon dioxide and loses water.

In short, by recognizing that fluxes of heat, mass, and momentum are all transport phenomena, we set the stage for an extraordinarily productive synthesis of ideas.

4   WHAT'S IN A NAME?

The preceding discussion outlines the perspective of this text—with the goal of informing ecology, we will use response functions (mechanistic where possible, measured when necessary) to explain how plants and animals work and interact. To streamline the prose, we need a name for this approach. Several possibilities come to mind. None is entirely satisfactory, but the phrase *ecological mechanics—ecomechanics* for short—seems to be the best of the lot,[1] and the term has a distinguished heritage; it was first used by Wainwright et al. (1976) in their classic treatise on mechanical design in organisms. Taoist philosophy warns, however: "all that is dark derives from the labeling of things" (Mitchell 1998), so it is best if I clarify my intent in choosing this label.

By using ecomechanics to describe this text's perspective, I do not mean to imply that this approach is absent from current-day research. As noted previously, mechanistic approaches are in fact widely used: functional ecology, biophysical ecology, bioenergetics, ecological physiology, biomechanics, and materials science (among others) all view the study of biological mechanisms as their central modus operandi in much the fashion I propose here.

In each case, however, the focus and range of these mechanistic approaches are restricted compared to my vision for ecomechanics. Biomechanics and functional ecology, for instance, have historically concentrated their efforts at the level of the organism rather than extending their mechanistic approach to higher levels of organization. Biophysical ecology and bioenergetics emphasize the transfer of heat and mass between organisms and their environment but de-emphasize the role of momentum transfer, that is, of forces. Ecological physiology focuses on the processes through which organisms survive but, other than at the cellular level, generally neglects the structures and materials that make those processes possible. And materials scientists seldom consider the physiological and ecological implications of their findings.

In proposing the term ecomechanics, my intent is to emphasize the commonality of research philosophy that unites these fields. I hope that this text will foster cooperation among disciplines and across levels of organization, from molecules to ecosystems.

5   WHAT'S LEFT OUT

To keep this text from becoming unwieldy, I have had to make tough choices as to what to include and what to leave out. Two criteria guided these decisions. First, the emphasis here is on the physical interactions among organisms and between organisms and their environment. Consequently, even though they clearly fall within ecomechanics' purview, those aspects of physiology that deal with organisms' internal working receive short shrift. Second, I have minimized discussion of ecomechanical

---

[1] To some, mechanics is narrowly defined as the study of forces and the motions they cause. Because the purview of ecological mechanics, as described here, includes many aspects of biology that do not involve forces, inclusion of the term *mechanics* is thus deemed inappropriate. However, this definition of mechanics is unnecessarily narrow. For example, the *Random House Dictionary* defines mechanics as the "branch of physics that deals with the action of forces on bodies and with motion, comprised of kinetics, statics, and kinematics." The phrasing of this definition and the inclusion of kinematics (the study of motion without regard to forces) implies that force is not essential to mechanics. As long as one takes an inclusive view of things in motion (e.g., heat, mass, momentum, energy, information), the appearance of mechanics in ecological mechanics and ecomechanics should cause no more of a problem than it does in the term quantum mechanics.

topics that are well covered in other texts. In particular, I have omitted coverage of sound and hearing (whose physics are explained by Fletcher, 1992) and light and optics (well explained by Johnsen, 2012) and have included only a brief overview of low-Reynolds-number fluid dynamics, a topic admirably covered by Kiørboe (2008) and Dusenbery (2009). As partial compensation for the remaining omissions, I have included in the appropriate sections directions to useful sources of information on the many worthy topics I have been forced to exclude.

6 A ROAD MAP

Our exploration of ecological mechanics is divided into four parts. This chapter and the two that follow (Part I) set the stage by providing an introduction to the ecomechanical perspective, basic principles of Newtonian physics, and the concept of response functions.

Part II develops the theme of heat, mass, and momentum transport, with chapters on diffusion, fluid mechanics, boundary layers, hydrodynamic forces, locomotion, and thermal biology.

Next, Part III deals with the structural response of organisms to the forces imposed on them—how organisms bend, twist, and break.

Lastly Part IV synthesizes information from the previous sections in an ecological context. We examine the role of variation in response functions, and how variation depends on the scale of measurement. We delve into the statistics of extremes, and explore how mechanistic response functions can be used to predict the patterns in which organisms assemble.

7 ONLINE SUPPLEMENTS

In my quest to keep the message simple, I have been forced to leave out many of the intriguing (but complex) details and all but a few of the derivations of equations. Fortunately, the Internet provides a means for those details and derivations to be available to the interested reader. Throughout the text, I note where further discussion of a topic can be found in an online supplement. (Supplements—along with problem sets, additions, and corrections—can be accessed through the book's Web page within the Princeton University Press site, http://press.princeton.edu/titles/10641.html.) Some of these supplements are quite short—the proof of an assertion or a brief annecdote— but others are extensive, amounting to whole chapters on subjects that could not be included in the text. I have also included problem sets so that you can test your understanding.

8 A NOTE ON NOTATION

In a subject this broad, finding unique symbols for variables is well nigh impossible. Due to the limited number of Greek and Roman letters, I have been forced to use some symbols more than once, but I have endeavored to minimize the repetition. To help you keep symbols straight, there is an index to the page where each is defined. The letter k is a special case—I employ it as utility variable, defined in the context of each local discussion. So be warned, unlike other symbols, k's identity changes frequently.

Now, on with the show.

Part I

BASIC CONCEPTS

Chapter 2

Basic Physics and Math

We begin our formal exploration of ecomechanics with a review of basic physics—so-called first principles that are valid everywhere, all the time, and on which the rest of this text is built. In the process, we review some of the basic mathematics that will be useful in our quest.

1 LOCATION

Our first task is to provide a means to locate objects in space. Any of a variety of coordinate systems could be used, but for our purposes the standard Cartesian system of three mutually perpendicular axes suffices (Figure 2.1A) . We orient the x- and y-axes horizontally and align them with directions most relevant to the subject at hand. The z-axis is vertical. Given this orientation, it remains only to specify the system's origin. This we do on a case-by-case basis, choosing the location most convenient for each particular application. Regardless of where we locate it, we assume that the origin is stationary relative to Earth's surface.

2 DIMENSIONS AND UNITS OF MEASURE

Having provided means for locating objects, we next desire to describe their physical character. There are three basic attributes of the physical universe from which all other characteristics of importance to ecomechanics can be measured—*time*, *length*, and *mass*.

Time, as we will see, is in some respects the most basic dimension. The Système International (SI) unit of time is the *second*, symbolized by s. The second was originally measured relative to Earth's rotation: 1/86,400 the length of an average day. Unfortunately, the length of an average day is slowly increasing, making Earth a lousy clock. In 1961, the second was redefined as the time encompassed by 9,192,631,770 oscillations of a specific wavelength of light emitted by a cesium-133 atom. Atomic clocks use this concept to count time with exquisite accuracy, varying by less than 10^{-9} s per day.

The SI unit for length is the *meter*, m. Initially defined in the late 1700s as one ten-millionth the distance from the North Pole to the equator along a meridian passing

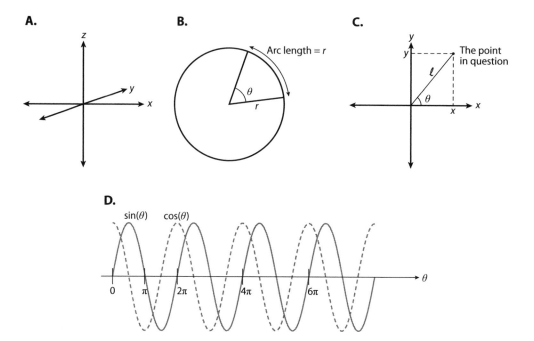

Figure 2.1 A. The Cartesian coordinate system. **B.** Angle (in radians) is arc length divided by radius. **C.** A point in the (x, y)-plane can be specified by either its angle relative to the x-axis (θ) and its distance from the origin (ℓ) or its (x, y)-coordinates. **D.** Sine and cosine waves are periodic with a period of 2π.

through Paris, the meter has been redefined several times since. Most recently (in 1983) its length was set relative to the speed of light: 1 m is the distance traveled by light in a vacuum in exactly 1/299,792,458 s. Accurate measurement of length thus depends on accurate measurement of time. The standard of length can be used to quantify area and volume; area has units of square meters (m²) and volume, cubic meters (m³).

The measurement of volume in turn allows us to measure mass. The SI unit of mass is the *kilogram* (kg), originally defined in theory as the mass of 1 cubic decimeter (0.001 m³) of pure water at its temperature of maximum density. In practice, the SI standard of mass is a block of platinum-iridium alloy residing at the International Bureau of Weights and Measures in Sèvres, France. Periodically, the standard kilogram is removed from its vault and used to verify the weight of copies, which then serve as standards for practical use.

This system worked well until, at the latest verification in 1988, it was discovered that the standard kilogram had lost mass relative to its primary copies. To circumvent the problems ensuing from such a variable standard, there is a movement afoot to replace the standard kilogram with a more universal standard based on counting the exact number of atoms in a mole. For example, a mole of silicon-28 atoms has, by definition, a mass of 0.028 kg. Thus, if one knew exactly how many atoms there were in a mole, one could count out that many atoms of silicon-28 and know that the resulting mass was exactly 28 g. Unfortunately, current estimates of the number of atoms in a mole (Avagadro's number) have an uncertainty of 30 parts per billion, slightly more than the uncertainty of 20 parts per billion associated with weighing the platinum-iridium standard.

Table 2.1 Factors for Conversion from Imperial to SI Units.

To Convert from	To	Multiply by
Feet	meters	0.3048
Yards	meters	0.9144
Miles	kilometers	1.60934
Miles per hour	meters per second	0.44704
Pounds	kilograms	0.45359
Pounds per square inch	pascals	6894.76
Kilocalories	joules	4190.02
Horsepower	watts	745.700
Degrees	radians	0.01745

Together, mass and volume define an object's density:

$$\text{density}, \rho = \frac{\text{mass}}{\text{volume}}. \tag{2.1}$$

Density has units of $\text{kg} \cdot \text{m}^{-3}$.

3   CONVERTING UNITS

Although SI units are the standard for this text, they are unfortunately not the standard for everyday use in the United States. To facilitate matters, Table 2.1 provides the factors through which a variety of non-SI units can be converted.

A word of advice: your scientific life will be simpler and more pleasant if at the very beginning of any analysis, you convert all measurements to standard SI units. The utility of this advice was brought home to the scientific community in spectacular and tragic fashion in September 1999, when Mars Climate Orbiter (an exploratory space craft) broke up on its approach to the red planet after its thrusters fired with inappropriate force. The software controlling the thrusters expected input in SI units (newtons, defined later), but the navigational software provided data in imperial units (pounds force), and the mismatch doomed the $125,000,000 project.

4   TRIGONOMETRY

In addition to the dimensions of time, length, and mass, it is necessary to include in our list of basic physical attributes a measure of angle, and for most purposes it is convenient to use *radians* (rad) rather than the more familiar degrees. Consider a circle with radius r on which I have measured a length along its circumference (an arc length) equal to its radius (Figure 2.1B). By definition, 1 rad is the central angle subtended by this arc. More generally,

$$\text{radians}, \theta = \frac{\text{arc length}}{\text{radius}}. \tag{2.2}$$

Because radians are a ratio of two lengths, they are themselves dimensionless.

While we are on the subject of angles, let's review some basic trigonometry. Consider an arbitrary point and its location relative to a set of Cartesian axes (Figure 2.1C). We can quantify the point's location in two ways. In one, we draw a line from the point to the origin. The length of this line (ℓ) and its angle relative to the x-axis (θ) specify the point's location. Alternatively, we can locate the point by specifying its x and y values. These methods are connected via the sine and cosine functions. By definition,

$$\sin \theta = \frac{y}{\ell}. \tag{2.3}$$

Thus,

$$y = \ell \sin \theta. \tag{2.4}$$

Similarly,

$$\cos \theta = \frac{x}{\ell}, \tag{2.5}$$

so,

$$x = \ell \cos \theta. \tag{2.6}$$

Given θ and ℓ, we can thus specify y and x.

Sines and cosines are *periodic* functions—they repeat themsleves at regular intervals (Figure 2.1D). For example, when $\theta = 0$, $\sin \theta = 0$, but because there are 2π rad in a circle, $\sin \theta$ is also zero when $\theta = 2\pi, 4\pi, 6\pi$, and so on. Similarly, when $\theta = 0$, $\cos \theta = 1$, but $\cos \theta$ is also 1 when $\theta = 2\pi, 4\pi, 6\pi$, and so forth.

5   TRANSLATION

Having settled on a coordinate system and standards for time, length, and angle, we have the ingredients with which to quantify motion. Motion can consist of movement from one location to another (*translation*), spinning in place (*rotation*), or a combination of the two. Let's begin with translation.

Suppose we take a particular rock on the forest floor as the origin of a Cartesian coordinate system with the x-axis extending north-south, the y-axis, east-west, and the z-axis, up-down. At noon (time t_1), we use a tape measure to ascertain that a particular sloth is browsing in a tree at location x_1, y_1, z_1 relative to this origin (Figure 2.2). Twenty-four hours later (time t_2), we locate the sloth again and note that it has moved to x_2, y_2, z_2. During the interval $\Delta t = t_2 - t_1$, the animal has changed its location relative to the axes of our coordinate system—translated—by the amounts

$$\Delta x = x_2 - x_1, \tag{2.7}$$
$$\Delta y = y_2 - y_1, \tag{2.8}$$
$$\Delta z = z_2 - z_1. \tag{2.9}$$

We can then use the Pythagorean theorem to quantify the total translational distance $\Delta \ell$ the sloth has been displaced:

$$\Delta \ell = \sqrt{\Delta x^2 + \Delta y^2 + \Delta z^2}. \tag{2.10}$$

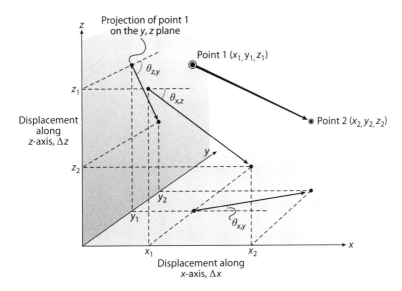

Figure 2.2 The location of a sloth at times t_1 (point 1) and t_2 (point 2). Projections of point 1 and point 2 on the (x, y)-, (x, z)-, and (y, z)-planes allow one to measure θ_{xy}, θ_{xz}, and θ_{yz}, respectively. (See text.)

From our measurements, we could also quantify the direction in which the sloth moved, specified by the angles θ_{xy}, θ_{yz}, and θ_{xz} (see Figure 2.2).

Because translation has both magnitude and direction, it is a *vector*, in contrast to *scalars* (such as mass), which have only magnitude. If this were a physics text, we would deal with translation as the vector it is, but for our purposes it is convenient to deal with distance and direction separately, and most of the time we can avoid dealing with direction althogether by appropriately orienting the axes of our coordinate system. When directional information is necessary, we express it by specifying the components of translation as they project on the axes of our coordinate system.

6   TRANSLATIONAL VELOCITY

How fast was the sloth traveling between our observations just given? The ratio of distance traveled to time of travel is the average translational speed $\bar{s}$ (here and throughout the rest of the text, an overbar on a variable indicates the average):

$$\bar{s} = \frac{\Delta \ell}{\Delta t}, \tag{2.11}$$

a relationship shown in Figure 2.3A. Note the term *average* in this definition. During the night (when darkness prevents observation) our sloth might well have perambulated through a circuitous path, traveling a distance much greater than $\Delta \ell$. Given our two observations, however, our best estimate of the animal's speed is the average just calculated. If we desire a more precise estimate, we must make measurements at shorter intervals. The shorter the interval, the more exactly our measure of average speed resembles the sloth's *instantaneous speed*, s. Taken to the

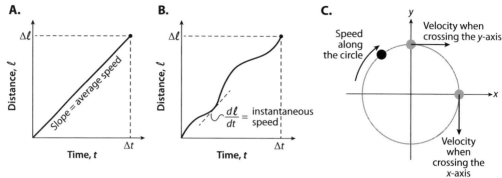

Figure 2.3 A. Average speed is displacement per time. **B.** Instantaneous speed is the slope of the displacement-time curve at a point. **C.** As an object moves in a circular path, speed is constant but direction changes.

limit, as Δt approaches zero,

$$s = \lim_{\Delta t \to 0} \frac{\Delta \ell}{\Delta t} = \frac{d\ell}{dt}. \tag{2.12}$$

In other words, instantaneous translational speed is the first derivative of displacement as a function of time, the slope of a graph of ℓ versus t (Figure 2.3B). Translational speed has SI units of meters per second $(m \cdot s^{-1})$.

The combination of speed and direction is the vector that defines *translational velocity*, with components in the x-, y-, and z-directions:

$$u = \frac{dx}{dt}, \tag{2.13}$$

$$v = \frac{dy}{dt}, \tag{2.14}$$

$$w = \frac{dz}{dt}. \tag{2.15}$$

The overall magnitude of velocity (i.e., speed) is

$$s = \sqrt{u^2 + v^2 + w^2}. \tag{2.16}$$

7 TRANSLATIONAL ACCELERATION

Just as displacement can change through time, so can velocity. The components of the *translational acceleration* vector are

$$a_x = \frac{du}{dt} = \frac{d^2x}{dt^2}, \tag{2.17}$$

$$a_y = \frac{dv}{dt} = \frac{d^2y}{dt^2}, \tag{2.18}$$

$$a_z = \frac{dw}{dt} = \frac{d^2z}{dt^2}, \qquad (2.19)$$

and the magnitude of translational acceleration is

$$a = \sqrt{a_x^2 + a_y^2 + a_z^2}. \qquad (2.20)$$

If velocity increases, the direction of acceleration is the same as that of velocity. However, if velocity decreases, acceleration acts in a direction opposite that of velocity. Translational acceleration has units of meters per squared seconds ($m \cdot s^{-2}$).

It is important to note that because velocity is a vector (having both magnitude and direction), an object can accelerate without changing its speed. Consider, for instance, an object moving clockwise along a circular path at a constant speed (Figure 2.3C). At the top of the circle, the object has maximal speed in the positive x-direction, but zero speed along the y-axis. A short time later, at the right side of the circle, it has zero speed along the x-axis and maximum speed in the negative y-direction. In short, even though the object moves at a constant overall speed, its speed along both the x- and y-axes changes, so according to equations 2.17 and 2.18, the object is accelerating. (For a more complete explanation of this phenomenon, see Supplement 2.1.)

8   ROTATION

To this point, we have considered only motions in which an object shifts from one location to another. Objects can also move by rotating—changing their orientation without changing their position. Earth rotates about an axis passing through its poles, the hands on a clock rotate around the center of the clock's face, and figure skaters are renowned for rotating in place on the tip of a skate blade.

To quantify rotation, we must first specify the axis about which an object rotates (e.g., Figure 2.4); rotation is measured in a plane perpendicular to this axis. We next pick a starting orientation for measuring the angle through which the object has rotated. As with the origin for a Cartesian coordinate system, we are free to pick any convenient orientation to specify $\theta = 0$. Having specified the axis of rotation and initial orientation, we can then quantify any change in θ through time.

If at time t_1 an object is oriented at angle θ_1 and at time t_2 it lies at angle θ_2, its *average angular velocity*, $\bar{\omega}$, is

$$\bar{\omega} = \frac{\theta_2 - \theta_1}{t_2 - t_1} = \frac{\Delta\theta}{\Delta t}. \qquad (2.21)$$

Taken to the limit as Δt approaches zero, *instantaneous angular velocity* ω is

$$\omega = \lim_{\Delta t \to 0} \frac{\Delta\theta}{\Delta t} = \frac{d\theta}{dt}. \qquad (2.22)$$

Similarly, the rate of change of angular velocity is *angular acceleration*, a_θ:

$$a_\theta = \frac{d\omega}{dt} = \frac{d^2\theta}{dt^2}. \qquad (2.23)$$

Because radians are dimensionless, angular velocity has units of s^{-1} and angular acceleration, s^{-2}.

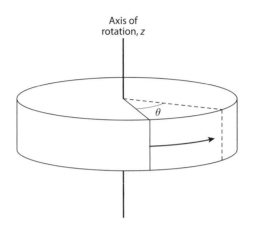

Figure 2.4 Rotation is measured relative to a defined axis. Here, rotation in the (x, y)-plane is measured relative to the z-axis.

Objects can translate and rotate at the same time. One can keep track of such combined motion by first specifying how the axis of rotation translates, and then by specifying how the object rotates around that translating axis.

9   FORCE AND TRANSLATIONAL ACCELERATION

In 1687, Isaac Newton (1643–1727) published his *Principia Mathematica*, perhaps the greatest single document in scientific history. In it, he outlined the basic laws that govern the motion of macroscopic objects and in the process invented calculus.[1] At the core of Newtonian physics are three laws; I present them first as they apply to translation.

1. *An object's translational velocity can change only through the imposition of a net force.*

This means that if an object is initially stationary with respect to our coordinate system, it stays still unless acted upon by a net force.[2] Similarly, if an object has an initial velocity, this velocity—speed, direction, or both—changes only in response to a net force. Thus, Newton's first law expresses the concept of *inertia*, the fact that unless subjected to a force, objects don't change their state of motion.

This law seems at odds with everyday experience. Give a soccer ball some initial velocity across a grassy field and it quickly comes to a standstill. Shove a brick across the floor and it similarly grinds to a halt. These sorts of experiments led Aristotle (384–322 BC) to assume that objects are in their natural state when stationary and that moving objects revert to that state. Part of Newton's genius was to see beyond this seductively simple assumption. Balls and bricks slow down because of the forces imposed by friction, a concept we will return to at the end of this chapter.

2. *The rate of change of translational velocity—that is, acceleration a—is proportional to net force F and inversely proportional to the mass m of an object. The direction of acceleration is the same as that of net force.*

---

[1] Gottfried Liebnitz (1646–1716) independently developed much of the same mathematics, and it is largely Leibnitz's notation that is used today.

[2] The term *net* in this context implies an imbalance in the forces applied to an object. If, for example, an object is subjected to two equal forces, one pulling it to the right, the other to the left, the forces balance each other and no *net* force is applied. Picture a tug-of-war between two evenly matched teams.

This relationship is encapsulated in one of the classic equations of physics:

$$a = \frac{F}{m},$$ (2.24)

or, in revised form,

$$F = ma.$$ (2.25)

It is in this latter arrangement that we will most often encounter Newton's second law, and if you remember only one equation from this text, pick this one. When mass is measured in kilograms and acceleration in meters per second squared, force has the units of *newtons* (N).

A common example provides some tangibility to this relationship. At the surface of the earth, objects are subjected to a gravitational acceleration g (9.8 m $\cdot$ s^{-2}). As a consequence, an object of mass m experiences a downward force known as *weight*:

$$\text{weight, } F_g = mg.$$ (2.26)

For instance, a 70-kg human being has a weight of 686 N.

3. *When two objects push or pull on each other, the force on one object due to the interaction is equal in magnitude but opposite in direction to the force on the other.*

This law codifies the concept of a *reaction force*. Imagine yourself pushing on a brick wall. As you push, you can readily feel the wall push back. The two forces are equal in magnitude and opposite in direction.

10   MOMENT AND ROTATIONAL ACCELERATION

Just as there are three laws for translational motion, there are three laws that govern how objects rotate, laws based on the rotational analogues of force and mass. The need for analogues arises because, when a rigid object spins, different parts of it move at different speeds. Each little bit of the object moves with the same angular velocity (Figure 2.5A), but the farther a bit of mass is from the axis, the larger the circle it travels (equation 2.2) and the greater its speed:

$$s = \omega R.$$ (2.27)

We need to take this spatially variable speed into account as we formulate the laws of rotational motion.

The rotational analogue of force is *moment*, M, defined as the product of force and the perpendicular distance, $\ell_\perp$, between the force's line of action and the axis of rotation (Figure 2.5B):

$$M = F\ell_\perp.$$ (2.28)

To specify perpendicular distance for a given axis and force, first draw the force vector, F. Then connect the axis of rotation to the vector with a line, forming a right angle. The length of this line is the perpendicular distance, also known as the *moment arm*, or *lever arm*. Moment has SI units of newton meters, (N $\cdot$ m). Because moment is the

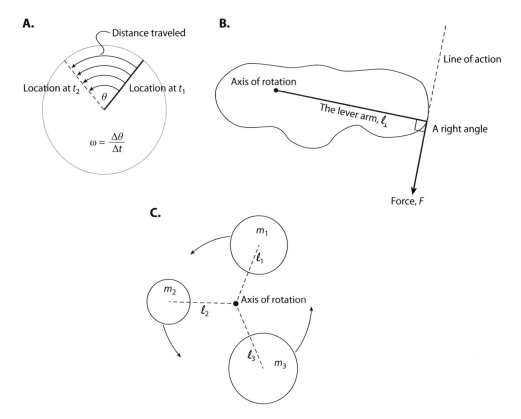

A. Distance traveled

Location at t_2 Location at t_1

θ

$\omega = \dfrac{\Delta\theta}{\Delta t}$

B.

Line of action

Axis of rotation

The lever arm, $\ell_\perp$

A right angle

Force, F

C.

m_1

ℓ_1

m_2

ℓ_2 Axis of rotation

ℓ_3 m_3

Figure 2.5 A. For a solid object, distance traveled during rotation increases linearly with distance from the axis of rotation. **B.** Moment is the product of force and the length of the lever arm. **C.** Three masses (fixed relative to each other) rotate about a central axis, an arrangement used to illustrate the moment of inertia. (See text.)

product of force and a lever arm, the same force can have different effects depending on where it is applied. To open a door (that is, to get it to rotate around its hinges), it is easier to apply a force at the knob (where the lever arm is largest) than near the hinges (where the lever arm is minimal).

The rotational analogue of mass is the *rotational moment of inertia*, J_m, which requires some explanation. The tendency for an object to resist rotation depends on both the object's total mass and how that mass is arranged relative to the axis of rotation. To be specific, if the object consists of several discrete masses $m_1, m_2, ..., m_n$ held rigidly at distances $\ell_1, \ell_2, ..., \ell_n$ from the axis (Figure 2.5C), the object's rotational moment of inertia is

$$J_m = \ell_1^2 m_1 + \ell_2^2 m_2 + ... + \ell_n^2 m_n = \sum_{i=1}^{n} \ell_i^2 m_i. \qquad (2.29)$$

In other words, the contribution of each mass to resisting spin is weighted by the *square* of its distance from the axis of rotation. As a consequence, masses farther from the axis have a much larger effect than masses near it.

If an object is composed of a continuous mass, its rotational moment of inertia is defined by the integral form of the summation in equation 2.29:

$$J_m = \int_0^{\ell_{max}} \ell^2 dm. \qquad (2.30)$$

Here, dm represents each infinitesimal mass element in the object, and ℓ_{max} is maximum distance from the axis. For an example (J_m of a cylinder), see Supplement 2.2.

Now that we have rotational analogues for force and mass, we can define relationships for rotation that parallel Newton's laws for translation.

1. *An object's angular velocity can change only through imposition of a net moment. In other words, if an object isn't rotating initially, it doesn't rotate unless acted upon by a moment. Similarly, if an object has an initial angular velocity, ω changes only if a moment is applied.*
2. *When a moment is imposed, angular acceleration, a_θ, is proportional to applied moment and inversely proportional to the object's rotational moment of inertia*

$$a_\theta = \frac{M}{J_m}. \qquad (2.31)$$

3. *When one object imposes a moment on another, the moment imposed on the first object due to their interaction is equal in magnitude but opposite in direction to the moment on the other.*

This third law poses a practical problem for astronauts. If during a spacewalk an astronaut attempts to tighten a nut clockwise with a wrench, the resulting moment causes him or her to rotate counterclockwise.

In summary, angular velocities and accelerations can be handled analogously to translational velocities and accelerations, the primary difference being that in the case of rotation, we must take into account the positions of force and mass relative to the axis of rotation.

11   MOMENTUM

Newton's second law of motion (equation 2.25) is central to much of physics. However, the law as we have formulated it is different from that envisioned by Newton himself. Recalling that acceleration (a) is the time rate of change of velocity (u, for instance), we can rewrite equation 2.25:

$$\begin{aligned} F &= ma \\ &= m\frac{du}{dt}. \end{aligned} \qquad (2.32)$$

If we allow for the possibility that mass as well as velocity might change through time, we must pull mass into the derivative:

$$F = \frac{d\,(mu)}{dt}. \qquad (2.33)$$

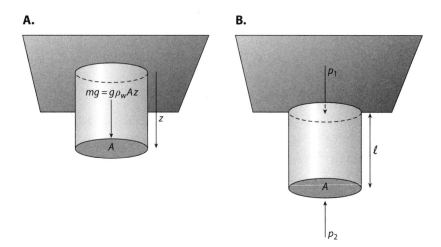

Figure 2.6 Pressure and buoyancy. **A.** A cylindrical column of water extending down from the water's surface. **B.** A cylindrical column of water within the water column.

Thus, force is equal to the time rate of change of the quantity mu. Because of its role in creating force, the product of mass and velocity is given its own name—*translational momentum*:

$$\text{translational momentum} = mu. \tag{2.34}$$

In fact, equation 2.33 is how Newton envisioned his second law, and it is more generally applicable than the common form ($F = ma$).

Just as there are rotational analogues of mass and force, there is a rotational analogue of translational momentum. *Angular momentum* is the product of rotational moment of inertia and angular velocity:

$$\text{angular momentum} = J_m\omega. \tag{2.35}$$

And, by analogy to equation 2.33,

$$M = \frac{d\,(J_m\omega)}{dt}. \tag{2.36}$$

Moment is the time rate of change of angular momentum.

12 PRESSURE

So far we have talked about force as if it were applied to an object at a specific point. However, we will often encounter cases in which imposed force is distributed across some area. In that case, it is convenient to speak in terms of *pressure, p*—force divided by the area over which it is applied:

$$p = \frac{F}{A}. \tag{2.37}$$

Pressure has units of $N \cdot m^{-2}$, also known as pascals (Pa) in honor of Blaise Pascal (1623–62).

In preparation for our discussion of fluid mechanics in Chapter 6, let's explore the concept of static pressure in a fluid. Consider a column of water extending down into the ocean (Figure 2.6A). The column has cross-sectional area A, and we desire to calculate the pressure acting a distance z below the water's surface. Water in the column has density ρ_w.

The volume of water in the column above depth z is Az, and it has mass $\rho_w Az$. Thus, the weight of this mass—that is, the weight pushing down on water at depth z—is

$$F = g\rho_w Az. \tag{2.38}$$

This weight is distributed over the area of the column, so the pressure p at depth z due to the water above is

$$p = \frac{F}{A} = g\rho_w z. \tag{2.39}$$

This *hydrostatic pressure* increases linearly with depth below the surface. The same physics applies to air. Atmospheric pressure at ground level is due to the weight of air above, and it decreases as one moves upward.

Although hydrostatic and atmospheric pressures are caused by the downward pull of gravity, they act equally in all directions. For example, a spherical balloon submerged in water stays spherical as depth and pressure increase. Lacking direction, pressure is a scalar.

13 BUOYANCY

Pressure imposed by air or water affects the gravitational force an object experiences. To see why, consider a cylinder of length ℓ and cross-sectional area A with volume $V = \ell A$. If the cylinder is made from a material with density ρ, its mass is ρV, and its weight is $F_g = g\rho V$.

Let's immerse this cylinder in a fluid of density ρ_f (Figure 2.6B) such that pressure at its top end is p_1, and—due to its greater depth—pressure at the bottom end is

$$p_2 = p_1 + g\rho_f\ell. \tag{2.40}$$

Multiplying pressure times area to calculate forces on the top and bottom (F_1 and F_2, respectively), we can calculate the net upward force—*the buoyant force*—acting on the cylinder:

$$
\begin{aligned}
F_{net} &= F_2 - F_1 \\
&= \left(p_1 + g\rho_f\ell\right)A - \left(p_1 A\right) \\
&= g\rho_f\ell A \\
&= g\rho_f V.
\end{aligned}
\tag{2.41}
$$

The buoyant force is proportional to volume, and it tends to offset weight (Archimedes principle), such that the *effective weight* of an object immersed in a fluid is

$$F_{g,\text{eff}} = F_{\text{net}} - F_g$$
$$= g\rho_f V - g\rho V$$
$$= g\left(\rho_f - \rho\right) V. \tag{2.42}$$

The quantity $\rho_f - \rho$ is the cylinder's *effective density*. If $\rho > \rho_f$, the cylinder is negatively buoyant, and the net force is downward. If $\rho < \rho_f$, the cylinder is positively buoyant, and net force is upward. If $\rho = \rho_f$, the cylinder is neutrally buoyant; its weight is just offset by the buoyant force. The density of most plants and animals is similar to that of water, so their effective weight in water is small. Air is much less dense than water, so the effective weight of terrestrial organisms is larger than that of their aquatic cousins, which, as we will see, has multiple consequences.

14   ENERGY

As insightful as Newton was about the mechanical workings of the universe, there is one crucial aspect of physics that escaped his genius. The concept of *energy* so central to our present understanding of mechanics was not clearly developed until the early 1800s, nearly a hundred years after Newton's death, and it was not fully fleshed out until the maturation of thermodynamics in the late 1800s.

Energy exists in many forms—mechanical, chemical, and electromagnetic, for instance. In this text, we focus on just two types of energy—mechanical energy (including heat) and electromagnetic energy (in the form of light).

14.1   Mechanical Energy I: Potential Energy

Mechanical energy W is defined as the capacity to do work, where work is in turn defined as the product of force and the distance through which that force moves its point of application:

$$W = F\ell. \tag{2.43}$$

Although both force and displacement have directions associated with them—and are, therefore, vectors—energy is nonetheless a scalar. The SI unit of energy is the joule, J, named in honor of James Prescott Joule (1818–89).

(Note that both mechanical energy and moment have dimensions of force times distance, but it is important to realize that they are distinct because the distances involved are different. In the case of energy, the relevant distance is that through which an object moves. For moments, it is instead the length of the moment arm.)

As a simple example of mechanical energy, consider the apparatus shown in Figure 2.7A, a bucket of water with mass m suspended by a rope from a pulley. By tugging on the rope with a force equal to the bucket's weight (mg), we can slowly winch the bucket upward against the pull of gravity. In the process of lifting the bucket through distance z, the work we do is

$$W = F_g z = mgz. \tag{2.44}$$

This energy is now available to do other kinds of work. For instance, we could attach a generator to the pulley's axle, and by letting the bucket fall back to earth, we could

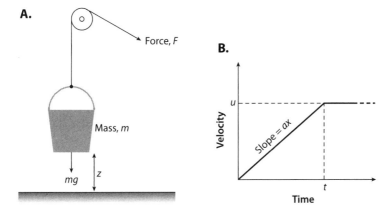

Figure 2.7 A. As a bucket is lifted, gravitational potential energy is stored. **B.** Calculating kinetic energy. (See text.)

generate electricity. Or the rope could be tied to a mill and the energy of the elevated bucket used to grind your morning coffee. Each of these scenarios demonstrates that by doing work on the bucket (lifting it against gravity), we store energy in the system that consequently has the potential for doing other types of work. This stored energy— $mg\,z$—is *gravitational potential energy*.

Gravitational potential energy is just one example of the many types of mechanical potential energy. For instance, in ecomechanics we will also encounter *elastic potential energy*, energy stored in the deformation of a material. A garage-door spring is an everyday example: as the door is closed, a beefy steel spring is stretched, and the energy stored in this deformation can subsequently be used to assist in opening the door. *Pressure-volume potential energy* is another example. When you push down on the handle of a bicycle pump, you apply pressure p to the gas in the pump, reducing its volume by an amount ΔV. Recalling that pressure is force per area, and that area and volume have units of length squared and length cubed, respectively, we see that

$$p\Delta V \propto \frac{F}{\ell^2}\ell^3 = F\ell. \tag{2.45}$$

In other words, the product of pressure and change in volume is equivalent to force times distance, that is, work.

Work done in compressing a gas has the potential to do other kinds of work. The pressure-volume work expended to compress the gas in a bottle of champagne can be used to propel the bottle's cork, for instance.

Two aspects of mechanical potential energy are worthy of additional note:

1. *One form of mechanical potential energy can be transformed into another.*

   For example, by shooting a champagne cork skyward against the tug of gravity, the pressure-volume energy in the bottle is transformed into gravitational potential energy.

2. *Mechanical potential energy is always relative.*

   The bucket-and-rope system of Figure 2.7A provides an instructive example. The magnitude of gravitational potential energy possessed by the bucket depends on where we choose to start measuring z.

In the situation described before, we set z to zero at ground level, implicitly assuming that a bucket on the ground could fall no further and would, therefore, have no potential energy. But what if, having lifted the bucket, we now dig a hole under it? By thus shifting the ground to a lower level, we increase the bucket's capacity to do work without doing any further work on the bucket itself. Similarly, if we were to place a stool under the bucket, reducing the distance it could fall, we would reduce its potential energy. In short, because mechanical energy is equal to the product of force and distance, when quantifying gravitational potential energy we need to specify the starting point from which distance is measured.

For pressure-volume potential energy, this dictum translates to specifying either the initial volume or pressure against which changes are quantified. For example, we have noted that a champagne bottle stores potential energy, as evidenced by the fact that it can blast a cork away from the bottle. But this is true only if pressure outside the bottle is less than that inside. If you submerge a champagne bottle deep enough in a river such that hydrostatic pressure outside the bottle is equal to gas pressure inside, the system no longer possesses the potential to do work on the cork.

This phenomenon was demonstrated in entertaining fashion for some unfortunate nineteenth-century English dignitaries who visited the site of a new bridge over a local river. Construction of the bridge's towers involved using caissons, chambers installed in the bottom of the river into which high-pressure air was pumped to allow workers to excavate for the towers' foundations. The dignitaries, invited into a caisson to celebrate the foundation's completion, opened a bottle of champagne preparatory to a victory toast. In the words of Leonard Hill, writing in his classic treatment of caisson disease (the bends):[3]

> On drawing the cork the champagne proved flat, and, thinking it bad, they drank none, all except one, who tasting liked the wine, and drank half the bottle, which he then corked and put in his pocket. During decompression in the air-lock the cork blew out with a loud report, and there was a great commotion, as one Councillor asserted he had been shot. This was nothing to the commotion made by the other who had drunk when the champagne effervesced in his stomach.

14.2 Mechanical Energy II: Kinetic Energy

Potential energy (either graviatational or pressure-volume) isn't tied to an object's motion: a bucket suspended motionless above the ground still has gravitational potential energy, and a stationary bottle of champagne still contains pressure-volume energy. By contrast, there is a type of energy for which motion is necessary. The application of force through a distance can be transformed into *kinetic energy* associated with an object's speed.

To see how this works, let's take a body of mass m, initially at rest, and apply a force F to it, acting along the x-axis. From equation 2.25 we know that application of this

---

[3] Hill, L. 1912. *Caisson Sickness, and the Physiology of Work in Compressed Air.* Edward Arnold, London.

force causes m to accelerate at constant rate a (Figure 2.7B). After time t we remove the force. How much work have we done on the object?

To calculate work—that is, energy—we need to know both force and distance; let's begin with force. Because acceleration is constant, the velocity of the mass accumulates linearly through time as shown, increasing from zero at the start of the experiment to u at time t. We can thus calculate acceleration a_x:

$$a_x = \frac{u}{t}, \tag{2.46}$$

and, knowing that $F = ma_x$,

$$F = m\frac{u}{t}. \tag{2.47}$$

How far did the object travel during force application? The answer can be obtained using calculus (Supplement 2.3), but in this case it is easy to make a simpler calculation. The average speed during the application of force is

$$\bar{s} = \frac{0 + u}{2} = \frac{u}{2}. \tag{2.48}$$

From equation 2.11, we know that total distance traveled is equal to average speed multiplied by the time of travel, t. Thus, distance x is

$$x = \frac{ut}{2}. \tag{2.49}$$

Mutliplying force times distance, we calculate the work required to get mass m up to speed:

$$W = \frac{mu}{t} \times \frac{ut}{2} = \frac{mu^2}{2}. \tag{2.50}$$

This is the object's *kinetic energy*.

As with momentum, kinetic energy is affected by both an object's mass and its speed. Despite its slow pace, a steamroller has substantial kinetic energy because it has a large mass. Despite its relatively small mass, an artillery shell in flight has substantial kinetic energy because it has a high speed. Because of their kinetic energy, both steamrollers and artillery shells are capable of doing work.

We postpone a detailed discussion until Chapter 11, but it is important to to note here that heat is a form of kinetic energy associated with the random motion of molecules.

14.3   Light

Light has inextricable characteristics of both waves and particles. For convenience, we will think of a beam of light as a collection of energetic particles—photons—but the energy transported by each of these quanta is set by the frequency, f, at which the particle's electromagnetic wave oscillates (measured in cycles per second, Hz):

$$\text{photon energy} = hf. \tag{2.51}$$

Here h is Planck's constant, $6.626 \times 10^{-34} \mathrm{J} \cdot \mathrm{Hz}^{-1}$. It is indicative of the deep wave-particle duality of light that the energy of a particle depends on the frequency of a wave.

Instead of thinking about light in terms of frequency, it is more common to think in terms of wavelength, λ. The two are related:

$$f = \frac{c}{\lambda}, \tag{2.52}$$

where c is the speed of light in a vacuum, approximately $3 \times 10^8 \mathrm{\,m} \cdot \mathrm{s}^{-1}$. Inserting equation 2.52 into equation 2.51, we see that

$$\text{photon energy} = \frac{hc}{\lambda}. \tag{2.53}$$

The shorter its wavelength, the greater a photon's energy. The wavelengths of light visible to us humans vary from approximately 400 nm for light we see as violet to approximately 700 nm for light we perceive as red (1 nm = 10^{-9} m). These wavelengths correspond to photon energies of 500 to 280×10^{-21} J.

Light is important to biology in many ways; for example, photons power the chemical reactions of photosynthesis. But we will not concern ourselves with biochemical energetics. Instead, our interest in light is confined to the heat energy (a form of kinetic energy) it can impart when absorbed. Sunlight, the primary source of energy on Earth, comprises photons with a spectrum of wavelengths ranging from the ultraviolet ($\lambda < 400$ nm), through the visible, far out into the infrared ($\lambda > 700$ nm). When exposed to full sunlight at sea level, a black surface absorbs all these wavelengths, gaining approximately 1000 J of heat energy per second for each square meter exposed. A large black iguana, for instance, with 0.1 m^2 of skin exposed to the sun, soaks up heat at a rate equal to that given off by a 100 W lightbulb. We will explore the details of sunlight and its absorption in Chapter 11.

15   CONSERVATION OF ENERGY

In physics, much of the utility of the concept of energy is based on the extraordinary fact that energy is conserved. Whenever energy in one form disappears, an equal amount reappears in some other form, a fact codified as the first law of thermodynamics. To return to a now-familiar example: when a cork is removed from a bottle of champagne, the pressure-volume energy of the system decreases but this decrease is just matched by an increase in the kinetic energy of the cork as it is propelled skyward. As the cork moves up, it is slowed by the tug of gravity, eventually coming to a halt. At that instant, it has lost all its kinetic energy, but now has an amount of gravitational potential energy equal to the kinetic energy it started with, and falling back to earth it could do work.[4] Similarly, light energy can be absorbed and turned into an equal amount of heat. But nowhere in these exchanges is any energy actually lost.

The fact that energy is conserved is philosophically unsettling. *Why* is energy conserved, you might ask? What underlying principle ensures that energy can be exchanged but never disappear? The answer is, we don't know. It is simply a fact that in every experiment that has ever been conducted, when energy decreases in one

---

[4] For simplicity, I have ignored the energy the cork would lose to its friction with air.

form, an equal amount appears in another form (or forms) (Feynman et al. 1963). This principle was open to question for a while in the early twentieth century, but Einstein's famous formula $E = mc^2$ saved the day by expanding the concept of energy to include its transformation into mass and vice versa (see Supplement 2.4).

16   FRICTION

Having just said that energy can be readily transformed, it is necessary to point out that some of these transformations are more easily accomplished than others. In particular, although any form of light or mechanical energy can readily be transformed into heat, it is much more difficult to reverse the process. In fact, according to the second law of thermodynamics, energy transformed into heat can *never* be fully recovered (Atkins 1984). This characteristic sets heat apart from all other kinds of energy, and leads to an important definition: For our purposes, *friction* is any process that converts mechanical energy into heat, the thought being that once converted into heat, energy has a reduced capacity to do mechanical work. As we have noted, it was friction that fooled Aristotle into thinking that the natural state of an object is to be stationary. Less familiar might be the friction that occurs as fluids move. Stir a cup of coffee and the resulting swirls gradually subside due to internal friction within the fluid. We will return to the friction of fluids—their *viscous* nature— in Chapter 5. In each of these cases, friction causes the system to wind down, not by making energy disappear (that's apparently impossible), but rather by converting it to the degraded form of heat.

For an object resting on a solid surface, friction comes in two forms: static and dynamic. The force required to get the object moving is

$$F_{sf} = k_{sf} F_n, \tag{2.54}$$

where k_{sf} is the *coefficient of static friction*, which varies depending on the nature of the surfaces in contact, and F_n is a normal force, the force pushing the object onto the surface. (For a horizontal surface, F_n is often the object's weight, mg.) Once the object is moving, less force is required to keep it moving:

$$F_{df} = k_{df} F_n, \tag{2.55}$$

where the *coefficient of dynamic friction*, k_{df}, is less than the coefficient of static friction. The difference between static and dynamic friction makes cross-country skiing possible.

17   POWER

One last aspect of energy requires our attention. Power is the rate at which energy changes through time:

$$\text{power} = \frac{dW}{dt}. \tag{2.56}$$

The units of power are joules per second, known as watts (W) after James Watt (1736–1819), one of the inventors of the steam engine.

To give this concept some personal significance, consider the power needed to propel your car. As you drive at $u = 30$ m/s (67 mi/h) along a level highway, the engine (via the tires) must continuously apply a force F of approximately 400 N to the tarmac to overcome the resistance of moving through the air. By applying a force through a distance, the engine does work, and because the distance traveled by the car increases steadily through time, work accumulates at a constant rate. The power expended by your engine is thus

$$\text{power} = \frac{\text{energy}}{\text{time}}$$
$$= \frac{F \times \text{distance traveled}}{\text{time of travel}}$$
$$= F \times \frac{\text{distance traveled}}{\text{time of travel}}$$
$$= F u. \tag{2.57}$$

In other words, force times speed is equal to power, in this case $400 \text{ N} \times 30 \text{ m} \cdot \text{s}^{-1} = 12,000$ W (16 horsepower). The extra power provided by a typical automobile engine is needed only to go up hills, accelerate while passing, and overcome various types of friction. We will use this same sort of calculation to measure the power needed for fish to swim and birds to fly.

18 SUMMARY

Physics begins with the ability to locate objects in space and proceeds to keep track of how location and orientation change through time. Coupling this information with Newton's laws of motion sets the stage for predicting the forces and moments that pervade the physical environment and impact plants and animals. Time and again through the remainder of this text, we will use the principles outlined in this chapter; you may wish to revisit this discussion periodically to refresh your memory.

Chapter 3

Response Functions

In the course of this text, we will characterize the many mechanisms through which plants and animals respond to each other and to their environment. We will see how the fluid-dynamic forces imposed on organisms depend on the velocity of air and water, and how the flexibility of trees and kelps depends on the molecular architecture of the materials from which they are constructed. We will investigate how body temperature responds to sunlight, how the dispersal of seeds, pollen, spores, and larvae is governed by the chaotic swirl of turbulent flow, how the rate at which predators capture prey is governed by prey density, and how species diversity depends on the time allowed for succession. Each of these relationships is a response function, and ecomechanics requires us to weave them together as we attempt to define the connections among molecules, cells, and physiology; between physiology and individual performance; and among individuals, populations, and communities.

Before we dive into ecological mechanics in earnest, it will be useful to train ourselves not to get lost in the details of individual response functions. When presented with a function's graph, it is important not to get bogged down in the units of the axes or the size of the error bars, but rather to notice the function's basic characteristics. Is it increasing or decreasing? Bounded or unbounded? Monotonic or modal? Paying attention to these basics highlights the function's fundamental message, which in turn helps to locate the function in the overall context of biology.

1   CONCEPTS AND TERMINOLOGY

Response functions come in three varieties: conceptual, empirical, and mechanistic. Conceptual functions describe—usually in a simple, abstract, and mathematically tractable form—how a system might function in some idealized world. They allow us to organize our thoughts and play with ideas; one seldom expects them to be an exact description of the real world.

By contrast, there is nothing abstract or idealized about empirical response functions. The result of laboratory investigations or carefully monitored field experiments, they describe empirical reality as revealed by measurements.

Mechanistic response functions bridge the gap between the two. By incorporating insight into *why* a system functions as it does, mechanistic functions use well-established concepts of physics, engineering, physiology, and behavior to model the performance of a system not only under current conditions, but also to predict how the system will behave under novel conditions.

In some cases, response functions are statements of *causality*. If temperature is x and the relative humidity is y, sweat will evaporate at rate z. However, response functions are not limited to descriptions of cause and effect. In many cases, biological response functions are instead statements of *potential*. If prey density is x, a predator can feed at rate y, but only if it chooses to do so. Prey density does not cause the animal to feed, instead it sets a limit as to how fast it can feed. For our purposes, functions of potential response can be just as useful as those of causal response.

And lastly, response functions can be statements of *probability*. If the intensity of wind's turbulence is x, the time it takes for pollen grains to encounter a flower is described by the probability distribution y: there might be a 10% chance that encounter time is less than 1 min, for instance, 17% chance less than 2 mins, and so on. In this case, the system's response to a particular set of circumstances is itself a function (the probability distribution) rather than a single value, but this is a defined response nonetheless.

2   MONOTONIC FUNCTIONS

The simplest response functions are monotonic: as the input variable (on the x-axis) increases, the response (on the y-axis) either increases (Figure 3.1A) or decreases (Figure 3.1B) but not both. Note that the rate of increase or decrease need not be constant. If the slope of the function steadily increases, the function *accelerates* (its graph is concave upward). If the slope becomes more negative, the function *decelerates* (its graph is concave downward). Note that a function can decelerate even though it is increasing (Figure 3.1A) or accelerate even though it is decreasing (Figure 3.1B). (The importance of acceleration and deceleration will become apparent later in this chapter when we deal with the calculation of averages.)

It is useful to split monotonic functions into two subcategories. Some functions are *unbounded*, such as those shown in Figure 3.1. As long as the input variable increases, the system continues to respond. For example, we will find in Chapter 9 that the hydrodynamic drag force F_D imposed on a large aquatic organism increases with increasing water velocity u without practical limit:

$$F_D = ku^2. \tag{3.1}$$

Here k is a coefficient that depends on water's density and the size and shape of the organism. Similarly, we will see in Chapter 12 that the higher an organism's body temperature (T_b) the greater the rate (H) at which it loses heat by emitting infrared radiation:

$$H = kT_b^4. \tag{3.2}$$

In this case, k is a coefficient that depends on the composition of the organism's integument.

Although common in purely physical processes such as these, unbounded responses are rare in biology. Life has limits, and as a result most physiological response functions are *bounded*. For example, the rate at which a plant photosynthesizes increases with increasing light intensity but only up to a point (Figure 3.2A). At that limit, the plant's photosynthetic machinery processes photons as fast as it can, and

A. Increasing functions

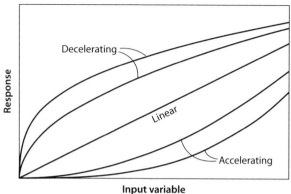

B. Decreasing functions

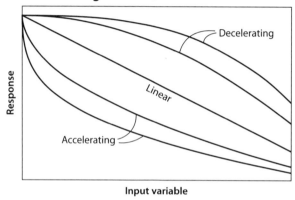

Figure 3.1 Monotonic functions can increase (**A**) or decrease (**B**). If a function is concave upward, it accelerates; if concave downward, it decelerates.

brighter light can do nothing to speed it further. Some physical responses are bounded as well. For instance, when a rain-forest snail moves from sunlight into shade, its body temperature decreases exponentially through time, but (barring evaporative cooling) it can never get colder than the surrounding air (Figure 3.2B).

For future reference, let's explore an example of a bounded monotonic function from population biology. In an elegant experiment in 1959, C. S. Holling, a Canadian entomologist and ecologist, blindfolded a student[1] and had her tap her finger on a table to locate randomly placed sandpaper discs. When a disc was located, the student removed it, set it aside, and searched for the next disc. The rate at which discs were thus "captured" was analyzed as a function of their spatial density on the table.

Watching the student, Holling noted that her time was divided between two activities: searching for discs and handling discs once they were located. From this basic observation, Holling derived a simple equation that accurately modeled how the student (a predator) responded to prey density (ρ_{prey}, the spatial density of discs). The

---

[1] Patricia Baic.

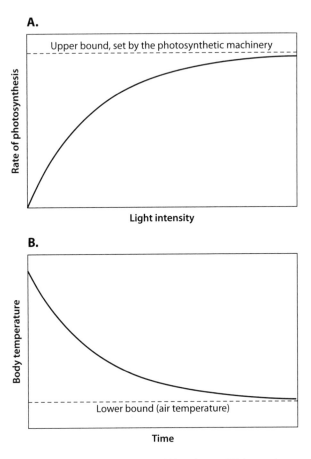

Figure 3.2 Nonlinear functions can have upper (**A**) or lower (**B**) bounds.

rate of prey capture C is:

$$C = \frac{k\rho_{\text{prey}}}{1 + \frac{k\rho_{\text{prey}}}{C_{\text{max}}}}. \tag{3.3}$$

Here k sets the slope of the curve at low prey density and C_{max} is the maximum rate of prey capture, the upper bound. This equation (the Holling type II functional response, Figure 3.3) accurately describes the dynamics of a wide variety of invertebrate predators.[2]

Biochemists deal with an analogous phenomenon at much smaller scale. The rate at which an enzyme can convert substrate to product depends on the time it takes for diffusion to deliver substrate molecules to the enzyme's active site, a time that depends on substrate concentration. Thus, delivery time plays a role analogous to searching time in Holling's experiment. Once reactant molecules have been delivered, it takes time for the enzyme to reconfigure, for the chemical reaction to occur, and

---

[2] If this is type II, what is type I? In a Holling type I functional response, capture rate increases linearly with increasing prey density up to an upper bound.

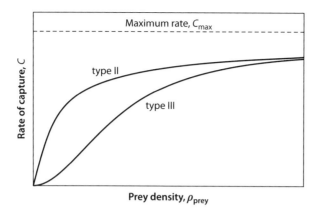

Figure 3.3 Holling's functional responses.

for the product to be released. In other words, an enzyme has a handling time in the same sense that a predator does. It should come as no surprise, then, that the Michaelis-Menten equation—which describes the kinetics of enzyme-mediated reactions (Hochachka and Somero 2002)—is mathematically identical to the type II functional response.

Type II and Michaelis-Menten functions model processes that depend on only two factors: searching (or delivery) time and handling time. Introduction of a third factor produces a more complicated—and often more realistic—response. For example, if there are hiding places available to prey and prey density is low, most prey can hide. With few prey out and about, the rate of prey capture is low. However, as prey density increases, fewer prey are able to find unoccupied shelters, and the rate of capture accelerates. Similar effects obtain if contacting prey in quick succession (which becomes more likely as prey density increases) allows the predator to learn how to capture prey more efficiently. These general scenarios can be described by a Holling type III reponse function:

$$C = \frac{k\rho_{prey}^2}{1 + \frac{k\rho_{prey}^2}{C_{max}}}. \tag{3.4}$$

As shown in Figure 3.3, this function is monotonic and bounded, but, unlike the type II function, type III is sigmoidal: it accelerates at low prey density (the curve is initially concave upward) before decelerating at higher prey density (the curve there is concave downward).

Again, there is an analogy at smaller scales. In many animals, oxygen is transported from the respiratory organs (the gills or lungs) to the tissues while bound to hemoglobin. Often, these oxygen-binding molecules form multi-unit structures; the oxygen-binding molecules in your blood, for instance, are composed of four, inter-acting hemoglobin units. When one hemoglobin unit binds to oxygen, it increases the likelihood that a second unit will bind. When two units are bound to oxygen, it increases the probability that a third will bind, and so forth. This inter-unit facilitation results in an oxygen binding curve (the fraction of hemoglobin in the bound state as a function of oxygen concentration) with a sigmoidal shape that can be modeled by an equation analogous to that for the Holling type III response (Prosser and Brown 1961).

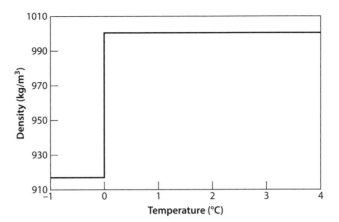

Figure 3.4 A threshold response: water's phase transition to ice.

2.1   Thresholds

The monotonic functions discussed so far vary relatively gradually with change in the input variable. In contrast, some responses undergo an abrupt shift as the input variable reaches a critical value, a shift that qualifies these *threshold responses* as a special type of montonic function.

Phase changes are classic threshold phenomena in physics. For instance, as temperature decreases from 4°C to 0°C, the density of water decreases very slightly. But as temperature descends through 0°C, density can abruptly and drastically decrease as liquid water becomes solid ice (Figure 3.4). A similar sort of phase transition can take place in a mucus gel, although in this case the transition is between collapsed and expanded states rather than between liquid and solid. As the ionic concentration of the fluid around it decreases through a threshold value, a collapsed gel abruptly expands, increasing its volume in milliseconds by more than an order of magnitude (Verdugo et al. 1987). This transition can help to explain how huge volumes of mucus can be stored in tiny secretory cells.

Threshold responses often arise in physiological processes; action potentials in nerves are a classic case. When a nerve cell is at rest, its interior has a negative charge (about −60 mV) relative to its exterior, and the cell is said to be polarized. If the degree of polarization is reduced by a millivolt or two from this resting potential, the slow flow of sodium ions out of the cell does not change. But if polarization is reduced by some threshold amount (typically 5 to 10 mV), voltage-gated sodium channels suddenly open in the cell's membrane—a threshold response. The resulting rapid outflow of sodium ions further reduces the cell's polarization, causing more sodium channels to open, and this positive feedback results in the firing of an action potential.

Thresholds can also be found in population dynamics. When the spatial density of a susceptible population is low, for example, a disease cannot effectively propagate. Above a threshold density, however, disease transmission suddenly becomes effective and an epidemic can ensue (Solé and Bascompte, 2006). Forest fires have analogous dynamics. When the density of trees is low, fires remain local; above a certain threshold density, fires can run rampant.

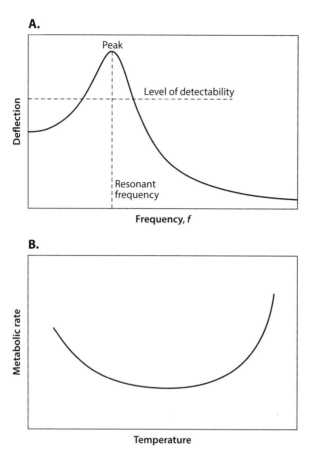

A.

B.

Figure 3.5 Unimodal response functions: **A**. Deflection of a sensory hair as a function of sound frequency. **B**. A mammal's metabolic rate as a function of temperature.

3   PEAKS AND VALLEYS
Unimodal Response Functions

As useful as monotonic functions are, they are too simple to describe the many biological processes that instead have a peak or valley where the function switches from increasing to decreasing, or vice versa.

Consider, for example, how the deflection of a sensory hair on a cricket's abdomen varies with the frequency of the force applied by sound waves (Figure 3.5A). We will investigate the mechanics of this system in Chapter 18, but a qualitative description suffices for present purposes. Due to the stiffness of the hair's base, the hair's deflection at low frequencies is small, too small for nerves to detect. At high frequencies, deflection is smaller still, a consequence of the hair's inertia. Near the hair's resonant frequency, however, the interaction between stiffness and inertia amplifies deflection to the point where it can be sensed. By adjusting its stiffness, each sensory hair can be tuned to detect a specific frequency (Casas and Dangles 2010).

Unimodal response functions are common in both physiology and ecology. As a muscle contracts, the force it can produce is maximal at an intermediate length.

Phytoplankton species diversity is maximal at intermediate nutrient levels (Irigoien et al. 2004), and species diversity in coral reefs, rain forests, and intertidal boulder fields is maximal at intermediate frequency of disturbance (Connell 1978; Sousa 1979).

The converse of peaks are valleys. For example, the metabolic rate of endothermic ("hot-blooded") animals varies with air temperature (Figure 3.5B). At low air temperature, metabolic rate is high as the animal shivers to maintain its internal temperature. At high ambient temperatures, the organism cannot cope by evaporative or convective cooling, and metabolic rate increases as body temperature rises. Thus, resting metabolic rate is minimal—a valley—at an intermediate range of temperatures, the organism's thermal neutral range (Schmidt-Nielsen 1997). Rates of DNA repair and programmed cell death vary with temperature in similar valley-type fashion (Yao and Somero 2012), and time to hatching in some insect larvae (which determines their exposure to parasites and predators) is minimal at intermediate temperatures (Potter et al. 2009).

Because peaks and valleys often define the circumstances under which organisms function best or worst, it is important to understanding how such functions arise. There is a wide variety of possible mechanisms, but three are particularly common.

3.1   The Difference Between Functions

The presence of a peak or valley often suggests that a response function is the product of competition between two underlying monotonic functions. Consider, for instance, the reproductive output of sea anemones as a function of body size.

In a classic investigation of the effects of size, Sebens (1982) studied feeding, growth, and reproduction in the sea anemone *Anthopleura elegantissima*, which inhabits wave-swept shores and makes its living by engulfing organisms (primarily mussels and snails) that fall on its oral disk after they have been dislodged by waves. The larger its gonads, the more sperm or eggs an anemone can produce. But constructing a gonad is possible only if the animal can acquire more energy (in the form of food) than it needs for maintenance metabolism. For anemones, the rate at which energy is expended in metabolism depends on body mass, which is porportional to the cube of the body's diameter (Figure 3.6A). By contrast, the rate at which anemones procure energy depends on the surface area of the oral disk, and therefore on the square of diameter. Thus, as the animal grows, its metabolic requirement for energy increases faster than its ability to feed.

At any size, the difference between energy intake and expenditure is the anemone's *scope for growth*, the surplus energy available to grow gonads. Scope for growth is maximal at an intermediate diameter (Figure 3.6B). (For a detailed mathematical description, see Supplement 3.1.)

According to this relationship, the greater the rate at which food is provided by the environment, the larger the optimal size of an anemone. One would thus expect to find larger anemones on wave-exposed shores (where mussels and snails are often dislodged) than on protected shores (where they are not). Conversely, the greater the metabolic rate of anemone tissue, the smaller the optimal size. Increased temperature—which would raise metabolic rate—should thus result in a decrease in the size of anemones.

Similar considerations help to explain why phytoplankton are small. The physics of diffusion dictates that the rate at which a phytoplankton cell can absorb the carbon

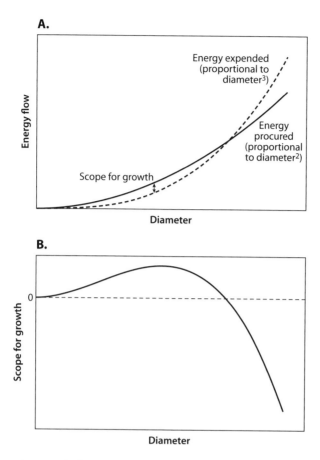

Figure 3.6 Scope for growth. Energy inputs and expenditures (**A**) and their difference (**B**).

needed for photosynthesis increases in direct proportion to the cell's diameter (see Chapter 4), while metabolic rate increases with the cell's volume. As a result, a phytoplankter's scope for growth is largest at a particular—small—cell size (Jumars 1993). Thus, if cell size is a heritable trait, increased fecundity selects for optimal body diameter.

3.2   The Derivative of a Function

Peaks or valleys can also result when one takes the derivative of a bounded monotonic response. Consider, for instance, the pattern of population growth described by the logistic function, an increasing, bounded relationship (Figure 3.7A):

$$N = \frac{1}{\left(\frac{1}{N_0} - \frac{1}{K}\right) e^{-rt} + \frac{1}{K}}. \tag{3.5}$$

Here N_0 is the small initial population size, N is subsequent size at time t, and r is the intrinsic rate of increase. When the population is small, growth is slow (the slope of the curve—its derivative—is gradual) because there simply aren't many individuals to reproduce. When the population is near carrying capacity, K, growth is again slow,

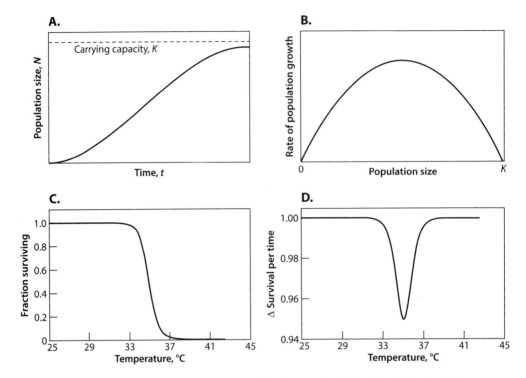

Figure 3.7 The logistic function (**A**) and its derivative (**B**). Survivorship (**C**) and its derivative (**D**).

in this case because resources are scarce.[3] The population grows fastest (the slope is steepest) when N is half the carrying capacity. Translating these thoughts to a graph of rate versus population size gives us a peaked function (Figure 3.7B).

Conversely, taking the derivative of a decreasing, bounded monotonic function leads to a valley-type response function. For example, the first derivative of the survivorship function of Figure 3.7C is the valley shown in Figure 3.7D.

3.3   Multiplying Functions

Alternatively, a peak- or valley-type response can result from the multiplication of two functions. Thermal reaction norms in marine invertebrates provide an example. There is some minimum temperature (T_{min}) below which the organism is not viable. As body temperature increases above this lower limit, potential metabolic rate $\mathcal{M}$ (and thereby performance) increases with temperature (Figure 3.8A). But metabolism can be limited by (among other factors) the rate of oxgen delivery, D, which decreases with increasing temperature. Realized performance can be modeled as the product of these two factors:

$$\text{performance}(T) = \mathcal{M}(T)D(T). \tag{3.6}$$

[3] The logistic equation has an interesting quirk. If it is applied to a population that reproduces periodically, rather than continuously, it can give chaotic results. For a full discussion of this phenomenon, consult Moon (1992).

A.

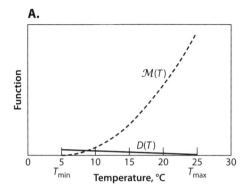

B.

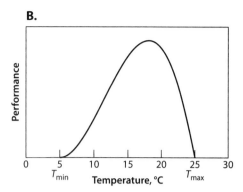

Figure 3.8 Components of thermal performance (**A**) and their product (**B**).

The resulting performance curve—a key ingredient in our understanding of the evolution of thermal physiology—increases gradually to a peak as temperature rises, and then decreases rapidly as temperature rises further (Figure 3.8B). (For the details of the mathematics, see Supplement 3.2.)

In summary, unimodal response functions commonly provide information regarding the conditions under which organisms perform best, and understanding the underlying basis for the existence of a peak or valley can provide insight into the salient mechanisms governing performance.

4   COMPLICATIONS

The simple response functions we have discussed so far suffice for some applications, but many biological systems require us to take additional factors into account.

4.1   Uncertainty

For example, the response functions just discussed are all deterministic. According to these relationships, if one specifies the value of the input variable, the function specifies the exact response of the system. In reality, this is seldom the case. There is almost always some chance involved in how a system responds, and this uncertainty can be incorporated into response functions in at least two ways.

First, one can separately account for a system's average response (a deterministic function of the sort we have discussed) and the nature of the variation around that average. This sort of two-part response function is common in investigations of turbulent flow (Davidson 2004). For example, the steeper the streambed, the faster the average flow ($\bar{u}$) in a mountain stream, a relationship that can be accurately described by a response function (Figure 3.9A). But at each average flow speed, there is some variation in speed due to turblence, and the pattern of these deviations (Δu) can be measured and expressed as a probability distribution (Figure 3.9B): there might be a 40% chance that the velocity measured at a random instant lies within 10 cm/s of the mean, for instance.

Quantification of response then proceeds in two steps. For a given value of the input variable, one first uses the average response function to calculate the expected

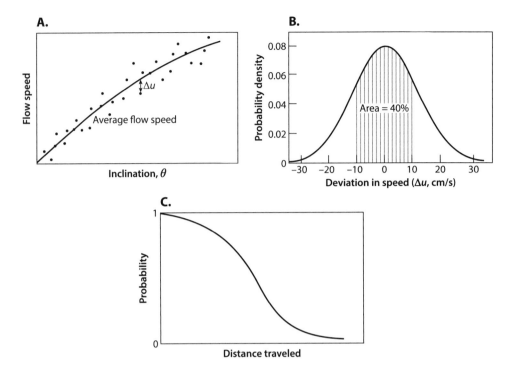

Figure 3.9 A. Hypothetical instantaneous velocity in a mountain stream (dots) is the sum of average velocity (solid line) and deviations picked at random from a distribution (**B**). **C.** Probability of traveling further than a given distance before hitting bottom.

response. A value is then chosen at random from the distribution of response variation and this deviation is added to the expected speed to give one random realization of the actual flow:

$$u(t) = \bar{u} + \Delta u(t). \tag{3.7}$$

A similar procedure can be used in many other situations where both a predictable average and chance variation are present.

In some cases, the system under consideration is driven by chance alone. Turbulence again provides an example. In turbulent flow, small particles (e.g., seeds, pollen, larvae, or spores) move in a random walk, and the distance they travel as a function of time can be described only statistically. (We'll examine this process in Chapter 4.) In cases such as these, the response in question is a measure of probability. For example, in the case of random walks, the response function specifies the exact probability that a seed or larva will travel more than a certain distance before first hitting the forest floor or seabed (Figure 3.9C).

One might object that probability is not really a response in the same sense that we have used the term up to now. But think about it this way. Let's suppose that a kelp plant releases 10^8 spores, and we know from our response function that there is a probability of 0.12 that a spore chosen at random will travel more than 10 m from its parent. We can thus calculate that, on average, 1.2×10^7 spores travel more than 10 m from the kelp. In this fashion, it is easy to translate the relationship between distance and probability into one that relates distance and number of spores hitting bottom, a function that fits comfortably within our notion of a response.

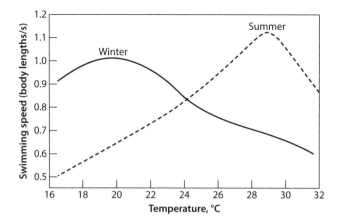

Figure 3.10 Crocodiles' optimal swimming temperature is higher in summer than in winter. Redrawn from Glanville and Seebacher (2006).

4.2   Change Through Time

It is often the case that biological response functions change through time as a result of physiological acclimatization, community dynamics, or evolution. For example, the temperature at which estuarine crocodiles swim fastest shifts with the seasons to match the temperature of the water (Figure 3.10; Glanville and Seebacher 2006). In such cases, it is necessary to specify the time at which a particular response function is to be applied, or to express response as an explicit function of time as well as of the input variable.

4.3   Hysteresis

The response functions we have discussed so far are independent of the direction of change in the input variable, but this need not be the case. A strand of spider's silk, for instance, is stiff when stretched. But much of the energy required to extend the material is dissipated as heat, so force follows a different path as the strand contracts (Figure 3.11A). This change in path (depending on whether the input variable is increasing or decreasing) is known as *hysteresis*. The area within the shaded loop is a measure of silk's ability to dissipate energy, an adaptive characteristic for a material that intercepts flying insects (see Chapter 14). Other examples abound. Species diversity in a community might remain high until physical stress reaches a threshold, but diversity might not recover until stress declines below a smaller value (Figure 3.11B). As light gets brighter, the rate of photosynthesis increases. But prolonged exposure to bright sunlight can damage a plant's photosynthetic apparatus. As a result, photosynthesis follows a different return trajectory as irradiance declines (Figure 3.11C).

5   JENSEN'S INEQUALITY
The Fallacy of the Average

As we explore ecomechanics, we will often use response functions to describe how an organism, object, or process behaves, not at any particular time or place, but

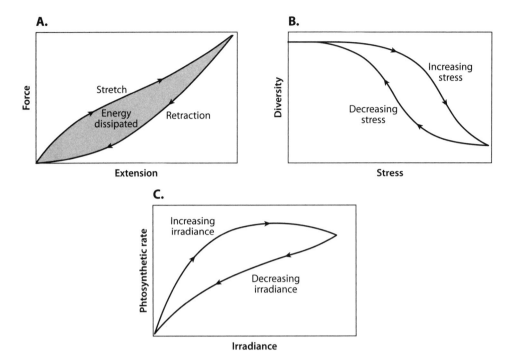

Figure 3.11 Hysteresis. **A.** A force-extension curve for a viscoelastic material. **B.** Plant diversity as a function of environmental stress. **C.** Photosynthesis as a function of light intensity.

on average. For example, in order to ripen its gonads, a predator must eat a certain amount of food. How long will it take until the predator is ready to reproduce? The answer is calculated easily if we know the average rate of injestion; we simply divide the total amount needed by the average rate at which food is obtained. But measuring that average can be tricky. According to a species' functional response (Figure 3.3), the rate of prey capture depends on the density of prey. However prey density can vary through time, and unless one is careful to take that variation into account, one can easily arrive at an erroneous estimate of mean predation rate and thereby a mistaken estimate of the time required to ripen a gonad. Understanding the reasons for the error is an important part of understanding how the world works.

6   LINEAR VERSUS NONLINEAR

Consider again the generic response functions shown in Figure 3.1. Some accelerate (they are concave upward), others decelerate (they are concave downward), and it is clear that acceleration and deceleration can happen to varying extents. By contrast, only if nowhere in its length does a function accelerate or decelerate is it a straight line. This criterion separates functions into two groups: *linear* functions (straight lines), and *nonlinear* functions (all others).

Because the definition of linearity is so narrow, the term nonlinear is ipso facto incredibly broad. As a colleague of mine once remarked, dividing functions into linear and nonlinear is analogous to dividing biology into bananas and nonbananas; there are few of the former and a multitude of the latter. Indeed, a glance back at the sampling of response functions discussed earlier in this chapter reveals that they are nearly all nonlinear, a trend that will hold up through the rest of this text. Because nonlinear functions are so pervasive in science, we need to understand their quirks. In particular, we need to understand how nonlinearity affects a function's average value.

7   THE EFFECTS OF NONLINEARITY

The average (or mean) of a set of n values is calculated by summing the values (x_1, x_2, ..., x_n) and dividing by n:

$$\bar{x} = \frac{1}{n} \sum_{i=1}^{n} x_i. \tag{3.8}$$

The average is used in two basic contexts. First, it provides a *descriptor* of a set of data. For example, average rainfall describes how much water is commonly available to plants. In the second context, the averge is used as a *predictor*. How much rainfall there has been in the past is a reasonable predictor of how much rainfall plants can expect in the year to come, hence an alternative term for the average is the *expectation*.[4] In biology, the average is used in both contexts.

Use of the average as a descriptor is the subject of inferential statistics, and we will not pursue it further. By contrast, using the average as a predictor is part and parcel of much of ecomechanics, and we had best know how it works. As I have intimated, the average can be a dangerous tool when applied to nonlinear functions.

Consider, for instance, the life of a desert cactus. Its environment is windy and hot, which imposes both physical and physiological stress. Take, for example, the force exerted by wind on a giant saguaro cactus. As we will see in Chapter 9, for large organisms and fast flows such as this, aerodynamic drag increases as the square of wind velocity u:

$$F_D = k A_{pr} u^2. \tag{3.9}$$

Drag is thus an accelerating, nonlinear function of velocity. Here A_{pr} is the area the cactus projects into flow, and the coefficient k depends on the saguaro's shape.

Let's suppose that that the size and strength of the cactus are an ontogenic response to the average drag the plant experiences: cacti in calm areas grow tall and weak; cacti in windy areas grow shorter and stronger. Therefore, to predict the size and strength of a saguaro in a particular environment, we would need to know the average drag.

If wind blows at constant speed, average drag is easily calculated. The situation is a bit more complex, however, if wind speed varies. Let's consider a situation in which velocity alternates between $0\,\mathrm{m \cdot s^{-1}}$ and $10\,\mathrm{m \cdot s^{-1}}$, a simple approximation of a gusty day. If the wind spends half its time at each velocity, its average speed ($\bar{u}$) is $5\,\mathrm{m \cdot s^{-1}}$, and we could use this average speed to predict average force:

$$\text{predicted average drag} = k A_{pr} \bar{u}^2 = 25 k A_{pr}. \tag{3.10}$$

---

[4] Technically, the expectation is the average as n becomes very large, but for our purposes the distinction is immaterial.

Alternatively, we can calculate average drag directly. When velocity is zero, drag is zero; when velocity is 10 m · s^{-1}, drag is $100kA_f$. Average force is thus.

$$\text{actual average drag} = \frac{0 + 10^2 kA_f}{2} = 50kA_{pr}. \tag{3.11}$$

In other words, the average drag we obtain when we take variation into account is twice that predicted from average speed alone. If we were to rely on force calculated from average speed, we would seriously underestimate the average drag imposed on the cactus and thereby be mistaken in our prediction of its size and strength.

Wind affects more than just force, however. A saguaro absorbs energy from the sun, and to keep from overheating it must shed some of this heat to the surrounding air. The faster the wind, the greater the cooling potential, but the relationship is again nonlinear (see Chapter 11 for details):

$$\text{cooling potential, } C = kA_s\sqrt{u}. \tag{3.12}$$

In contrast to drag, cooling potential is a decelerating function of wind speed. In this case, A_s is the cactus's surface area and k is a coefficient dependent on its size and shape.

Given the gusty scenario just described, we could predict average cooling potential using average wind speed (5 m · s^{-1}):

$$\text{predicted average cooling potential} = kA_s\sqrt{\bar{u}} = 2.24kA_s. \tag{3.13}$$

However, if we take variability into account,

$$\text{actual average cooling potential} = \frac{0 + \sqrt{10}kA_s}{2} = 1.58kA_s. \tag{3.14}$$

The potential for the cactus to cool in gusty winds is nearly 30% *less* than we would calculate using average wind speed.

In summary, if we were to use average wind speed as a predictor from which to estimate a cactus's morphology and thermal stress, we could seriously miscalculate its ability to survive.

These simple calculations convey the essence of a mathematical principle known as *Jensen's inequality*, named for J. L. Jensen (1859–1925), a French mathematician who published a seminal paper on the subject in 1906. The principle can be summarized as follows. Consider a function $g(x)$. Let's assume that there is some variation in x and that x has mean $\bar{x}$. Jensen's inequality asserts that if $g(x)$ is nonlinear:

$$g(\bar{x}) \neq \overline{g(x)}. \tag{3.15}$$

In other words, unless g is linear, the function of the mean $(g(\bar{x}))$ does not equal the mean of the function $(\overline{g(x)})$. You'll be doing yourself a favor if you repeat that last sentence a few times until the concept soaks in.

Now, if $g(\bar{x}) \neq \overline{g(x)}$, $g(\bar{x})$ is either greater than or less than $\overline{g(x)}$. In our drag example, the function of the mean was less than the mean of the function; in the cooling example it was more. How can we tell by looking at a function which it will be? It is here that the terms accelerating and decelerating come to the fore, providing two simple rules:

1. If $g(x)$ accelerates (it's concave upward), the function of the mean is less than the mean of the function, $g(\bar{x}) < \overline{g(x)}$.

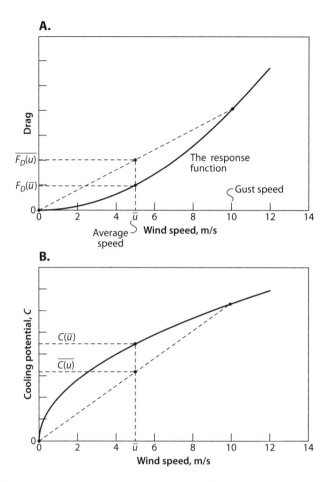

A. Average drag in a gusty wind is higher than the drag in a steady wind.
B. By contrast, average cooling potential is lower.

Figure 3.12

2. If $g(x)$ decelerates (it's concave downward), the function of the mean is greater than the mean of the function, $g(\bar{x}) > \overline{g(x)}$.

These rules can be varified graphically. Consider wind-imposed drag (Figure 3.12A). As specified by equation 3.9, drag accelerates with increasing wind speed, and the speeds used in our example (0 and 10 m · s^{-1}) are noted. Drag at average velocity—that is, $F_D(\bar{u})$—is the point on the curve corresponding to the average speed, 5 m · s^{-1}. By contrast, average drag—$\overline{F_D(u)}$—is the midpoint of a line connecting the drag curve at 0 and 10 m · s^{-1} (see Supplement 3.3 for a proof of this assertion). Because the function accelerates, the average of the function is greater than the function of the average.

The same logic can be applied to saguaro thermal dynamics (Figure 3.12B). Because cooling potential decelerates, the average of the function is less than the function of the average.

Now consider the curve shown in Figure 3.13, a thermal performance curve of the sort described earlier in the chapter. Near its peak, the function is concave downward. Therefore, Jensen's inequality tells us that, if temperature varies, the

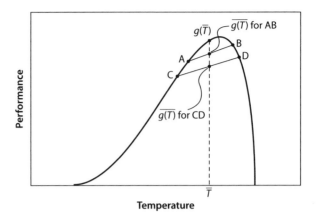

Figure 3.13 The difference between $\overline{g(T)}$ and $g(\overline{T})$ increases with increasing variation in T. (See text.)

average performance must be less than the function of the corresponding average temperature. But another important conclusion can be garnered by changing the difference between temperatures; that is, by adjusting the amount of variation. Temperatures A and B vary relatively little, and their midpoint (the average of the function) is only slightly lower than the function of the average. Temperatures C and D vary more, and their midpoint deviates farther from $g(\overline{T})$. In short, the more variation there is in T, the greater the effect of Jensen's inequality, a conclusion that applies not just to temperature variation, but to variation in general. We will expand on this conclusion in Chapter 19 where we take up the subject of variation in greater detail.

One last note before we move on. Jensen's inequality applies only to nonlinear functions. What happens if $g(x)$ is linear? In that case, $g(\overline{x}) = \overline{g(x)}$, and it doesn't matter whether you calculate the function of the average or the average of the function. Supplement 3.3 provides a mathematical proof.

8   CONCEPTS AND CONCLUSIONS

This chapter has explored response functions as a concept, the intent being to retrain our mental eye to look at the qualitative characteristics of a response rather than its quantitative details. Is the response increasing or decreasing? Accelerating or decelerating? Bounded or unbounded? What is the biological relevance of a peak or valley, and what are the underlying processes that result in a maximum or minimum? Is there hysteresis? The ability to discern these characteristics—and the habit of asking these questions—will be valuable as we use response functions to understand how plants and animals interact with each other and their environment.

Most environmental and biological response functions are nonlinear, however. As a result, we have seen that it is dangerous to use average conditions (e.g., average wind speed, average prey concentration) to predict the average response (e.g., average drag, average temperature, average rate of prey capture). This is the message of Jensen's inequality, a concept so fundamental and important that it should be—but almost never is—taught in introductory science classes.

Jensen's inequality has never been more important than it is in these times of rapid climate change. Predictions of future conditions are almost always couched as averages—average air temperature, average rainfall, average storm intensity. A thorough understanding of Jensen's inequality will be a critical part of our effort to predict how biology—with all its variability—will respond to the future's average conditions.

To this end, keep your eyes open as we explore the principles of ecologial mechanics. Whenever we discuss a nonlinear function—and there will be many—a red flag should shoot up in your mind and you should ask yourself: how would variation affect the average response? We will tackle this question directly in Chapters 19 to 21 where we deal with scale transition theory and its applications.

Part II

THE MECHANICS OF TRANSPORT

Chapter 4

Diffusion
Random Motion and
Its Consequences

With this chapter, we begin an investigation of transport, both from an engineering perspective (the movement of heat, mass, and momentum) and from a biological point of view (the movement of whole organisms). Transport phenomena have profound biological importance. The movement of heat into and out of organisms determines their temperature, which in turn can affect their metabolic rate, growth rate, survivorship, and even their sex. Transport of mass (e.g., oxygen, carbon dioxide, and nutrients) often limits the maximum size of plants and animals. Transport of momentum creates forces, which affect almost every aspect of life: when and where plants and animals can survive, the shapes in which they grow, and how they move. And the movement of whole organisms affects their ability to find food and mates, provides escape from predators, and governs the spread of populations. In short, the principles of transport underlie much of our understanding of biology.

We begin our overview of these principles with an exploration of random motion. This might seem an odd place to start, but, as we will see, the random motion of objects causes them to disperse—a process known as *diffusion*—which plays a key role in many forms of transport. This chapter outlines the principles of diffusion. My primary intent is to provide a basis for the use of these principles in later chapters, but I include one example of the utility of these principles: a prediction of the maximal size of phytoplankton.

1 RULES OF A RANDOM WALK

Random motion is, by definition, unpredictable. For example, as the larvae of benthic invertebrates waft randomly in a turbulent ocean current, no one can foresee exactly when a particular larva will encounter the seafloor and be able to settle. But what might at first appear to be the bane of prediction turns out to be an incredible boon. When the movement of individuals is random, the transport of *groups* of individuals is predictable. We can, for instance precisely estimate the *average* time required for larvae to hit bottom.

To see how this works, we start with a thought experiment in which we track the motion of particles through time as they move randomly along a line. This is a far cry from the three-dimensional motion of larvae, but the principles of this one-dimensional experiment can be generalized to address more realistic scenarios.

The experiment begins when we place a group of particles on the x-axis, along which they are free to slide independently. Each particle moves according to the following rules:

1. At time $t = 0$, the particle is at the origin, $x = 0$.
2. Subsequently, the particle moves at constant speed u.
3. Every τ seconds—the *characteristic interval*—we flip a coin to determine the particle's course. Heads and the particle continues in the same direction, tails and it reverses direction. This process ensures that, on average, half the time the particle moves in the positive x-direction and half the time in the negative x-direction. Note that in interval τ each particle travels distance $\delta = u\tau$.

These rules will eventually lead us to two important conclusions:

1. Random motion causes a group of particles to disperse, that is, to diffuse.
2. The distance they disperse typically increases not in proportion to time but rather to the *square root* of time.

This second fact runs counter to intuition and has numerous biological consequences. It explains, for instance, how nerves can rely on the diffusion of neurotransmitters to convey information and why trees need vascular tissue to deliver glucose to their roots.

To understand how the rules of a random walk lead to these conclusions, let's warm up by calculating the average position of a group of particles after each has taken a given number of steps.

2 AVERAGE POSITION

Let's assume that our group consists of n particles, which we track individually: at time t particle i ($i = 1, 2, ..., n$) is at location $x_i(t)$. According to our rules, position at time t differs from postion at time $t - \tau$ by distance δ:

$$x_i(t) = x_i(t - \tau) \pm \delta. \tag{4.1}$$

(The $\pm$ sign is a reminder that the particle spends half its time moving in each direction.) Given equation 4.1, the average displacement of the n particles at time t can be expressed in terms of their displacement at $t - \tau$:

$$\bar{x}(t) = \frac{1}{n} \sum_{i=1}^{n} x_i(t)$$

$$= \frac{1}{n} \sum_{i=1}^{n} [x_i(t - \tau) \pm \delta]$$

$$= \frac{1}{n} \sum_{i=1}^{n} [x_i(t - \tau)] + \frac{1}{n} \sum_{i=1}^{n} \pm \delta. \tag{4.2}$$

Because particles' direction changes randomly, positive travel tends to cancel out negative travel, and

$$\frac{1}{n} \sum_{i=1}^{n} \pm \delta \approx 0. \tag{4.3}$$

Thus, the mean position at t is equal to the mean position at $t - \tau$:

$$\overline{x}(t) = \frac{1}{n} \sum_{i=1}^{n} [x_i(t - \tau)]$$

$$= \overline{x}(t - \tau). \tag{4.4}$$

Given that $\overline{x}(0) = 0$, this means that $\overline{x}(\tau)$ also is 0, as is $\overline{x}(2\tau)$, and so forth. In other words,

$$\overline{x}(t) = 0. \tag{4.5}$$

On average, a group of particles goes nowhere.

3   DISPERSION
The Variance of Position

However, this doesn't mean that every particle remains at the origin. Rather, it implies only that for each particle displaced to the right, there is likely to be a particle displaced an equal distance to the left. This effect is illustrated in Figure 4.1A, where I have tracked the paths of five particles through five hundred characteristic intervals. Even though their average location remains near the origin, the particles disperse. Our job now is to quantify this spread.

The trick is to track how far a particle is from the origin without regard to whether it has moved right or left; a particle at $x = -1$ has been displaced just as far as a particle at $x = 1$. Taking a cue from statistics, we erase the effect of sign by squaring the value for each location; that is, $(-1)^2$ is the same as $(+1)^2$. In other words, we keep track of particles' average spread by calculating the average square of their distance from the origin. Because the origin is also the average location (equation 4.5), by calculating average squared distance, we are also calculating particles' mean square deviation from the average, their *variance*, σ_x^2.

With this idea in mind, we return to the calculation of averages. Because

$$x_i(t) = x_i(t - \tau) \pm \delta,$$

the square of location at time t is

$$x_i^2(t) = [x_i(t - \tau) \pm \delta]^2. \tag{4.6}$$

Expanding the square, we see that

$$x_i^2(t) = [x_i^2(t - \tau)] \pm [2x_i(t - \tau)\delta] + \delta^2. \tag{4.7}$$

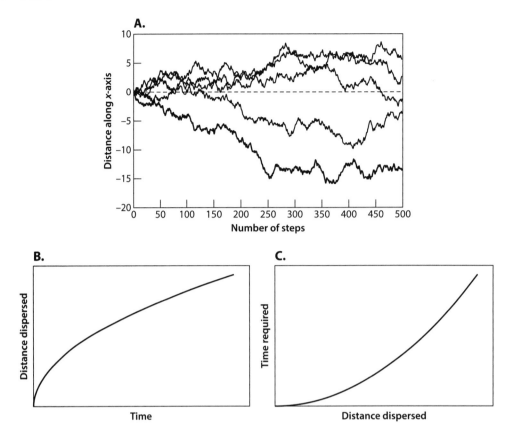

Figure 4.1 Random walks. **A.** As they step randomly, five particles disperse, but their average position stays near the origin. **B.** Dispersion increases as the square root of time. **C.** Time required to disperse a given distance increases as the square of distance.

(At this point it is easy to get confused with the notation. Recall that $x_i(t - \tau)$ is the displacement of the ith paticle at time $t - \tau$. The square of this value is $x_i^2(t - \tau)$, not $x_i^2(t - \tau)^2$. That is, we square the magnitude of location, but that value still applies to time $t - \tau$.) Given this expression for a single particle, we then average over all n particles to calculate σ_x^2:

$$\sigma_x^2(t) = \frac{1}{n} \sum_{i=1}^{n} \left\{ \left[x_i^2(t - \tau) \right] \pm \left[2x_i(t - \tau)\delta \right] + \delta^2 \right\}$$

$$= \frac{1}{n} \sum_{i=1}^{n} x_i^2(t - \tau) + \frac{2}{n} \sum_{i=1}^{n} \pm x_i(t - \tau)\delta + \frac{1}{n} \sum_{i=1}^{n} \delta^2. \tag{4.8}$$

The first term here is the average squared deviation at time $t - \tau$:

$$\frac{1}{n} \sum_{i=1}^{n} x_i^2(t - \tau) = \overline{x^2}(t - \tau) = \sigma_x^2(t - \tau). \tag{4.9}$$

The second and third terms can be simplified. Because the direction of motion varies randomly,

$$\frac{2}{n}\sum_{i=1}^{n}\pm x_i(t-\tau)\delta \approx 0. \tag{4.10}$$

In the third term, we add δ^2 to itself n times and then dividing by n, which just gives us δ^2:

$$\frac{1}{n}\sum_{i=1}^{n}\delta^2 = \delta^2. \tag{4.11}$$

Thus, the end result is

$$\sigma_x^2(t) = \sigma_x^2(t-\tau) + \delta^2. \tag{4.12}$$

In other words, σ_x^2 at time t is equal to σ_x^2 at time $t-\tau$ plus δ^2. Because δ^2 is always positive, displacement increases through time.

How fast? Given that all particles start at the origin,

$$\sigma_x^2(0) = 0, \tag{4.13}$$
$$\sigma_x^2(\tau) = 0 + \delta^2 = \delta^2, \tag{4.14}$$
$$\sigma_x^2(2\tau) = \delta^2 + \delta^2 = 2\delta^2, \tag{4.15}$$

and so forth. So after N characteristic intervals,

$$\sigma_x^2(N\tau) = N\delta^2. \tag{4.16}$$

Now, $N\tau = t$, the total time from the beginning of the experiment, and thus $N = t/\tau$. Substituting this equality into equation 4.16:

$$\sigma_x^2(t) = \frac{t}{\tau}\delta^2. \tag{4.17}$$

Finally, recalling that $\delta = u\tau$, we conclude that

$$\sigma_x^2(t) = \tau u^2 t. \tag{4.18}$$

Mean square displacement (variance) increases in direct proportion to time and characteristic interval, and in proportion to the square of particle speed.

We could stop here and use σ_x^2 as an index of how far the group has spread. But σ_x^2 has units of distance squared, an awkward measure of dispersion. So for general use, let's quantify the average spread of particles using the the square root of σ_x^2—the *standard deviation*—which has units of distance:

$$\sigma_x(t) = u\sqrt{\tau t}. \tag{4.19}$$

Thus, dispersion increases with the square root of time (Figure 4.1B), the nonintuitive fact I alluded to at the beginning of the chapter.

4   THE DIFFUSION COEFFICIENT

In our attempt to understand diffusion, we have so far employed a mechanistic approach: exploring the logical consequences of a simple set of rules. Alternatively, we can work from an empirical perspective. Rather than predicting dispersion from first principles, we can directly measure the rate at which particles spread and use that information to quantify a characteristic parameter of the process, the *diffusion coefficient*, $\mathcal{D}$. $\mathcal{D}$ is defined as half the temporal rate of change of the variance in particle location:

$$D = \frac{1}{2}\frac{d\sigma_x^2}{dt}.$$ (4.20)

(The reason for the factor of $1/2$ will become apparent shortly when we discuss Fick's equation.) Variance has units of m^2, so the diffusion coefficient has units of $m^2 \cdot s^{-1}$.

Having measured $\mathcal{D}$, we can use it to predict particle spread. Integrating equation 4.20 to solve for σ_x, we find that

$$\sigma_x(t) = \sqrt{2\mathcal{D}t}.$$ (4.21)

The larger the diffusion coefficient, the faster particles spread, and (as we have already concluded) dispersion increases as the square root of time.

We can invert this relationship to calculate the time required for a group of particles to disperse to a given σ_x:

$$t(\sigma_x) = \frac{\sigma_x^2}{2\mathcal{D}}.$$ (4.22)

Time required increases as the square of distance (Figure 4.1C).

These relationships between $\mathcal{D}$, dispersion, and time can be extended to two- and three-dimensional random walks. If, while particles step in the x-direction they use the same rules to step independently in the y-direction (as animals might when moving on the ground), their spread, σ_r (the standard deviation of r, the radial distance from the origin), can be calculated using the Pythagorean theorem:

$$\sigma_r(t) = \sqrt{\sigma_x^2(t) + \sigma_y^2(t)}$$ (4.23)

$$= \sqrt{2\mathcal{D}t + 2\mathcal{D}t}$$

$$= \sqrt{4\mathcal{D}t}.$$ (4.24)

The time it takes to spread to σ_r is

$$t(\sigma_r) = \frac{\sigma_r^2}{4\mathcal{D}}.$$ (4.25)

Similarly, in three dimensions (the dispersion of marine larvae, for instance):

$$\sigma_r(t) = \sqrt{\sigma_x^2(t) + \sigma_y^2(t) + \sigma_z^2(t)} = \sqrt{6\mathcal{D}t},$$ (4.26)

$$t(\sigma_r) = \frac{\sigma_r^2}{6\mathcal{D}}.$$ (4.27)

In summary, having measured the diffusion coefficient for a group of particles, we can use that coefficient to calculate how the group disperses as a function of time.

Although $\mathcal{D}$ is typically calculated from empirical measurements, it can be related to our random-walk model. Given that $\sigma_x^2(t) = t\tau u^2$ (equation 4.18),

$$D = \frac{1}{2}\frac{d(t\tau u^2)}{dt}$$

$$= \frac{\tau u^2}{2}.$$ (4.28)

Thus, when it is feasible to measure u and τ directly, they can be used to calculate $\mathcal{D}$. $\mathcal{D}$ increases with an increase in either τ or u, but it is more sensitive to an increase in u.

5   WHAT DRIVES RANDOM WALKS?

We are now in a position to relate diffusion to the real world. Notice that in the preceding calculations we never defined the scale at which we were working: Speeds could be millimeters per year or kilometers per second, the characteristic interval could be picoseconds or centuries, the particles could be molecules or giraffes; it doesn't matter. As long as particles obey our rules, they spread in proportion to the square root of time. What ties our equations to reality is the magnitude of the diffusion coefficient, $\mathcal{D}$, which as we have just seen depends on u and τ, speed and characteristic interval, respectively. In turn, u and τ depend on what is driving the random walk.

5.1   Molecular Diffusion

Let's begin with small molecules such as oxygen, carbon dioxide, glucose, and mineral nutrients whose diffusion is driven by the random motion of thermal agitation. As we will see in Chapter 11, the average speed at which a small molecule moves (m · s^{-1}) is set by the ratio of its temperature (T) to its mass (m):

$$u = 6.43 \times 10^{-12}\sqrt{\frac{T}{m}}. \tag{4.29}$$

The higher the temperature, the faster the molecule moves; the larger its mass the slower it moves. Here T is absolute temperature (kelvins, K) and the factor of 6.43×10^{-12} is a result of expressing temperature on the Kelvin scale (see Chapter 11). Mass is in kilograms; for example, an oxygen molecule has a mass of 5.3×10^{-26} kg, and at room temperature (293 K), it has a surprisingly high speed: $478 \, \text{m} \cdot \text{s}^{-1}$.

However, molecules' characteristic intervals are very short. In air an oxygen molecule moves for only approximately 1.6×10^{-10} s before impacting another gas molecule and heading off in a random direction. At a speed of $478 \, \text{m} \cdot \text{s}^{-1}$, this means that oxygen molecules travel approximately $0.08 \, \mu\text{m}$ between impacts, a distance known as the *mean free path*. In water, the medium is so tightly packed that the characteristic interval is 10,000–fold shorter, corresponding to a mean free path of 10^{-11} m.

Inserting these values for u and τ into equation 4.28, we find that the diffusion coefficients of small molecules in air are approximately $10^{-5} \, \text{m}^2 \cdot \text{s}^{-1}$. In water, $\mathcal{D}$ is 10,000-fold smaller, approximately $10^{-9} \, \text{m}^2 \cdot \text{s}^{-1}$.

Several biologically meaningful conclusions can be drawn from these values. First, let's consider diffusion in water (Table 4.1). Despite water's small $\mathcal{D}$, diffusion can be an effective means of transport over short distances, a fact essential for nerve conduction and muscle contraction. When a nerve impulse reaches the end of an axon, it is transmitted to the next cell not by electrical means but rather by the diffusion of small molecules. For a nerve attached to a muscle, for instance, it takes only $1.25 \, \mu\text{s}$ for the neurotransmitter (acetylcholine) to bridge the 50-nm gap between cells.

Table 4.1   The Time Required for Oxygen to Diffuse a Given Distance at 20°C.

Distance (m)	Seconds	Hours	Years
Air			
10^{-9}	5×10^{-14}		
10^{-6}	5×10^{-8}		
10^{-3}	0.05		
10^{-2}	5		
1		14	
10		1400	
100			16
1000			1600
Water			
10^{-9}	5×10^{-10}		
10^{-6}	5×10^{-4}		
10^{-3}	500		
10^{-2}		14	
1			16
10			1600
100			160,000
1000			16,000,000

By contrast, at larger distances diffusion in water is incredibly ineffective. A muscle cell in the interior of a larval fish is separated from the body's surface by a millimeter or so of watery tissue, a distance that would require 500 s for a diffusing oxygen molecule to traverse. As a consequence of this sluggish transport, even small fish need a circulatory system to deliver oxygen to their muscles. The situation is much more dire for trees. It would take nearly 160,000 y for glucose molecules produced in a redwood's crown to diffuse the 100 m to the tree's roots, making it obvious why trees rely on the active transport of their phloem to deliver glucose to the roots.

Life can be different in air, where $\mathcal{D}$ is 10,000-fold larger. For example, an oxgen molecule can travel a millimeter in only 0.05 s, and insects have evolved to take advantage of this relatively rapid transport. They have a network of small tubes—tracheae—connecting muscles to the outside air, and diffusion alone is often sufficient to deliver oxygen, although some active pumping of air may be needed when the animals are active.

5.2   Brownian Motion

Large molecules (e.g., proteins) and small organisms (e.g., viruses, bacteria, and phytoplankton) are too large to fit our model for the random motion of small molecules. Rather than themselves moving at a constant speed between discrete changes in direction, these particles move in response to random impacts from the medium's small molecules. As a result, although these relatively large particles perform randoms walks (known as *Brownian motion*[1]), the analysis of their diffusion requires a different approach. Basing his analysis on the force it takes to move an object through a viscous fluid, Einstein (1905) calculated the Brownian diffusion coefficient for a sphere of radius r:

$$\mathcal{D} = \frac{\mathcal{R}T}{\mathcal{N}} \frac{1}{6\pi\mu r}. \tag{4.30}$$

Here $\mathcal{R}$ is the universal gas constant ($8.31\,\mathrm{J\,mol^{-1} \cdot K^{-1}}$), T is absolute temperature, $\mathcal{N}$ is Avagadro's number (6.02×10^{23}), and μ is the fluid's viscosity (approximately $10^{-3}\,\mathrm{Pa \cdot s}$ for water at room temperature; see Chapter 5).

Equation 4.30 immediately leads to a robust conclusion: diffusion in water by Brownian motion is too slow to have much practical biological importance. For a virus 10 nm in radius, $\mathcal{D}$ is a mere $2 \times 10^{-11}\,\mathrm{m^2 \cdot s^{-1}}$, 100-fold smaller than that of an oxygen molecule. A bacterium 1 μm in radius has a $\mathcal{D}$ 100-fold smaller still ($2 \times 10^{-13}\,\mathrm{m^2 \cdot s^{-1}}$), and the $\mathcal{D}$ of a phytoplankton cell with $r = 10\,\mu$m is 10 times smaller than that ($2 \times 10^{-14}\,\mathrm{m^2\,s^{-1}}$). Thus, even though they undergo random walks, for virtually all practical purposes passive viruses, bacteria, and phytoplankton don't diffuse. The viscosity of air is approximately 2% that of water, so the Brownian diffusion coefficient in air is roughly 50 times that in water, but still too small to have much practical value.

The consequences of negligible Brownian diffusivity vary among organisms. For aquatic viruses, it means that it is the active movement of their hosts rather than diffusion of the virus that mediates initial virus-host contact. For phytoplankton, which support photosynthesis by absorbing bicarbonate and mineral nutrients, the minuscule diffusivity of the plankton isn't a problem. Even though these cells are sedentary, bicarbonate and nutrients (being small molecules) diffuse rapidly, and deliver themselves. For bacteria, which need to move to the detritus particles they eat, the fact that passive bacteria don't diffuse has likely acted as a factor guiding them to evolve a capacity for locomotion.

5.3   Locomotory Diffusion

Organisms of all sizes can propel themselves, giving them the potential to perform random walks and, thereby, to disperse. If we can measure u and τ for a particular organism, we can predict its rate of dispersal.

Consider for instance the hypothetical spread of a newly mutated gene in squirrels. The gene appears in a single female and is passed to her progeny, which, after a year with their parents, mate and head off in random directions. Each mated pair travels 2 km before settling down to repeat the process. How fast do the young squirrels (and thereby the mutant gene) spread? This is a simple case of two-dimensional diffusion where "particles" (squirrel pairs) have a τ of 1 y and a u of $2\,\mathrm{km \cdot y^{-1}}$. The resulting diffusion coefficient ($\mathcal{D} = \tau u^2/2$) is $2\,\mathrm{km^2 \cdot y^{-1}}$.

Knowing $\mathcal{D}$, we can calculate the gene's spread. Using the standard deviation (σ_r) as the radius of the gene's dispersion, the area covered by the mutant squirrels is $\pi\sigma_r^2$.

---

[1] Named for the botanist Robert Brown, who described it in 1828.

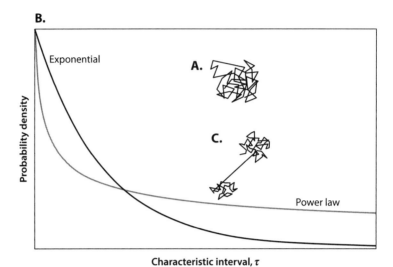

Figure 4.2 A. A random walk with Poisson-distributed τ. **B.** Comparing the distribution of τ between the exponential Poisson distribution and the power-law Lévy distribution. **C.** A Lévy random walk.

From equation 4.25 we know that $\pi \sigma_r^2 = 4\pi \mathcal{D}t$, so the rate at which the gene occupies new area is

$$\frac{d(4\pi \mathcal{D}t)}{dt} = 4\pi \mathcal{D} = 25.1 \text{ km}^2 \cdot \text{y}^{-1}. \tag{4.31}$$

Because σ_r increases with the square root of time, the area covered by two-dimensional diffusion of this sort increases directly with time. Similar calculations can be applied to many organisms (Okubo and Levin 2001).

One must be careful when estimating u and τ, however. Consider, for instance, the dispersion of bacteria. Unlike molecules and squirrels, for which τ is constant, bacteria sample their environment by swimming with variable τ, heading in one direction for awhile before tumbling and heading off in a random direction (Figure 4.2A). The interval between tumbles varies dramatically. In any given period, there is a constant probability that the cell will tumble, leading to an exponential distribution of interval times—the likelihood that a given intertumble interval has length τ decreases as $e^{-\tau}$ (Figure 4.2B; see Supplement 4.1). As this graph shows, the interval between tumbles is usually short, but there are occasional long runs, and these boost the diffusion coefficient by a factor of two relative to that of a random walk with constant τ (Visser and Thygesen 2003).

Food is scarce and patchy for marine bacteria, and they rely on their locomotory diffusion to find the detritus particles that they digest. Without the boost from the exponential distribution of τ, life could be grim. But given the high speed at which marine bacteria swim (up to $400\,\mu\text{m} \cdot \text{s}^{-1}$) and the exponential distribution of intervals, the diffusion coefficients for bacteria are approximately $10^{-9}\,\text{m}^2 \cdot \text{s}^{-1}$, approximately the same as those of small molecules, and Kiørboe (2008) has shown that this diffusivity is sufficient to explain the rate at which bacteria colonize organic detritus particles.

More extreme effects can be had if the distribution of τ varies as τ^{-k} (where k is a constant) rather than $e^{-\tau}$; that is, as a power function rather than an exponential function (Figure 4.2B). Random walks with power-law distributions of this sort are known as *Lévy flights* and they have been observed in the dispersal of a variety of animals, everything from snails to albatrosses to deer (Viswanathan et al. 2011). For $1 < k \leq 3$, Lévy dispersion increases not with the square root of time, but rather with time to an exponent greater than $1/2$, a process known as *superdiffusion* (Viswanathan et al.):

$$\sigma \propto t^{1/(k-1)} \tag{4.32}$$

Furthermore, the pattern of movement changes (Figure 4.2C). For example, when k is approximately 2, a particle shuffles back and forth, exploring in detail a local region of its surroundings, and then at random times travels to a new region, where detailed exploration begins anew. Under certain circumstance, this type of diffusive behavior can be an optimal strategy for finding food or mates.

5.4 Turbulent Diffusion

Roughly speaking, turbulence is the chaotic motion of a fluid. It is often useful to think of turbulence as the random motion caused by the combined action of eddies within eddies, each swirling in a random direction with a random speed. As turbulent eddies interact, they carry particles along unpredictable paths. It is important to note that it is not the movement of particles through the medium that matters; instead it is movement of the medium itself—the turbulence—that drives the random walk.

Biologically relevant turbulence is often created as air or water moves past a solid surface (see Chapter 6). Unfortunately, the resulting randoms walks are more complicated than those we have encountered so far. For starters, the statistics of these walks depends on the direction of movement. Parallel to a boundary, particles are free to move long distances and diffusion is relatively rapid; horizontal turbulent diffusion at the surface of oceans and lakes is a good example. By contrast, perpendicular to a boundary movement is constrained by the surface, and diffusion is relatively slow.

Predictions of turbulent dispersion are further complicated by the fact that in turbulent flow the diffusion coefficient increases through time. When particles are closely packed, only small eddies contribute to their spread; big eddies move the entire compact group without affecting their spacing. However, as particles are spread by small eddies, larger and larger eddies can contribute to dispersion, and the rate of dispersion increases.

Because of these (and other) complications, it is very difficult to make precise quantitative predictions regarding turbulent diffusion. However, if we are willing to settle for order-of-magnitude answers, useful conclusions can be drawn. To that end, let's focus on z-directed motion toward or away from a solid horizontal boundary. For instance, it is this diffusion that often accounts for the vertical distribution of phytoplankton, spores, larvae, and pollen. If we can specify the u and τ that characterize particle movement along the z-axis, we can calculate a turbulence version of the diffusion coefficient for comparison to molecular and locomotory diffusion.[2]

---

[2] McNair et al. (1997) present a method for making more-precise calculations, but discussion of the method is beyond our purview.

First, u. The average speed of particles as they are buffeted by turbulence near a solid boundary can be characterized by the *shear velocity* u_*, an index of turbulence intensity (Chapter 6). Typically, u_* is 10% of $\bar{u}$, the average speed of the wind or current.

Next, τ: particles in turbulent flow don't take discrete steps, but empirical observations suggest that particles change directions more often the closer they are to the boundary. In essence, their mean free path is proportional to z. If we assume that the constant of proportionality is on the order of 1, we can estimate the characteristic interval

$$\tau \approx \frac{z}{u_*}, \tag{4.33}$$

where z is distance from the boundary.

With estimates of u_* and τ in hand, we use equation 4.28 to calculate the turbulent diffusion coefficient for motion along the z-axis:

$$\mathcal{D}_z \approx \frac{1}{2}\tau u_*^2 = \frac{1}{2}z u_*. \tag{4.34}$$

Using our rule of thumb that u_* is a tenth of $\bar{u}$:

$$\mathcal{D}_z \approx \frac{z\bar{u}}{20}. \tag{4.35}$$

For example, a meter above the sea floor in a $0.2\,\text{m} \cdot \text{s}^{-1}$ current, algal spores have a turbulent diffusivity of approximately $10^{-3}\,\text{m}^2 \cdot \text{s}^{-1}$, 11 orders of magnitude greater than their Brownian diffusion constant and 6 orders of magnitude greater than the $\mathcal{D}$ of a swimming bacterium. A meter above the ground in a gentle breeze ($2\,\text{m} \cdot \text{s}^{-1}$), pollen grains have a $\mathcal{D}_z$ on the order of $10^{-2}\,\text{m}^2 \cdot \text{s}^{-1}$, 10 orders of magnitude greater than that of Brownian diffusion. Rough as these estimates are, they nonetheless indicate that turbulence transports particles much more effectively than does molecular or Brownian diffusion.

As noted before, turbulence can also be created in the surface layer of lakes and oceans, driven by the interaction of wind with the water's surface. In the middle of a lake or ocean, the horizontal flow of water is constrained by neither solid boundaries nor the air-water interface. In these circumstances, velocities in turbulent eddies can reach meters per second, while eddies can travel for seconds to hours before changing directions. As a result, horizontal turbulent diffusivities can have coefficients of 0.1 to $1000\,\text{m}^2 \cdot \text{s}^{-1}$, and these large diffusivities can rapidly disperse aquatic propagules.

It is important to note that the turbulent diffusion described here applies to the dispersion of macroscopic particles as they move with the fluid. To a first approximation, this motion does not affect the diffusion of small molecules in the vicinity of each particle. For example, as a phytoplankton cell is transported by a large oceanic eddy, it doesn't move relative to the water around it, and the diffusive delivery of bicarbonate and nutrients continues as if the cell were in still water (Kiørboe 2008).

5.5   Drivers of Random Walks: A Summary

It may be helpful at this point to recap our results. Random walks can be driven by thermal agitation, Brownian motion, locomotion, and turbulence. The diffusion of

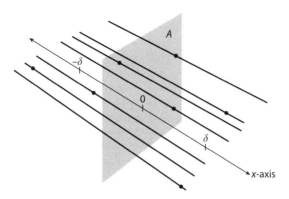

Figure 4.3 The system used to derive Fick's law. Each particle steps randomly along an axis parallel to the *x*-axis and is free to move through area *A*.

small molecules by thermal agitation can be effective at small scales, but is useless at large scales. Brownian diffusion is seldom if ever effective for living things, but locomotion and turbulence can be effective means of transport over a wide range of scales.

6 FICK'S EQUATION

So far we have investigated diffusion in terms of dispersion. There are many cases, however, in which it helps to translate this viewpoint into a measure of bulk transport—how fast can material be delivered or removed by random motion? How fast does carbon dioxide diffuse into leaves, for instance? How fast does water evaporate from a seaweed? (The explanation of diffusion that follows here is based on the classic presentation of Berg (1983).)

We again examine the movement of individual particles as they slide along a line, but rather than dealing with multiple particles on a single axis, let's consider multiple axes parallel to each other, each with a single particle positioned at random (Figure 4.3). We erect a hypothetical area *A* perpendicular to the lines, and for each line place its origin at the intersection with this area. Particles can step back and forth unimpeded through *A*, and we desire to keep track of the *flux* of particles, the net number per time moving through *A*.

To calculate flux, we again appeal to the rules of a random walk. As before, each particle moves a distance $\delta = u\tau$ every τ seconds. This simplifies matters considerably. Only those particles that lie within δ of area *A*—particles initially located between $-\delta$ and $+\delta$—can cross through the area in their next step, so these are the only particles we need to take into account.

Consider the situation at the start of our experiment. Among the particles within δ of *A*, there are n_L to its left; τ seconds later, half these particles will have stepped to the right and thus will have crossed through our hypothetical area. By the same token, there are n_R particles to the right of *A*, and at time τ half of them will have stepped to the left through *A*. The net number of particles transiting *A* to the right is

$$\text{net number moving to the right} = \frac{1}{2}n_L - \frac{1}{2}n_R. \qquad (4.36)$$

For reasons that become apparent in a moment, we rearrange these terms:

$$\text{net number moving to the right} = -\frac{1}{2}(n_R - n_L).\tag{4.37}$$

(The factor of $\frac{1}{2}$ here is the reason for the $\frac{1}{2}$ in equation 4.20.)

Flux is defined as the net number of particles crossing A per time per area, so to calculate flux (symbolized by $\mathcal{F}_x$), we divide equation 4.37 by time τ and A:

$$\mathcal{F}_x = -\frac{1}{2}\frac{n_R - n_L}{\tau A}.\tag{4.38}$$

Now for a mathematical sleight of hand. We multiply the right side of this expression by 1 (disguised as δ^2/δ^2) and rearrange:

$$\mathcal{F}_x = -\frac{\delta^2}{2\tau}\frac{1}{\delta}\left(\frac{n_R}{A\delta} - \frac{n_L}{A\delta}\right).\tag{4.39}$$

We then note that the product of A (an area) and δ (a length) is a volume. Thus, $n_R/(A\delta)$ is the number of particles per volume, a *concentration*, C. Specifically, it is the concentration of particles immediately to the right of area A. Similarly, $n_L/(A\delta)$ is the concentration of particles to the left of area A.

It is useful to assign specific locations to these concentrations: n_R is the number of particles located anywhere between 0 and δ. If the particles are uniformly distributed, their average position is $\delta/2$, and we use this as a characteristic location for concentration to the right of A. That is, we express the concentration to the right of A as $C(\delta/2)$. Similarly, concentration to the left of A is $C(-\delta/2)$. Using this notation,

$$\mathcal{F}_x = -\frac{\delta^2}{2\tau}\frac{C(\delta/2) - C(-\delta/2)}{\delta}.\tag{4.40}$$

There are only two steps left before we reach our goal. First, we substitute $u\tau$ for δ in the first fraction, with the result that

$$\frac{\delta^2}{2\tau} = \frac{\tau u^2}{2} = \mathcal{D}.\tag{4.41}$$

Then there is the term

$$\frac{C(\delta/2) - C(-\delta/2)}{\delta},\tag{4.42}$$

the difference in two values (the concentrations) divided by the distance between them (δ). Harken back to your days in introductory calculus, and you will realize that if we allow δ to approach zero this term becomes dC/dx, the gradient of concentration. Thus, our final simplification is

$$\mathcal{F}_x = -D\frac{dC}{dx}.\tag{4.43}$$

The flux of particles (net transport per time per area) is proportional both to the diffusion coefficient and the gradient in concentration: the steeper the gradient and the larger the diffusion coefficient, the greater the flux. The negative sign tells us that transport is from areas of high concentration to those of low concentration.

This is *Fick's first equation of diffusion*,[3] and it provides entry into the practical use of diffusion theory in much of biology, physics, and engineering. Note that in deriving this equation, we haven't needed to specify what is driving the random walk. Consequently, Fick's equation applies to all modes of diffusion.

Before we put Fick's equation to work, it is useful to note that the same logic we applied to calculate flux along the x-axis can be used in other directions, so

$$\mathcal{F}_y = -D\frac{dC}{dy}, \tag{4.44}$$

$$\mathcal{F}_z = -D\frac{dC}{dz}. \tag{4.45}$$

Similar logic allows us to calculate the radial flux toward or away from a central point, with results that look very similar:

$$\mathcal{F}_\ell = -D\frac{dC}{d\ell}. \tag{4.46}$$

In this case, ℓ is radial distance from the center.

7   THE MAXIMUM SIZE OF PHYTOPLANKTON

Let's put Fick's equation to use. Consider, for instance, a common problem in biology—a sphere of radius r (a reasonable model of a cell) is immersed in water with a certain concentration of a particular solute molecule, and we desire to know the rate at which the cell can absorb the solute. To give this scenario some tangibility, let's use a single-celled phytoplankton (a dinoflagellate, for instance) as our cell, and nitrate and ammonium (the dinoflagellate's source of nitrogen for growth and reproduction) as the solutes of interest. We assume that the volume of water surrounding the cell is so large that the concentration of nutrients (C_∞) is maintained in the bulk water regardless of any fluxes in the immediate vicinity of the phytoplankter. At what rate does the delivery of nitrogenous nutrients allow the dinoflagellate to grow and divide?

To answer this question, we start by assuming that every nutrient molecule that arrives at the cell's surface (a distance $\ell = r$ from the cell's center) is immediately absorbed. Consequently, the concentration at $\ell = r$ is zero. Given this condition, a steady state is soon reached in which concentration at radius $\ell\ (\geq r)$ is

$$C(\ell) = C_\infty \left(1 - \frac{r}{\ell}\right). \tag{4.47}$$

(For a derivation, see Supplement 4.2.) We can use this relationship as an entry into Fick's equation.

The derivative of equation 4.47 with respect to ℓ is

$$\frac{dC(\ell)}{d\ell} = \frac{C_\infty r}{\ell^2}. \tag{4.48}$$

---

[3] Named for Adolf Eugen Fick (1829–1901) a German-born physiologist. Fick has another law named in his honor, which describes the temporal rate of change of concentration, but we will not have an opportunity to use it.

Inserting this expression into the radial version of Fick's equation (equation 4.46) we see that

$$\mathcal{F}_r(\ell) = -DC_\infty \frac{r}{\ell^2}. \tag{4.49}$$

At the cell's surface ($\ell = r$),

$$\mathcal{F}_r(r) = -\frac{DC_\infty}{r}. \tag{4.50}$$

That is, the flux of nitrate or ammonium to the phytoplankter is radially inward (the negative sign tells us this) and is inversely proportional to the cell's radius. The larger the diffusion coefficient and the greater the bulk concentration of nutrients, the faster the flux, as you might expect. But the larger the cell, the lower the flux, which you might not expect.

Now flux is the number of particles delivered per time per area. To calculate the total rate at which nutrients are delivered to our phytoplankter (the *current*, moles per second), we need to multiply $\mathcal{F}_r$ by the cell's surface area, $4\pi r^2$:

$$\text{current} = 4\pi DC_\infty r. \tag{4.51}$$

(The negative sign disappears because we define current as positive inward.) This conclusion should seem strange. Even though the cell's surface area increases with the square of radius, current increases with radius only to the first power.

This disproportionate scaling has been used by Kiørboe (2008) to predict the maximum size of phytoplankton. These unicellular plants are typically denser than seawater, and thus they sink. As a consequence, the average depth of a group of cells increases through time, although turbulence can cause the group to disperse, some being swept upward while others are carried down. The phytoplankton population can be maintained in the well-lit surface layer of the ocean only if those cells that are by chance dispersed upward reproduce fast enough to counter the continuous loss of cells that results from the average rate at which the population sinks; for typical oceanic tubulence they must divide approximately once per day. (We will explore this line of reasoning in greater depth in Chapter 24.)

However, before it can reproduce, a phytoplankter must double its nitrogen content to have sufficient nutrients to make a new cell. (Nitrogen is the limiting nutrient in much of the ocean.) If C_N is the concentration of nitrogen in a cell (approximately 1500 moles per m³), the amount of nitrogen that must be absorbed prior to cell division is

$$\text{nitrogen needed} = \frac{4}{3}\pi r^3 C_N. \tag{4.52}$$

Dividing nitrogen needed by the rate at which it is delivered, we can solve for the time t (in seconds) required to accumulate a new cell's worth of nitrogenous nutrients:

$$t = \frac{\frac{4}{3}\pi r^3 C_N}{4\pi DC_\infty r}$$

$$= \frac{r^2 C_N}{3DC_\infty}. \tag{4.53}$$

We then solve for the maximum size of cell that can reproduce at a given rate:

$$r_{max} = \sqrt{\frac{3\mathcal{D}tC_\infty}{C_N}}. \tag{4.54}$$

Setting t to 1 day, we can solve for the maximum viable cell radius. At 10°C, the diffusion coefficient of nitrate and ammonium is approximately $1.5 \times 10^{-9}\,\mathrm{m^2 \cdot s^{-1}}$. In a typical midlatitude area of the ocean, in winter nitrate is the primary source of nitrogen for phytoplankton, with a concentration of $0.01\,\mathrm{mol \cdot m^{-3}}$. Plugging in these values, we estimate that the maximum winter-time radius of a phytoplankter is 51 μm. In summer, nitrate is often exceedingly scarce, and ammonium (at a concentration of $10^{-4}\,\mathrm{mol \cdot m^{-3}}$) is the only source of nitrogen. In this case, r_{max} is only 5.1 μm. These sizes are quite close to those observed in the field. Thus, diffusion neatly accounts for the tenfold seasonal variation in phytoplankton size that has long been noted by biological oceanographers.

These calculations have substantial evolutionary and ecological importance. Single-celled phytoplankton account for nearly half of earth's primary productivity, and the carbon they fix forms the basis for the pelagic marine food web. The fact that they are constrained to such small sizes makes for fundamental differences between terrestrial and marine ecosystems. Terrestrial plants have a tremendous range of sizes—from diatom scum to towering sequoias—and the size range of herbivores is similarly broad—from fruit flies to elephants. In large part because phytoplankton are so small, the size range of oceanic herbivores is much more constrained—from ciliates at the low end (barely larger than the phytoplankton they eat) to a few fish larvae at the high end. These constraints in turn affect the design of predators, with consequences that cascade up the food chain.

8 CONCEPTS AND CONCLUSIONS

There are four important messages to take away from this chapter:

1. Random motion of particles causes them to disperse, a process known as diffusion.
2. In general, dispersion distance increases with the square root of time, allowing diffusion to be an effective means of transport at small scales, but compromising its viability at large scales.
3. "Small" and "large" depend on what drives a random walk. Molecular diffusion is important for distances less than a few millimeters; locomotory and turbulent diffusion can be important at much larger distances.
4. Particles diffuse down a concentration gradient at a rate that depends on the diffusion coefficient.

We have explored one example of how these facts can be applied in a biological context, but this is only the beginning. Throughout the rest of this book, diffusive processes will appear, at times in the most unlikely circumstances. When they do, you might want to return to this short list of messages to refresh your understanding.

Chapter 5

An Introduction to Fluid Mechanics

In this chapter, I introduce the basic principles of fluid dynamics. Building on the concept of viscosity, we will learn how to tell solids from fluids, and how to use Reynolds numbers to characterize patterns of flow. In particular, Reynolds numbers allow us to predict whether flow will be orderly or turbulent. We then use the fact that energy is conserved to formulate Bernoulli's equation, a relationship that tells us how pressure varies along streamlines. These concepts will have frequent application in chapters to come. But before we can begin, we need terminology to describe the manner in which materials change shape.

1 SHEAR

In Chapter 13 we will undertake a thorough investigation of the ways in which materials can be deformed—including tension and compression—but for present purposes we are interested only in one mode of deformation, *shear*. To see how shear works, consider a small cube of material with sides of length dz (Figure 5.1). If we apply forces parallel to opposite faces, acting in opposite directions, the cube deforms by an amount dx. The ratio of dx to dz—a measure of proportional deformation—is known as *shear strain*, γ, an index of how much the shape of the cube is changed:

$$\gamma = \frac{dx}{dz}. \qquad (5.1)$$

The behavior of γ provides the means to differentiate between solids and fluids.

2 WHAT IS A FLUID?

Consider two hypothetical experiments performed on a cube of material like that shown in Figure 5.1. In the first experiment, the cube is made of rubber, a classic example of a solid. When loaded in shear, the rubber deforms. The larger the applied force, the greater the shear strain, and, to a good approximation, the amount of deformation does not depend on how long the force is imposed. We apply the force,

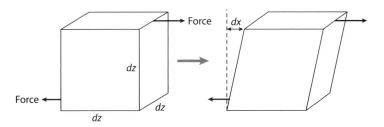

Figure 5.1 When loaded in shear, a material deforms.

measure γ, and get a cup of coffee. When we come back in an hour, γ hasn't changed. Thus, for a pure solid,

$$F = k_s \gamma. \tag{5.2}$$

The coefficient of proportionality, k_s, (with units of newtons) is a measure of the cube's shear stiffness. In engineering parlance, an object that abides by equation 5.2 is said to be *elastic*.

The second experiment is a repeat of the first, except that the cube is made of water rather than rubber. When a force is applied, the cube is again sheared, and if water were solid, the cube would deform a certain amount and then stop. Instead, as a fluid, water flows, deforming at rate $d\gamma/dt$. The larger the force applied, the faster the deformation. In other words, for a fluid, it is the *rate* of deformation—rather than the amount—that is proportional to force. Couched in terms of shear strain, for a pure fluid,

$$F = k_f \frac{d\gamma}{dt}. \tag{5.3}$$

Here the coefficient k_f (with units of newton seconds) is a measure of the fluid cube's *viscosity*. An object that behaves according to equation 5.3 is said to be *viscous*, and it is the contrast between elasticity and viscosity that separates solids from fluids.

The line between the two is not as distinct as one might desire. Most solids have some viscous characteristics. If a force is applied to a rubber cube for a day rather than an hour, the cube's deformation will in fact increase fractionally, revealing the material's latent viscous tendencies. And some biological fluids—mucus comes to mind—display a bit of elasticity. Materials with this dual nature are said to be *viscoelastic*. We will deal with solids and viscoelastic materials in Chapter 13, but for the moment our focus is on pure fluids where force is proportional solely to strain rate.

3   LIQUIDS AND GASES

Fluids come in two varieties—liquids and gases—a distinction based on two aspects of their density:

1. Liquids are more dense than gases. The density of water (the liquid of interest to ecomechanics) is approximately $1000 \, \text{kg} \cdot \text{m}^{-3}$, whereas the density of air (the pertinent gas) is approximately $1.2 \, \text{kg} \cdot \text{m}^{-3}$, 833-fold less.

Table 5.1 Physical Properties of Air (at 1 Atmosphere Pressure), Freshwater, and Seawater (Data from Denny 1993).

$T^\circ C$	ρ $kg \cdot m^{-3}$	μ 10^{-6} Pa·s	ν 10^{-6} m²·s⁻¹
		Air	
0	1.293	17.18	13.3
10	1.247	17.68	14.2
20	1.205	18.18	15.1
30	1.165	18.66	16.0
40	1.128	19.14	17.0
		Freshwater	
0	999.87	1790	1.79
10	999.73	1310	1.31
20	998.23	1010	1.01
30	995.68	800	0.80
40	992.22	650	0.66
		Seawater ($S = 35‰$)	
0	1028	1890	1.84
10	1027	1390	1.35
20	1025	1090	1.06
30	1022	870	0.85
40	1018	710	0.70

2. The density of liquids is more or less constant, whereas the density of gases varies dramatically with changes in pressure and temperature. For example, water 10 m below the surface of a lake or ocean is subjected to a pressure twice that at the surface, but its density increases by only 0.005%. By contrast, a bubble of air at a depth of 10 m has half the volume—and therefore twice the density—it would have at the surface. Across the biologically relevant temperature range (roughly 0°C to 40°C), water's density varies by only 1%, while the density of air varies by 13% (Table 5.1).

The differences in density between water and air are due to the different manner in which their molecules interact. Molecules in a liquid are sufficiently attracted to each other to maintain themselves in close (but not rigid) physical contact, and this tight packing leads to a high density. By contrast, gas molecules have negligible mutual attraction, and thermal agitation spreads them out, resulting in a low density.

Throughout the rest of this text we will explore the contrasts between water and air, but here at the beginning we instead concentrate on their similarity—they are both fluids.

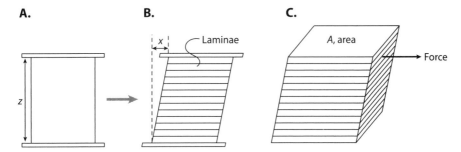

A.   **B.**   **C.**

Figure 5.2 A. The no-slip condition allows two solid plates to gain purchase on the fluid between them. **B.** As the top plate moves, fluid is sheared, behaving like a stack of laminae. **C.** Force required to shear a fluid is proportional to the fluid's area.

4   DYNAMIC VISCOSITY

As we have seen, the difference between fluids and solids becomes apparent when they are deformed. Everyday experience makes us familiar with how to deform solids: how to stretch a rubber band, break a stick, or bend a paper clip. But how does one grab a fluid to shear it?

Consider the simple system shown in Figure 5.2A: a volume of fluid sandwiched between two horizontal plates. What happens if we slide the plates parallel to each other in opposite directions? One might easily suppose that they could glide over the fluid, leaving it undisturbed. Instead, physics dictates that fluid in contact with a solid surface acts as if it were stuck to that surface, a property known as the *no-slip condition*. It doesn't matter whether the fluid is air or water, nor does the composition of the solid surface make a difference. At the interface between fluid and solid, there is no appreciable slip.[1] This less-than-intuitive phenomenon provides us with a mechanism to gain purchase on the fluid; as we slide the plates, fluid at their surfaces moves with them, and the fluid in between is sheared.

It is convenient to think of the fluid between the plates as a stack of infinitesimally thin layers (*laminae*), akin to a deck of cards. As the fluid shears, each layer slides relative to the layers above and below (Figure 5.2B), and it is this layer-to-layer slippage that controls fluid's response to a shearing force. In reality, there are no discrete laminae in a fluid, but, for simple patterns of fluid flow such as this, the analogy to a deck of cards is useful, and these orderly motions are therefore known as *laminar flow*. (The analogy loses its utility when flow is turbulent, but we postpone that discussion to the next chapter.)

Now, the defining characteristic of a fluid is that force is proportional to the rate of deformation:

$$F = k_f \frac{d\gamma}{dt} = k_f \frac{d\,(dx/dz)}{dt}. \tag{5.4}$$

---

[1] The validity of this assertion was a matter of debate until well into the twentieth century (see Vogel 1994), and in some systems (rarified gases and liquids in microfluidic chambers) a minute amount of slip has been observed. But the slip is so slight that, for our purposes, it is functionally nonexistent.

Rearranging the differentials, we find that

$$F = k_f \frac{d\,(dx/dt)}{dz}.$$ (5.5)

But dx/dt is u, velocity in the x direction, so

$$F = k_f \frac{du}{dz}.$$ (5.6)

In other words, for a fluid, force is proportional to the *velocity gradient, du/dz.*

F in equation 5.6 is the force required to deform a particular cube of fluid with sides of length dz. Thus, k_f applies only to a sample with these particular dimensions. But intuition and everyday experience tell us that the amount of force required to impose a certain velocity gradient depends on the sample's size; it takes more force to stir a whole vat of soup than it does to stir a cup of tea. To be specific, the force required to create a given velocity gradient in a fluid depends on the area, A, over which force is applied (Figure 5.2C). Thus, if we desire to quantify the properties of the fluid itself, rather than merely the properties of a particular sample, we need to normalize force to area:

$$\frac{F}{A} = \frac{k_f}{A}\frac{du}{dz}.$$ (5.7)

At this point, it is convenient to introduce two new symbols. Shear force per area is known as *shear stress, τ,* with units of $N \cdot m^{-2}$ (that is, Pa):

$$\text{shear stress, } \tau = \frac{F}{A},$$ (5.8)

and the ratio of k_f to a sample's area is the fluid's *dynamic viscosity, μ* ($N \cdot s \cdot m^{-2}$, or Pa $\cdot$ s):

$$\mu = \frac{k_f}{A}.$$ (5.9)

Thus

$$\tau = \mu \frac{du}{dz}.$$ (5.10)

This equation (Newton's law of friction) provides a basic description of how fluids behave: under the action of an applied shear stress, a velocity gradient is established, and the ratio of stress to gradient is set by the dynamic viscosity of the fluid. Or we can turn this logic around: wherever we find a velocity gradient in a viscous fluid, a shear stress exists.

For a given velocity gradient, shear stress is much greater in water than in air because water is much more viscous. For example, at 20°C water's dynamic viscosity is 56 times that of air (Table 5.1). Water and air also differ in how temperature affects their viscosity. The viscosity of water decreases dramatically with increasing temperature; μ at 40°C is only 38% of μ at 0°C. By contrast, air's viscosity *increases* with temperature, albeit only slightly; μ at 40°C is 11% higher than that at 0°C. (For an informal explanation of why viscosity varies with temperature differently in water and air, see Supplement 5.1.)

5   VISCOUS FRICTION

We have seen how shear can be modeled as the sliding of fluid laminae relative to each other. In this perspective, a fluid's viscosity can be thought of as the friction that must be overcome to keep laminae moving. The more viscous the fluid, the greater the tendency for each lamina to stick to the layers above and below, and the more force required to maintain a given strain rate.

In Chapter 2, however, we noted that friction converts useful mechanical energy to heat. Force is required to slide laminae relative to each other, but because this force is used to overcome the fluid's friction, energy expended in deforming the fluid is converted to heat, and is thereby lost as a means for doing mechanical work. To put it another way: the existence of a velocity gradient in a viscous fluid implies the loss of mechanical energy to heat.

6   REYNOLDS NUMBER

The fact that shear results in a loss of mechanical energy provides insight into the pattern of fluid motion. Consider the flow that results when you stir cream into coffee. As long as you actively stir, the cream swirls, and when you stop stirring, swirling continues for several seconds. By contrast, try stirring cream into a cup of honey. Swizzle as you might, you really can't get any eddies going, and when you stop stirring, flow ceases almost instantly. Why the difference?

There are two major factors that contribute to these contrasting patterns of flow. The first is the fluid's inertia. Once stirred up, even a small volume of coffee (a fluid particle, if you will) has substantial momentum, which tends to keep it moving. In large part, it is the unrestrained momentum of fluid particles that produces the turbulent eddies that swirl the cream. Because coffee is a viscous fluid, the internal friction of the sheared liquid gradually dissipates the particles' momentum, and flow eventually stops. But because coffee has a relatively low viscosity, inertial forces dominate and dissipation takes considerable time.

By contrast, honey is 10,000 times as viscous as coffee, so viscous forces dominate. While actively stirred, a particle of honey can have the same momentum as a particle of coffee, but its movement is restrained by viscous friction. As a result, eddies never really develop, motion remains laminar, and as soon as you stop stirring, flow ceases.

These observations suggest that it is the ratio of inertial to viscous forces that determines the pattern of fluid motion—turbulent eddies versus laminar flow. This ratio is known as the *Reynolds number*, Re:[2]

$$\mathrm{Re} = \frac{\text{inertial force}}{\text{viscous force}}. \tag{5.11}$$

Because Re is the ratio of two forces, it is dimensionless.

It is all well and good to propose that Re can serve as an index of the pattern of flow, but how can we know a priori what inertial and viscous forces act in particular circumstances? To answer this question, let's consider a hypothetical situation: a small solid cube with sides of length ℓ suspended motionless in fluid of density ρ, which

---

[2] Named for Osborne Reynolds (1842–1912), a prominent British fluid dynamicist.

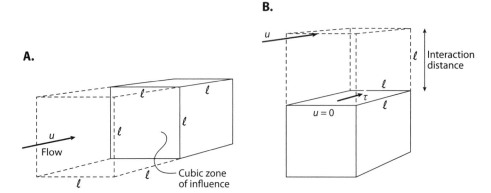

Figure 5.3 A solid cube immersed in flow is used to define the components of the Reynolds number. **A.** Inertial force. **B.** Viscous force. (See text.)

flows past the cube at velocity u (Figure 5.3). By characterizing the pattern of flow around the cube, we can quantify the inertial and viscous forces acting on the fluid, thereby allowing us to calculate the Re of the flow.

First, the inertial force. As fluid approaches the cube's upstream face, the presence of the cube deflects flow to either side. For simplicity, let's assume that the solid cube's influence on flow extends upstream to a volume equal in size and shape to that of the solid cube itself (Figure 5.3A). As fluid moves into this zone of influence it is redirected, causing it to lose much of its x-directed velocity, and this change in velocity affects the fluid's kinetic energy.

Consider, for instance, a cubic fluid particle with sides of length ℓ destined to flow into the zone of influence. Before it arrives, the fluid particle has kinetic energy proportional to its mass $(\rho \ell^3)$ times the square of its velocity (u^2). As the particle's flow is redirected by the solid cube—that is, as work is done on it by the solid—the fluid loses much of its x-directed kinetic energy. In other words,

$$\text{change in energy} \propto \rho \ell^3 u^2. \tag{5.12}$$

Recall that work is force times distance. Therefore, dividing this change in energy (= work) by the length of the interaction zone (ℓ) allows us to calculate the inertial force acting on the fluid. To a first approximation,

$$\text{inertial force} \propto \frac{\rho \ell^3 u^2}{\ell} = \rho \ell^2 u^2. \tag{5.13}$$

Now, the viscous force. Due to the no-slip condition, as fluid flows past the solid cube, a velocity gradient is established adjacent to each lateral face, imposing a shear stress (Figure 5.3B):

$$\tau = \mu \frac{du}{dy}. \tag{5.14}$$

Lacking specific information about the precise dimensions of the velocity gradient, we again suppose that the extent of the cube's influence is proportional to its size, such that velocity u is reached a distance on the order of ℓ away from the cube's face. As a result, the velocity gradient is approximately u/ℓ, and the shear stress is

$$\tau \propto \mu \frac{u}{\ell}. \tag{5.15}$$

This stress (force per area) acts on the area of the cube's lateral face (ℓ^2), so

$$\text{viscous force} \propto \tau \ell^2 = \mu u \ell. \tag{5.16}$$

We are now in a position to calculate the Reynolds number:

$$\text{Re} = \frac{\text{inertial force}}{\text{viscous force}} = \frac{\rho \ell^2 u^2}{\mu u \ell} = \frac{\rho u \ell}{\mu}. \tag{5.17}$$

The greater the density and velocity of the fluid and the larger the object, the higher the Reynolds number. The greater the fluid's viscosity, the lower the Reynolds number.

In our derivation of Re, we used the length of the solid cube's sides as a handy measure by which to characterize its size and thereby the spatial scale of its interaction with the fluid. Other simple objects have similarly obvious characteristic lengths—the diameter of a sphere, for instance. However, for many biological objects, it is not clear what length to choose. For wind flowing past a cactus, one could reasonably pick either the plant's width or its height, which can be quite different. As a consequence, for the same density, viscosity, and flow speed, the Re one calculates for flow around a cactus differs depending on which characteristic length one chooses. In cases such as this, one choice is as good as another, but it becomes necessary to specify explicitly what characteristic length is being used.

Note that equation 5.17 involves the ratio of dynamic viscosity to density, a ratio that occurs so often in fluid dynamics it is given its own name and symbol—*kinematic viscosity*, ν:

$$\nu = \frac{\mu}{\rho}. \tag{5.18}$$

Kinematic viscosity has units of $m^2 \cdot s^{-1}$. (Note for future reference that the units of ν are the same as those of a diffusion coefficient. In fact, we will find in Chapter 6 that ν is the diffusion coefficient for momentum.) Given this definition, we can restate equation 5.17:

$$\text{Re} = \frac{u \ell}{\nu}. \tag{5.19}$$

Relative to its density, air is more viscous than water; as a result its kinematic viscosity is 7- to 24-fold higher than that of water, depending on temperature (Table 5.1). Thus, for a given flow speed and object size, Re in air is 7- to 24-fold lower than it is in water.

As I have suggested, Re is an index of the pattern of fluid flow. When inertial forces are small compared to viscous forces (that is when Re $\ll 1$), flow is orderly and laminar. Conversely, when inertial forces far outweigh viscous forces (when Re $\gg 1$), fluid has a tendency to form turbulent swirls and eddies. Note that these conclusions hold whether the fluid in question is air or water.

Two pertinent biological examples provide some feel for this concept. Consider a limpet attached to an intertidal rock; the animal is roughly $\ell = 1$ cm long. Seawater at a typical ocean temperature of 10°C has a kinematic viscosity of $1.35 \times 10^{-6}\ m^2 \cdot s^{-1}$, and the velocity of water in a wave as it surges past the limpet is typically $10\ m \cdot s^{-1}$.

Thus, the flow around the limpet has the following Re:

$$Re = \frac{10^{-2} \text{ m} \times 10 \text{ m} \cdot \text{s}^{-1}}{1.35 \times 10^{-6} \text{ m}^2 \cdot \text{s}^{-1}} = 7.4 \times 10^4. \tag{5.20}$$

Because this Re vastly excedes 1, we know without having to look that the wave-induced flow around the limpet is likely to be turbulent.

By contrast, consider a pollen grain as it sinks through air at a velocity of $2 \text{ cm} \cdot \text{s}^{-1}$. The grain has a diameter of 40 μm, and the kinematic viscosity of room-temperature air is $1.5 \times 10^{-5} \text{m}^2 \cdot \text{s}^{-1}$. The Reynolds number for flow around the grain is thus

$$Re = \frac{(40 \times 10^{-6} \text{ m}) \times (2 \times 10^{-2} \text{ m s}^{-1})}{1.5 \times 10^{-5} \text{ m}^2 \text{ s}^{-1}} = 5.3 \times 10^{-2}, \tag{5.21}$$

a value sufficiently small that we can guess with some assurance that flow around the pollen grain is laminar.

A note on precision: recall that our derivation of Re was based on rough guesses for interaction distance and velocity gradient. As a consequence, our expression for the ratio of inertial to viscous forces is at best an approximation. If we are honest, we need to take this intrinsic lack of precision into account. For example, when we calculate a Reynolds number of 7.4×10^4 for a limpet, we should interpret this to mean that the ratio of inertial to viscous forces is roughly 10^5. Similarly, the Reynolds number of a pollen grain (5.3×10^{-2}) tells us that the ratio of forces in that situation is approximately 0.1.

7   CONSERVATION OF MASS AND VOLUME

At the small scale of individual organisms and at the velocities encountered in biology, local changes in temperature and pressure are sufficiently small that we can safely ignore the changes they incur in the density of air and water. Consequently, the volume of a mass of fluid is conserved during flow, a property known as the *principle of continuity*.

Continuity provides us with a useful tool. Consider, for instance, the rigid pipe shown in Figure 5.4A. At its inlet, the pipe has cross-sectional area A_1, but some distance downstream area decreases to A_2 as the pipe necks down. Let's assume that the pipe is filled with fluid whose density (and, therefore, volume) is constant. What happens if we inject a new volume V of fluid into the pipe's inlet?

Volume is the product of area and length, so to make room for injected volume V, fluid already in the fat section of the pipe must move a distance

$$x_1 = \frac{V}{A_1}. \tag{5.22}$$

But because the pipe is rigid and volume is conserved, as fluid is injected at the upstream end, an equal volume must exit downstream. Thus, fluid in the narrow section of pipe moves a distance

$$x_2 = \frac{V}{A_2}. \tag{5.23}$$

A. New volume, V

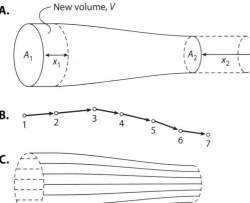

B.

C.

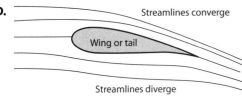

D.

Streamlines converge

Wing or tail

Streamlines diverge

Figure 5.4 A rigid pipe with decreasing cross-sectional area illustrates the principle of continuity: **A.** Volume entering equals volume exiting. **B.** A streamline. **C.** A streamtube. **D.** Flow around a wing or tail.

The ratio of x_2 to x_1 is thus

$$\frac{x_2}{x_1} = \frac{\left(\frac{V}{A_2}\right)}{\left(\frac{V}{A_1}\right)} = \frac{A_1}{A_2}. \qquad (5.24)$$

Because A_1 is greater than A_2, x_2 is greater than x_1.

Suppose that it takes t seconds to inject volume V. The distances x_1 and x_2 are therefore traversed in time t, with the result that

$$u_1 = \frac{x_1}{t} = \frac{\left(\frac{V}{t}\right)}{A_1}, \qquad (5.25)$$

$$u_2 = \frac{x_2}{t} = \frac{\left(\frac{V}{t}\right)}{A_2}, \qquad (5.26)$$

$$\frac{u_2}{u_1} = \frac{A_1}{A_2}. \qquad (5.27)$$

Because $A_2 < A_1$, the velocity of water at the pipe's outlet (u_2) is greater than velocity at its inlet (u_1). This is why it is so satifying to put your thumb over the end of a garden hose. Water flows into the hose (a rigid pipe) at a constant rate, and velocity in the bulk of the hose is relatively slow. But, by constricting the hose's outlet with your thumb—thereby reducing A_2—the principle of continuity requires the water to speed up, allowing you to squirt the person next to you.

8   STREAMLINES

The preceding example involved a rigid tube, and the conclusion reached can be applied directly to rigid biological pipes such as the xylem tubes in terrestrial plants. However, when combined with the concept of *streamlines*, the principle of continuity can be applied in a more general context.

Figure 5.4B represents a slice through a steadily moving fluid where the arrows define a streamline, the path a fluid particle takes as it flows: a particle starting at point 1 moves to point 2, then to point 3, and so forth. Streamlines such as this have three important properties:

1. The path defined by the streamline is constant through time.
2. Particles can speed up and slow down as they travel along the stream-line, but at any given point, the speed of each particle is the same as it passes through.
3. Particles introduced at the upstream end of the streamline travel precisely along it, never deviating. In more techical terms, particles on a streamline always move tangent to the line.

Properties 1 and 2 constitute a definition of what we mean by *steady flow*, and it is important to keep in mind that streamlines as we have defined them here exist only when flow is steady.

Let's take a moment to examine in more detail what the existence of streamlines implies. First, *streamlines can never cross*. If they could, a particle arriving at the intersection could subsequently take two different paths, violating rule 3. As a corollary to the fact that streamlines can't cross, we can deduce that *flow can never cross a streamline*. In order for a fluid particle to cross a streamline, its direction of flow would have to deviate from the streamline, and that isn't allowed.

The streamline shown in Figure 5.4B is just an example. In similar fashion, we could trace streamlines from every starting position in the fluid, and the pattern of these streamlines would provide a detailed and useful description of the fluid's steady motion.

The utility of streamlines is demonstrated in Figure 5.4C, which shows a set of streamlines that chart the outline of a *streamtube* resembling the pipe of Figure 5.4A. Even though this streamtube has no physical walls, its effect is exactly the same as if it were a rigid pipe. Because flow can't cross streamlines, fluid can't move into or out of the streamtube through its walls. And because the position of streamlines doesn't change through time, the streamtube acts as if it were rigid. The similarity between pipe and streamtube allows us to make accurate quantitative predictions about the velocity of fluid in the streamtube. As the streamtube contracts, velocity increases just as it would in a rigid pipe.

This analysis provides us with a convenient shorthand method for deducing the pattern of fluid flow. Take, for example, the situation shown in Figure 5.4D, a flowwise cross section through the fluid moving from left to right around a bird's wing. Streamlines, which are evenly spaced upstream of the wing, converge on the top side of the structure and diverge on the bottom. Without knowing anything else about the flow (and barring flow into or out of the page), we can immediately conclude that flow over the top of the wing, where streamlines converge, is fast relative to flow under the bottom, where streamlines diverge. Were we to measure the relative spacing between streamlines, we could accurately quantify the ratios of fluid velocity (equation 5.27).

9 CONSERVATION OF ENERGY
Bernoulli's Principle

As we discussed in Chapter 2, conservation of energy is a basic tenet of physics, and it leads to one of the most useful relationships in fluid mechanics—Bernoulli's principle. As we will see, among many other uses Bernoulli's principle allows us to translate the pattern of flow past wings into the forces that result. However, before we can dig into the the the practical use of Bernoulli's principle, we first need to introduce the concept of an ideal fluid.

For present purposes, a fluid is ideal if, as it flows, it doesn't lose energy to friction. At first glance, this would seem to pose a serious problem. Air and water, the fluids that we care about, are both viscous and therefore prone to frictional loss of energy. But think back for a moment to equation 5.10: the shear stress required to keep a fluid moving is equal to the product of dynamic viscosity and the velocity gradient. As a consequence, if the velocity gradient is zero, shear stress—and consquently frictional loss of energy—vanishes even though the fluid is viscous. Furthermore, the only mechanism we have encountered so far to create a velocity gradient in a fluid is the no-slip condition at a solid-fluid boundary. Thus, in the absence of solid surfaces—or in fluid sufficiently far removed from a solid that its presence doesn't matter—substantial velocity gradients aren't readily formed, and fluids behave approximately as if they are inviscid. Thus, in practical terms, as long as we stay far away from solid walls, we are free to treat air and water as if they were ideal, and to assume that a fluid particle's energy remains constant as it flows. This assumption allows us to draw the following conclusions.

A fluid particle's energy comes in three forms:

1. If it has mass m and flows at velocity u, the particle has kinetic energy $\frac{1}{2}mu^2$.
2. Elevation z relative to some reference height gives the particle gravitational potential energy mgz.
3. And lastly, if a net pressure p is applied to the particle over area A, the particle does work pAx as it moves through distance x. The product of area and distance is volume V; thus the energy associated with net pressure is $pAx = pV$.

For an ideal fluid, the sum of these three energy components is a constant, k:

$$pV + mgz + \frac{1}{2}mu^2 = k \tag{5.28}$$

This is one form of Bernoulli's principle, named in honor of Daniel Bernoulli (1700–82), who first derived it in 1738. (For a formal derivation, see Supplement 5.2.)

When dealing with fluids it is often convenient to deal with mass per volume—that is, density—rather than mass per se. To this end, we divide equation 5.28 by volume, arriving at an alternative expression for the conservation of energy. Noting that mass per volume is density, ρ:

$$p + g\rho z + \frac{1}{2}\rho u^2 = k. \tag{5.29}$$

Each of the terms in equation 5.29 has units of energy per volume, which (upon working through the math) is the same as force per area, that is, pressure. First there is p, the pressure imposed by external forces. Next is $\rho g z$, the static pressure associated with the weight of the fluid. And, lastly, there is $\frac{1}{2}\rho u^2$, a term known as the *dynamic pressure*, pressure due to the fluid's motion.

At this point, a small problem might be niggling at you. It is all well and good to say that the sum of p, $\rho g z$, and $\frac{1}{2}\rho u^2$ is constant, but how can we employ this relationship if we don't know that constant's magnitude? There are indeed times when one might

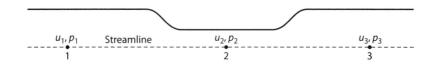

Figure 5.5 Illustrating Bernoulli's principle: flow at three points in a streamtube of variable height but constant width, here shown in cross section.

need to know k explicitly, but more commonly it is sufficient simply to know that k is indeed constant.

Consider, for example, a cross section through a hypothetical streamtube of varying height (Figure 5.5). Let's explore how we can use Bernoulli's equation to deduce the fluid's status at the three points shown on the streamline. Although we do not know what k is at any of these points, we know that it is the same for all. Thus,

$$p_1 + \rho g z_1 + \frac{1}{2}\rho u_1^2 = k = p_2 + \rho g z_2 + \frac{1}{2}\rho u_2^2. \tag{5.30}$$

In this simple case, $z_1 = z_2$, so

$$p_1 + \frac{\rho u_1^2}{2} = p_2 + \frac{\rho u_2^2}{2}, \tag{5.31}$$

and working through the algebra, we conclude that

$$p_1 - p_2 = \rho \frac{u_2^2 - u_1^2}{2}. \tag{5.32}$$

We now invoke our knowledge of streamlines. Because streamlines converge in the vicinity of point 2, u_2 must be greater than u_1. As a consequence, the right side of equation 5.32 is positive. But that implies that p_1 (upstream pressure) must be greater than p_2 (downstream pressure).

The same logic can be applied to points 2 and 3, spanning the portion of the streamtube where flow broadens out:

$$p_2 - p_3 = \rho \frac{u_3^2 - u_2^2}{2}. \tag{5.33}$$

Because u_3 is less than u_2, p_3 must be greater than p_2. In short, *in an ideal fluid there is a trade-off between velocity and pressure: where velocity is high, pressure is low, and vice versa.*

This trade-off makes intuitive sense when viewed from the perspective of a fluid particle. As it approaches a constriction in flow, the high pressure behind and low pressure ahead cause the particle to speed up. As flow broadens out, the low pressure behind and high pressure ahead slow it down.

10   CONCEPTS, CONCLUSIONS, AND CAVEATS

Air and water—life's media—figure prominently in our exploration of ecomechanics. Air is a gas; water, a liquid; and their densities and viscosities are therefore vastly different. Despite these contrasts, both air and water are fluids, and we have emphasized this conceptual similarity:

- The central characteristic of a fluid is its viscosity, a stickiness that causes sheared liquids and gases to dissipate energy.
- Reynolds number provides a quantitative index of the ratio of inertial to viscous forces, a ratio that governs the pattern of flow.
- The principle of continuity and the concept of streamlines provide tools for quantifying the pattern in which the velocity of a fluid—any fluid—changes through space.
- And, Bernoulli's principle tells us how these changes in velocity are inversely correlated with changes in pressure.

Most of these conclusions apply universally to fluids and, therefore, to both air and water. Bernoulli's equation, however, applies only to steady flow far removed from solid surfaces. Flow in the real world is often unsteady, and in the vicinity of organisms it is necessarily near a solid surface (the plant or animal's integument). In these cases, Bernoulli's equation provides valuable insight but only a rough estimate of actual pressure. Our task in the next three chapters is to figure a way to apply these basic principles to the complex flows characteristic of nature.

For a more detailed introduction to fluid dynamics, consult Vogel (1994), from which you can move on to the bibles of the subject (e.g., Batchelor 1967; Happel and Brenner 1973).

Chapter 6

Boundary Layers I
Equilibrium Layers

Near solid surfaces, viscosity inhibits flow, leading to the creation of a region of relatively slow-moving fluid known as the *boundary layer*. There is, for instance, a boundary layer around each of a pine tree's needles, which is, in turn, encased in a thicker boundary layer surrounding a branch, and the entire tree is surrounded by the large-scale boundary layer of Earth's atmosphere. Similar hierarchies pertain to a cricket's sensory hairs, which operate in the body's boundary layer, which lies within the boundary layer of the forest floor. Boundary layers are everywhere, and life is embedded in them.

Some boundary layers have enough space and time to equilibrate with the surrounding flow. Equilibrium boundary layers of this sort can be found in rivers, streams, and ocean currents, for example, or on open savannahs in a steady breeze. By contrast, where flow changes rapidly (e.g., wave-, gust-, or sound-driven oscillations) or interaction distance is limited (e.g., flows around individual organisms), boundary layers do not have a chance to equilibrate.

To cope with the broad scale and differing characteristics of boundary-layer flows, our discussion will extend across two chapters. In this first chapter, we explore the mechanics of equilibrium boundary layers. Their constancy allows for relatively simple analysis and sets us up for the second chapter, in which we deal with dynamic boundary layers. In each chapter we first deal with orderly, laminar flow and then progress to the effects of turbulence.

Our goal is to quantify several characteristics of boundary layers. What is the pattern of flow within them? What shear stress is required to create and maintain them? How do these layers of retarded flow affect the transport of mass and momentum? And, of course, what are the biological consequences? As we will see, boundary layers affect the forces imposed on plants and animals and play a major role in determining the transport of materials to and from organisms.

1   BASIC CONCEPTS

We begin with a review of three key concepts:

1. Momentum is the product of mass and velocity (mu). For fluids, which don't come in discrete chunks, it is convenient to deal with momentum per volume (ρu) rather than momentum itself.
2. Dynamic viscosity μ is the stickiness of a fluid, its resistance to being sheared. Because of viscosity, force is required to maintain a velocity gradient (du/dz) in a fluid. If force F acts over area A, Newton's law of friction tells us that

$$\frac{F}{A} = \tau = \mu \frac{du}{dz}. \qquad (6.1)$$

   The greater the viscosity or the steeper the velocity gradient, the larger the shear stress τ.
3. The no-slip condition mandates that directly at a solid surface, there is no movement of fluid relative to the surface.

Together, these concepts provide the basis for undersanding boundary layers.

2   LAMINAR BOUNDARY LAYERS
An Introduction

Consider Figure 6.1, a flat plate oriented parallel to flow. In panel A, fluid upstream of the plate moves with uniform velocity u_∞ relative to the stationary plate. In panel B, the plate moves with velocity u_∞ relative to stationary fluid. In both cases, because there is no velocity gradient upstream of the plate, there is no shear stress in the upstream fluid.

The situation changes abruptly, however, as fluid encounters the solid. At the plate's leading edge, relative velocity goes to zero in the fluid lamina adjacent to the boundary (layer 0), instantly creating a difference in velocity between it and layer 1, the next lamina out. Coupled with the fluid's viscosity, this steep velocity gradient imposes a shear stress on the fluid (equation 6.1), slowing it down if the plate is stationary and speeding it up if it is the plate that moves. (At the same time, viscous friction imposes a shear stress—a viscous drag—on the solid surface. We'll return to this drag in Chapter 8.)

The interaction between the solid surface and layers 0 and 1 is just the beginning, however. Consider, for instance, flow past the stationary plate (panel A). As flow slows in layer 1, a velocity gradient is established between it and layer 2. The friction associated with this gradient slows layer 2, creating a velocity gradient between layers 2 and 3, and so forth. As time passes and fluid moves downstream along the plate, the region of reduced relative flow—the boundary layer—extends farther and farther away from the solid boundary, and velocity at its outer edge blends asymptotically into the mainstream (Figure 6.1C).

For the moving plate, the process proceeds in reverse. As fluid accelerates in layer 1, a velocity gradient is established between it and layer 2. The friction associated

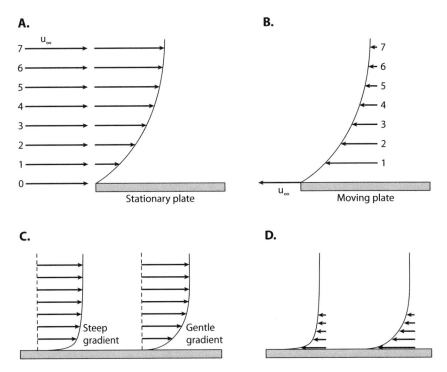

Figure 6.1 A. Fluid flowing past a stationary plate is impeded. **B.** A plate moving through a stationary fluid drags fluid with it. **C** and **D.** The velocity gradient at the surface is maximal at the leading edge.

with this gradient accelerates layer 2, which creates a velocity gradient between layers 2 and 3, and so forth. As time passes and the plate moves through the fluid, the boundary layer extends farther and farther away from the solid boundary, and velocity at its outer edge asymptotically approaches zero (Figure 6.1D).

Given sufficient time and distance, boundary layers can extend until they reach some obstruction (the fluid's surface, for instance), after which flow settles into an equilibrium. More commonly, however, the boundary layer reaches the trailing edge of the object and is shed as a *wake*, a region of disturbed flow in the object's lee. We will deal with nonequilibrium boundary layers of this sort in Chapter 7, but for the moment we focus on equilibrium layers.

In an equilibrium boundary layer, viscous forces in the fluid have equilibrated with the shear stress on the adjacent solid surface, so there is no net force acting on the fluid. As a consequence, although velocity may vary with z (the distance from the boundary), at any particular distance x downstream from a leading edge velocity at a given z is constant. We can use this equilibrium to predict the spatial variation in shear stress and the resulting pattern of flow in the boundary layer.

A mountain stream provides a handy example of equilibrium flow (Figure 6.2A). We align our coordinate system with the tilted streambed such that the x-axis is parallel to both the bed and the water's surface and the z-axis extends perpendicular to the bed. The force required to maintain steady flow is provided by the acceleration of gravity, which—because the streambed is tilted—has a component pulling water downstream. To simplify matters, rather than analyzing the situation in terms of x-directed force itself, we formulate the problem in terms of force per volume, Φ.

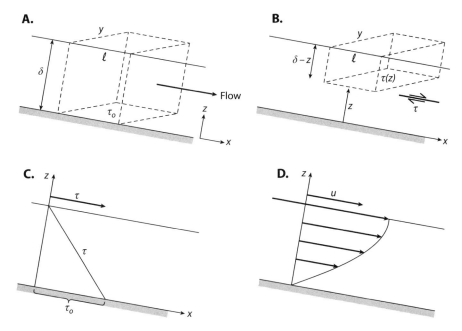

Figure 6.2 An equilibrium boundary layer in a stream of depth δ. **A.** Gravity pulls water parallel to the tilted stream bed (in the x-direction). **B.** For each volume of water (delineated by the dashed lines), the downstream force is equaled by shear stress on the volume's lower surface (z above the streambed). **C.** Shear stress decreases linearly with increasing z. **D.** Velocity increases nonlinearly with increasing z.

Let's investigate the x-directed force acting on a volume ℓ long and y wide, extending from the streambed to the surface, a distance δ:

$$F = \text{volume} \times \frac{\text{force}}{\text{volume}} = \ell y \delta \Phi. \tag{6.2}$$

This force imposes a boundary shear stress (τ_0) on the water-streambed interface:

$$\tau_0 = \frac{F}{\text{contact area}}$$

$$= \frac{\ell y \delta \Phi}{\ell y}$$

$$= \delta \Phi. \tag{6.3}$$

By rearranging terms, we can define Φ in terms of τ_0 and the stream's depth:

$$\Phi = \frac{\tau_0}{\delta}. \tag{6.4}$$

Next, we examine the force acting on a plane ℓ long by y wide a distance z above the streambed (Figure 6.2B). Between this plane and the water's surface, there is a volume $\ell y (\delta - z)$. The downstream force acting on this volume is $\ell y (\delta - z) \Phi$, which applies a shear stress on fluid in the plane a distance z above the bed:

$$\tau(z) = \frac{\ell y (\delta - z) \Phi}{\ell y} = \Phi (\delta - z) \tag{6.5}$$

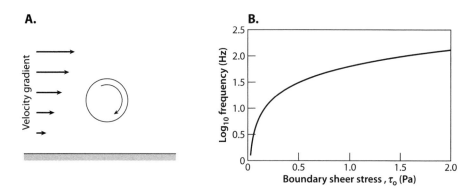

Figure 6.3 A. A spherical particle rotates in a boundary layer. **B.** The rate of rotation increases with boundary shear stress.

But $\Phi = \tau_0/\delta$ (equation 6.4), so

$$\tau(z) = \frac{\tau_0}{\delta}(\delta - z)$$

$$= \tau_0 \left(1 - \frac{z}{\delta}\right). \tag{6.6}$$

This relationship—the goal of our first task—is shown in Figure 6.2C: shear stress decreases linearly from its maximum at the streambed (τ_0) to zero at the water's surface.

Our next job is to determine the velocity gradient associated with this stress gradient. This is easily done. We know from equation 6.1 that

$$\frac{du}{dz} = \frac{\tau}{\mu}. \tag{6.7}$$

Thus, in our stream

$$\frac{du}{dz} = \frac{\tau_0}{\mu}\left(1 - \frac{z}{\delta}\right). \tag{6.8}$$

Like the shear stress, the velocity gradient decreases from a maximum at the streambed to zero at the surface.

Given this linear variation in velocity gradient, how does velocity itself vary with z? To answer this question, we integrate equation 6.8 to solve for u:

$$u(z) = \frac{\tau_0}{\mu}\left(z - \frac{z^2}{2\delta}\right). \tag{6.9}$$

This expression is graphed in Figure 6.2D. Velocity is zero at the substratum (as it must be) and in this particular case it increases parabolically to a maximum at the water's surface. Different circumstances can alter the exact shape of the velocity profile, but its general character is captured in this example: a gradual, nonlinear approach to maximum (that is, mainstream) velocity.

3 MOMENT AND ROTATION

In addition to the stress it places on the solid substratum, shear in a boundary layer affects solid objects suspended in flow. Consider a solid sphere embedded in a boundary layer's velocity gradient (Figure 6.3A). As a result of the no-slip condition, fluid in contact with the sphere acts as if it is stuck to the sphere. However, because fluid velocity varies with distance from the substratum, the bottom of the sphere is stuck to fluid moving at a lower velocity than that stuck to its top. The resulting viscous interaction applies a moment, causing the sphere to rotate (Kessler 1986):

$$\text{moment} = 4\pi \mu r^3 \left(\frac{du}{dz} - 4\pi f \right). \tag{6.10}$$

Here r is the sphere's radius and f is its rotational velocity (cycles per second, Hz). Under the influence of this moment, the sphere spins faster and faster until $4\pi f = du/dz$, at which point net moment is zero. Thus, the equilibrium spinning rate of a sphere in a velocity gradient is

$$f_{eq} = \frac{1}{4\pi} \frac{du}{dz}. \tag{6.11}$$

The steeper the velocity gradient, the faster the sphere tumbles. Note that f_{eq} depends only on du/dz; it is independent of μ. So, for a given velocity gradient, particles spin as fast in air as they do in water.

We know from equation 6.8 how du/dz varies as a function of boundary shear stress and location in a stream's water column. Substituting this relationship for du/dz into equation 6.11,

$$f_{eq} = \frac{1}{4\pi} \frac{\tau_0}{\mu} \left(1 - \frac{z}{\delta} \right), \tag{6.12}$$

which leads to the conclusion that the rate of rotation is maximal at the substratum:

$$f_{eq,max} = \frac{1}{4\pi} \frac{\tau_0}{\mu}. \tag{6.13}$$

Even for small shear stresses, passive particles spin at tens of cycles per second (Figure 6.3B).

These rapid rotation rates make it extremely difficult for small swimming organisms (such as larvae) to control their movements near a substratum. Unless they use their appendages to continuously apply a moment counter to that imposed by the velocity gradient, they lose control over their orientation and tumble arse over teakettle.

Flow-induced rotation can also pose a problem for external fertilization. In the boundary layer where sperm and eggs are released, eggs spin rapidly, creating a steep velocity gradient in their vicinity. In this local velocity gradient, a sperm's tail acts like a weather vane, forcing the sperm to swim tangentially around the egg and making it difficult for the two gametes to contact each other (e.g., Denny et al. 2002; Guasto et al. 2011; Zimmer and Riffell 2011).

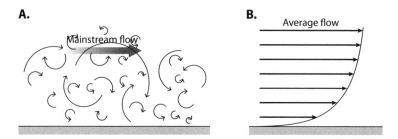

Figure 6.4 A. Turbulent eddies complicate the pattern of boundary-layer flow. **B.** However, consistent patterns can be discerned when flow is averaged over time.

4   TURBULENT BOUNDARY LAYERS
An Introduction

As we have encountered it so far, a boundary layer consists of an orderly gradient of velocities extending away from a solid surface. But quite often flow in a boundary layer can be turbulent. The presence of eddies creates a chaotic distribution of flow speeds and directions whose random motions can have substantial biological consequences. How can we reconcile this swirling with the orderly concept of a boundary layer?

At any instant, we can't. Imagine a snapshot of the velocities near a flat plate as a turbulent mass of water flows over it (Figure 6.4A). Water immediately in contact with the plate must be stationary due to the no-slip condition, but otherwise flow varies in a seemingly unpredictable fashion as turbulent eddies stretch and divide, forming smaller eddies that likewise deform and split in an ever-changing cascade.

Order can be discerned, however, if we average velocities through time (Figure 6.4B). Temporally averaged velocity increases in an orderly fashion with distance from the substratum, and it is this gradient of average velocity that defines the turbulent boundary layer. Making sense of turbulent boundary layers thus requires us to combine the disparate aspects of their nature: chaotic flows that carry fluid particles on random walks and the orderly pattern of these flows when averaged over time.

In our introduction to laminar boundary layers, we envisioned a wave of shear stress that starts at the boundary and moves outward into the fluid. An alternative perspective—one particularly useful for turbulent layers—tracks the inward transport of fluid momentum as it is absorbed by the solid surface. To visualize this process, consider the following analogy: Imagine two trains traveling on parallel tracks (Figure 6.5). Each train comprises a series of boxcars loaded with sandbags such that the mass of the two trains is the same, but train 1 moves faster than train 2, so its momentum is greater. So far, there is nothing in this scenario that implies that one train has any effect on the other.

Now let's suppose that men in each box car start tossing sandbags from their train to the other. For every sandbag thrown from train 1 to train 2, one is thrown from 2 to 1, so that the mass of each train is constant. But the sandbags thrown from train 1 have higher velocity in the direction of travel, so when they land in train 2, they increase its net momentum. Conversely, the sandbags thrown from train 2 have lower velocity, so as they land in train 1 its net momentum is reduced. In short, as sandbags

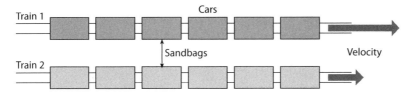

Figure 6.5 Momentum is exchanged by turbulent eddies, illustrated here by an analogy to two trains traveling on parallel tracks at different speeds. (See text.)

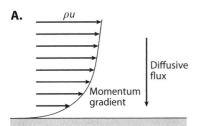

Figure 6.6 **A**. Momentum flows down the average velocity gradient. **B**. Roughness elements on an otherwise smooth plate.

are exchanged, they tend to equalize the trains' momentum; it's as if the faster train is pulling the slower train forward and the slower train is dragging the faster one back. The more rapidly sandbags are exchanged, the greater the effect.

An analogous process occurs in turbulent boundary layers (Figure 6.6A). Averaged over time, fluid at distance z from a boundary acts like a train, and the turbulent exchange of mass with fluid at larger or smaller z transports mass (and thereby momentum) between fluid layers, creating a shear stress. Note that, because mass is exchanged by turbulence-driven random walks, momentum is transported diffusively.

5   THE ONSET OF TURBULENCE

As you have likely guessed, tubulence complicates the prediction of the velocity gradient in boundary-layer flow. To cope with these complications, we approach the subject in stages, beginning with a discussion of when and where turbulence is found.

As fluid moves over a smooth solid surface, instabilities can arise in the boundary layer, locally increasing the boundary shear stress and causing turbulent eddies to form. There is an extensive literature exploring the mechanics of turbulence generation on smooth plates (see Schlichting and Gerstens 2000), but it has little relevance to biology because there are very few truly smooth surfaces in nature. Instead, we focus our attention on how turbulence arises on the rugose surfaces typical of the sea bed, the ground, and organisms.

Consider a smooth plate coated with *roughness elements* (Figure 6.6B). In laboratory experiments, these are typically sand grains of near-uniform size; in the real world, they could be almost anything: surface texture on a rock, denticles on a shark's skin, or trees in a forest. In each case, the relevant size of roughness elements is indexed by their average height, D. Numerous experiments have shown that the transition from a laminar to a turbulent boundary layer can be predicted using the *roughness Reynolds number*:

$$\text{Re}_* = \frac{\sqrt{\frac{\tau_0}{\rho}} D}{\nu}. \tag{6.14}$$

where ν is again the fluid's kinematic viscosity and ρ is its density. If Re_* is greater than approximately 60, flow in an equilibrium boundary layer is turbulent (Schlichting and Gerstens 2000).

We can rearrange equation 6.14 to solve for the roughness-element height required to ensure that the boundary layer is turbulent:

$$D_{\text{crit}} = 60\nu \sqrt{\frac{\rho}{\tau_0}}. \tag{6.15}$$

The higher the boundary shear stress, the smaller the critical roughness height.

To put this relationship in a biological context, we need to know what shear stresses matter to plants and animals. A variety of benchmarks could be chosen, but let's use the stress required to move small particles. For example, on desert floors, river beds, and ocean bottoms, benthic organisms must cope with the movement of sand. In water, a typical quartz sand grain 1 mm in diameter begins to move when τ_0 is approximately 0.5 Pa (Middleton and Southard 1984). Lacking the buoyancy they feel in water, sand grains in air need more shear stress to get moving (0.75 Pa). Shear stress can also affect organisms directly; a τ_0 of 1 to 2 Pa can dislodge marine larvae and algal spores as they attempt to settle (Koehl and Hadfield 2004). Thus, a range of τ_0 from 0.5 to 2 Pa serves to separate flows that are too benign to move even small particles from those that are likely to have significant biological effects.

In seawater at 10°C (where ν is 1.35×10^{-6} m²· s⁻¹), this range of τ_0 corresponds to critical roughness heights from 1.8 to 3.7 mm, really quite small. At 10°C, air's kinematic viscosity is 11 times that of water, but its density is 850-fold less, so (for the same range of shear stress) roughness elements in air need to be only 0.7 to 1.4 mm high to initiate turbulence in a boundary layer. In short, when stresses are large enough to move sediments, spores, and larvae, the presence of even minimal roughness ensures that equilibrium boundary layers are turbulent.

Let's return for a moment to the definition of the roughness Reynolds number. In order for Re_* to be dimensionless, $\sqrt{\tau_0/\rho}$ must have units of velocity. This combination of variables is known as the *friction velocity*, or *shear velocity*, u_*, a term I briefly mentioned in Chapter 4. It is an important descriptor of turbulent flow:

$$u_* = \sqrt{\frac{\tau_0}{\rho}}. \tag{6.16}$$

Technically, u_* (pronounced "you star") is a mathematical convenience: boundary shear stress expressed as a velocity by relating it to ρ. Informally, however, u_* serves as an indicator of the pattern in which instantaneous fluid speed in a boundary layer deviates from the long-term mean. As we will see, it incorporates aspects of both the magnitude of deviations and their correlation in space and time. Rewriting equation 6.14 in terms of u_*, we find that

$$\text{Re}_* = \frac{u_* D}{\nu}, \tag{6.17}$$

which has a more familiar form for a Reynolds number.

Laboratory and field measurements lead us to suspect that u_* ranges from approximately 5% of average mainstream velocity ($\bar{u}_\infty$) in benign flows to more that 20% of $\bar{u}_\infty$ in very energetic turbulence. For convenience, I will use a value in the middle of this range as a rule of thumb:

$$u_* \approx 0.1\bar{u}_\infty. \tag{6.18}$$

Thus,

$$\mathrm{Re}_* \approx \frac{\bar{u}_\infty D}{10\nu}, \tag{6.19}$$

which we can use to estimate the velocity required to ensure that the boundary layer is turbulent. Setting Re_* to its critical value of 60 and solving for D,

$$D_{\mathrm{crit}} \approx \frac{600\nu}{\bar{u}_\infty}. \tag{6.20}$$

For seawater moving at $\bar{u}_\infty = 0.1$ m·s^{-1}, roughness elements need to be only 8 mm high to ensure that the boundary layer is turbulent. If $\bar{u}_\infty$ is 1 m·s^{-1}, D_{crit} is a mere 0.8 mm. In a 1-m·s^{-1} breeze, terrestrial roughness elements need to be 9 mm high to ensure turbulence, but in a 10-m·s^{-1} wind, they need to be only 0.9 mm high. Again the take-home message is: relatively minor roughness leads to boundary-layer turbulence for flow speeds that are common in both water and air.

6 TURBULENT SHEAR STRESS

Given the broad circumstances in which boundary layers are turbulent, it behooves us to ask of them the same questions we asked of laminar layers: What is the boundary shear stress? How does τ vary with distance from the substratum? What is the velocity profile? We approach these questions by first taking a step back and examining how momentum diffuses in laminar flow.

Recall that our deck-of-cards analogy led us to conclude that

$$\tau = \mu \frac{du}{dz}.$$

Multiplying the right side of this equation by ρ/ρ (1 in disguise),

$$\tau = \frac{\mu}{\rho} \frac{d\,(\rho u)}{dz}. \tag{6.21}$$

Noting that kinematic viscosity $\nu = \mu/\rho$, we see that

$$\tau = \nu \frac{d\,(\rho u)}{dz}. \tag{6.22}$$

Now recall that ρu is momentum per volume. Thus, viscous shear stress in a laminar boundary layer is governed by ν and the gradient in momentum (which, in turn, is set by the velocity profile).

The form of this relationship is similar to that of Fick's equation for the transport of mass (Chapter 4):

$$\mathcal{F}_z = -D\frac{dC}{dz}, \tag{6.23}$$

where $\mathcal{F}_z$ is mass flux in the z-direction, $\mathcal{D}$ is the molecular diffusion coefficient, and dC/dz is the concentration gradient. Thus, by analogy, shear stress (the flux of momentum across the direction of flow) is caused by the diffusion of momentum down the velocity gradient, with ν playing the role of the diffusion coefficient. Recall from Chapter 5 that ν indeed has the dimensions of a diffusion coefficient: $m^2 \cdot s^{-1}$.

When the boundary layer is turbulent, the same general concept applies, but in this case it is the gradient of *average* momentum that matters, and shear stress is augmented as momentum is transported by the random motion of fluid in eddies. In turbulent flow, the rate of diffusion can be modeled using ϵ, the *eddy* or *turbulence diffusivity*, a turbulence equivalent of kinematic viscosity. Thus, by analogy to equation 6.23, shear stress due to turbulence is

$$\tau_{turb} = \epsilon \frac{d\,(\rho\bar{u})}{dz} \tag{6.24}$$

Equation 6.24 raises an obvious question: how large is the shear stress caused by turbulence relative to that caused by viscosity? To answer this question, we introduce the concept of Reynolds stress.

7   REYNOLDS SHEAR STRESS

As we have just seen, shear stress in a turbulent boundary layer is caused in part by the fluid's random motion. To incorporate these random fluctuations in speed into our thinking, we model velocity in the x-direction as the sum of mean velocity ($\bar{u}$) and the instantaneous deviation from that mean (Δu):

$$u = \bar{u} + \Delta u. \tag{6.25}$$

Similar logic applies to vertical velocity w:

$$w = \bar{w} + \Delta w. \tag{6.26}$$

With this notation in place, we can explore the consequences of combined vertical and horizontal velocity fluctuations. As we will see, if Δu and Δw are correlated, they can transport momentum and thereby cause a shear stress that augments that due to molecular viscosity. The correlation between Δu and Δw is expressed as the temporal average of their product, $\overline{\Delta u \Delta w}$, a value known as the *covariance*, and the stress caused by the covariance is *the Reynolds shear stress*:

$$\tau_{Re} = -\rho\,\overline{\Delta u \Delta w}. \tag{6.27}$$

(For a derivation, see Supplement 6.2.)

To see how the covariance behaves, let's investigate how its components interact. If, at any instant, Δu and Δw have opposite signs, their product $\Delta u \Delta w$ is negative. If Δu and Δw have the same sign, $\Delta u \Delta w$ is positive. When Δu and Δw vary randomly and independently of each other—as they would in mainstream flow—negative values on average cancel positive values, and the covariance is zero. Thus, in mainstream flow, fluctuations in velocity typically do not contribute to shear stress.

In a turbulent boundary layer, however, the time-averaged velocity gradient induces a correlation between Δu and Δw. Near a boundary, average vertical velocity is approximately zero, so if Δw of a fluid particle is positive, for that moment the particle (a miniature analogue of a sandbag) moves away from the substratum.

But, by moving outward in the boundary layer, the particle moves to an area of higher average u. Introduction of this relatively slow moving fluid momentarily reduces the x-directed speed of the fluid (the train) into which it has moved, resulting in a momentary decrease in u, that is, a negative Δu. Thus, on average, in a boundary layer, when Δw is positive Δu is negative, and their product, $\Delta u \Delta w$, is negative.

Similarly logic applies if Δw is negative. By moving closer to the substratum, a fluid particle augments the x-directed velocity of the fluid into which it moves, momentarily inducing a positive Δu. With Δw negative and Δu positive, $\Delta u \Delta w$ is again negative.

Thus, *because of the velocity gradient* there is, on average, a negative correlation between Δu and Δw, and the resulting exchange of momentum induces a Reynolds shear stress that adds to the stress imposed by molecular viscosity. This leads us to a turbulence version of Newton's law of friction (equation 6.1):

$$\tau = \text{viscous shear stress} + \text{Reynolds shear stress}$$

$$= \mu \frac{d\bar{u}}{dz} - \rho \overline{\Delta u \Delta w}. \tag{6.28}$$

Now to the main point of this exercise. How large is $-\rho \overline{\Delta u \Delta w}$ compared to $\mu du/dz$? Consider, for instance, flow in a mountain stream. If the boundary layer is laminar, we know from evaluating equation 6.9 at $z = \delta$ that

$$\tau_0 = \frac{2 \mu \bar{u}_\infty}{\delta}, \tag{6.29}$$

where I have substituted $\bar{u}_\infty$ for $u(\delta)$. For example, if mainstream flow is 1 m·s^{-1}, μ is 1.3×10^{-3} Pa·s, and the stream is 1 m deep, viscous shear stress is 0.0026 Pa.

If, instead, boundary-layer flow is turbulent, velocity fluctuations in each direction are—according to our rule of thumb—10% of mainstream velocity. If these fluctuations are perfectly correlated, $\overline{\Delta u \Delta w} = \overline{\Delta u} \times \overline{\Delta w}$ and the Reynolds shear stress (equation 6.27) is $0.01 \rho \bar{u}^2 = 10$ Pa, almost 4000 times that due to viscosity.

In reality, the correlation between Δu and Δw is unlikely to be perfect, but Reynolds stress can be large nonetheless. If the correlation is only 10% of its maximum value, for instance, Reynolds shear stress in this example is still nearly four hundred times viscous shear stress. In short, turbulent diffusion is much better than molecular diffusion at transporting momentum. To answer our original question, ϵ is typically much larger than ν.

We can put equation 6.28 to immediate use by revisiting Figure 6.2C, where we found that, in an equilibrium boundary layer, total shear stress decreases linearly from its maximum at the solid boundary to zero at the stream's surface. A quick glance back at that analysis reveals that we never assumed that flow was laminar—at any distance from the streambed we assumed that the downstream pull of gravity was just offset by the upstream tug due to shear stress, but we never specified the origin of that stress. Thus, regardless of the mechanism creating it, shear stress decreases linearly from the bottom to the top of an equilibrium boundary layer. If flow is turbulent, however, equation 6.28 tells us that most of the required shear stress is due to eddy diffusivity.

This leads us to the situation shown in Figure 6.7. Total shear stress varies linearly as required. However, through much of the boundary layer, this total is due almost entirely to eddy viscosity. Only amongst the roughness elements—where flow is in intimate contact with solid surfaces—does molecular viscosity contribute noticeably.

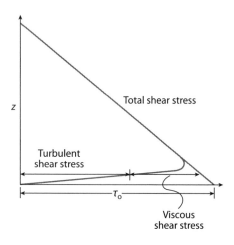

Figure 6.7 Shear stress in a turbulent boundary layer decreases linearly with distance from the substratum.

8   THE TURBULENT VELOCITY PROFILE

With this shear-stress profile in hand, our next task is to describe the turbulent velocity profile. Unfortunately, this can't be accomplished in the same straightforward fashion we used for laminar boundary layers. In laminar flow, momentum is transported by molecular viscosity, which (as an intrinsic property of the fluid) doesn't vary with distance from the solid surface. By contrast, in turbulent flow momentum is transported by eddies, whose interaction with the solid boundary causes ϵ to vary as a function of z. As a result, it is difficult to predict the turbulent velocity profile from first principles, and we must rely on empirical measurements.

Traditionally, the description of the velocity profile of an equilibrium turbulent boundary layer begins with a discussion of the layers adjacent to smooth surfaces (see Supplement 6.3). But smooth in this context refers to surfaces with roughness elements far too small to trigger turbulence ($D \ll D_{\text{crit}}$). Such surfaces exist in nature, but they are so rare that we can safely ignore them. The discussion that follows concerns the sort of rough surfaces that are much more common for organisms and their environment.

On a rough surface, the boundary layer can be divided into two sublayers: the true boundary layer which extends outward from the top of the roughness elements, and the interstitial boundary layer of flow amongst the roughness elements.

8.1   Above the Roughness Elements

We can predict the general shape of the turbulent velocity profile above the roughness elements through an analogy to the diffusion of gases. Recall from Chapter 4 that the diffusion coefficient of gases is the product of molecules' speed and the distance they travel between collisions (their mean free path). I propose that eddy diffusivity in a turbulent boundary layer can be modeled as being proportional to the product of u_* (used an index of the speed at which fluid particles move in their random walks) and z, the distance eddies travel before encountering the solid boundary. Thus, including the appropriate coefficient of proportionality,

$$\epsilon = \kappa u_* z. \tag{6.30}$$

Here κ is von Kármán's constant (dimensionless), which is generally accepted to be approximately 0.4. Substituting this expression into equation 6.24 (and noting that ρ is constant and can therefore be pulled out of the differential) we see that boundary shear stress due to turbulence is

$$\tau_0 = \kappa u_* z \rho \frac{d\bar{u}}{dz}. \tag{6.31}$$

Now, from equation 6.16, we know that $\tau_0 = \rho u_*^2$, so, when the dust clears,

$$\frac{d\bar{u}}{dz} = \frac{u_*}{\kappa z}. \tag{6.32}$$

We can now separate variables and integrate to solve for $\bar{u}$ as a function of z:

$$\bar{u}(z) = \frac{u_*}{\kappa} \ln(z) + k, \tag{6.33}$$

where k is the constant of integration. If we suppose that in fact $\bar{u}$ goes to zero at height z_0 above the boundary, $k = -\frac{u_*}{\kappa} \ln(z_0)$, and

$$\bar{u}(z) = \frac{u_*}{\kappa} \ln\left(\frac{z}{z_0}\right), \tag{6.34}$$

a relationship known as a *law of the wall*. Note that $\bar{u}$ increases with the logarithm of z; consequently, this portion of the boundary layer is commonly known as the *log layer*. For closely packed, rounded roughness elements, $z_0 \approx D/30$, but for biological roughness (e.g., vegetation) $z_0 \approx D/5$ (Kaimal and Finnigan 1994).

Before we finish, we need to take into account one more aspect of reality. In the absence of roughness, it is obvious where we should locate the origin of our z-axis— it should start at the solid boundary. But when roughness elements are present, they raise the virtual location of the solid boundary a distance z_1, known as the *zero plane displacement*. This displacement can be incorporated into the law of the wall:

$$\bar{u}(z) = \frac{u_*}{\kappa} \ln\left(\frac{z - z_1}{z_0}\right). \tag{6.35}$$

For many rough surfaces, z_1 is in the range from $0.6D$ to $0.8D$, but its value varies with the shape and distribution of roughness elements.

The velocity profile predicted by equation 6.35 is shown in Figure 6.8A. As demanded by equation 6.32, the velocity gradient $(d\bar{u}/dz)$is gentle except near the tops of the roughness elements, testament to the fact that tubulence is very good at transporting momentum, and thereby at homogenizing the velocity among different regions of flow. The bulk of the velocity gradient is confined to a narrow layer near the roughness elements, and it is the steepness of this gradient that accounts for the high shear stresses that accompany turbulent boundary layers.

Equation 6.35 is useful in characterizing the velocity gradient in steady aquatic flows (e.g., creeks, rivers, tidal currents). For example, Eckman and others (1981) used equation 6.35 to characterize the flow around worm tubes on a muddy-bottomed tidal flat. The flow patterns around these tubes have a substantial effect on the local bottom topography and, thereby, on the community structure of the organisms living in the mud; Nowell and Jumars (1984) review the literature on the subject. Equation 6.35 has also seen wide use in predicting flows over grasslands and forests subject to steady winds (Kaimal and Finnigan 1994).

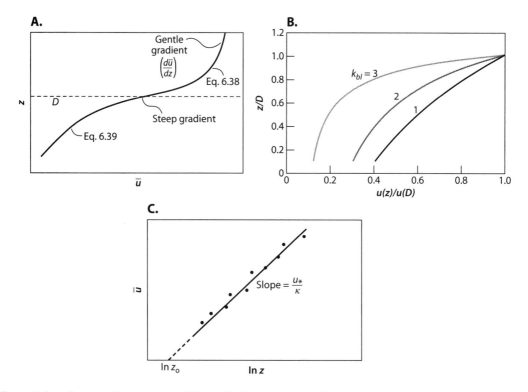

Figure 6.8 A. The velocity gradient is different for flow above and below the tops of the roughness elements. **B**. Flow within an array of roughness elements depends on k_{bl}. (See the text.) **C**. Calculating shear velocity, u_*. (See text.)

8.2   Among the Roughness Elements

So far we have investigated the turbulent velocity profile above the peaks of roughness elements. What happens down among the elements? What is average flow like within a mussel bed, grassy field, or forest? Knowledge of these velocities can be extremely useful, for instance, when trying to predict the flux of gases and particulate food.

As a general rule, average velocity decreases exponentially with distance below the top of the roughness elements (Figure 6.8A; Campbell and Norman 1998):

$$\bar{u}(z) = \bar{u}(D)\exp\left[k_{bl}\left(\frac{z}{D} - 1\right)\right].\tag{6.36}$$

The coefficient k_{bl} determines how rapidly velocity decreases as one moves toward the substratum (Figure 6.8B), and it varies with the shape of the roughness elements. A few representative values for k_{bl} are given in Table 6.1. As this table suggests, k_{bl} has been investigated most thoroughly in an agricultural context, where the roughness elements are leafy plants. The more tightly packed the plants and the leafier each one is, the higher k_{bl} is.

Equation 6.36 applies from $z = D/10$ to $z = D$. Near the base of the roughness elements, the solid boundary establishes its own local boundary layer, which—depending on whether it is laminar or turbulent—has a velocity profile described by equation 6.9 or 6.35.

Table 6.1 Measured Values of the Roughness Coefficient. Data from Campbell and Norman (1998).

Plant	k_{bl}
Immature corn	2.8
Oats	2.8
Wheat	2.5
Corn	2.0
Sunflower	1.3
Christmas trees	1.1
Larch trees	1.0
Citrus orchard	0.4

Equation 6.36 should be used with caution. It serves well as a general rule but can be misleading in particular cases. In coniferous forests and kelp beds, for example, foliage is tightly packed near the top of the roughness elements, forming a dense canopy, but the trunks and stipes below offer little resistance to flow. Under these circumstance, flow among the roughness elements may be channeled into the tunnel beneath the canopy, complicating the velocity profile. For a more thorough discussion of flow among roughness elements, consult Kaimal and Finnigan (1994).

9   MEASURING u_*

Although friction velocity is a central concept in the study of turbulent flow, the intensity of turbulence—of which u_* is an index—varies greatly from one situation to the next, depending on both the nature of mainstream flow and the size and shape of roughness elements. As a result, it is difficult a priori to predict u_* with any accuracy; this is why I have resorted to estimating u_* using a general rule of thumb. For anything other than the kind of rough, back-of-the-envelope calculations we have made in this chapter, it is necessary to measure u_* directly.

There are two ways to accomplish this. The first is applicable to flows in which turbulence is sufficiently intense that Reynolds shear stress far outweighs stress from molecular viscosity. Under these circumstances, equation 6.28 tells us that

$$\tau_0 \approx -\rho \overline{\Delta u \Delta w}. \tag{6.37}$$

Recalling that $u_*^2 = \tau_0/\rho$, we can then surmise that

$$u_* \approx \sqrt{-\overline{\Delta u \Delta w}}. \tag{6.38}$$

Thus, if we can measure Δu and Δw in the boundary layer and calculate their correlation, we can estimate u_*. (To be accurate, we would need to measure the correlation at a few different heights near the substratum and extrapolate to what the correlation would be at the boundary.) Measurement of Δu and Δw involves first measuring u and w for a lengthy period. From these data one then calculates $\overline{u}$ and $\overline{w}$, after which one can calculate Δu and Δw and their correlation. These measurements require specialized apparatus: hot-wire anemometers, acoustic or laser Doppler velocimeters, or laser light sheets and high-speed cameras.

If one lacks this technology, an alternative method is available, requiring only the ability to measure $\bar{u}$ as a function of z. Harking back to equation 6.35, we see that when $\bar{u}(z)$ is plotted as a function of $\ln(z)$, the slope of the line is u_*/κ, which can be estimated by calculating the linear regression of the data (Figure 6.8C). With u_*/κ in hand and the value of κ known (it is approximately 0.4), u_* is thus obtained.

There is an art to this procedure, however (Kaimal and Finnigan 1994). Velocity measurements taken too far above the substratum (beyond the proximal 20% of the boundary layer) may deviate from a logarithmic profile, as will measurements too close to the tops of roughness elements. It is therefore necessary to base one's analysis only on the subset of measurements that show a logarithmic relationship. Accurate estimate of u_* requires multiple measurement of $\bar{u}$ at different heights above the boundary, ideally with the z location of those points spaced logarithmically.

10   SWEEPS AND BURSTS

By concentrating on the average gradient in a turbulent boundary layer—as predicted by the logarithmic profile of equation 6.35 or the exponential profile of equation 6.36—it is easy to lose sight of the instantaneous aspects of flow. As we saw in Figure 6.4A, at any instant fluid motion near the boundary can differ from average flow. Indeed, experiments show that energetic eddies (called *sweeps*) are created by shear at the top of the roughness elements, and they periodically dip down to touch the substratum, imposing velocities that can momentarily equal or even exceed those in the mainstream. In response to a sweep, a *burst* of turbulence can be formed at the substratum, causing fluid to be ejected up into the boundary layer. These episodic events can have substantial biological consequences.

John Crimaldi and his coworkers have made a careful study of turbulent flow within a bed of benthic clams. As you might expect from equation 6.36, average velocity is low within the bed and decreases with decreased spacing between clams. Benthic ecologists had assumed that this reduction in velocity serves as a hydrodynamic refuge, allowing clam larvae to settle and grow amongst conspecific adults. But Crimaldi and his colleagues (2002) reached the opposite conclusion. Although average flow velocity decreases in the bed, maximum velocity *increases*. Because clam shells act as roughness elements, they increase both the intensity and frequency of sweeps, decreasing the probability that larvae can successfully settle. In short, the dynamic interaction of clam shells with flow makes it less likely that larvae can hide amongst their elders.

This hydrodynamic constraint could well have ecological consequences. Larvae can use the presence of conspecific adults as a clue that the local environment is amenable to growth and reproduction—if mom and pop survived here, it should be okay for me. Aggregative settlement can thus have survival value. This advantage could be negated, however, if the presence of adults makes it physically difficult for larvae to settle.

This is not to say that all aggregations are inhospitable. The clam beds used by Crimaldi and his colleagues were sparsely populated. Laboratory experiments (e.g., Nowell and Church 1979; Friedrichs et al. 2000) show that when roughness elements are closely spaced (covering one-twelfth of the substratum or more), flow tends to skim over the tops of the roughness rather than extend down into the interstices. Many intertidal and shallow subtidal rocky shores are densely covered with rugose plants and animals. It seems likely that these substrata are characterized by skimming flow and thus can potentially serve as a refuge from high water velocities. But the incursion of sweeps into dense assemblages—mussel beds, for instance—has not been studied, due in large part to the difficulty of measuring flow accurately amongst tightly packed organisms.

Just as larval settlement can be affected by sweeps, larval dispersal can be affected by bursts. One might easily suppose that interstitial species living within a clam bed, where average flow is slow, might have trouble dispersing their larvae. But, as sweeps impinge on the substratum, the consequent ejection of fluid in a burst can carry larvae out into the free stream.

The random motions that account for Reynolds stress in a boundary layer also transport CO_2 and water vapor in plant canopies (Kaimal and Finnigan 1994), so sweeps and bursts can affect the exchange of these important gases. Just as the presence of a clams complicates the delivery of larvae to the seafloor, trees complicate the exchange of gases in a forest. On the one hand, trees act as roughness elements, augmenting the intensity of turbulence at their upper edges. On the other hand, friction with leaves and branches saps the energy of the resulting sweeps as eddies swirl toward the substratum. Whether overall exchange is increased or decreased by trees depends on their size, shape, and packing, and robust generalities are again difficult to draw. For a review of theory, see Kaimal and Finnigan (1994). We will investigate aerial boundary layers in greater depth in Chapter 11 as we explore convective heat transfer and evaporation.

11   CONCEPTS, CONCLUSIONS, AND CAVEATS

The viscosity of air and water interacts with the no-slip condition to produce the boundary layer, a velocity gradient adjacent to any solid surface, and this region of retarded flow has several important effects:

- Transport of momentum down the velocity gradient imposes shear stress on the solid. We will see in Chapters 8 and 9 that this shear stress accounts for most of the drag on small organisms at low speeds and on streamlined organisms at high speeds.
- Turbulence transports momentum across the boundary layer much more effectively than does a fluid's dynamic viscosity. As a result, shear stresses are greater when the boundary layer is turbulent.
- Turbulence is triggered by roughness elements, ensuring that among equilibrium boundary layers, turbulence is the rule rather than the exception.
- Flow among roughness elements is complicated.

It is important to keep in mind that this chapter has dealt only with equilibrium boundary layers, layers that have had sufficient time and space for shear stress to equilibrate with the forces driving flow. Ocean currents and steady winds acting over large stretches of real estate can produce equilibrium boundary layers, as can flow in streams and rivers. But truly equilibrium boundary layers are relatively rare. It is much more common that space, time, or both are insufficient for flow to equilibrate. In these cases, the relationships developed in this chapter may be applicable in a qualitative sense, but they are guaranteed to be quantitatively inaccurate. To complete our investigation of boundary layers, we need to account for their dynamics, and this is the subject of the next chaper.

As always, I must warn you that this introduction to equilibrium boundary layers has only scratched the surface of the subject. For the complete story, I advise you to consult the bibles on the subject: Schlicting and Gerstens (2000) and Kaimal and Finnegan (1994).

Chapter 7

Boundary Layers II
Dynamic Layers

In the last chapter, we focused on equilibrium boundary layers of the sort associated with streambeds, benthic marine habitats, and spacious swaths of savannah or forest. However, boundary layers also commonly occur in circumstances where there is insufficient time or space for flow to equilibrate. A typical patch of forest floor or rocky substratum, for instance, consists of a series of individual surfaces, each with its own leading edge and a length of at most a few meters. Even if fluid flows steadily over these topographical features, it doesn't have time to come to equilibrium before exiting downstream, and if there are gusts or oscillations in the flow, equilibrium is nearly impossible. For the same reasons, the boundary layers surrounding individual organisms (or parts of organisms) seldom, if ever, equilibrate. In this chapter we explore the characteristics of these nonequilibrium, dynamic boundary layers.

1   BOUNDARY-LAYER THICKNESS DEFINED

When flow first encounters a solid surface, a velocity gradient diffuses into the fluid adjacent to the boundary, and velocity asymptotically approaches that of the mainstream, u_∞. This gradual change in velocity makes it difficult to locate a specific point at which the boundary layer stops and mainstream flow begins. Thus, if we desire to divide the fluid into boundary layer and mainstream, we are forced to pick a more or less arbitrary cutoff, some characteristic distance from the substratum that we will call the *boundary layer thickness*, δ. Traditionally, this cutoff is set at the point where flow reaches 99% of mainstream velocity; that is, δ is the z for which $u(z) = 0.99\overline{u}_\infty$ (Figure 7.1A).

Arbitrary as it is, this definition has heuristic value. Beyond this cutoff, there is very little gradient in velocity, and in the absence of a velocity gradient, fluid behaves as if it were inviscid. By contrast, in the boundary layer, flow *is* influenced by viscosity, with all that portends: shear stress, momentum flux, and the resulting temporal and spatial changes in velocity. Thus, boundary-layer thickness defined in this fashion divides flow into an outer or free-stream layer, where the effects of viscosity are negligible, and the boundary layer, where viscosity matters.

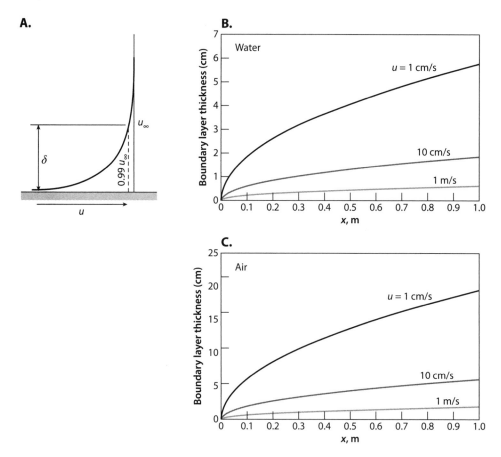

Figure 7.1 A. Boundary-layer thickness is the distance from a solid surface at which flow reaches 99% of mainstream velocity. **B** and **C.** Laminar boundary layer thickness increases as the square root of distance from a leading edge; the faster the mainstream flow, the thinner the layer (**B**, water; **C**, air).

2   THICKNESS OF DYNAMIC LAMINAR BOUNDARY LAYERS

Having defined δ, our next job is to predict its magnitude, which we accomplish using an analogy to molecular diffusion. Recall that when molecules diffuse, the average distance they travel is

$$\sigma = \sqrt{2\mathcal{D}t}. \tag{7.1}$$

where $\mathcal{D}$ is the diffusion coefficient and t is time (Chapter 4). In the last chapter, we concluded that boundary layers also grow through a diffusive process—the diffusion of momentum. Thus, by analogy, we expect δ to increase in proportion to the square root of the time over which flow interacts with a solid substratum. Recalling that kinematic viscosity, ν, is the diffusion coefficient for momentum, we hypothesize that

$$\delta \propto \sqrt{\nu t}, \tag{7.2}$$

a guess that turns out to be accurate. Careful analysis of the details of viscous flow (Blasius 1908) provides the coefficient of proportionality:

$$\delta = 5\sqrt{vt}. \tag{7.3}$$

The longer the period of contact, the thicker the boundary layer.

It is often more convenient to reformulate this relationship in terms of the distance traveled by fluid as it interacts with a solid. Noting that distance x from the leading edge can be expressed as $u_\infty t$, we can define t in terms of x and u_∞:

$$t = \frac{x}{u_\infty}. \tag{7.4}$$

Substituting this expression for t in equation 7.3:

$$\delta = 5\sqrt{\frac{vx}{u_\infty}}. \tag{7.5}$$

For a given mainstream flow, boundary-layer thickness increases with the square root of distance from the leading edge. For a given distance, δ is inversely proportional to the square root of mainstream velocity.

Equation 7.5 is graphed in Figure 7.1B and C for various velocities in both water and air, respectively. For example, a giant-kelp frond a meter long bathed in a 10-cm $\cdot$ s^{-1} current can grow a boundary layer 1.8 cm thick. By contrast, for a beech leaf 3 cm wide in a 1-m $\cdot$ s^{-1} breeze, δ is at most 3.4 mm.

In a final variation on the theme of equation 7.3, we can calculate a Reynolds number (the *local Reynolds number*, Re_x) using x, the distance from the leading edge, as our characteristic length:

$$Re_x = \frac{u_\infty x}{v}. \tag{7.6}$$

In terms of Re_x, v is

$$v = \frac{u_\infty x}{Re_x}. \tag{7.7}$$

Substituting this expression into equation 7.5 and rearranging variables, we see that

$$\frac{\delta}{x} = \frac{5}{\sqrt{Re_x}}. \tag{7.8}$$

For any position on an object, the higher the local Reynolds number, the thinner the boundary layer.

Equation 7.8 suggests that for the high Reynolds numbers characteristic of large organisms, boundary-layer thickness is a small fraction of an organism's length, consistent with the examples cited before. However, the reverse can be true for small parts of organisms. Consider, for instance, the sensory hairs that a mantis shrimp (a stomatopod) uses to detect odors (Figure 7.2A and B). These cylindrical structures are tiny—20 μm in diameter—and they are embedded in the boundary layer of the shrimp's antennule, where they are subjected to a flow of approximately 1 cm $\cdot$ s^{-1}. At this speed, flow around the hairs has a local Reynolds number of approximately 0.1. As water flows from a hair's leading edge to its side, it travels a distance $\pi \ell/4$, where ℓ is the hair's diameter. Using equation 7.8, we calculate that the boundary layer on the side of a sensory hair is approximately 250 μm thick, 12.5 times the hair's

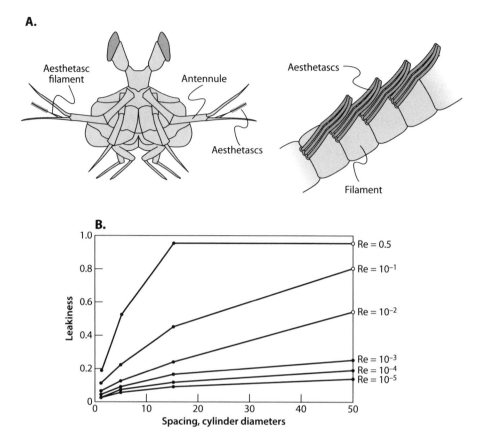

Figure 7.2 A. Mantis shrimps' antennules bear rows of hairlike aesthetascs (redrawn from Meade and Koehl 2000). **B**. Reynolds number and the spacing between aesthetascs determine whether flow goes between the aesthetascs (high leakiness) or around them (low leakiness). Redrawn from A. Y. Cheer and M.A.R. Koehl. 1987. Paddles and rakes: Fluid flow through bristled appendages of small organisms. *J. Theor. Biol.* 129:17–39, with permission from Elsevier.

diameter.[1] Thus, if two hairs are closer together than 25 diameters, their boundary layers overlap, and flow between them is inhibited—water tends to go around the array of hairs rather than through. In fact, the shrimp's sensory hairs are only 5 to 10 diameters apart (Meade and Koehl 2000), suggesting that there is considerable overlap in their boundary layers. Delivery of odor molecules might thus be jeopardized.

Mantis shrimp avoid this problem by flicking their antennules forward, increasing their speed relative to the surrounding water—and hence their Re_x—by roughly an order of magnitude (Meade and Koehl 2000). This 10-fold increase in Re_x causes a 3.2-fold decrease in boundary-layer thickness, opening a gap in the hydraulic dam between sensory hairs and allowing water to flow through. At the end of a flick, the

---

[1] An exact calculation of boundary-layer thickness for a cylinder at low Reynolds number is a complex problem (Steinmann et al. 2006), so we make do with this approximation.

antennule is pulled back slowly, allowing the animal time to sense odor in the water now trapped between hairs.

Although the specifics of hair diameter and spacing differ, a wide variety of crustacea use flicking motions to sniff the fluid around them (Koehl 2011), effectively using the Re_x-dependent thickness of the boundary layer as a means to discretely sample the environment's aroma.

The same physics apply in air. Male silk-moths use their plumose antennae to sense the pheromone released by females up to a mile away. Tightly packing sensory sensilla into the antennae increases the animal's sensitivity but potentially reduces flow through the structure. Vogel (1983) found that, although 43% of an antenna's profile was open space, only 8% to 18% of the air impinging on the structure passed through it. The faster moths fly, the higher the Re_x of the sensilla and the greater the effective porosity of the antenna.

Viscous interaction between sensory hairs can be important in air as well as water. The sensory hairs on a cricket's abdomen are spaced so closely together that, when subjected to the flow of a sound wave, the movement in one hair affects the movement of adjacent hairs, and this aerodynamic linkage can change the system's response to the sound (Casas et al. 2010).

3   SHEAR STRESS IN DYNAMIC LAMINAR BOUNDARY LAYERS

Having quantified the thickness of a dynamic laminar boundary layer, we are now in a position to estimate the shear stress it imposes on a solid surface. In Chapter 9 we will see how this stress—friction drag—resists fish and other aquatic creatures as they swim. (In this case, mainstream velocity is the speed at which the animal moves through stationary water.)

To a first approximation, the gradient in velocity across the boundary layer is

$$\frac{du}{dz} \approx \frac{u_\infty}{\delta}. \tag{7.9}$$

Given this approximation, shear stress at the fluid-solid interface is

$$\tau_0 \approx \mu \frac{u_\infty}{\delta}. \tag{7.10}$$

Thus, using equation 7.5 for δ, we estimate that

$$\tau_0 \approx \mu \frac{u_\infty}{5\sqrt{\frac{\nu x}{u_\infty}}}$$

$$\approx 0.2 \mu u_\infty^{3/2} \sqrt{\frac{1}{\nu x}}. \tag{7.11}$$

Calculations by Blasius (1908)—which take into account the nonlinear nature of the velocity gradient—show that actual boundary shear stress is slightly higher:

$$\tau_0 = 0.332 \mu u_\infty^{3/2} \sqrt{\frac{1}{\nu x}}. \tag{7.12}$$

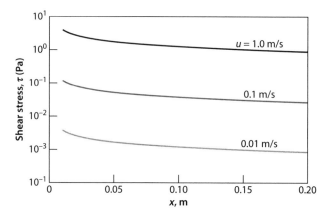

Figure 7.3 Shear stress decreases with distance from a leading edge (x).

This expression is graphed in Figure 7.3 for a variety of mainstream flows in water (i.e., swimming speeds). The faster the flow, the higher the shear stress, and stress decreases gradually with increasing distance from the leading edge (e.g., from the front of a fish).

Equation 7.12 can be rearranged to highlight the roles that viscosity and density play in determining shear stress. Recalling that $\nu = \mu/\rho$, we see that

$$\tau_0 = 0.332 u_\infty^{3/2} \sqrt{\frac{\mu\rho}{x}}. \tag{7.13}$$

The higher the viscosity, the higher the shear stress, as one might expect, but density is just as important. The higher the density, the greater the fluid's mainstream momentum and hence the greater the diffusive rate of momentum transport, i.e., shear stress. For a given u_∞ and x, shear stress in water is approximately 180 to 300 times that in air.

4   MASS TRANSPORT

We turn now to another biological process affected by boundary layer thickness: the transport of mass. Photosynthesis provides an example.

Photosynthesis uses light energy to convert inorganic carbon into sugars that can then be used for metabolism and growth. In air, where the primary source of carbon is carbon dioxide (CO_2) the basic reaction of photosynthesis is

$$6CO_2 + 6H_2O \xrightarrow{\text{light}} C_6H_{12}O_6 + 6O_2. \tag{7.14}$$

Carbon dioxide combines with water to produce glucose and oxygen. In seawater, where the primary source of inorganic carbon is the bicarbonate ion, HCO_3^-,

$$6HCO_3^- + 6H_2O \xrightarrow{\text{light}} C_6H_{12}O_6 + 6O_2 + 6OH^-. \tag{7.15}$$

Bicarbonate combines with water to yield glucose, oxygen, and hydroxyl ions.

Let's explore the consequences of these reactions. To photosynthesize, leaves and seaweeds' fronds must be supplied with carbon dioxide or bicarbonate from the surrounding fluid, and they need to rid themselves of oxygen and (for seaweeds) hydroxyl ions. Lacking any form of active ventilation, plants rely on diffusion for the transport of these materials, and the rate of transport (moles per area per time) is governed by Fick's equation:

$$\mathcal{F}_z = -\mathcal{D}\frac{dC}{dz},$$ (7.16)

where z increases with distance from the leaf or frond. The steeper the concentration gradient dC/dz, the faster supplies are delivered and wastes are removed.

It is here that boundary layers come into play. The relatively stagnant fluid of the boundary layer separates mainstream fluid (with concentration C_∞) from the surface of the leaf or frond, where concentration is C_b. Thus, to a first approximation, the concentration gradient for transport of material to or from the plant is

$$\frac{dC}{dz} \approx \frac{C_\infty - C_b}{\delta_c}.$$ (7.17)

Here δ_c is the thickness of the *concentration boundary layer*, the distance above the leaf or frond at which C is 99% of C_∞. When dC/dz is positive, $\mathcal{F}_z$ is negative, and mass flows into the plant; when dC/dz is negative, $\mathcal{F}_z$ is positive and it flows out. If we can predict the thickness of the concentration boundary layer, we can predict the flux of molecules.

To predict δ_c, we make the assumption that diffusion of mass is analogous to diffusion of momentum. We know that δ is set by the diffusivity of momentum (ν), whereas mass transport is set by the diffusivity of molecules ($\mathcal{D}$), so we can guess that the ratio of δ to δ_c is set by the ratio of ν to $\mathcal{D}$. This dimensionless value is known as the *Schmidt number*:

$$\mathrm{Sc} = \frac{\nu}{\mathcal{D}}.$$ (7.18)

Aerial Schmidt numbers are on the order of 1, while Scmidt numbers in water are on the order of 1000. To see why, we employ an analogy. Imagine a cue ball moving across a pool table when only a few other balls are present, a situation analogous to a gas, where the mean free path is large. Odds are that the cue ball's momentum will arrive at the opposite rail at the same time as the cue ball itself. By analogy, in air, momentum and mass diffuse at nearly the same rate, so $\mathrm{Sc} \approx 1$ (Table 7.1). Now imagine a cue ball traveling on a crowded table (an analogue of water, where molecules are crowded). As the cue ball travels, its momentum is quickly transfered to another ball, which bounces off another ball, and so forth. The cue ball's momentum quickly arrives at the opposite rail, even though the cue ball itself has barely moved. By analogy, the diffusion of momentum in water is much faster than the diffusion of molecules, leading to Schmidt numbers on the order of 1000.

Scaling arguments and empirical measurements (Incropera and DeWitt 2002) show that

$$\delta_c = \frac{\delta}{Sc^{1/3}}.$$ (7.19)

Thus, in water, concentration boundary layers are an order of magnitude thinner than velocity boundary layers; in air they are nearly the same size (Table 7.1).

Table 7.1 Schmidt Numbers for Various Substances in Seawater and Air. Values Calculated from Data in Denny (1993).

T°C	Sc			$Sc^{1/3}$		
	O_2	CO_2	HCO_3^-	O_2	CO_2	HCO_3^-
In Seawater						
0	1859	1600		12.29	11.70	
10	877	925	868	9.57	9.74	9.54
20	505	599		7.96	8.43	
30	318	409		6.83	7.42	
	O_2	CO_2	H_2O	O_2	CO_2	H_2O
In Air						
0	0.743	0.957	0.636	0.906	0.985	0.860
10	0.743	0.953	0.631	0.906	0.984	0.858
20	0.744	0.954	0.624	0.906	0.981	0.855
30	0.744	0.941	0.615	0.906	0.980	0.850
40	0.749	0.939	0.614	0.908	0.979	0.850

We can now use equation 7.19 to estimate mass flux. From equation 7.5 we know that in laminar boundary layers, $\delta = 5\sqrt{vx/u_\infty}$. Thus,

$$\delta_c = \frac{5\sqrt{\frac{vx}{u_\infty}}}{Sc^{1/3}}. \tag{7.20}$$

Substituting this value into equation 7.17,

$$\frac{dC}{dz} \approx 0.2\,(C_\infty - C_b)\,Sc^{1/3}\sqrt{\frac{u_\infty}{vx}}. \tag{7.21}$$

Multiplying by $\mathcal{D}$ gives us the inward flux:

$$\mathcal{F}_z \approx 0.2\mathcal{D}\,(C_\infty - C_b)\,Sc^{1/3}\sqrt{\frac{u_\infty}{vx}}. \tag{7.22}$$

For a plant it matters less what the flux is at a particular spot (x) than what flux is averaged over the whole leaf or frond. If the leaf or frond has length ℓ in the direction of flow, average flux is

$$\overline{\mathcal{F}_z}(\ell) \approx 0.2\mathcal{D}\,(C_\infty - C_b)\,Sc^{1/3}\sqrt{\frac{u_\infty}{v}}\left(\frac{1}{\ell}\int_0^\ell \sqrt{\frac{1}{x}}\,dx\right)$$

$$= 0.4\mathcal{D}\,(C_\infty - C_b)\,Sc^{1/3}\sqrt{\frac{u_\infty}{v\ell}}. \tag{7.23}$$

Several aspects of this relationship should mesh comfortably with your intuition. Flux of molecules across the boundary layer depends on the Schmidt number and the molecular diffusivity of the molecules involved, values set by the medium. Transport also depends on the difference in molecular concentrations between the mainstream and the organism's surface, which for leaves is determined by the concentration of CO_2 in air and for seaweeds by the concentration of bicarbonate in seawater.

Surprisingly, however, flux across the boundary layer—that is *perpendicular* to the leaf or frond—also depends on u_∞, velocity *parallel* to the structure. This less-than-intuitive interaction is due to the fact that velocity parallel to the leaf or frond imports fresh fluid from the mainstream and thereby controls the concentration boundary layer's thickness (equation 7.5).

And lastly, flux depends on the length of the leaf or frond. Short structures experience higher fluxes than do long structures.

To grasp the biological importance of equation 7.23, let's examine photosynthesis in giant kelps. C_∞, the concentration of bicarbonate ions in seawater at a typical pH of 8.0, is 1.6 mol $\cdot$ m^{-3}, and let's assume that a kelp frond is capable of absorbing every bicarbonate ion that reaches its surface, so that $C_b = 0$. The Schmidt number for bicarbonate in 10°C seawater is 868 and its diffusivity is 1.6×10^{-9} m$^2 \cdot$ s^{-1}. At 10°C the kinematic viscosity of seawater is 1.31×10^{-6} m$^2 \cdot$ s^{-1}, and let's assume a steady current of 10 cm $\cdot$ s^{-1}. For a frond 1 m long in the direction of flow, bicarbonate is delivered at an average rate of 2.7×10^{-6} mol $\cdot$ m$^{-2} \cdot$ s^{-1}

By contrast, Gerard (1986) determined that in bright light, fronds of the giant kelp *Macrocystic pyrifera* can use bicarbonate at rates up to 6.5×10^{-6} mol m$^{-2} \cdot$ s^{-1}. Thus, for a blade this size in flow this slow, diffusive transport of bicarbonate across the boundary layer limits the maximum rate of photosynthesis. If the boundary layer remains laminar, a velocity of 58 cm $\cdot$ s^{-1} would be required to deliver bicarbonate at the maximum rate the plant can use it. This may be one reason why these kelps occur only where wave-driven water velocities periodically augment current velocity. (However, we will see later that the boundary layer may become turbulent well below this speed.)

Delivery is only part of the story, however. As a frond uses bicarbonate in photosynthesis, it releases hydroxyl ions, one hydroxyl ion for each bicarbonate ion taken up. Given the rate of bicarbonate delivery in 10-cm $\cdot$ s^{-1} flow (2.7×10^{-6} mol $\cdot$ m$^{-2} \cdot$ s^{-1}), what is the OH$^-$ concentration at the frond's surface? Concentration of OH$^-$ in pH-8 seawater is 10^{-3} mol $\cdot$ m^{-3}. We can calculate the OH$^-$ concentration at the frond surface by solving equation 7.23 for C_b using the same values as above for Sc and v and 1.9×10^{-9} m$^2 \cdot$ s^{-1} for D and setting flux equal to the known rate of bicarbonate delivery. The answer is 1.35 mol $\cdot$ m^{-3}, 1350 times that in the surrounding seawater. It is possible that this high hydroxyl ion concentration—which corresponds to a pH of 11.1 (see Supplement 7.1)—could interfere with the alga's ability to process carbon in slow flow.

A similar calculation can be conducted for the oxygen liberated by the kelp's photosynthesis, one oxygen molecule per carbon atom fixed. Oxygen concentration in saturated seawater at 10°C is approximately 0.25 mol $\cdot$ m^{-3} and its diffusivity is 1.9×10^{-9} m$^2 \cdot$ s^{-1}. Crunching through the math, we calculate that the concentration of oxygen at the blade's surface is 1.1 mol $\cdot$ m^{-3}, a fourfold increase above that in bulk seawater. One of the key enzymes in photosynthesis (Rubisco) performs poorly in the presence of high concentrations of oxygen, and this local increase may, therefore, have a negative effect on the efficiency of photosynthesis in slow flow, again a reason for kelps to live where waves augment the current.

What about gas exchange for leaves? The concentration of CO_2 is air is currently 400 parts per million, which translates to 0.017 mol $\cdot$ m^{-3}, approximately 1/100 the concentration of bicarbonate in water. In light of this substantial reduction in concentration, one might suppose that transport across the boundary layer could constrain the flux of carbon to leaves. But mass transport also depends on molecular diffusivity, Schmidt number, and kinematic viscosity (equation 7.22):

$$\mathcal{F}_z \propto \frac{\mathcal{D}Sc^{1/3}}{\nu^{1/2}}. \tag{7.24}$$

Recalling that $Sc = \nu/\mathcal{D}$, this expression can be simplified:

$$\mathcal{F}_z \propto \frac{\mathcal{D}^{2/3}}{\nu^{1/6}}. \tag{7.25}$$

The ratio of $\mathcal{D}^{2/3}$ to $\nu^{1/6}$ is roughly 300-fold greater in air than in water, more than enough to offset the low aerial concentration of CO_2. Thus, for leaves the boundary layer is seldom the limiting factor in the delivery of carbon dioxide (Lambers et al. 2008).

The consequences of boundary-layer diffusion are not peculiar to fronds and leaves. Similar considerations control diffusive delivery of oxygen to a fish's gill, and, as we will see in Chapter 11, boundary-layer thickness also controls the rate of convective heat exchange between an organism and its environment and the rate that it loses water by evaporation.

5   THE DYNAMIC LAMINAR VELOCITY PROFILE

Having defined the thickness of a laminar boundary layer, our next task is to describe the profile of velocity within the layer. A precise answer was provided by Blasius (1908), who noted that the shape of the velocity profile is constant under many circumstances, differing only in its magnitude. This similarity of shape is revealed when speed at a given height in the boundary layer—expressed as a fraction of mainstream velocity—is plotted against the dimensionless variable, ϑ:

$$\vartheta(z) = z\sqrt{\frac{u_\infty}{2\nu x}}. \tag{7.26}$$

This universal laminar profile for flow past a flat plate is shown in Figure 7.4A (values are tabulated in Supplement 7.2). For a given value of x, the larger z is, the larger ϑ is, and the higher the velocity. For a given z, the larger x is, the smaller ϑ is and the lower the velocity.

Unfortunately, Blasius's velocity profile cannot be expressed as a simple equation. Instead, it is common to rely on approximations to the exact solution. A variety are available, but two serve well for most practical purposes (Massey 1989; Figure 7.4B). In the proximal half of the boundary layer ($z < \delta/2$), velocity increases approximately linearly with distance from the substratum:

$$\frac{u(z)}{u_\infty} \approx 1.6\frac{z}{\delta}. \tag{7.27}$$

where δ is specified by equation 7.3. For the full boundary layer, another expression must be used. A simple, reasonably accurate approximation is

$$\frac{u(z)}{u_\infty} \approx \frac{2z}{\delta} - \left(\frac{z}{\delta}\right)^2. \tag{7.28}$$

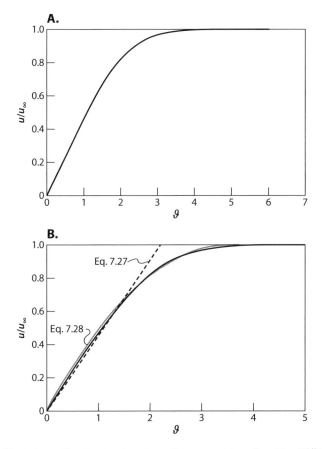

Figure 7.4 A. Blasius's profile of a laminar boundary layer ($\vartheta = z(u_\infty/(2vx))^{1/2}$). **B**. The Blasius profile compared to two approximations.

It is important to remember that equation 7.27 and 7.28 are only approximations. Their resemblance to Blasius's exact solution can be judged from Figure 7.4B. Note, for instance, that equation 7.28 assumes that, at $z = \delta$, velocity in the boundary layer equals u_∞ rather than $0.99u_\infty$ as mandated by the definition of δ.

6   DYNAMIC TURBULENT BOUNDARY LAYERS

As we have seen, the orderly flow in a laminar boundary layer devolves into turbulence when the damping effect of viscosity can no longer restrain the tendency for the fluid's inertia to form eddies. This transition occurs either when mainstream flow is fast enough, or the boundary layer is thick enough, so that fluid along its outer edge isn't sufficiently under the steadying influence of the solid boundary.

Engineers have studied the transition to turbulence for flow over smooth flat plates (e.g., airplane wings). However, as I noted in Chapter 6, objects in the natural world are seldom smooth enough for these studies to have much relevance. Information about the turbulent transition on rough surfaces is scant, but Schlichting and Gerstens (2000) suggest that, in a growing boundary layer, roughness elements play a role in triggering turbulence if they exceed a critical height:

$$D_{\text{crit}} \approx \frac{200\nu}{u_\infty}. \tag{7.29}$$

For example, in water with a mainstream velocity of $10 \, \text{cm} \cdot \text{s}^{-1}$, D_{crit} is approximately 2.5 mm, so the ruffles and ridges characteristic of many kelp blades are likely to trigger turbulence in the blades' boundary layers, which can increase the rate of diffusive transport (see below). In air moving at $2 \, \text{m s}^{-1}$, D_{crit} is only approximately 1.5 mm, so even minor roughness makes a difference for a leaf in a gentle breeze. In short, the boundary layers of all but the smoothest of real-world surfaces are likely to be strongly influenced by roughness elements, which makes it difficult to draw robust generalities about when and where dynamic boundary layers transition to turbulence.

Information about the growth of turbulent boundary layers over rough surfaces is similarly difficult to come by. However, O'Riordan and her colleagues (1995) suggest that:

$$\frac{\delta}{x} = \frac{0.37}{\text{Re}_x^{0.2}}. \tag{7.30}$$

Comparison to equation 7.8 shows that for $\text{Re}_x < 6000$, laminar boundary layers are thicker than turbulent layers. At higher Reynolds numbers, however, turbulent boundary layers can be substantially thicker than their laminar counterparts. For example, in the surf zone of wave-swept shores, water can move at $10 \, \text{m} \cdot \text{s}^{-1}$ for a distance of 10 m as a wave moves inshore, resulting in a Re_x of 7.4×10^7 and a δ of 10 cm for the turbulent boundary layer. If flow remained laminar under these circumstances, δ would be only 0.6 cm thick. For a $10\text{-m} \cdot \text{s}^{-1}$ wind whipping across 10 m of rough rock, Re_x is 6.6×10^6, and the turbulent δ is 16 cm compared to 2 cm for the corresponding laminar layer.

7   MASS FLUX IN TURBULENT BOUNDARY LAYERS

Viewed naively, these comparisons suggest that at high Re_x the greater thickness of turbulent boundary layers could inhibit mass flux to surfaces. However, in this respect, boundary-layer thickness is a misleading metric. Mass flux is limited by the retarded flow near a boundary, across which material can be transported only by molecular diffusion. However, for laminar and turbulent boundary layers of the same thickness, the region of retarded flow (flow less than 50% of mainstream, say) is much thinner in the turbulent layer (Figure 7.5A). Conversely, a turbulent boundary layer can be much thicker than its laminar counterpart and still have the same effect on mass transport (Figure 7.5B). In short, just because a turbulent boundary layer is thick does not mean that it inhibits mass flux. This conclusion mirrors our discussion in Chapter 6, where we found that turbulent boundary layers transport momentum much more effectively than do laminar layers.

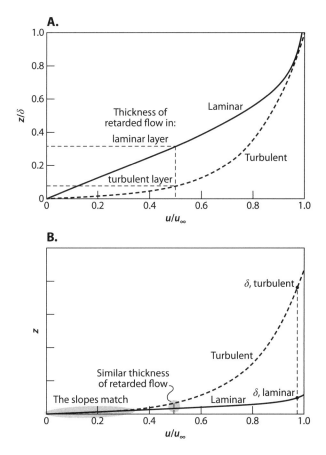

Figure 7.5 Turbulent and laminar velocity profiles compared. **A.** Layers of the same thickness. **B.** Layers with the same boundary shear stress.

8   OSCILLATING FLOW

So far we have dealt with mainstream flows that are either constant or intermittant. There are cases, however, in which mainstream flow oscillates predictably. The ebb and flow of tides creates currents that reverse four times a day, and ocean waves produce flows that oscillate with periods of 4 to 20 s. Flow in sound waves reverses hundreds of times every second. The same principles apply to boundary layers in these oscillating flows—boundary-layer thickness increases with time and distance from a leading edge—but because flow speed isn't constant, our simple approach doesn't readily yield quantitative answers. In dealing with this problem, we rely instead on a solution devised by G. G. Stokes (Schlichting and Gersten 2000).

Stokes calculated that, to a first approximation, the velocity near a solid substratum in contact with oscillating fluid is

$$u(z, t) = [u_0 \cos(2\pi f t)] - \left\{ u_0 \exp\left(-z\sqrt{\frac{\pi f}{\nu}}\right) \cos\left(2\pi f t - z\sqrt{\frac{\pi f}{\nu}}\right) \right\}. \quad (7.31)$$

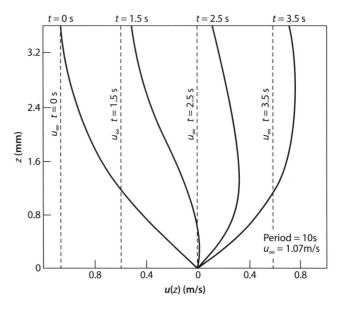

Figure 7.6 Velocity profiles in oscillating flow. (See text.)

Here u_0 is the amplitude of velocity (the highest speed in a cycle) and f is the frequency of oscillation (Hz). This behavior is depicted in Figure 7.6.

Equation 7.31 contains a strange fillip. Note the term $z\sqrt{\frac{\pi f}{\nu}}$ in the second cosine expression; it introduces a phase shift in boundary-layer velocity relative to that of the oscillating mainstream. For instance, when $z\sqrt{\frac{\pi f}{\nu}} = \pi/2$, velocity in the boundary layer is maximal when mainstream velocity is zero.

The fact that, at times, velocity in the boundary layer can exceed that in the mainstream makes it impossible to apply our usual definition of boundary-layer thickness. Instead, for oscillating flows we define δ as the z for which the retardation of flow speed in a cycle (the term in braces in equation 7.31) is maximally 1% of u_0. This occurs when z is such that

$$\exp\left(-z\sqrt{\frac{f\pi}{\nu}}\right) = 0.01. \tag{7.32}$$

Given that $\ln(0.01) \approx -4.61$, this means that

$$\delta \approx 4.61\sqrt{\frac{\nu}{\pi f}}. \tag{7.33}$$

The lower the frequency, the thicker the boundary layer. For example, ocean waves with a frequency of 0.1 Hz induce a boundary layer 1 cm thick; tidal oscillations with a frequency of 2.2×10^{-5} Hz are associated with boundary layers 0.65 m thick.

At the opposite end of the frequency spectrum, sound waves drive oscillating flows at frequencies measured in tens to thousands of cycles per second. For a

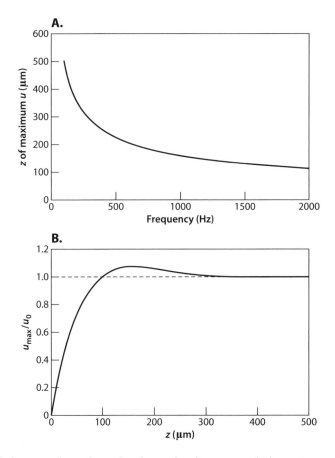

Figure 7.7 A. In an oscillating boundary layer, the distance at which maximum velocity occurs decreases with increasing frequency. **B.** Velocity in an oscillating boundary layer can exceed that in the mainstream ($f = 1000\,\text{Hz}$ in this example).

100-Hz sound wave in air, δ is 1 mm. For a 1000-Hz signal, δ is a mere 0.3 mm. This frequency-dependent boundary-layer thickness allows crustacea and insects to tune their hearing. As we have seen, these organisms use hairs on the surface of their bodies to sense sound-driven flow. A short hair is sheltered from low-frequency flow by its relatively thick sound-generated boundary layer but is exposed to high-frequency oscillations because their boundary layer is thin. Thus, the length of a hair affects the sounds it can sense. (We will return to this topic in Chapter 18.)

The presence of the term $z\sqrt{\frac{\pi f}{\nu}}$ in equation 7.31 makes for yet another odd twist. At a certain height above the substratum (a height that depends on frequency, Figure 7.7A), maximum velocity in an oscillation can be approximately 7% higher than mainstream velocity (Figure 7.7B; see Supplement 7.3). To achieve maximum sensitivity to flow, the sensory hairs of insects and crustacea should have their ends near this height, which is somewhat less than δ.

9 CONCEPTS, CONCLUSIONS, AND CAVEATS

In this chapter we investigated how boundary layers grow through space and time:

- Laminar boundary layer thickness increases as the square root of distance from a leading edge or the square root of time over which flow interacts with a solid surface.
- Because the kinematic viscosity of air is larger than that of water, aerial boundary layers are thicker than aquatic layers (all other factors being equal).
- Boundary-layer thickness plays a major role in the diffusive transport of molecules to and from surfaces.

We have seen how these factors influence the photosynthetic capabilities of leaves and seaweeds and the ability of mantis shrimp to smell and of insects to hear. The principles outlined in this and the preceding chapter will find common use in the chapters to come.

However, it is important to keep in mind that this foray into boundary-layer physics is only an introduction, and the information herein must therefore be used with caution. For instance, the equations presented here for the growth of boundary layers assume that the solid surface is parallel to mainstream flow; any tilt or curvature in the surface affects the boundary layer. In these cases, our equations can provide qualitative insight, but they are no longer quantitatively accurate.

Special care should be taken when dealing with turbulent boundary layers. Depending as they do on the chaotic motion of fluids, they can be very sensitive to the precise circumstances: the size, shape, and packing of roughness elements and the intensity of turbulence in flow prior to contact with the solid boundary. Care also must be taken when dealing with large-scale boundary layers in air. Even slight changes in temperature and humidity can affect the density of air, leading to buoyancy effects that complicate predictions. To cope with these and other details, you should consult the bibles of boundary layers: Schilchting and Gerstens (2000) and Kaimal and Finnigan (1994).

Chapter 8

Fluid-Dynamic Forces I
Introduction and Low-Reynolds-Number Flows

When fluid moves relative to an organism, forces are imposed. These fluid-dynamic forces dictate the power required to swim or fly, the rate at which phytoplankton and pollen sink, and the likelihood of trees and corals surviving storms. Fluid-dynamic forces have guided the evolution of many biological structures, from sensory hairs to fins and wings, leaves, kelp blades, and limpet shells. In this and the next chapter, I provide a brief overview of fluid-dynamic forces to establish themes that will play out in the rest of the text.

There are three categories of force we need to consider:

1. When fluid and organism move relative to each other, *drag* acts in the direction of relative flow. For example, when fluid is stationary, drag resists the movement of organisms. Conversely, when fluid flows, drag pushes stationary objects downstream (Figure 8.1A).
2. By contrast, *lift* acts perpendicular to flow (Figure 8.1B). Lift is familiar as the force that keeps birds and airplanes aloft, where it acts upward in opposition to gravity. For our purposes, however, lift isn't so narrowly defined. Any fluid-dynamic force acting perpendicular to flow—be it up, down, or sideways—is, by our definition, lift.
3. If the speed of fluid relative to an organism changes through time, an additional force—the *acceleration reaction*—is imposed, adding to or subtracting from drag depending on whether speed is increasing or decreasing (Figure 8.1C). The sum of drag and the accleration reaction is the total in-line force.

We will explore lift and the acceleration reaction in Chapter 9, but it is convenient to start our discussion of fluid dynamic forces with drag, in particular, drag at a low Reynolds number. Owing to the steadying influence of viscosity, low-Re flows (for our purposes, Re < 0.1) are sufficiently orderly that they are amenable to exact mathematical description. Furthermore, forces at low Re are relatively simple, with drag playing such a dominant role that we can generally neglect the other forces. Lift,

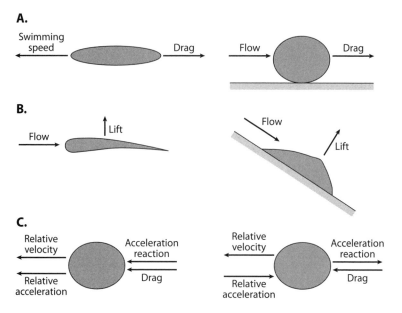

Figure 8.1 Fluid-dynamic forces: **A.** Drag. **B.** Lift. **C.** Acceleration reaction.

although possible, does not seem to contribute in any important way at low Re, and accelerational forces are important in only a few circumstances that we will discuss in Chapter 9.

1   DRAG AT LOW REYNOLDS NUMBER

Recall that

$$\text{Re} = \frac{u\ell_c}{\nu}, \tag{8.1}$$

where u is relative velocity, ℓ_c is length along the axis of flow, and ν is kinematic viscosity. For Re to be less than 0.1 (our criterion for low Re), $u\ell_c$ must be less than 0.1ν. But ν is small (10^{-6} m$^2 \cdot$ s^{-1} in water, 1.5×10^{-5} m$^2 \cdot$ s^{-1} in air), so the requirement that $u\ell_c < 0.1\nu$ places severe limits on u and ℓ_c. For instance, if u is a mere 1 mm $\cdot$ s^{-1}, an aquatic plant or animal must be less than 100 μm long for Re to qualify as low. For air moving at 1 mm $\cdot$ s^{-1}, ℓ_c must be less than 1.5 mm. In short, when we talk about forces at low Re, we are primarily discussing drag on the bacteria, gametes, zooplankton, phytoplankton, and pollen of the world.

How big is this drag? Recall from Chapter 7 that, at the interface between a solid object and flowing fluid, viscosity and the no-slip condition interact to create a velocity gradient, du/dz, which imposes a *viscous drag*, F_V:

$$F_V = \mu A \frac{du}{dz}. \tag{8.2}$$

Here μ is the dynamic viscosity of the fluid and A is the area of contact between fluid and organism. From our derivation of Re in Chapter 6, we know that du/dz is proportional to u/δ, where δ is boundary-layer thickness. At high Re, where inertial

Table 8.1   Force Coefficients for Flow Parallel or Perpendicular to an Object's Major Axis.

	ℓ/d	k_v Flow $\parallel$	Flow $\perp$
Circular disk	1:50	6.383	4.255
Oblate spheroid	1:4	6.174	4.857
Oblate spheroid	1:3	6.012	4.950
Oblate spheroid	1:2	5.794	5.074
Sphere	1:1	5.317	5.317
Prolate spheroid	2:1	4.896	5.608
Prolate spheroid	3:1	4.762	5.858
Prolate spheroid	4:1	4.734	6.093
Circular cylinder	50:1	6.598	10.448

forces come into play, δ is proportional to the square root of the length of the organism (Chapter 7), but at low Re δ is directly proportional to length itself. Length, in turn, is proportional to the square root of the organism's surface area. In short, at low Re:

$$\frac{du}{dz} \approx \frac{u}{\delta} \propto \frac{u}{\sqrt{A}}. \tag{8.3}$$

Incorporating this thought into equation 8.2 and including a coefficient (k_v) to quantify the proportionality, we arrive at a robust prediction for the general behavior of viscous drag at low Reynolds numbers:

$$F_V = k_v \mu u A^{1/2}. \tag{8.4}$$

Viscous drag is directly proportional to velocity and viscosity and to the square root of area.

Mathematicians have calculated k_v for disks and spheroids, shapes similar to pollen, spores, phytoplankton and many aquatic larvae (Table 8.1; Happel and Brenner 1973; Vogel 1994). k_v varies among shapes but not as much as one might suppose. For a sphere, $k_v = 5.32$. For a thin circular disk oriented broadside to flow—which one might expect to experience much more drag than a sphere—k_v is only 20% higher (6.38). Even more surprising, for the same disk held with its faces parallel to flow—which one would intuitively expect to experience very little drag—k_v is only 20% lower than that of a sphere (4.26).

The general applicability of equation 8.4 allows us to make several important calculations. Consider, for instance, the role of phytoplankton in controlling Earth's temperature. Carbon dioxide is a greenhouse gas, and its concentration in the atmosphere is controlled in part by phytoplankton, which absorb CO_2 in the process of photosynthesis, and then sequester the fixed carbon by sinking out of the surface mixed layer into the ocean's interior. Earth's temperature is controlled in large part by the net rate at which carbon is drawn out of the atmosphere, so it is therefore sensitive to the rate at which phytoplankton sink, and the rate of sinking is governed by the interaction between weight and drag. We will analyze this process in greater detail in Chapter 24, but let's have a preliminary look now.

As a phytoplankter descends, its speed initially increases in response to gravity. But the faster it sinks, the greater the viscous drag it feels (equation 8.4), and it quickly reaches an equilibrium speed—the *terminal velocity*—at which drag equals weight. Let's assume that phytoplankton are approximately spherical. In that case, particle volume V is a function of A, the sphere's surface area:

$$V = \frac{1}{6\sqrt{\pi}} A^{3/2}. \tag{8.5}$$

Weight (F_g) varies with volume:

$$F_g = g \left(\rho_b - \rho_f \right) V$$
$$= \frac{g \left(\rho_b - \rho_f \right)}{6\sqrt{\pi}} A^{3/2}, \tag{8.6}$$

where ρ_b is the particle's density and the term $\rho_b - \rho_f$ accounts for the buoyant effect of fluid density, ρ_f (Archimedes principle).

Given this expression for weight, we can set it equal to drag (equation 8.4) and solve for terminal sinking speed, u_t:

$$\frac{g \left(\rho_b - \rho_f \right)}{6\sqrt{\pi}} A^{3/2} = k_v \mu u A^{1/2}, \tag{8.7}$$

$$u_t = \frac{g \left(\rho_b - \rho_f \right)}{6\sqrt{\pi} k_v \mu} A. \tag{8.8}$$

Thus, sinking speed increases in direct proportion to a particle's surface area—big phytoplankton sink much faster than small phytoplankton. (Note that turbulence can move phytoplankton at speeds higher than teminal velocity, but because the direction of turbulent motion is random, the resulting average displacement is zero. Thus, for sequestration it is the steady, slow sinking rate that matters.)

In Chapter 4 we saw that in nutrient rich waters phytoplankton are typically 50 μm in radius ($A = 3.1 \times 10^{-8}$ m^2). Assuming a cell density of 1080 kg $\cdot$ m^{-3}, these diatoms and dinoflagellates sink at 0.28 mm per second, nearly 25 m per day.[1] Under these circumstances, it takes only 4 days for cells to sink the 100 m required to exit the surface mixed layer, and carbon sequestration by phytoplankton can therefore be effective.

By contrast, in the nutrient-poor waters characteristic of much of the ocean, single-celled phytoplankton are typically 5 μm in radius ($A = 3.1 \times 10^{-10}$ m^2) and sink 100 times slower, a mere 25 cm per day. At this rate, it would take a phytoplankter over a year to sink 100 m. Because phytoplankton sink so slowly in nutrient-poor waters, the rate of carbon sequestration there could potentially be ineffective, with cascading effects for Earth's climate.

The same physics that govern the sinking of phytoplankton apply to other particles as well. Detritus particles, for instance, are stirred up in the boundary layer over muddy sediments and are then sorted out by their sinking rates in a process similar

---

[1] Sinking rates can be modified by cell shape—the more asymmetrical the cell (that is, the large k_v is), the slower it sinks—and some phytoplankton can regulate their sinking by modifying their density. The sinking rate of diatoms deviates from the general rule: big diatoms sink slower than one would expect. The reason for this discrepancy is explained in Supplement 8.1.

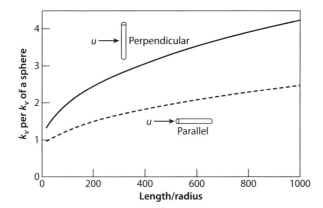

Figure 8.2 The force coefficient (k_v) for elongated ellipsoids at low Re, expressed relative to the force coefficient of a sphere.

to that by which wheat is separated from chaff, but in reverse: dense mineral particles (which have no food value) sink fast; less dense organic particle (which are nutritious) sink slower. The benthic suspension-feeding worm *Spio setosa* takes advantage of this spatial separation. By extending its feeding apparatus 4 to 5 cm above the bottom, it increases the fraction of organic material in its diet (Muschenheim 1987).

Sinking speed has also affected the coevolution between plants and animals. Many terrestrial plants use wind to carry pollen from one cone or flower to another, but wind-borne transport is effective only while pollen grains stay aloft. Due to air's low density, aerial particles sink twenty-fold faster than particles in water (equation 8.8). Plants have evolved to cope with this problem by making their pollen grains as small as possible: four cells in the conifers, a mere two cells in the flowering plants. However, even at these small sizes pollen sinks at approximately 2 cm · s^{-1}, quickly falling to the ground. It seems likely that this rapid sinking was an impetus for the evolution of flowers that attract birds, bats, and insects and employ them as pollinators.

Not all low-Re biological structures are sufficiently spherical for the k_v values listed in Table 8.1 to be valid. In particular, many important low-Re biological structures (e.g., sensory hairs and flagella) are cylindrical with a length ℓ much greater than their radius r. Force coefficients for such cylinders depend on the ratio of ℓ to r and the cylinder's orientation. For a cylinder perpendicular to flow (Figure 8.2):

$$k_v = \frac{2\sqrt{2\pi \frac{\ell}{r}}}{\ln(\frac{\ell}{r}) + 0.193}. \tag{8.9}$$

Thus, for a given surface area, the longer the cylinder, the greater the drag it feels (equation 8.4), and its drag is always larger than that of a sphere of equal surface area.

By contrast, a cylinder held with its length parallel to flow experiences roughly half the drag of a rod held perpendicular to flow, and, for relatively small values of ℓ/r, drag can be approximately equal to that of a sphere:

$$k_v = \frac{\sqrt{2\pi \frac{\ell}{r}}}{\ln(\frac{\ell}{r}) - 0.807}. \tag{8.10}$$

As we will see in Chapter 10, the difference in drag between cylinders parallel and perpendicular to flow provides the mechanism by which flagella propel micro-organisms.

2 LOW-RE ECOMECHANICS
Sensory Ecology and Predator-Prey Interactions

In Chapter 3 we encountered predators and prey in abstract terms as we modeled their interaction using Holling type II or III response functions. For some predators, low-Re fluid mechanics can enhance this broad-brush theoretical approach by accounting for the details of predator-prey dynamics.

These details have been best worked out in copepods (small marine crustacea) and a few insects (crickets, in particular). This might seem a paltry list, but copepods comprise 70% to 80% of all marine zooplankton, and insects are by far the most common terrestrial animals, so the detailed information from these few model species can be applied to the majority of all predator-prey interactions.

The exoskeletons of crustacea and insects are typically hairy, and many of those hairs (the flow sensilla of crustaceans and the filiform hairs of insects) are used to detect motion in the surrounding fluid (either air or water). These motions can potentially convey information about the location of prey or the approach of predators and are, therefore, a key element in predator-prey interactions.

The sensory hairs of crustacea and insects all have a common structure (Figure 8.3A), and the mechanics of flow-sensing hairs have been studied in great detail (Humphrey and Barth 2007). The base of the hair is embedded in the cuticle and is connected to a sensory neuron. As fluid moves, it imposes a drag on the hair, which rotates around its base, stimulating the nerve. The mechanical properties of the hair (primarily the rotational stiffness and viscosity of its base) interact with the hair's mass, the mass of fluid that behaves as if it moves with the hair (the "added mass"; see Chapter 9), and the viscosity of the surrounding fluid to form a damped resonant system whose characteristics govern the hair's frequency response (a topic we will cover in Chapter 18). Short hairs have high resonant frequencies, and therefore respond best to high-frequency motions of the fluid. Long hairs have low resonant frequencies, and respond best to low-frequency stimulation.

Sensory hairs are exquisitely sensitive (Casas and Dangles 2010). A copepod's sensillum needs to be deflected by a mere 10 nm to initiate an action potential. For a sensillum 0.1 mm long, this amounts to a rotation of a mere 10^{-4} rad (0.006°). Crickets' filiform hairs can sense air velocities as low as 0.03 mm $\cdot$ s^{-1}. The energy imparted to the hair by this flow is less than a tenth that of a single photon!

Sensitivity is a two-edged sword, however. It can provide information about the approach of a predator or the location of prey, but at the same time it opens the organism to a flood of information about irrelevant aspects of the environment. Indeed, some sensory hairs are sensitive enough to respond to their own Brownian motion.

In summary, sensory hairs can provide a steady stream of information—some of it useful, some of it not—about the movement of fluid in the vicinity of an animal. Let's explore two examples of how that information affects predator-prey interactions.

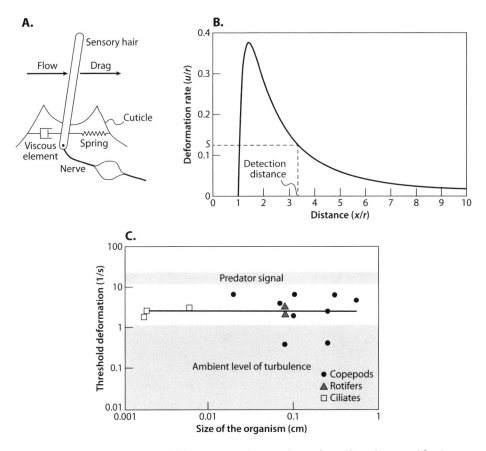

Figure 8.3 A. A schematic model for a sensory hair (redrawn from Humphrey and Barth (2007)). **B**. Deformation rate in the fluid. **C**. The threshold of a copepod's sensitivity is too high to sense ambient levels of turbulence but low enough to sense predators (redrawn from Kiørboe (2008)).

2.1   Copepods

Light is rapidly attenuated by seawater, so copepods cannot rely on sight to detect food or enemies. Indeed, copepods' eye spots are incapable of forming an image. Instead, copepods rely on their ability to sense flow to provide information about the approach of predators or prey. In particular, they rely on their ability to sense velocity gradients.

As water shears in the velocity gradient surrounding a predator or prey, it moves relative to a copepod, and drag is imposed on sensory sensilla. At the low Reynolds numbers where these interactions take place, the flow around objects can be precisely modeled (see Supplement 8.2), and we can use this theory to characterize the signal presented by predators and prey as they approach a copepod.

Let's suppose that the copepod is stationary in the fluid and a spherical object of radius r (a model for predator or prey) swims toward it at speed u. As the sphere moves through the water, it deforms the fluid in front of it, and Kiørboe (2008) shows that it is the rate of change of shear strain in the object's "bow wave" $(d\gamma/dt)$ that copepods can potentially detect, allowing them to take appropriate action. Strain rate $d\gamma/dt$ in the water varies with distance x from the sphere (Figure 8.3B; Kiørboe 2008):

$$\frac{d\gamma}{dt} = \frac{3}{2}ur\frac{(x^2 - r^2)}{x^4}.$$

(8.11)

Deformation rate rises gradually as x decreases (that is, as the sphere approaches the copepod) and then decreases rapidly to zero at $x = r$. If, as the sphere approaches, the strain rate reaches the copepod's threshold of sensitivity, the animal can detect the sphere's presence in advance of its arrival. If one knows the copepod's sensitivity, $S = (d\gamma/dt)_{crit}$, one can solve for the maximum detection distance:

$$d_{max} = r\sqrt{\frac{3u}{4rS}\left(1 + \sqrt{1 - \frac{8rS}{3u}}\right)}.$$

(8.12)

Laboratory experiments have shown that sensilla can sense a strain rate as low as $10\,\mathrm{s}^{-1}$ (Kiørboe 2008).

The larger the sphere and the faster it moves, the farther away it can be sensed. If the sphere is a predator trying to eat the copepod, it is likely to be relatively large (i.e., r is large). As a consequence, unless it swims slowly, its presence can be sensed at a relatively great distance. Predatory fish larvae know this. As they cruise along, if they spot a copepod they immediately slow down and approach slowly until they are within striking range (Munk and Kiørboe 1985).

By contrast, an approaching prey item (such as a swimming dinoflagellate) is likely to have a small radius and a slow speed. As a result, the velocity signature of prey can be sensed only at short distances, and it may instead be the prey's diffusive chemical signal that is detected first (Chapter 4).

This tidy picture could be complicated by noise. If turbulence or the Brownian motion of sensillae is sufficient to trigger a response from the copepod, it would be faced with the task of separating signal from noise as it attempts to detect a predator. Fortunately, the signals produced by predators are sufficiently stronger than those produced by Brownian motion or turbulence that copepods and other zooplankton can, in an evolutionary sense, adjust their sensitivity to filter out the background noise and still have an effective prey detector (Figure 8.3C).

Low-Reynolds-number fluid mechanics plays a role in many other aspects of aquatic predator-prey relationships; for a review, see Riisgård and Larsen (2010).

2.2   Crickets and Spiders

When it comes to predators, crickets face the same problems as copepods. Crickets have difficulty seeing predators that approach from behind, and to escape they rely on their ability to sense the air currents produced by their enemies.

Crickets' flow sensors are found on two conical appendages (called cerci; singular, cercus) at the posterior end of the the abdomen. Each cercus is coated with sensory hairs of varying lengths. As noted before, these hairs are incredibly sensitive, and several strategies have evolved to allow the cricket to filter signal from noise.

In part, the filter is mechanical. As air moves over a cercus, a boundary layer is formed (Chapter 7) such that the fluid's velocity depends on distance from the insect's surface. Thus, the length of hairs affects not only their resonant frequency, but also the speed of the fluid with which they interact. Because the fluid motions are often oscillatory or impulsive, the shape of the boundary layer can be complex (e.g., Steinmann et al. 2006), but, through a combination of theory and direct measurement, a clear picture of how a fluid stimulus is applied to crickets' sensory hairs is beginning to emerge.

Low frequencies produce relatively thick boundary layers: not as thick as long hairs but thick enough to reduce the flow speeds imposed on short hairs. As a result, long hairs are exposed to the flow of low-frequency sounds and can detect them, but short hairs are shielded from low-frequency flow. By contrast, high frequencies fall above the resonant frequencies of long hairs (and, therefore, cannot effectively stimulate them), but these frequencies have thin boundary layers, allowing flow to stimulate short hairs effectively. Thus, each length of hair responds best to a particular frequency, providing a wealth of sensory information about environmental flows. Further filtering and analysis is carried out by the nervous system, which ultimately informs the animal when it is time to make a run for it. Crickets are an ideal organism for these studies because it is possible to record nervous activity from an intact animal in its natural environment (Dupuy et al. 2012).

Wolf spiders prey on crickets, and like a predator approaching a copepod, they produce a "bow wave" that preceeds them. One might suppose that crickets' sensitivity to these motions would serve as an effective burglar alarm. Unfortunately for the crickets, spiders have evolved behaviors to avoid detection. Jérôme Casas and his colleagues measured the fluid motions caused by moving spiders and constructed a device to mimic them. They then used the device to test the response of crickets. When the spider mimic moved very slowly, crickets couldn't detect it. When the mimic moved very fast, crickets could detect it but didn't have time to escape. At velocities in between, crickets escaped. Spiders take this information into account. Casas and his colleagues (2008) observed spiders as they attacked. In most of the encounters, spiders either moved very slowly or very fast, thereby avoiding the speeds at which crickets can escape.

Similar sensory investigations have documented the mechanics of interactions between water spiders (in this case, the prey) and their frog predators (Suter 2003), and between praying mantises and the bats that hunt them (Triblehorn and Yager 2006).

In summary, low-Re fluid dynamics can form a bridge between sensory physiology and behavioral ecology, allowing for a detailed mechanistic understanding of predator prey interactions.

3   CONCEPTS, CONCLUSIONS, AND CAVEATS

Our brief discussion of low-Re fluid-dynamic forces has revealed several general themes:

- For microorganisms, it is fluid viscosity rather than density that governs force, and the force that dominates is viscous drag. The viscosity of water is 30 to 100 times that of air (depending on temperature), so for a given velocity, low-Re viscous drag is much greater in water than air.

- For phytoplankton and other low-Re particles, sinking rate is directly proportional to the object's surface area—therefore, small phytoplankton sink slowly and are ineffective at sequestering carbon.
- The fluid-dynamic forces associated with low-Re flows provide sensory cues in predator-prey interactions.

As always, you should keep in mind that the mechanisms of low-Re fluid forces presented in this chapter have been described in broad strokes. We will deal with some of the finer points in Chapter 10, but if you are interested in digging deeper into the nuts and bolts of low-Re fluid-dynamic forces, I advise you to consult a text specifically on the subject. Dusenbury (2009) is an excellent place to start, Happel and Brenner (1973) is the bible, and Kiørboe (2008) beautifully illuminates the ecomechanics.

Chapter 9

Fluid-Dynamic Forces II
High Reynolds Numbers

At low Reynolds numbers, fluid-dynamic forces are relatively simple. Lift and the acceleration reaction are generally negligible, and drag is open to precise mathematical analysis. Life at high Re is more complex. Lift and the acceleration reaction often make substantial contributions to the overall force, and turbulence introduces an element of chaos into the mix. In this chapter, we explore the nature of these complexities, an investigation that comes in four parts. First we develop the theory of drag, lift, and the acceleration reaction, and find that each is associated with a shape-dependent coefficient. We then survey biology to see how these coefficients vary among plants and animals, which leads us to develop rules of thumb as to their relative magnitudes. We finish with a look at how high-Reynolds-number fluid-dynamic forces can be used to explain the size limits of seaweeds and the temporal pattern of species diversity in intertidal boulder fields.

1 FORCES

1.1 Drag at High Reynolds Number

In the last chapter, we saw that viscous drag is the preeminent force imposed on organisms at low Re. At high Re, the picture is quite different. For many macroscopic creatures the dominant force is another form of drag—*pressure drag*.

The mechanism of pressure drag can be explained using Bernoulli's principle. Consider the idealized situation shown in Figure 9.1: uniform flow moving from left to right where it encounters a rigid vertical plate. Streamlines diverge about the center of the plate as flow makes its way around the obstruction, but for simplicity, let's focus on the horizontal streamline exactly at the plate's center. Well upstream of the plate—at point 1—the fluid has speed u. Fluid moves along the center streamline until it hits the plate at point 2, where it comes to an abrupt halt ($u = 0$). Bernoulli's equality for these two points is, thus,

$$p_1 + \frac{1}{2}\rho_f u^2 = p_2. \qquad (9.1)$$

Figure 9.1 Flow around a flat plate. (See text.)

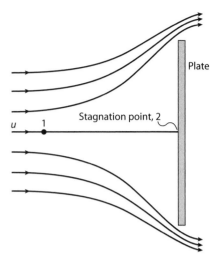

Plate

Stagnation point, 2

u 1

In other words, the pressure at the *stagnation point* where the streamline hits the rigid plate (p_2) is greater than the general pressure acting in the fluid (p_1) by an amount equal to the dynamic pressure, $\frac{1}{2}\rho_f u^2$.

Although it is not so simple to calculate the pressure at other points on the plate's upstream face, it seems reasonable to assume that they, too, have an excess above background of the same order of magnitude as the dynamic pressure. Thus, based on our understanding of Bernoulli's principle, we can guess that when fluid runs into an unyielding object, pressure is locally increased on the object's upstream face by roughly $\frac{1}{2}\rho_f u^2$. If the upstream face of the plate has projected area A_f (the area of the shadow the plate would cast for light arriving in the direction of flow[1]) and we assume that pressure on the downstream face remains at ambient pressure, p_1 (a dubious assumption we will return to shortly), the increased pressure on the upstream face exerts a downstream force—pressure drag, F_D:

$$F_D \approx \frac{1}{2}\rho_f u^2 A_{pr}. \tag{9.2}$$

To make this approximation exact, we can include a dimensionless coefficient—the *drag coefficient*, C_D—that adjusts equation 9.2 for the exact pattern of flow in the object's vicinity and, thereby, for the actual distribution of pressure:

$$F_D = \frac{1}{2}\rho_f u^2 A_{pr} C_D. \tag{9.3}$$

Because C_D depends on the pattern of flow, we expect it to depend on the Reynolds number, and we can guess that it also depends on the object's shape. A bluff object (such as our plate) has the potential to bring upstream flow to a near halt and, therefore, has a C_D of approximately 1. In contrast, a streamlined object—a fish for example—does not cause streamlines to divert radically and has a C_D less than 1.

In practice, equation 9.3 is used to define C_D. An object of known A_{pr} is placed in flow of known u and ρ_f and F_D is measured. From these data, C_D is calculated:

$$C_D = \frac{2F_D}{\rho_f u^2 A_{pr}}. \tag{9.4}$$

As noted before, F_D (and, therefore, C_D) depends on the pattern of flow around an object, but what physical mechanism determines that pattern? The explanation is best approached indirectly. Consider a cyclist poised on a hill, prepared to coast down into the valley below (Figure 9.2A). As she shoves off, the biker's momentum is low,

---

[1] This concept applies to all shapes, not just flat plates. For example, the surface area of a sphere with radius r is $4\pi r^2$, but its projected area, the area of its shadow, is πr^2.

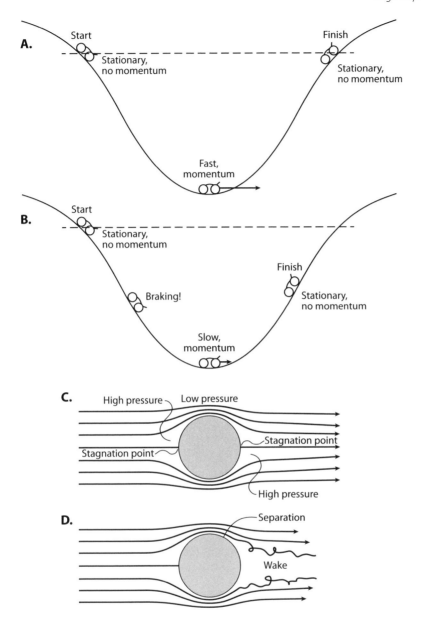

Figure 9.2 A and **B**. An analogy to the loss of momentum in a boundary layer. (See text.)
C and **D**. The symmetrical pattern of inviscid flow around a cylinder (**C**) is altered by the loss
of momentum in a viscous boundary layer (**D**).

but driven by gravity she picks up speed, and her momentum reaches a maximum
as she arrives at the valley floor. Having acquired this momentum, she can trade it
for altitude as she glides up the next hill, allowing her to coast back to the the same
elevation at which she started. The result is different, however, if the cyclist applies
her brakes as she travels downhill (Figure 9.2B). The friction of the brakes converts
gravitational potential energy into heat rather than momentum. As a consequence,

the cyclist hasn't acquired a full measure of speed when she reaches the valley floor, and therefore she can't coast as far up the next hill.

Now consider the flow of an inviscid fluid past a horizontal circular cylinder (Figure 9.2C). Directly upstream, streamlines diverge, which tells us that velocity is low, and (from Bernoulli's equation) we can infer that pressure is high. By contrast, because streamlines pinch together along the top and bottom of the cylinder, velocity is high, and pressure is low. In other words, as fluid flows laterally along the upstream face of the cylinder, it moves from an area of high pressure to an area of low pressure, speeding up in the process, and (like our cylist coasting downhill) it gains momentum. Along the downstream face, this process plays out in reverse. Streamlines diverge, velocity decreases, and the fluid's momentum allows it to coast "uphill" into the consequently increasing pressure. Because the fluid is inviscid, no momentum is lost to friction, and, as a result, the pressure downstream is equal to that upstream: no upstream-downstream pressure difference, no pressure drag. So in the absence of viscosity, $C_D = 0$.

We reach a different conclusion, however, when viscosity and the resulting boundary layer are taken into account (Figure 9.2D). As a viscous fluid flows along the upstream half of the cylinder, momentum is lost to heat through friction in the boundary layer. As a result, when fluid arrives at the downstream side of the cylinder it cannot travel as far uphill into the adverse pressure gradient as it otherwise would. Instead, fluid nearest the surface (which has lost the most momentum) is brought to a halt before it reaches the cylinder's trailing edge. This stationary fluid obstructs the fluid behind it, causing flow to deviate from the path it would have taken in the absence of viscosity. The boundary layer *separates* from the cylinder and rolls up, forming eddies that move downstream as a wake.

Pressure in the wake is therefore approximately equal to the pressure at the separation point, which—being on the lateral side of the cylinder—is lower than the pressure acting on the cylinder's upstream face. Indeed, for bluff bodies like cylinders and flat plates, wake pressure is slightly lower than ambient hydrostatic pressure. Thus, the loss of momentum in the boundary layer—and the resulting flow separation—produces an upstream-downstream pressure difference that leads to drag. The farther upstream flow separates on the downstream face: the wider the wake, the lower its pressure, and the greater the pressure drag.

Before leaving the subject of drag, there is one more aspect that requires discussion. For streamlined objects (e.g., fish and birds), flow has little tendency to separate, and drag is therefore due almost entirely to viscous friction. In such cases, it is more appropriate to use the entire wetted surface area of the object, A_w, in an alternative definition of the drag coefficient:

$$C_{DW} = \frac{2F_D}{\rho_f u^2 A_w}. \tag{9.5}$$

Because wetted area is larger than projected area—in many cases, quite a bit larger—this alternative definition leads to a smaller value for the drag coefficient than that calculated using A_{pr}. The difference between C_D and C_{DW} can lead to confusion when making comparisons among species and between experiments. For a thorough discussion of the many ways in which drag coefficients can be defined, consult Vogel (1994). Unless otherwise noted, all drag coefficients used in this text are based on projected area.

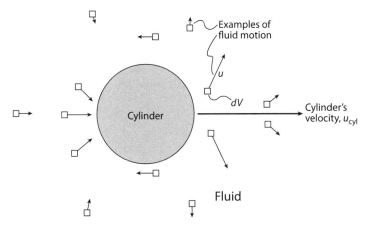

Figure 9.3 Flow around a moving cylinder, seen here in cross section.

1.2   Acceleration Reaction

To this point, we have investigated the in-line force—drag—imposed as fluid moves past an object at constant speed. When fluid accelerates, an additional in-line force is imposed. To understand how this force—the *acceleration reaction*—works, we again approach the subject indirectly.

Consider an infinite sea of stationary, inviscid fluid in which we have immersed an infinitely long horizontal cylinder moving perpendicular to its axis at speed u_{cyl} (Figure 9.3). Because the fluid has no viscosity, no velocity gradient is established around the cylinder, and no fluid is dragged along by friction. Nonetheless, the cylinder's motion makes the fluid move: fluid in front of the cylinder must get out of the way, and other fluid must move to fill the space left behind. The speed of the displaced bits of fluid varies with distance from the cylinder. A fluid element—an infinitesimal volume of fluid, dV—near the upstream side of the cylinder is shoved aside at a speed approaching that of the cylinder itself, while fluid elements elsewhere move more slowly.

Recall that the momentum of a bit of fluid is the product of its speed and mass or, equivalently, the product of speed, density, and volume. Thus, as it moves, each fluid element has momentum $u\rho_f dV$, where u varies from one element to the next. If we were to sum up all these individual momenta, we could calculate the total momentum imparted to the fluid by the cylinder's motion. This task was accomplished by Charles H. Darwin (grandson of *the* Charles Darwin), who arrived at an intriguing result:

$$\text{total momentum} = \int_{\infty} u\rho_f dV = u_{cyl}\rho_f V_{cyl}, \tag{9.6}$$

where V_{cyl} is the volume of the cylinder itself. In other words, although the momentum imparted by the cylinder is distributed throughout the fluid, its combined magnitude is equal to the momentum of a mass of fluid ($\rho_f V_{cyl}$) traveling at the same speed as the cylinder, u_{cyl}. Thus, even in the absence of viscosity, fluid behaves as if a mass of fluid—the *added mass*—were dragged along with the cylinder.[2]

---

[2] Because fluid moves out of the way as an object moves and then fills in behind, there is some net movement of fluid in the object's direction of motion. It has been proposed that this

As long as the cylinder moves at constant speed, the presence of the added mass isn't noticeable; fluid momentum is constant and therefore no forces are involved. But as soon as the cylinder's speed changes, the fluid's momentum changes, and a force must be imposed. In other words, because the added mass acts as if it moves with the cylinder, any acceleration of the cylinder involves acceleration of the added mass, which requires a force.

We are now in a position to calculate the force required to accelerate a cylinder of mass m and volume V in an inviscid fluid of density ρ_f. Even if the fluid weren't there, a force F_M would be required to accelerate the cylinder's mass at rate a:

$$F_M = ma. \tag{9.7}$$

Accelerating the added mass requires an additional force, the *added mass force*:

$$\text{added mass force, } F_{AM} = \rho_f V_{\text{cyl}} a. \tag{9.8}$$

The overall force required to accelerate the cylinder in an inviscid fluid—the *acceleration reaction*—is thus

$$\begin{aligned} F_A &= F_M + F_{AM} \\ &= ma + \rho_f V_{\text{cyl}} a \\ &= \left(m + \rho_f V_{\text{cyl}} \right) a. \end{aligned} \tag{9.9}$$

Perhaps it is surprising, but to a good approximation the presence of viscosity does not change the situation, and the concepts of added mass and added mass force are applicable to real as well as inviscid fluids.

Lest we get carried away with the beauty of this concept, we need to note several complications. First, we have seen that for a cylinder, the volume of fluid that acts as if it moves with the object—the *added volume*—is equal to the cylinder's volume V_{cyl} (equation 9.6). But this is true only for infinitely long cylinders. For other objects—the undulating fins of a fish, for example, or the waving leaves of a palm tree—the ratio of added volume to object volume can be different. To acknowledge this variability, we define a new term: the *added mass coefficient*, C_A:

$$C_A = \frac{\text{added volume}}{\text{object volume}}. \tag{9.10}$$

Like C_D, C_A is dimensionless. We can incorporate C_A into our equation for the force required to accelerate an object in a stationary fluid (equation 9.9):

$$F_A = \left(m + C_A \rho_f V \right) a. \tag{9.11}$$

Here $C_A \rho_f V$ (rather than $\rho_f V_{\text{cyl}}$) is the added mass. For future reference, we can rewrite this equation, noting that the mass of an object is equal to its volume times its density (ρ_b):

$$F_A = \left(\rho_b + C_A \rho_f \right) V a. \tag{9.12}$$

As with C_D, the magnitude of the added mass coefficient depends on the pattern in which fluid flows around an object. (Representative examples of C_A are shown later in Figure 9.8.)

---

movement of fluid—Darwinian mixing—might contribute substantially to the overall mixing of the ocean as animals migrate vertically in the water column (Katija 2012).

So far, we have considered the force necessary to accelerate an object in a stationary fluid. When dealing with sedentary organisms, we are just as likely to encounter a situation in which fluid accelerates past a stationary object (e.g., wave-driven flows past corals, wind gusts past trees). What force is imposed in this case? You might suppose that, all things being relative, the force on a stationary organism in accelerating flow would be equal to that of an accelerating organism in stationary fluid. But you would be wrong.

To see why, we approach the subject circuitously, beginning with a review of the buoyant force resulting from hydrostatic pressure. Recall from Chaper 2 that hydrostatic pressure p increases with depth beneath the water's surface such that the change in pressure between two points separated by vertical distance z is

$$\Delta p = \rho_f g z. \tag{9.13}$$

Acted upon by this hydrostatic pressure gradient, a stationary object with volume V feels a buoyant force:

$$\text{buoyant force} = \rho_f V g. \tag{9.14}$$

Here the product $\rho_f V$ is the mass of fluid displaced by the object, a quantity known as the *equivalent mass*.[3] Note that a stationary object in a stationary fluid does not effect the fluid's momentum, so gravitational buoyancy does not involve any added mass effects.

Through a similar line of reasoning (see Supplement 9.1), one can show that when a fluid accelerates at rate a, the acceleration induces a pressure gradient between points separated by distance x along the axis of acceleration:

$$\Delta p = \rho_f a x. \tag{9.15}$$

(Note that, unlike gravity, a can act in any direction.) Just as the vertical hydrostatic pressure gradient leads to gravitational buoyancy, this acceleration-induced pressure gradient leads to a *virtual buoyancy*, a buoyant force acting in the direction of acceleration:

$$\text{virtual bouyancy} = \rho_f V a. \tag{9.16}$$

Thus, when a stationary object is immersed in an accelerating fluid, virtual bouyancy is imposed.

Virtual buoyancy is only part of the story, however. Unlike the gravity-induced buoyant force acting on a stationary object in a stationary fluid—for which there are no added mass effects—a stationary object submerged in accelerating fluid involves an added mass. As an accelerating fluid flows past a stationary object, the object changes the pattern of flow, slowing some fluid particles, accelerating others, and thereby changing their momentum. Because the object changes the fluid's momentum, the fluid imposes a force (known as the *added mass force*) on the object. In effect, the

---

[3] Equivalent mass is different from added mass. Equivalent mass is the mass of a volume of fluid equal to the volume of the object. Fill a bucket to the very top with water and then submerge an object in the bucket. The mass of water that sloshes over the rim of the bucket is the object's equivalent mass. By contrast, added mass cannot be so easily measured. Instead, added mass is a calculated quantity: the mass that, when traveling at the same relative speed as an object, has the same momentum as the that imparted to the fluid by the object.

object is accompanied by an added mass, $C_A \rho_f V$, and the force required to keep this added mass stationary in the midst of accelerating flow is

$$\text{added mass force} = C_A \rho_f V a. \tag{9.17}$$

The overall force imposed on a stationary object by an accelerating fluid is the sum of virtual buoyancy and added mass force:

$$F_A = \rho_f V a + C_A \rho_f V a$$
$$= (1 + C_A) \rho_f V a. \tag{9.18}$$

The value $1 + C_A$ is sometimes given its own name: the *inertia coefficient, C_M*.

To recap, we have two equations for the force acting when an object accelerates relative to a fluid. If the object itself accelerates and the fluid is stationary (equation 9.12),

$$F_A = \left(\rho_b + C_A \rho_f \right) V a. \tag{9.19}$$

If the object is stationary and the fluid accelerates,

$$F_A = \left(\rho_f + C_A \rho_f \right) V a. \tag{9.20}$$

The contrast between these two equations is centered on the densities of fluid and object. When the object moves, both ρ_f and ρ_b matter; when the object is stationary, only ρ_f comes into play. If both fluid and object accelerate, life gets a bit more complicated (see Supplement 9.2).

Note that in both equations 9.19 and 9.20, the accelerational force imposed on an object is proportional to the object's volume. This is in contrast to drag, which is proportional to the object's area. The ratio of acceleration reaction to drag is thus proportional to volume/area—that is, to the linear dimension of the organism. This explains why F_A is generally negligible relative to F_D in microorganisms. Conversely, F_A can be larger than F_D in large organisms such as massive corals. The primary exception to this logic involves the small hairs animals use to sense sound (Fletcher 1992). While the ratio of F_A to F_D for sensory hairs is proportional to the linear dimension of the hair (as it is for any object), it is also proportional to the ratio of acceleration to velocity^2 (see "Rules of Thumb" later in the chapter). Sound waves impose low velocities but high accelerations. Therefore, although sensory hairs are small, the drag and acceleration reaction they experience are comparable in magnitude.

1.3   Total In-line Force

Drag and acceleration reaction add to form the total in-line force. For example, the total in-line force on a stationary object is

$$F_{IL} = F_D + F_A$$
$$= \left[\frac{1}{2} \rho_f u^2 A_{pr} C_D \right] + \left[(1 + C_A) \rho_f V a \right]. \tag{9.21}$$

This expression is commonly referred to as the *Morison equation*, in honor of J. R. Morison and colleagues (1950), who first employed it.

1.4   Lift

To understand the mechanism of lift, consider a horizontal circular cylinder in a viscous fluid, oriented with its axis perpendicular to flow (Figure 9.4A). As we have seen, fluid speeds up as it flows laterally past the cylinder and then slows down as it leaves the cylinder behind. Per Bernoulli's equation, where fluid moves fast, pressure is low, but because the cylinder is symmetrical, the low pressure on one lateral side is just offset by that on the other, and there is no net force imposed perpendicular to flow.

Now imagine what happens if the cylinder rotates clockwise around its axis (Figure 9.4B). Viscosity and the no-slip condition conspire to drag fluid along with the cylinder, and this rotation-induced flow (technically, *circulation*, Γ) adds to fluid velocity on top of the cylinder (lowering the pressure) and subtracts from velocity below (increasing pressure). In this fashion, rotation breaks the flow's symmetry, and the cylinder experiences a side-to-side pressure difference, in this case tending to push the cylinder toward the top of the page. Acting over the cylinder's lateral projected area, A_{pl} (its length times its diameter), this pressure difference creates lift. The magnitude of lift is proportional to the product of fluid density, relative velocity of the fluid in the x-direction, circulation, and area:

$$F_L \propto \rho_f u \Gamma A_{pl}. \tag{9.22}$$

This phenomenon is easily demonstrated with a narrow strip of paper. Tossed into the air, the strip rotates around its long axis (Figure 9.4C) and the resulting lift causes it to glide sideways as it falls. The seeds of some trees make use of this trick. They sport wings that act like the paper strip, and lift carries them away from their parent tree (Vogel 1994). This kind of rotation-induced lift also plays an important role in baseball where it is responsible for the curve in a curve ball. As a pitcher releases the ball, he imparts a substantial spin, and the resulting lift baffles batters by pulling the ball to the side.

Consider again the pattern of flow around a rotating cylinder (Figure 9.4B). We used this pattern to infer the distribution of pressure that leads to lift, but there is an alternative way to arrive at the same conclusion. As fluid approaches the cylinder from upstream, it moves parallel to the x-axis and therefore has no y-directed momentum. But the cylinder's rotation redirects streamlines such that downstream of the cylinder flow has a component in the y-direction. In short, circulation causes the cylinder to impart y-directed momentum to the fluid, which (according to Newton's second law) it can do only by applying a force. If the cylinder applies a force to the fluid, the fluid imposes an equal but opposite force on the cylinder. This is lift.

Let's apply this new perspective to a flat plate held at an angle θ to flow. From what we know so far, we would expect flow around the plate to resemble that shown in the first part of Figure 9.4D with distinct stagnation points both upstream and downstream. But this pattern of flow seems unlikely. For example, fluid flowing along the bottom of the plate would have to make an abrupt turn at the plate's trailing edge to approach the downstream stagnation point. In reality, flow adjusts to avoid this sort of abrupt shift. The presence of the sharp trailing edge induces circulation around the plate, displacing the stagnation points such that the upstream point coincides with the plate's leading edge and the downstream point coincides with the trailing edge. The shift at the trailing edge is particularly important because it allows fluid to move tangent to the plate's surface as it exits the plate. As a result,

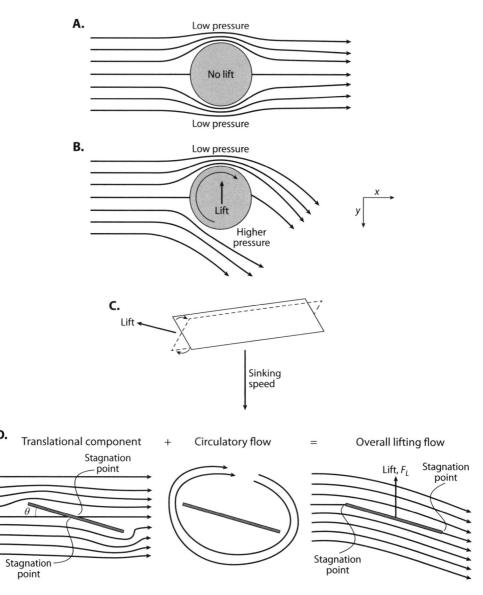

Figure 9.4 Flow around a cylinder (**A**) is altered when the cylinder rotates (**B**), giving flow a component along the y-axis. **C.** As a strip of paper falls, it rotates about its long axis, creating lift. **D.** Flow around a tilted plate: circulation adds to translational flow, producing lift. **E** and **F.** A hemicylinder (shown in cross section, **E**) and a limpet (**F**) experience lift. (See text.)

streamlines downstream of the plate have y-directed velocity similar to that produced by a rotating cylinder, and lift is created. (For a more thorough explanation of this counterintuitive phenomenon, see Vogel (1994).)

The strength of circulation induced by a tilted plate varies in direct proportion to the relative velocity between plate and fluid, which allows us to predict the magnitude of lift. From equation 9.22 we know that lift is proportional to $\rho_f u \Gamma A_{\text{pl}}$, but for our plate $\Gamma \propto u$. Thus,

$$F_L \propto \rho_f u^2 A_{\text{pl}}. \tag{9.23}$$

The precise magnitude of F_L depends on the pattern of flow, which in turn depends on the tilt of the plate (its angle of attack, θ). To account for this variation, we incorporate into our equation a dimensionless *lift coefficient*, C_L, and for symmetry with our expression for drag, we throw in a factor of $\frac{1}{2}$:

$$F_L = \frac{1}{2}\rho_f u^2 A_{\text{pl}} C_L. \tag{9.24}$$

This equation serves as a definition of the lift coefficient. If an object with planform area A_{pl} experiences a lift of F_L when exposed to velocity u:

$$C_L = \frac{2F_L}{\rho_f u^2 A_{\text{pl}}}. \tag{9.25}$$

The magnitude of C_L for a platelike object depends on the precise pattern of flow around the object and, therefore, on θ.

Lift of this sort isn't limited to flat plates. The wings of birds, bats, and insects come in a wide variety of shapes, but these *airfoils* are united in the fact that each has a sharp trailing edge, and each produces lift. The tail fins of dolphins, whales, and some fish have shapes that create circulation in a similar fashion and act as *hydrofoils*.

For objects that create lift via circulation, there is an intrinsic link between lift and drag. By redirecting flow, circulation increases y-directed momentum at the expense of x-directed momentum, and the rate of change of x-directed momentum imposes a drag. We will return to this *induced drag* when we discuss flight in Chapter 10.

There is one more mechanism that can produce lift, in this case associated with objects adjacent to a solid surface. To see how this works, we slice a cylinder in half along its axis and place the sliced surface of the resulting hemicylinder on a solid substratum (Figure 9.4E). We then allow an inviscid fluid to flow past the apparatus. Flow over the top of the hemicylinder is similar to that around the upper half of a whole cylinder in the mainstream. However, we assume that the gap between hemicylinder's lower surface and the substratum is too narrow to allow appreciable flow, although fluid in the gap can transmit pressure. Given these assumptions, both the upstream and downstream ends of the gap are exposed to fluid at stagnation pressure, and this high pressure is transmitted to the hemicylinder's basal surface. By contrast, Bernoulli's principle ensures that the pressure imposed on the hemicylinder's top surface is relatively low, and the resulting pressure difference imposes a net force—lift—tending to pull the hemicylinder away from the surface.

This kind of lift can tear the roofs off buildings in hurricanes. If windows in the building are closed, air inside remains at roughly ambient pressure, and wind whistling over the roof creates a difference in pressure between inside and outside. If the speed of the wind is sufficient, the roof lifts off.

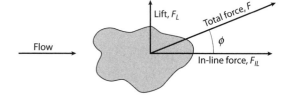

Figure 9.5 Total fluid-dynamic force is the vector sum of lift and in-line forces.

The same phenomenon can be observed under less stressful conditions. Semitractor trailers carrying grain or vegetables often have a boxlike structure with an open top. During transport, the top of the load is covered with a canvas tarpaulin. As the truck cruises the highway, the wind passing over the top of the trailer creates a pressure difference between outside and inside, and the tarp bulges upward.

A limpet is an excellent example of how this principle operates in nature (Figure 9.4F). The edge of its conical shell is closely applied to the substratum. As water flows past the shell, the average pressure around the shell's edge is at ambient pressure or slightly lower, and this pressure is transmitted to the foot, guts, and any water trapped beneath the shell. Fluid flowing over the top of the shell is at a pressure considerably lower than that underneath, and the stage is set for the imposition of lift. Lift has been measured on limpets and limpet models (Denny 2000), and at times it can be greater than the drag imposed on the animal.

1.5 Overall Force

To this point, we have treated in-line and lift forces separately, but in nature they occur simultaneously. The magnitude of the overall force, F, is the Pythagorean sum of in-line and lift forces:

$$F = \sqrt{F_{IL}^2 + F_L^2} \tag{9.26}$$

$$= \sqrt{(F_D + F_A)^2 + F_L^2}. \tag{9.27}$$

Different ratios of lift to in-line force lead to different angles at which the overall force is applied (Figure 9.5):

$$\phi = \arctan\left(\frac{F_L}{F_{IL}}\right). \tag{9.28}$$

2 FORCE COEFFICIENTS

To this point, we have treated C_D, C_L, and C_A in the abstract. However, before we can use fluid dynamics as a tool in ecological mechanics, we need to account for the details. How does the shape of an organism affect the fluid-dynamic forces it encounters? How do C_D, C_L, and C_A vary with Reynolds number? We will examine each coefficient in turn. We then finish up with two instructive examples in which fluid-dynamic forces provide ecomechanical insight: the limits to size in a competitively dominant species of seaweeds, and the control of algal diversity in an intertidal boulder field.

2.1   Drag Coefficient

We begin our investigation of C_D with two particularly simple examples—smooth circular cylinders and smooth spheres—in steady laminar flow. Although our findings for cylinders and spheres are rarely directly applicable to organisms, they provide a useful outline for understanding the general characteristics of the drag coefficient.

We start with the cylinder. As we have seen, pressure drag is associated with the separation of flow on the downstream side of an object, which leads to the formation of a wake, and this holds true for cylinders. As Reynolds number increases from 1 to approximately 500, separation produces more or less discrete eddies, but above Re of 500, shedding of vortices becomes less discrete, and to casual observation the wake appears chaotic or turbulent.

As long as turbulence is confined to the wake, drag imposed on the cylinder increases in proportion to the square of velocity, and the drag coefficient is constant at 1.0 to 1.2 for Reynolds numbers of 10^3 to 10^5 (Figure 9.6A). This is one of many instances in which C_D is greater than 1, implying in this case that the fore-aft pressure difference across the cylinder can be up to 20% greater than the dynamic pressure, $\frac{1}{2}\rho_f u^2$. This isn't due to an exceptionally large pressure on the upstream face; it's difficult to produce a pressure greater than dynamic pressure. Instead, C_D is greater than 1 because pressure in the wake (set by the pressure at the separation point) is less than ambient hydrostatic pressure.

At a Reynolds number of approximately 10^5, an important transition takes place as the cylinder's boundary layer—which at lower Re was laminar—becomes turbulent. As we have seen (Chapter 7), the shape of the velocity gradient in a turbulent boundary layer is such that relatively high velocities reach closer to a solid surface than they do in a laminar boundary layer. Thus, when the boundary layer adjacent to a cylinder becomes turbulent, fluid in the layer maintains higher velocity—and therefore greater momentum—as it flows past the cylinder. This increased momentum allows boundary-layer fluid to travel farther "uphill" against the adverse pressure gradient in the cylinder's lee, moving the separation point downstream.

In turn, the downstream displacement of the separation point has two consequences: it decreases the wake's size, and it increases the wake's pressure. Both act to decrease the upstream-downstream pressure difference applied to the cylinder, thereby reducing C_D. This reduction is both abrupt and dramatic—C_D decreases by a factor of 4; from 1.2 to approximately 0.3. After the boundary layer has transitioned to turbulence, a further increase in Re does not result in any gross change in the patten of flow and only a slight increase in C_D. (In contrast to its effects on pressure drag, the transition from a laminar to a turbulent boundary layer increases friction drag, but the increase in overall drag is relatively small.)

The drag coefficients of smooth spheres parallel those of cylinders. Over a broad range of Reynolds numbers, C_D is relatively constant (0.4 to 0.47), but near Re = 10^5, C_D abruptly decreases, also by a factor of approximately 4. The drop in C_D is so large that an increase in Re from just below transition to just above it actually results in a decrease in drag, not just a decrease in the drag coefficient. This is the reason why golf balls have dimples. A well-driven smooth golf ball travels at a velocity corresponding to Re just a bit too low to reach the transition to a turbulent boundary layer. However, if the ball has dimples, their irregularities are sufficient to trigger transition. Drag is consequently reduced, and the ball's carry is enhanced.

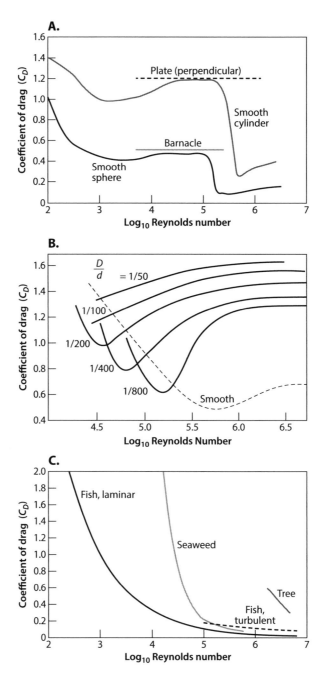

Figure 9.6 A. Drag coefficients, C_D, vary depending on Reynolds number and the shape of an object. **B.** The C_D for a cylinder is sensitive to the relative size of roughness elements (redrawn from Sarpkaya 1976). **C.** C_D for streamlined and flexible objects.

Because of the importance of smooth cylinders and spheres in engineering, their C_D-versus-Re curves are the ones most often reported in books on fluid mechanics, and, for bluff objects, they nicely illustrate the potential shifts in C_D as Re increases. However, from a biological standpoint these curves are less useful and may actually be misleading. Only rarely does a plant or animal closely resemble a cylinder or sphere, and those that do are seldom smooth. What effects do roughness and changes in shape have on the nature of the C_D-versus-Re curve?

Figure 9.6B shows the C_D-versus-Re behavior near the turbulent transition for cylinders of varying roughness, where roughness is measured as the ratio of D (the mean diameter of the roughness elements) to d (the cylinder's diameter). As roughness increases, the Re at which the turbulent transition occurs decreases, and the transition is less abrupt. More important for our purposes, the decrease in C_D associated with the transition is drastically reduced. For cylinders with roughness elements 1/50 the diameter, the effect of the turbulent transistion on C_D is quite minor. For a biological cylinder 1 cm in diameter, for instance, a ratio of $D/d = 1/50$ amounts to roughness elements of only 200 μm high. Most biologists would consider such a "rough" structure to be quite smooth. Thus, the natural roughness of biological cylinders renders them more or less immune to the abrupt changes in C_D accompanying turbulent transition, and the same is likely to hold true for other shapes. Turbulence in mainstream flow mimics the effect of roughness, lowering the critical Re and diminishing the reduction in drag. In short, for biological objects, the transition from a laminar to a turbulent boundary layer is unlikely to have substantial effect on C_D. Drag on an acorn barnacle, for instance, resembles that on a rough sphere: C_D is constant at 0.5 with no transition (Figure 9.6A).

I know of only one exception to this general pattern—a limpet shell that exhibited a classically abrupt transition in C_D (Denny 1989). At Re $= 10^5$, drag on this particular individual decreased by nearly half as its boundary layer became turbulent. One might suppose that a shell shape that reduces drag would be sufficiently advantageous that all limpets would have it. But it turns out that lift rather than drag is the worrisome force for limpets, and the shell that had the reduced C_D did not have a reduced C_L.

As informative as cylinders and spheres can be, it is dangerous to assume that their drag behavior applies to other shapes. As a pertinent example, let's explore the behavior of thin flat plates, beginning with a plate oriented with its flat surfaces perpendicular to flow (e.g., Figure 9.1). Because of the plate's sharp edges, flow around the plate must differ from that around a cylinder or sphere. Even at quite low Re, fluid moving from the upstream side of the plate to the downstream side can't make the abrupt 180° turn at the plate's periphery; consequently, flow separates at the plate's edge. With the separation points thus effectively fixed, transition to a turbulent boundary layer cannot substantially affect the wake's size or pressure, and consequently the drag coefficient for a flat plate oriented perpendicular to flow is virtually constant at 1.2 for all Re greater than 1000 (Figure 9.6A).

By contrast, when a thin plate is oriented parallel to flow, its presence causes very little disturbance. If it were not for viscosity, a very thin plate wouldn't affect flow at all (regardless of Re) and C_D would be zero. But, as we have seen, real fluids are viscous, so frictional forces are inevitable. (For continuity with our discussion of boundary layers, as we calculate the viscous drag on a plate at high Re, we assume that only one side of the plate is exposed to flow. If both sides are exposed, viscous drag will be twice as large.)

From Chapter 6, we know that if flow in the boundary layer is laminar, viscous force per area on a flat plate parallel to flow is

$$\tau_0 = 0.332 \mu u_\infty^{3/2} \sqrt{\frac{1}{\nu x}},$$

where x is distance from the plate's leading edge. Through some mathematical manipulations (Supplement 9.3), this equation can be massaged to yield two important relationships. First, for a plate with width y and length ℓ, total viscous drag (F_V) on one side of the plate is

$$F_V = 0.664 \left(\rho_f \mu\right)^{1/2} u_\infty^{3/2} y \ell^{1/2}. \tag{9.29}$$

Second, if y is a fixed fraction k of ℓ, the plate's area A is $k\ell^2$, and $y\ell^{1/2} = k^{1/4} A^{3/4}$. Thus

$$F_V = 0.664 k^{1/4} \left(\rho_f \mu\right)^{1/2} u_\infty^{3/2} A^{3/4}. \tag{9.30}$$

In other words, in laminar flow at high Reynolds numbers, viscous drag is proportional to $u_\infty^{3/2} A^{3/4}$. This proportionality is intermediate between what we deduced for viscous drag at low Re (where $F_V \propto u^1 A^{1/2}$) and pressure drag at high Re (where $F_D \propto u^2 A^1$).

Although equation 9.29 aptly describes the factors that contribute to viscous drag (also known as friction drag), it is unusual to see it presented in this form. Instead, equation 9.29 is traditionally shoehorned into the model used to describe pressure drag. That is, one sets equation 9.29 equal to the standard model for drag at high Re (equation 9.3), and absorbs all the characteristics of friction into a viscous drag coefficient, C_{DV}:

$$0.664 \left(\rho_f \mu\right)^{1/2} u_\infty^{3/2} y \ell^{1/2} = \frac{1}{2} \rho_f u_\infty^2 A C_{DV}. \tag{9.31}$$

Note that in this case we use $A \, (= y\ell)$, the wetted area of one side of the plate, rather than the plate's area projected in the direction of flow. Solving for C_{DV} (Supplement 9.4), we find that

$$C_{DV} = \frac{1.33}{\sqrt{\frac{\rho_f u \ell}{\mu}}}. \tag{9.32}$$

The denominator here should look familiar: it's the Reynolds number, calculated using ℓ (the plate's length) as the characteristic length. Thus, for a smooth plate parallel to flow,

$$C_{DV} = \frac{1.33}{\sqrt{Re}}. \tag{9.33}$$

In short, the viscous drag coefficient decreases with increasing Reynolds number.

For plates with a turbulent boundary layer (Re greater than approximately 10^5), the analogous equation is

$$C_{DV} = \frac{0.072}{Re^{0.2}}. \tag{9.34}$$

To gain some feel for how these relationships apply to biology, let's consider a fish gliding through water, whose streamlined smooth body acts approximately like a flat plate, ensuring that most of the drag it feels is viscous drag. For a typical fish, wetted area A_w is approximately 15 times frontal area (A_{pr}, the area projected into flow), and Webb (1975) suggests that pressure drag is approximately 60% as large as viscous drag. Thus, the total drag on the fish is

$$F_D = \left[\frac{1}{2}\rho_f u^2 \left(15 A_{pr}\right) C_{DV}\right] + 0.6 \left[\frac{1}{2}\rho_f u^2 \left(15 A_{pr}\right) C_{DV}\right]$$

$$= 12\rho_f u^2 A_{pr} C_{DV}. \tag{9.35}$$

Setting this expression equal to the standard model for pressure drag, we can calculate a fish's C_D:

$$12\rho_f u^2 A_{pr} C_{DV} = \frac{1}{2}\rho_f u^2 A_{pr} C_D,$$

$$C_D = 24 C_{DV}. \tag{9.36}$$

Thus, if the boundary layer is laminar, the drag on a fish is

$$C_D \approx \frac{32}{\sqrt{\text{Re}}}. \tag{9.37}$$

If the boundary layer is turbulent,

$$C_D \approx \frac{1.7}{\text{Re}^{0.2}}. \tag{9.38}$$

These estimates are graphed in Figure 9.6C. C_D decreases with increasing Re, reaching values below 0.1 by Re $= 10^5$. Note that the C_D calculated here is for a rigid fish. If the fish bends its body as it swims (which most fish do), calculating C_D becomes difficult, a topic we will address briefly in Chapter 10.

The drag coefficient can also change if an organism's shape changes. For instance, leaves and algal fronds can reconfigure in flow, assuming more streamlined postures with concomitantly reduced drag coefficients. Two representative examples—trees and seaweeds—are shown in Figure 9.6C. As a result of reconfiguration, C_D is proportional to $\text{Re}^{-0.5}$ to Re^{-1}, meaning that for these shape shifters, pressure drag is proportional to u^1 to $u^{3/2}$ rather than u^2. In this respect, the drag on reconfigurable bluff objects mimicks that of streamlined objects, albeit for different reasons. For a thorough account of reconfiguration's ability to reduce drag, see Vogel (1994).

In a few cases, reconfiguration can increase drag. For example, kelp fronds sometimes flap in flow much as flags flap in the wind, a mode of reconfiguration that increases C_D. To combat this effect, some wave-exposed kelps develop longitudinal corrugations in their fronds that, by increasing the frond's flexural stiffness, reduce flapping and maintain a low C_D (Rominger and Nepf 2014). (See Chapter 18 for a discussion of flexural stiffness.)

2.2   Lift Coefficient

Driven by the aeronautical industry, engineers have made a tremendous effort to understand the lift of airplane wings and propellors, and their efforts have been mirrored by biologists studying animal flight. In general, an airfoil's lift coefficient increases with increasing angle of attack up to a point (Figure 9.7). Beyond this critical angle, lift decreases, drag increases, and the wing *stalls*. In steady flow, birds, bats, and insects typically have maximum C_Ls ranging from 0.6 to 1.5 at angles of $25°$ to $35°$

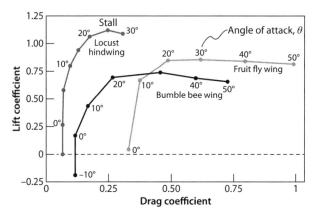

Figure 9.7 The relationship between lift coefficients and drag coefficients (a polar plot) for insect wings varies with the angle of attack (redrawn from Biewener (2003)).

(Vogel 1994; Alexander 2003). (As we will see in the next chapter, lift coefficients can be higher in the unsteady flows created as wings flap.)

For objects that get their lift from proximity to the substratum, lift coefficients are lower. For example, Denny and Blanchette (2000) measured values of 0.2 to 0.4 for limpets.

Little information is available as to how C_L varies as a function of Reynolds number; the few measurements available suggest that, within a species, it varies negligibly.

2.3   Added-Mass Coefficient

Added-mass coefficients have been calculated or measured for a variety of standard engineering shapes in mainstream flow (Figure 9.8), but these values must be used with care in biological contexts. They may apply for organisms suspended in a fluid (e.g., fish and birds) but are likely to be misleading for objects near boundaries. For example, the theoretical C_A for a sphere is 0.5, but values as high as 4.0 have been measured for spheres near solid surfaces. Furthermore, the values shown here are for objects with smooth surfaces, which organisms seldom have; roughness tends to increase C_A.

Table 9.1 provides measured C_As for a few aquatic organisms. Some general patterns are evident. For the animals listed (which are rigid), the more streamlined the creature (that is, the lower its drag coefficient), the lower its added mass coefficient, although this rule of thumb explains only about half the variation in C_A.

Seaweeds provide a stark contrast. When ocean waves break on a rocky shore, intertidal seaweeds are subjected to accelerations that can exceed $100 \text{ m}^2 \cdot \text{s}^{-1}$ (Gaylord 1999), and one might expect that these algae would compensate by having low C_As. Instead, seaweeds's C_As are surprisingly high, with typical values ranging from 2 to 8, indicating that these flexible organisms affect an unusually large volume of fluid relative to their own volume. In some cases, the large volume can be explained by the three-dimensional morphology of the alga. As the seaweed reconfigures in flow, water is loosely confined in the interstices between blades. In these cases, the added mass can be interpreted as an actual, physical mass of fluid. In other cases, the large added mass is likely due to the flapping and fluttering of the alga, an added mass in the traditional sense. Gaylord (1997) quantified the effect of blade morphology on the added mass coefficients of seaweeds through the use of a "flatness index,"

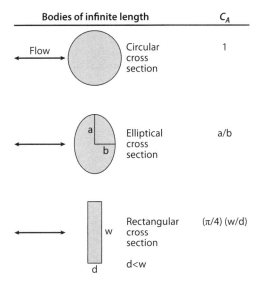

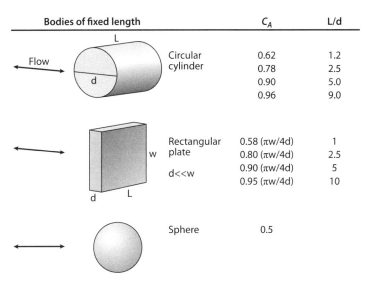

Figure 9.8 Added mass coefficients (C_A) for standard shapes.

FI. Noting that ellipsoidal objects have little surface area relative to their volume, whereas flat objects have large surface area per their volume, Gaylord suggested that the dimensionless ratio

$$FI = \frac{(\text{maximum projected area})^{3/2}}{\text{volume}} \tag{9.39}$$

is an appropriate index of morphology. A sphere has the lowest possible flatness index, 1.33; seaweeds have indices ranging from 20 to 150. In general, the added mass coefficient of seaweeds increases with increasing flatness index:

$$C_A = 0.18 FI^{0.76}. \tag{9.40}$$

This relationship explains about 40% of the observed variation in seaweeds' C_A.

Table 9.1 Drag and Added-Mass Coefficients for some Animals and Plants. Data from Denny and colleagues (1985) and Gaylord (1997).

Animals		C_D	C_A
Lottia gigantea (head on)	limpet	0.22	0.37
Lottia gigantea (tail on)	limpet	0.22	0.30
Lottia gigantea (broadside)	limpet	0.48	0.43
Lottia pelta	limpet	0.45	0.68
Lottia digitalis	limpet	0.52	0.84
Semibalanus cariosus	acorn barnacle	0.50	0.31
Balanus glandula	acorn barnacle	0.50	0.73
Thais canaliculata	predatory snail	0.67	0.72
Strongylocentrotus purpuratus	sea urchin	0.67	1.56
Seaweeds			
Egregia menziesii	feather boa kelp	0.12	4.52
Cryptopleura ruprechtianum	seaweed	0.13	8.02

Few reliable data are available regarding the effect of Reynolds number on the added mass coefficient of organisms. Perhaps the best information at hand is that for rough cylinders. The rougher the cylinder's surface, the smaller the effect of Re, and Sarpkaya (1976) found that C_A of a cylinder with roughness elements 1/50 its diameter varied by only 24% over a Re range from 5×10^4 to 10^6. Most biological objects are much rougher than this, and presumably their added mass coefficients vary even less with changes in Re.

3   RULES OF THUMB

In this survey of high-Re fluid-dynamic forces, I have treated them all as if they were of equal biological importance. However, there are cases in which one or several of these forces are negligible relative to the others, allowing us to propose simplified rules of thumb for the analysis of forces imposed on plants and animals. When comparing drag, lift, and acceleration reaction, I use pressure drag—the force that is often the largest—as a standard.

3.1   Friction Drag

Combining the equations for friction drag on a flat plate (equations 9.32 and 9.33) with the equation for pressure drag (equation 9.3), we arrive at the conclusion that

$$\frac{\text{friction drag}}{\text{pressure drag}} \approx 1.33 \frac{A_w}{A_{pr}} \frac{1}{C_D \sqrt{\text{Re}}} \tag{9.41}$$

or

$$\frac{\text{friction drag}}{\text{pressure drag}} \approx 0.07 \frac{A_w}{A_{\text{pr}}} \frac{1}{C_D \text{Re}^{0.2}}, \tag{9.42}$$

depending on whether the boundary layer is laminar or turbulent, respectively.

For example, a limpet has a wetted area approximtely four times its projected area and a drag coefficient of roughly 0.4. At Re $= 10^5$, friction drag amounts to less than approximately 10% of pressure drag for either a laminar or turbulent boundary layer. Indeed, at high Re, friction drag is comparable to pressure drag only if wetted area A_w is large compared to projected area A_{pr}, as is the case for fish.

3.2   Acceleration Reaction

For a stationary object in an accelerating fluid, acceleration reaction is $(1 + C_A) \rho_f V a$, and its ratio to pressure drag is

$$\frac{\text{acceleration reaction}}{\text{pressure drag}} = \frac{1 + C_A}{C_D} \frac{2a}{u^2} \ell_c, \tag{9.43}$$

where volume per projected area is represented by a characteristic length, ℓ_c.

Noting from Table 9.1 that shapes with high drag coefficients usually have high added-mass coefficients, we estimate that the ratio $(1 + C_A)/C_D$ is likely to be on the order of 3 for rigid, bluff biological shapes. Thus, acceleration reaction is comparable to pressure drag only when $6a\ell_c$ is approximately equal to u^2.

Consider, for example, a cylindrical massive coral 1 m in diameter and 1 m high ($\ell_c = 0.79$ m) in typical wave-driven subtidal flows: $u = 1$ m $\cdot$ s^{-1} and $a = 1$ m $\cdot$ s^{-2}. In this case, acceleration reaction is approximately 4.5 times pressure drag. Indeed, in their study of disturbance on coral reefs, Massel and Done (1993) concluded that it is acceleration reaction (rather than drag) that is primarily responsible for overturning massive corals. By contrast, for a small coral the same shape but only 0.1 m high and wide, acceleration reaction is only 45% as large as drag.

Flexible objects can present a different picture. For the seaweed *Cryptopleura ruprechtianum*, $(1 + C_A)/C_D = 69$, so acceleration reaction can be comparable to pressure drag even when acceleration and length are relatively small. For instance, for a frond 10 cm long and an acceleration of 10 m $\cdot$ s^{-2}, acceleration reaction is larger than drag unless $u > 12$ m s^{-1}.

3.3   Lift

In steady flow, the ratio of lift to pressure drag depends only on the relative force coefficients and the ratio of the areas over which force is applied:

$$\frac{\text{lift}}{\text{pressure drag}} = \frac{C_L}{C_D} \frac{A_{\text{pl}}}{A_{\text{pr}}}. \tag{9.44}$$

For the wings of birds, bats, and insects, lift per drag is typically near 2, although it can be as high as 10 for some gliding birds. For limpets, lift per drag ranges from 1 (for high-spired species) to 4 (for low-spired species). For bluff objects (a cactus or massive coral), lift per drag is approximately zero.

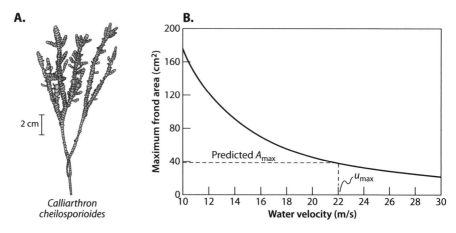

Figure 9.9 A. *Calliarthron cheilosporioides.* **B.** Maximum frond area of *Calliarthron* is limited by water velocity.

4   ECOMECHANICAL EXAMPLES

4.1   Maximum Size of Seaweeds

We will employ the principles outlined in this and the preceding chapters throughout the rest of this book. But a preview of the manner in which lift and drag can be used in ecomechanics may help these concepts to soak in.

Intertidal seaweeds are subjected to high water velocities as waves break on the shore, and area exposed to flow increases as seaweeds mature, increasing the drag imposed by a given velocity. For some species, the strength of the plant does not keep pace with the force imposed, limiting the size to which individuals can grow. Patrick Martone and I (Martone and Denny 2008) explored this relationship in the erect coralline alga, *Calliarthron cheilosporioides*, a branched red seaweed that dominates space in the low intertidal zone on exposed shores of California (Figure 9.9A). Like most seaweeds, *C. cheilosporioides* is flexible, allowing it to reconfigure in flow. As a result, its drag coefficient decreases with increasing Reynolds number:

$$C_D \approx 1.1 \text{Re}_f^{-0.29}. \tag{9.45}$$

Here Re_f is the *frond Reynolds number*:

$$\text{Re}_f = \frac{u\sqrt{A}}{\nu}, \tag{9.46}$$

where A is the maximum projected area of the frond, essentially half its wetted area, and $\sqrt{A}$ serves as a characteristic length for this complicated and variable shape. Rearranging equation 9.46, we see that

$$A = \frac{\nu^2 \text{Re}_f^2}{u^2}. \tag{9.47}$$

Incorporating equations 9.45 and 9.47 into the standard model for drag (equation 9.3), we arrive at an expression that relates water velocity and frond area to the force imposed:

$$F_D = \frac{1}{2}\rho u^2 AC_D \qquad (9.48)$$

$$= 0.55\rho v^2 Re_f^{1.71}. \qquad (9.49)$$

As a *Calliarthron* frond grows, area is added at the distal ends of branches, but the size of the basal portion of the plant does not change. As a result, large fronds have the same strength as small fronds, approximately 20.0 N. Setting this breaking force equal to applied drag and solving for critical frond Reynolds number, we find that fronds will break when $Re_f = 1.1 \times 10^6$. Inserting this value into equation 9.47 allows us to predict the relationship between water velocity and the maximum area to which a frond can safely grow:

$$A_{max} = \frac{\left[\left(1.1 \times 10^6\right) v\right]^2}{u^2}. \qquad (9.50)$$

The kinematic viscosity of seawater at 15°C (typical for *Calliarthron*) is 1.21×10^{-6} m$^2 \cdot$ s^{-1}; thus

$$A_{max} = \frac{1.77}{u^2}. \qquad (9.51)$$

This expression in graphed in Figure 9.9B. The faster the flow, the smaller the maximum frond size. At the site where we collected our specimens, we recorded a maximum water velocity of 22 m$\cdot$s^{-1}, corresponding to a maximum frond area of 37 cm^2. Average maximum frond size at this site was 41 cm^2, quite close to the predicted value, suggesting that the size of this ecologically important seaweed is limited by the interaction between frond structure—the fact that strength does not increase with frond area—and the maximum velocity found at a particular location.

The ability to predict maximum frond area provides a lens through which the low intertidal community can be viewed. In this habitat, erect coralline algae play much the same role that trees do in a forest: they provide both food and shelter for animals, and they compete for light with other plants. Knowing how fluid dynamics limits the size of these ecosystem engineers provides insight into the current structure of low-intertidal communities and allows us to predict how the forest will change with shifts in wave exposure.

4.2   Overturning Cobbles: The Control of Diversity

In a classic ecological study, Wayne Sousa documented the interaction between algal succession and disturbance in an intertidal boulder field (Sousa 1979). He noticed that small cobbles were nearly devoid of seaweeds, medium-sized rocks sported a diverse algal community, and large boulders were dominated by just a few species. This size-dependent pattern of diversity is similar to the time-dependent succession one observes on bedrock. Starting with bare rock, algae recruit, and species diversity initially increases through time as new individuals settle and grow. However, as the surface of the rock fills up, species begin to compete for light and space. With further passage of time, a few competitively dominant species begin to take over, and species diversity declines.

Sousa proposed that the frequency at which cobbles and boulders are overturned governs the state of succession in algal communities: overturning a rock wipes it clean of algae, and starts succession anew. He hypothesized that small cobbles are overturned frequently, before succession can advance very far, so they have low diversity. Medium-sized rocks are overturned less frequently, rarely enough to allow diversity to reach a peak, but frequently enough to prevent competitive dominants from taking over. Large boulders are overturned very infrequently, so they exhibit the final, low-diversity stage of succession.

To demonstrate the legitimacy of this concept, Sousa anchored small- and medium-size boulders to keep them from overturning. When the frequency of disturbance was thus decreased, diversity was low even on medium-size cobbles.

Sousa's idea makes wonderful sense provided that the smaller the cobble, the more often it is overturned. Field observations support this assumption, but we are now in position to provide a mechanistic explanation.

Consider the situation shown in Figure 9.10A. We model a cobble as an oblate ellipsoid (a shape like an M&M candy) with radius r and density ρ_r. The height of the cobble, z, is a fraction k of its radius; that is, $z = kr$. The cobble is butted up against its downstream neighbor such that, in order to overturn, it must pivot about a point at its downstream edge. In this situation, any drag or acceleration reaction acting on the cobble pushes it against its neighbors, but because these forces act in line with the pivot, they cannot overturn the cobble. By contrast, lift can impose an overturning moment, and it is there that we focus our attention.

Actually, there are two moments of concern. The weight of the cobble, F_g, acting through the rock's center of mass, applies a moment tending to rotate the cobble counterclockwise. The weight of the cobble is

$$F_g = \rho_e \left(\frac{4}{3}\right) k\pi r^3 g, \qquad (9.52)$$

where ρ_e is the boulder's effective density—rock density (ρ_r) minus fluid density (ρ_f)—and $\frac{4}{3}k\pi r^3$ is the boulder's volume. The magnitude of the weight-imposed moment is thus

$$\begin{aligned} M_g &= F_g r \\ &= \rho_e \left(\frac{4}{3}\right) k\pi r^4 g. \end{aligned} \qquad (9.53)$$

In the absence of flow, this moment is offset by the reaction force of the boulder resting on the substratum. In steady flow, however, lift provides a countermoment. At velocity u, the magnitude of lift is

$$\begin{aligned} F_L &= \frac{1}{2}\rho_f u^2 C_L A_{\mathrm{pl}} \\ &= \frac{1}{2}\rho_f u^2 C_L \left(\pi r^2\right). \end{aligned} \qquad (9.54)$$

If we assume that lift acts through the center of the boulder, the moment imposed by lift is

$$M_o = F_L r \qquad (9.55)$$

$$= \frac{1}{2}\rho_f u^2 C_L \left(\pi r^3\right). \qquad (9.56)$$

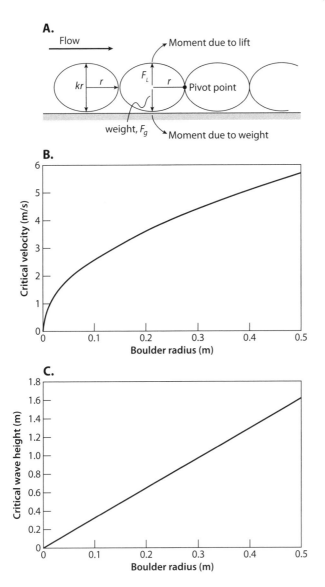

Figure 9.10 **A**. A simple model for the moments acting on a cobble in flow. Critical overturning velocity (**B**) and critical wave height (**C**) increase with cobble size.

A critical condition is reached when $M_w = M_o$. If the lift-induced moment is any larger, there is a net moment rotating the boulder clockwise, and it flips over. Thus, we can calculate the critical velocity at which the boulder flips by setting $M_g = M_o$ and solving for u:

$$u_{\text{crit}} = r^{1/2} \sqrt{\frac{8 \left(\frac{\rho_r}{\rho_f} - 1 \right) kg}{3C_L}}. \qquad (9.57)$$

The denser the rock, the larger k, or the lower the lift coefficient, the faster water must flow to overturn the cobble. But for our purposes, the main message of this calculation is that the velocity required to overturn a cobble increases with its size; to be specific, with the square root of its radius. In short, bigger boulders require larger velocities to overturn them. Ergo, in the nearshore environment where high velocities are encountered less often than low velocities, big boulders should indeed be flipped less often than small boulders.

What sort of velocities are we talking about? An average rock has a density of approximately 2700 kg · m^{-3} and the density of seawater is 1025 kg · m^{-3}. Data from limpets (some of which are shaped more or less like rounded cobbles) suggest that C_L is approximately 0.2. Let's suppose that our boulder has a height-to-radius ratio of 0.3. Given these conditions, the velocities required to overturn the cobble are shown in Figure 9.10B. A velocity of just over 2.5 m · s^{-1} will overturn a small cobble with a radius of 10 cm; a speed of 5.7 m · s^{-1} is needed to overturn a boulder 50 cm in radius.

We can take this analysis a step further. Rudimentary wave theory allows us to estimate the maximum water velocity imposed on surf-zone cobbles as a function of wave height h, (see Supplement 9.5):

$$u = \sqrt{2gh}, \tag{9.58}$$

which we can rearrange to solve for the wave height required to overturn boulders of a given size (Figure 9.10C):

$$h = \frac{u_{\text{crit}}^2}{2g}$$

$$= \frac{8 \left(\frac{\rho_r}{\rho_f} - 1 \right) kr}{6C_L}. \tag{9.59}$$

A wave only 0.3 m high is sufficient to overturn a cobble 10 cm in radius, but a wave more than 1.6 m high is needed to overturn a 50-cm-radius boulder. High waves occur less often than low waves, so small cobbles are overturned more often than large cobbles.

These calculations should be taken with a large grain of salt. The value used here for the lift coefficient is merely a guess, and we have assumed that a cobble interacts with its neighbors only at the pivot point, which will seldom be the case for rocks tossed together in nature. Nonetheless, this exercise illustrates the ability of fluid dynamics to provide a mechanistic explanation of an ecologically important process. With refinement, a model of this sort has the potential to accurately predict how overturning frequency—and, thereby, algal diversity—should vary as a function of wave climate.

5   CONCEPTS, CONCLUSIONS, AND CAVEATS

In this chapter, we dealt with some of the specifics of drag, acceleration reaction, and lift, allowing us to make a few general conclusions:

- At high Re, pressure drag is commonly the largest fluid-dynamic force imposed on organisms, and its magnitude is governed by the size of the wake. The shape of streamlined objects allows fluid to flow past gracefully, leaving a small wake and incurring little drag. By contrast, bluff objects cause flow to separate, producing a large wake with low pressures and, thereby, substantial drag. Turbulence in the boundary layer can maintain fluid's momentum as it moves past a bluff object, delaying separation and decreasing the size of the wake somewhat, but drag is still high compared to that of streamlined objects.

- For the bluff, rough objects found in nature, C_D is essentially constant for Reynolds numbers ranging from 10^3 to 10^7, affirmation of the fact that drag is proportional to the product of dynamic pressure and projected area.
- By contrast, for both flexible and streamlined objects, C_D decreases with increasing Re, albeit for different reasons. For flexible objects, C_D decreases because the object reconfigures when drag is imposed, becoming more streamlined. For streamlined objects, C_D decreases because drag is proportional to $u^{2/3}$ rather than u^2.
- For rigid objects, added mass coefficients are roughly correlated with drag coefficients—high-drag objects typically have substantial added mass. Bladelike flexible objects are different—they can have low drag coefficients but large added masses.
- Lift on an airfoil or hydrofoil varies with the angle of attack, reaching a maximum at an angle of $25°$ to $35°$. This variation allows animals to adjust the lift of their wings and tails. Even bluff objects on a substratum can experience lift.
- Added mass and lift coefficients are surprisingly independent of Re.
- Rules of thumb allow us to predict when and where each force coefficient dominates.

All these conclusions need to be treated with caution, however. Subtle influences on flow can make substantial differences in the forces imposed, and evolution is adept at finding innovative ways to adjust drag, lift, and added mass coefficients. For example, birds often have a special winglet (the alula) at the leading edge of their wings. As the angle of attack increases to the point where the wing is likely to stall, the alula separates from the rest of the wing, forming a leading edge slot that helps to direct flow along the wing, delaying stall. Many airplanes make use of this concept: as they come in to land, a metal version of an alula is extended in front of the wing, augmenting lift at the low speed necessary for a safe landing.

Important fluid-dynamic effects are still being discovered. For example, Frank Fish and his colleagues found that the tubercles on the leading edge of humpback whales' pectoral fins augment the lift these appendages create by allowing them to use higher angles of attack before they stall. Left to themselves, human engineers would likely never dream of increasing the efficiency of an airplane's wing by adding ugly bumps to its leading edge. But evolution maintains an open mind when it comes to strange experiments, and now that the tubercle effect has been noticed, it is under active consideration for airplane wings, wind-turbine blades, and surfboard skegs (Fish et al. 2011).

Surprises such as these constitute a warning: it is always advisable to measure C_D, C_A, and C_L directly at the Reynolds numbers appropriate to a given situation.

Chapter 10

Locomotion

In this chapter, I focus on two aspects of locomotory physics and physiology: the *speed* with which organisms move and the metabolic *cost* of transport. Both have particular importance in ecology. Speed plays a vital role in how animals catch prey and avoid predators, and given a fixed time frame—the length of a season or the duration of a low tide, for instance—speed determines the area over which animals can disperse, forage, seek mates, and find a place to hide. Thus, the more we know about the physics that determines speed, the better we can predict individuals' interactions with each other and with their environment. Similarly, cost plays a central role in the evolution of foraging behavior and life-history strategies: for example, it might be physically possible for an animal to forage far afield or migrate to distant breeding grounds, but if the cost of the journey exceeds the metabolic or reproductive benefits, it would be better to stay put.

Locomotion is far too large a subject to cover in depth in a single chapter. Instead, my intent is to introduce the principles of locomotion so that you can grasp the potential of this mechanistic approach. There are times, however, when the interaction of locomotory principles becomes so complex that simple predictions of speed and cost are not possible. In these cases, we will rely on empirical measurements.

We start with a brief overview of the physics of locomotion and then move on to a consideration of the motors that power movement—axonemes and muscles. With these basics under our belt, we explore each of the major types of locomotion in turn—crawling and burrowing, swimming, flying, and running. Information about individual modes of movement is then synthesized in an examination of their relative speeds and costs. We conclude with brief sections on jumping and climbing and a final synthesis in which the mechanics and energetics of locomotion are used to explain why some animals migrate and others don't.

1 THE PHYSICS OF LOCOMOTION

From a physical perspective, locomotion is all about *thrust*, the propulsive force created by organisms. Thrust has several functions. For example, it can be used to accelerate an organism: rapid acceleration by prey can be the key to avoiding predators, and rapid acceleration by a predator can be an effective tactic when catching prey. Often, however, thrust is used to *avoid* acceleration. For example, to fly

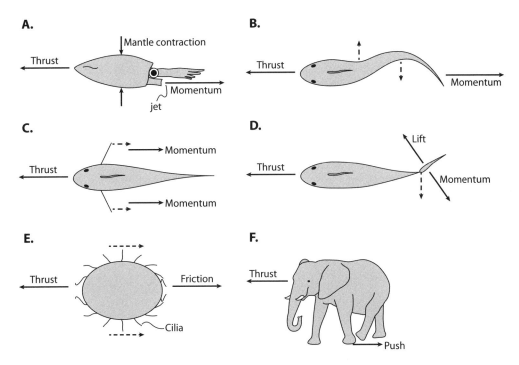

Figure 10.1 Modes of thrust generation. **A.** Jetting. **B.** Propagation of a wave along the body. **C.** Rowing using drag. **D.** Lift. **E.** Ciliary beating. **F.** Pushing on a solid substratum.

at a constant speed a bird needs sufficient thrust to avoid both accelerating downward under the pull of gravity and decelerating from the resistance of drag. Because the physics of thrust plays such a central role in locomotion, we start with a brief overview of how thrust is produced.

1.1 Sources of Thrust

Recall from Chapter 2 that force is equal to the time rate of change of momentum:

$$F = \frac{d\,(mu)}{dt},\qquad(10.1)$$

where m is mass and u is velocity. Foward thrust can thus be created by increasing the rearward velocity of some mass, usually a mass of fluid. Consider the following examples:

- As the hollow body of an animal squeezes, a jet of water can be expelled rearward. The rate at which fluid mass is ejected $(kg \cdot s^{-1})$ times the speed of flow in the jet $(m \cdot s^{-1})$ determines the rate at which momentum is transported and, thereby, the magnitude of thrust $(kg \cdot m \cdot s^{-2} = newtons)$. Jellyfish, scallops, and squids all swim using this type of reciprocating jet propulsion (Figure 10.1A).
- Many fish swim by undulating their bodies. The mechanism by which thrust is created is less straightforward than that of jet propulsion—it involves both added mass and pressure components

(Videler 1993)—but the result is the same. As the undulatory wave moves rearward, momentum is imparted to the surrounding water, propelling the fish forward (Figure 10.1B).

- Momentum can also be imparted to a fluid by rowing, the dragging of an appendage backward through a fluid. When a fish's pectoral fin is flapped, for instance, its pressure drag and acceleration reaction propel water backward (Figure 10.1C). Rowing is the primary form of thrust for a wide variety of animals: copepods, aquatic insects, and water striders are prominent examples.

- Alternatively, thrust can be created by moving an appendage perpendicular to the axis of transport. For example, the swinging of a tuna's tail (a hydrofoil) or the flapping of a bird's wing (an airfoil) imparts rearward momentum to fluid, and the lift produced therefore has a component in the direction of travel—a thrust (Figure 10.1D). Thrust via lift is common in both air and water: in addition to fish and birds, whales, dolphins, seals, bats, and many insects propel themselves with lift.

When fluid momentum is created by reciprocating motion, as it is for each of these propulsive strategies, it is often accompanied by the formation of a vortex ring: a toroidal volume of circulating fluid. When a jellyfish jets, for instance, shear between the jet and the adjacent fluid forms a vortex that travels rearward, entraining more fluid and augmenting the momentum produced by the animal. Vortices can also be shed as flapping wings and tails change directions. As fish and birds move slowly, they leave a series of discrete vortex rings in their wake, and the analysis of these rings' momentum has been used to assess the efficacy of thrust production (Videler 1993).

So far we have described thrust in terms of momentum, an approach most useful when applied to organisms moving at Reynolds numbers well above 1 where, by definition, inertial forces outweigh viscous forces. At lower Re, thrust is provided by cilia and flagella, whose interaction with the surrounding fluid has more to do with viscosity than with inertia (Figure 10.1E). In this viscous regime, creation of thrust by jetting, pressure drag, acceleration reaction, or lift is ineffective.

Lastly, organisms can produce thrust by pushing or pulling against a solid, immovable object. For example, when an elephant walks, its legs push back against the the ground, and the reaction to this traction pushes the beast forward (Figure 10.1F).

1.2   Levers

As an animal runs, its body moves forward only as fast as its legs move back. Similarly, birds and insects fly roughly as fast as their wingtips move through the air, and rowing animals get thrust from drag only if the backward speed of an appendage exceeds the body's forward velocity. Thus, to run, fly, or swim effectively, animals often need high-speed appendages, a goal achieved through the use of levers (Figure 10.2).

In its simplest form, a lever is a rigid rod that pivots about an axis. The angle θ through which the rod rotates is

$$\theta = \frac{\ell}{x}, \tag{10.2}$$

where ℓ is the length of the arc traveled by a point on the rod, and x is the lever arm (the point's distance from the pivot). The two arms of a lever are often different

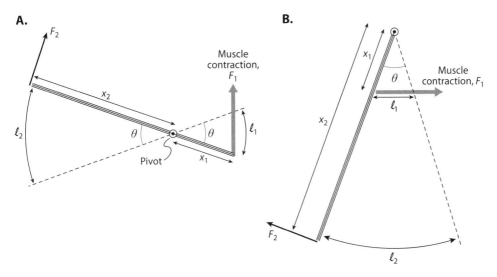

A.

F_2

Muscle contraction, F_1

x_2

l_2

θ θ l_1

Pivot x_1

B.

x_1

θ

Muscle contraction, F_1

l_1

x_2

F_2

l_2

Figure 10.2 The mechanics of levers. **A.** Muscle attachment inboard of the pivot (e.g., an insect wing). **B.** Muscle attachment outboard of the pivot (e.g., the legs of reptiles or mammals).

lengths, but as long as the rod is rigid they rotate through the same angle. For example, in Figure 10.2A,

$$\theta = \frac{\ell_1}{x_1} = \frac{\ell_2}{x_2}. \tag{10.3}$$

In a typical locomotory system, the pivot point is near the proximal end of a rodlike appendage (an insect's wing, for instance, Figure 10.2A) such that the distance from the pivot to the attachment point of a muscle, x_1, is less (often much less) than the appendage's outboard length, x_2. Because $x_2 > x_1$, a small displacement (ℓ_1) of the muscle results in a large displacement (ℓ_2) of the appendage:

$$\ell_2 = \ell_1 \frac{x_2}{x_1}. \tag{10.4}$$

The ratio x_2/x_1 is the system's *displacement advantage*. The same principles apply if the pivot is at the lever's inboard end, as it often is in a leg or fin (Figure 10.2B).

Using similar logic, we can quantify the relative speeds of any two points on a lever. If the inboard point moves at speed u, the speed of the outboard point is amplified by the displacement advantage:

$$u_2 = u_1 \frac{x_2}{x_1}. \tag{10.5}$$

This allows the relatively slow contraction of an inboard muscle to get the business end of a locomotory appendage up to speed. By varying x_2 relative to x_1, displacement advantage can be adjusted to match the required motions of a particular appendage to the contractile characteristics of a particular muscle.

Velocity amplification comes with a trade-off, however. Force F_1 imposed on an inboard point on the lever applies a moment $F_1 x_1$ about the pivot. If the lever rotates at constant angular velocity, this inboard moment is equal and opposite to the moment applied at an outboard point, F_2x_2. In other words,

$$F_2 = F_1 \frac{x_1}{x_2}. \tag{10.6}$$

Thus, the force acting at the outboard end (drag, for instance) is less than that applied at the inboard end by the inverse of the displacement advantage. In a lever system, what one gains in speed, one loses in force.

Recalling that power is the product of force and speed (Chapter 2), we can calculate the power required to move the inboard point:

$$P_1 = u_1 F_1. \tag{10.7}$$

Similarly, at the outboard point,

$$P_2 = u_2 F_2. \tag{10.8}$$

Substituting equations 10.5 and 10.6 for u_2 and F_2, we see that

$$P_2 = u_1 \frac{x_2}{x_1} F_1 \frac{x_1}{x_2} = u_1 F_1 = P_1. \tag{10.9}$$

In short, the trade-off between speed and force ensures that the power associated with rotation of a lever is equal at both ends.

1.3   Power

Like thrust, power plays a central role in locomotion. For example, as noted before, when an animal moves horizontally at constant speed, sufficient thrust must be produced to just equal drag. As a result, travel requires power:

$$\text{power} = \text{thrust} \times \text{speed}. \tag{10.10}$$

The more drag it has to overcome and the faster it moves, the more power an organism must supply.

Energy is also expended through time in keeping a creature aloft even when the organism has no vertical velocity. For example, at high Re the upwardly directed thrust needed to offset weight is accompanied by downward transport of momentum. In creating this momentum, force propels fluid and power is expended. When Re is low, upward force is created by moving an appendage downward against the friction of viscosity, and again power is expended. Only if an organism is neutrally buoyant or resting on the seafloor or ground can it maintain altitude without power.

2   SOURCES OF POWER
The Motors

Power is thus a necessary component of locomotion. At the biochemical level, adenosine triphosphate (ATP) is the near-universal currency for energy transport in living cells: approximately 50 kJ · mol^{-1} is stored in the bond that connects ATP's terminal phosphate group to the body of the molecule. When this bond is cleaved, the energy released can be used to power locomotion. However, ubiquitous as it is as a source of *chemical* energy, ATP by itself is useless as a source of *mechanical* energy. To power movement, energy released by the hydrolysis of ATP must be converted

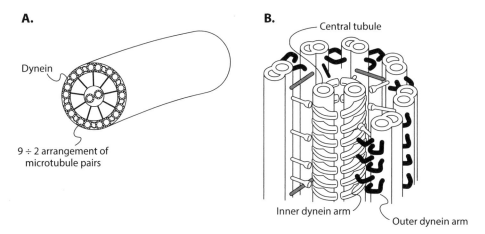

Figure 10.3 The axoneme. **A.** A cross section through a cilium or flagellum. **B.** Details of the dynein arms (redrawn from Holberton (1997)).

into motion in one of two ways: by the translation of microtubules relative to each other, or by the translation of actin filaments relative to myosin filaments.

2.1   Dynein and the Axoneme

Many micro- (and even some macro-) organisms move using cilia and flagella. Within each cilium or flagellum there is a ring of nine double microtubules surrounding two central, single tubules (Figure 10.3A), a highly conserved arrangement that forms the *axoneme*. A variety of molecules link microtubules within the axoneme, but for our purposes the important linkages are short protein bridges connecting the nine double microtubules to each other (Figure 10.3B). In addition to functioning as a cross bridge between microtubules, each of these *dynein arms* can act as an ATPase. The energy released by cleaving an ATP is used to bend the dynein arm, causing it to walk along the adjacent microtuble.

In a cilium, the coordinated walking of multiple arms causes microtubules to slide relative to each other, lengthening one side of the axoneme, shortening the other, and thereby bending the cilium. Reversing the direction of sliding reverses the curvature, bending the cilium back. The process is similar in a flagellum except that the sliding of tubules alternates sides along the axoneme's length, causing waves of bending to travel along the structure. The frequency of cyclic bending in both cilia and flagella can be as high as 40 Hz, although a rate of 10 to 20 Hz is more typical.

Both theoretical considerations (Wu 1977) and empirical measurements (Silvester and Sleigh 1984) suggest that at 10°C axonemes can produce approximately 4×10^{-10} watts per meter $(W \cdot m^{-1})$ of cilium or flagellum. Cilia are approximately 0.25 μm in diameter, so a meter of cilium has a volume of 4.9×10^{-14} m^3. Ciliary power production thus works out to roughly 8150 $W \cdot m^{-3}$ of axoneme, or (assuming the density of a cilium is 1080 kg $\cdot$ m^{-3}) approximately 7.5 $W \cdot kg^{-1}$ (Table 10.1). Power output of cilia and flagella increases by 20% to 40% for every increase of 10°C (Podolsky and Emlet 1993).

In a few exceptional cases, individual cilia gang together to form large, compound structures. For example, each comb of a comb jelly (a ctenophore) is constructed from a row of up to 100,000 cilia connected side by side to form a paddle (Barlow and Sleigh 1993). But such oddities are rare; in most instances axonemes act as individual motors. Rather than being attached to some locomotory appendage, the membrane coated axoneme *is* the locomotory appendage.

2.2   Actin and Myosin

By contrast, individual actin-myosin motors are commonly combined into larger structures—muscles—that are in turn attached to separate appendages: legs, wings, and fins. Actin monomers are large globular proteins that self-assemble to form long helical filaments. Myosin monomers also assemble into filaments, but by combining several types of monomers, myosin assembly results in a thicker filament from which globular heads extend (Figure 10.4A). The heads of myosin filaments attach to adjacent actin filaments at defined binding sites.

When the system is activated, individual myosin heads forcefully bend and then detach from actin (one ATP molecule is hydrolyzed to ADP for each bend-detach cycle), causing a myosin filament to walk its way along its actin neighbors. The rate of myosin's bending and recovery is determined by the rate at which ATP can be bound, hydrolyzed, and released, a rate that varies depending on the molecular structure of the myosin heads.

Actin and myosin filaments are typically bundled together to form highly ordered compound structures. This morphology has been studied most intensively in skeletal muscles, in which actin filaments interdigitate with myosin filaments, and are held in position by structures known as Z-discs (Figure 10.4B). In this arrangement, the heads on one end of each myosin filament face the opposite direction from those on the other. As a result, when myosin heads walk along adjacent actin filaments, Z-discs are pulled toward each other. The combination of Z-discs, actin, and myosin—together known as a *sarcomere*—thus functions as a contractile motor.

The modular nature of sarcomeres allows the construction of motors of a wide variety of sizes, shapes, and dynamics. Within a sarcomere, if actin and myosin filaments are long, many myosin heads can be engaged in pulling adjacent Z-discs together, and the force of contraction is potentially large (Figure 10.4C). The *rate* of contraction is slow, however, limited by the speed at which individual myosin heads can walk. Alternatively, the same overall length of contractile apparatus can be constructed from several short sarcomeres connected in series (Figure 10.4D). In this case, there are fewer myosin heads tugging on each actin filament, so the force of contraction is less, but the speed of contraction increases. If only the first in a linear series of sarcomeres contracts, the whole apparatus moves at the rate of that individual sarcomere. If a second sarcomere contracts as well, it adds its rate to that of the first, and the overall rate of contraction is doubled. Add in a third sarcomere, and the rate triples. In sum, although having several short sarcomeres in series reduces contraction force, it increases contraction rate.

Recall yet again that power is the product of force and speed. Consequently, long sarcomeres (which are forceful but slow) and short sarcomeres in series (which are weak but fast) tend to produce the same power.

Sarcomeres are commonly grouped together in three dimensions to form *myofibrils*, where the length of each sarcomere and the number of them attached in series determine the force and overall rate of contraction. Myofibrils are in turn grouped

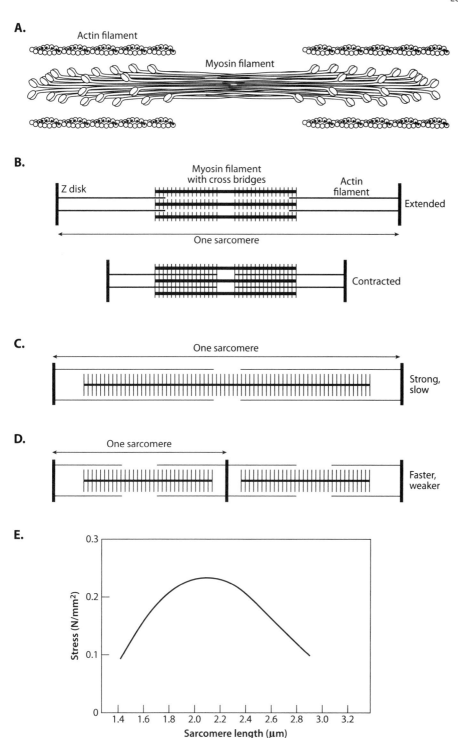

Figure 10.4 The actin-myosin motor. **A.** Actin and myosin filaments (redrawn from Schmidt-Nielsen (1997)). **B.** The sarcomere of a striated muscle filament. **C** and **D.** Sarcomere length affects contraction speed and force. (See text.) **E.** The force produced by a sarcomere is maximal at intermediate extension.

together to form muscle fibers, and muscles fibers can then be packaged to form muscles, in which the number of fibers acting in parallel determines the overall force of contraction. In this fashion, muscles can be tailored to specific locomotory needs. For a more thorough introduction to the flexible design of muscles, consult Biewener (2003).

Contractile force varies with sarcomere extension (Figure 10.4E). At maximum extension, few myosin heads can connect to actin, and force is minimal. As the sacomere contracts, more and more myosin heads connect and force rises, but only up to a point. When the sarcomere is so short that actin fibers run into the Z-discs, the fibers are displaced radially, contact between actin and myosin is disrupted, and force decreases.

When working in conjunction with locomotory appendages, muscles change length as the appendage moves. At times, contraction of the muscle moves the appendage, at other times the appendage's inertia stretches the muscle, and the dynamics of this oscillating system affects muscle performance. The details of this interaction are a bit complex and beyond the purview of this chapter, but Supplement 10.1 offers a brief overview.

In some cases, the primary function of a muscle is to change length very little or not at all. For example, by stiffening at the proper time during a running stride, the muscles in your calves allow your achilles tendons to stretch and store elastic energy. In large part, it is this stored energy—rather than the contraction of the muscle— that powers your next stride. As we will see, energy storage is a common theme in locomotion.

Power output of muscles varies from species to species and even from muscle to muscle within an individual. Muscles designed for slow contraction—in which the rate of myosin-head stepping is sluggish—may produce only 10 $W \cdot kg^{-1}$ at 10°C, a third more than the mass-specific power of cilia (Table 10.1). Although they aren't powerful, slow muscles are very efficient at transducing chemical energy to mechanical energy, and they are used when little power is required for locomotion. By contrast, some fast muscles—in which the myosin ATPase has a rapid turnover— can produce more than 100 $W \cdot kg^{-1}$, 13 times the power of cilia. Such extraordinary values are typically confined to the flight muscles of insects and birds, however, which operate at temperatures of 35°C to 40°C. The fast muscles of fish, which typically function at 10°C, produce 15 to 75 $W \cdot kg^{-1}$ with an average of approximately 40 $W \cdot kg^{-1}$ (James et al. 1995), about five times the mass-specific power of cilia.

Like the axoneme, the power output of muscle is temperature dependent, increasing by 50% to 100% with each increase of 10°C within a muscle's working temperature range.

This brief introduction has barely touched the highlights of muscle physiology. For a more thorough entrée into the rich literature of muscle mechanics, please consult Vogel (2001), Alexander (2003), and Biewener (2003).

2.3   Comparing Power

It is instructive to compare biological and synthetic motors (Table 10.1).

Axonemes and cool muscles come close to the power output of common electric motors but kilogram for kilogram produce only a fraction of the power of petroleum-fueled engines. Hot muscles do better. At 35°C, a kilogram of a dove's flight muscle

Table 10.1 Mass-specifc Sustained Power Output from a Variety of Biological and Synthetic Motors. Data from Vogel (2001).

Motor	$W \cdot kg^{-1}$
Axoneme (10°C)	7.5
Slow muscle (10°C)	10
Fast muscle (10°C)	40
Electric motor	45
Automobile engine	91
Motocycle engine	227
Fast muscle (35°C)	230
Aircraft engine (piston)	318
Aircraft engine (turbine)	1134

edges out a kilogram of motorcycle engine in producing sustained power (Tobalske et al. 2003), and a quail's flight muscles can produce 360 to 460 $W \cdot kg^{-1}$ in short bursts, briefly matching the output of a piston aircraft engine (Askew and Marsh 2002).

3 DRIFTING

Passive Locomotion

We will see later in this chapter how muscle power factors into the speed and cost of locomotion, but we begin our overview instead with locomotion in its most passive form. Travel is simple for plants and animals that hitch a ride with ambient flow.

With only a few exceptions, organisms are denser than the fluid around them, endowing them with a tendency to sink. Tossed into air or water, plants and animals will eventually hit bottom. As a consequence, the time a passive drifter spends in flow—and, therefore, the distance it travels—depend on two factors: its its sinking speed and its initial height.

The slower a particle sinks, the longer it can ride a current, and there are several ways that sinking rates can be reduced. Some aquatic organisms reduce their density by storing low-density lipids, exchanging heavy ions for light ones, or using gas bubbles or swimbladders. For others, altitude is maintained by swimming, a topic we'll address shortly. In many cases, however, simply being small suffices. As we deduced in Chapter 8, spores and larvae sink at rates of a few microns to a few millimeters per second, so slowly that they can travel considerable distance before falling out of a current. For example, zygotes of the brown seaweed *Sargassum nuticum* have a radius of 70 μm and sink at a speed of only 0.5 mm $\cdot$ s^{-1}(Norton and Fetter 1981). Released at a height of about a meter, they take 2000 s to hit bottom. Even in a sluggish current of 10 cm $\cdot$ s^{-1}, the propagules would travel more than 200 m before striking the seafloor. The spores of the giant kelp *Macrocystis pyrifera* are even smaller (only 3.5 μm in radius), and they sink at a mere 0.0012 mm $\cdot$ s^{-1} (Gaylord et al. 2002). Released only half a meter above the seafloor, they can nonetheless ride a current for nearly 5 days.

In air, pollen grains sink considerably faster (2 to 2.5 cm · s^{-1}, Niklas 1982). Released from a flower a meter above the ground, pollen would fall to earth in only 40 to 50 s, giving it little chance to fulfill its procreative duties. As noted previously, pollen's rapid sinking presumably provided an impetus for flowers' coevolutionary interaction with insects, birds, and bats. Additionally, many trees produce seeds that either have a large drag relative to their weight (e.g., dandelion seeds) or rotate to produce lift (e.g., maple seeds) as a means to prolonging their time in the air and thereby traveling as far as possible.

Because these calculations ignore the role of turbulence, they need to be taken with a grain of salt. To a first approximation, turbulence doesn't change the average settling time, but it drastically increases the variance of settling times, a phenomenon we will revisit in Chapter 19.

Plants and animals have evolved a variety of strategies to maximize the initial height of passive drifters. Some of these are so simple that we take them for granted: the male cones of a pine tree—from which pollen is released—are in the tree's crown, high above the ground. Other strategies are quite involved. Some fungi and plants maximize the initial height of their pollen and spores by launching them upward (Martone et al. 2010). Bunchberry dogwood (*Cornus canadensis*), for instance, uses a trebuchet-like mechanism to accelerate its pollen grains to a speed of 3 m · s^{-1}, lofting them 2.5 cm into the local boundary-layer flow. Spagnum moss does even better. Dehydration of its spore capsules increases their internal air pressure, which, when released, shoots spore-containing vortex rings 10 cm upward.

Despite its simplicity and low cost, passive locomotion has a substantial downside: drifters have little or no control over their trajectory; they go where the current takes them. There are circumstances in which drifting animals can affect their destination, but the requisite conditions are relatively rare. For example, in estuaries the current is typically seaward near the surface and landward near the bottom, allowing larvae to control their horizontal travel by varying their vertical position (e.g., Kunze et al. 2013).

4   CRAWLING
Adhesive Locomotion

In contrast to drifters, crawling animals sacrifice the simplicity of passive locomotion for control over their position. By adhering to the substratum with mucus, crawling animals can resist gravitational and fluid-dynamic forces—opening spatial niches that would otherwise be unavailable—but they must then contend with the challenges of walking on glue. Adhesive locomotion has been studied most thoroughly in a few gastropods (snails, slugs, limpets, etc.), but the same principles apply to other crawling organisms, such as flatworms.

4.1   Muscular Adhesive Locomotion

A typical gastropod sticks to the substratum with a thin layer of mucus produced by glands in the foot. The animal slides across this adhesive, using muscles to propagate waves of contraction or extension along its foot (Figure 10.5A). A portion of the foot moves forward in waves, and the remainder—the interwaves—are stationary relative

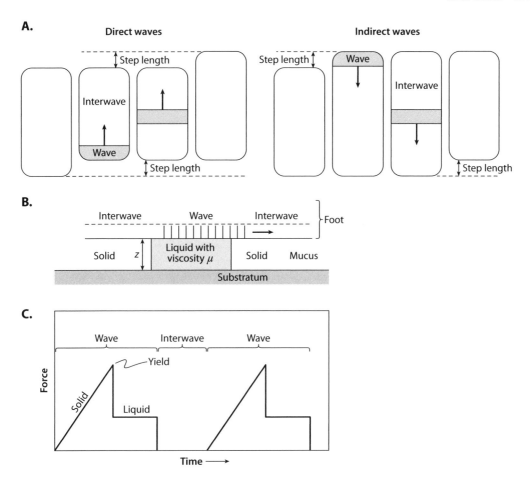

Figure 10.5 Snails crawl by passing waves along the foot. **A.** Waves can move in the direction or opposite the direction of travel. **B.** Cross section through the foot of a crawling snail. **C.** Shearing force during the passage of a wave. (See text.)

to the ground, providing the anchor against which muscles propel waves forward (Denny 1981; Lai et al. 2010).

The dynamic interplay between waves and interwaves can be affected by the properties of the mucus sandwiched between foot and substratum (Figure 10.5B). In a terrestrial slug, mucus under interwaves acts as a low-modulus solid (Denny and Gosline 1980). As a wave impinges on an interwave, mucus is sheared beyond its yield point and, under a wave, acts as a viscous liquid (Figure 10.5C). After the passage of the wave, the mucus rapidly "heals," returning to its solid behavior. In this fashion, interwaves provide continuous adhesion of foot to substratum while waves move against the reduced resistance of liquified mucus.

Miller (1974) conducted a broad survey of muscular crawling in gastropods, documenting an average speed of $1.6 \, \text{mm} \cdot \text{s}^{-1}$. The world speed record is $18.8 \, \text{mm} \cdot \text{s}^{-1}$, recorded by Donovan and Carefoot (1997) for the escape response of an abalone.

Power (P) per foot area (A) required for muscular pedal locomotion is set by the thickness (z) and viscosity (μ) of the mucus layer under the moving portions of the foot, and by the animal's speed, u (Figure 10.5B; Supplement 10.2):

$$\frac{P}{A} = \mu \frac{u^2}{z}.$$ (10.11)

The faster the animal crawls, the more power required.

What does this mean for a crawling gastropod? A typical mucus layer is 10 μm thick, and for a terrestrial slug, viscosity is 5 Pa · s (Denny and Gosline 1980). Inserting these values into equation 10.11, we find that for a representative crawling speed of 1.6 mm · s^{-1}, 1.3 W are required for each square meter of foot surface.

Now, a typical 4-g snail has approximately 1 cm^2 of foot area (Miller 1974), so 1.3×10^{-4} W would be required to move at the average snail's pace. The snail has approximately 1 g of pedal muscle mass, of which perhaps a third is contracting at any time. If this muscle supplies energy at a rate of 10 W · kg^{-1} (a reasonable value), the snail would need only 3.9% of its muscular capacity to power crawling. Thus, it doesn't appear that muscle power limits the speed of adhesive locomotion.

Why then are snails so slow? Because their form of crawling doesn't allow for amplification of speed by levers, snails may be limited by the intrinsic speed of muscle contraction. A consideration of the kinetics of wave motion in the foot (Supplement 10.3) suggests that the relative contraction rate—the strain rate—of pedal muscles during the passage of a wave is

$$\frac{\text{change in length}}{\text{resting length}} \approx \frac{u}{nk_\ell^2 \ell},$$ (10.12)

where u is again the overall speed of travel, n is the number of waves on the foot, ℓ is foot length, and k_ℓ is the fraction of ℓ occupied by an individual wave. An abalone with $n = 1$, $\ell = 0.1$ m, and $k_\ell = 0.3$ requires a strain rate of 2.1 s^{-1} to travel at its maximum escape speed of 18.8 mm · s^{-1} .

Maximal strain rate (the intrinsic rate of contraction; see Supplement 10.1) has not been measured for gastropod pedal muscles, but we can compare the strain rate calculated here to the maximum rate found in another molluscan muscle—the adductor muscle of a scallop. This muscle (the part of the scallop you eat) is specialized for rapid contraction and, in the absence of an external load, shortens at a strain rate of 3 s^{-1} (Marsh and Olson 1994). Given that gastropod pedal muscles work against a frictional load during locomotion—and therefore cannot reach their maximum rate of contraction—the rate calculated here (2.1 s^{-1}) is likely to be close to the maximum possible. So a snail's pace is probably constrained by how fast its muscles can contract.

4.2   Ciliary Adhesive Locomotion

Flatworms and many species of snails use cilia to crawl. The ciliated epithelium of the body or foot is connected to the substratum by a thin layer of mucus, which provides both a moist surface over which the animal can slide and a viscous medium through which cilia can row (Figure 10.6A).

The rowing stroke of a cilium is divided into two parts. During the effective stroke, the cilium acts as a more or less rigid rod, pivoting rapidly on its base (Figure 10.6B). The angle through which the cilium rotates varies among organisms, from 90° to nearly 180°. Having finished its effective stroke, the cilium then returns to its starting position by propagating a wave along its length, minimizing drag by keeping as close to the body as possible (Figure 10.6C). The return stroke also takes approximately

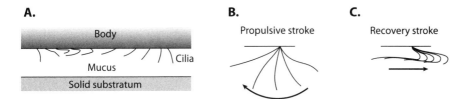

Figure 10.6 Cilia row through a layer of mucus (**A**), extending rigidly during the propulsive stroke (**B**), but flexing during the recovery stroke (**C**).

twice as long as the effective stroke, which reduces the cilium's speed and further reduces its drag. Drag reduction is important because it is the difference in drag between effective and return strokes that allows cilia to produce net thrust.

Theory developed by Blake and Sleigh (1974) suggests that the speed of a ciliated surface relative to the adjacent mucus is half to three quarters the cilia's tip velocity. For example, at a beat frequency of 20 Hz, each power stroke takes 0.017 s, such that for a stroke angle of 180°, cilia pivot at 190 rad · s^{-1}. For typical cilia 15 μm long, this translates to a tip speed of 2.8 mm · s^{-1} and a crawling velocity of 2.1 mm · s^{-1}, near the mean value of 2.8 mm · s^{-1} measured by Miller (1974) for the ciliary locomotion of snails. It thus seems likely that the speed of ciliary locomotion is governed by the speed of ciliary rotation.

Increasing cilia length could potentially increase snails' speed, but there are practical limits. The bending moment imposed by drag increases as the cube of a cilium's length, while the cilum's radius (an index of its ability to resist bending) stays the same (Supplement 10.4). As a result, long cilia would have a tendency to bend backward during the power stroke, negating their ability to provide thrust. In apparent response to this mechanical constraint, most cilia are only 10 to 15 μm long, and 100 μm is probably the useful limit (Sleigh and Blake 1977).

Speed could also be increased by increasing ciliary beat frequency, but beat frequency is limited by the rate at which dynein arms can walk along microtubules. As noted previously, few cilia can beat at greater than 40 Hz.

On average, snails that crawl using cilia move a bit faster than snails that use muscles (2.8 versus 1.6 mm · s^{-1}; Miller 1974), but maximum ciliary speed (6.7 mm · s^{-1}) is much slower than maximum muscular speed (18.8 mm · s^{-1}).

5 BURROWING

The physics of burrowing is similar to that of crawling, the primary differences being that mucus may not be present and the organism moves through the substratum rather than on it. The resulting increases in friction form both the resistance against which the animal must move and the anchor against which it can push.

Worms—the archetypical burrowing organisms—burrow by passing peristaltic waves along their bodies, a process based on the fact that the body as a whole has a fixed internal volume. Waves of muscular contraction shift the disposition of that volume, fattening some segments of the body while slenderizing the remainder. The fat segments push against the sides of the burrow, providing a frictional anchor from which the thin segments can force their way forward (see Alexander (2003) for a detailed explanation).

For many years, it was generally assumed that tremendous force was needed to ram the anterior end of a burrowing worm through the substratum. This may well be true for worms borrowing through soil, but recently Kelly Dorgan and her colleagues applied the theory of fracture mechanics (Chapter 15) to worms burrowing through mud and found that the force required to extend a burrow can be surprisingly low. Mud and other soft sediments behave as elastic solids through which a crack—the burrow—can be extended. A burrowing worm uses its sharp proboscis as a wedge to create a large force per area (a stress concentration) in the mud ahead. Because this stress is localized, only a small force is required. To extend the burrow, the worm need supply only the energy to create new surface area of mud, and given the low cohesive strength of sediment, this energy is relatively small (Dorgan et al. 2005, 2007).

Because the kinetics of burrowing are so closely tied to the mechanical properties of sediment—which vary from place to place—it is difficult to make general predictions about the speed of burrowing and its relationship to power.

6   WALKING AND RUNNING

Animals walk and run using legs, which provide three basic advantages: legs lift the organism off the ground, thereby reducing the frictional resistance to locomotion; by the same token, they allow animals to step over objects that would otherwise impede their movement; and, by acting as levers, legs allow high speeds of transport.

Let's begin with the effects of friction. The same basic physics that apply to crawling apply to walking or running—horizontal velocity creates drag on the body, which must be offset by thrust. However, because legs keep the body from dragging on the ground, animals that walk and run need to overcome only the resistance of moving through water or air. To those of us accustomed to moving though air, this drag is barely noticeable. Even at 4-min-mile pace (6.7 m · s^{-1}), the cost to a human being of overcoming drag in still air is only 7% of the overall cost of running (Schmidt-Nielsen 1997). However, because water is 833 times as dense as air, drag is a considerably larger concern for animals in water.

Crabs provide a heuristic example. Martinez (2001) found that, averaged over a stride, drag on crab legs is negligible: drag incurred as the leg is moved forward is offset by the drag-based thrust created as the leg moves backward. As a result, it is drag on the body that matters. Thus, the power per active muscle mass required to move at speed u is

$$\frac{\text{power}}{\text{active muscle mass}} = \frac{\text{body drag} \times \text{speed}}{\text{active muscle mass}} \tag{10.13}$$

$$= \frac{\left(\frac{1}{2}\rho_f u^2 C_D A_{\text{pr}}\right) \times u}{m}. \tag{10.14}$$

For the amphibious shore crab *Grapsus tenuicrustatus*, the drag coefficient (C_D) of the body in water is 0.47, and a typical individual has 7 g of muscle mass and a projected area (A_{pr}) of 17 cm^2. At any time during locomotion, about half of this mass is actively contracting so m is 3.5 g. The density of seawater (ρ_f) is 1025 kg · m^{-3}. Plugging these values into equation 10.14, we find that power per mass is 117u^3. Blickhan and Full (1987) measured the maximum power output of a shore crab to be 1.2 W · kg^{-1} of total body mass. Given that only about 10% of the body mass of a crab is locomotory

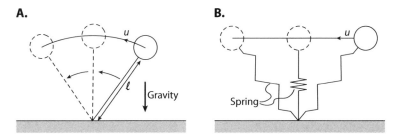

Figure 10.7 The dynamics of walking (**A**) and running (**B**) (See text.)

muscle (a fact apparent to anyone who has paid top dollar for several pounds of live crab to obtain a distressingly small meal), this implies that crab muscle produces approximately 12 $W \cdot kg^{-1}$, in line with estimates for other invertebrates. Using 12 $W \cdot kg^{-1}$ for power per muscle mass and knowing that power per mass is $117u^3$, we predict that the theoretical top speed for this animal is 0.47 $m \cdot s^{-1}$, reasonably near the 0.40 $m \cdot s^{-1}$ Martinez measured. In water, it seems that running speed is limited by the ability of muscles to produce the power needed to overcome drag.

Predicting speed in air is less straightforward. Due to air's low density, drag is much less important, but the lack of buoyancy becomes an issue. In water, organisms are buoyed by the medium's density, so gravitational work is negligible, but in air animals work against gravity as the body's center of mass is repeatedly raised and lowered.

There are two general patterns in which walking and running in air interact with gravity. As an animal walks, each leg acts like an inverted pendulum (Figure 10.7A). A foot is planted and the body then vaults over the leg, losing kinetic energy on the way up as the body slows, but at the same time storing energy as gravitational potential energy. This stored energy allows the animal to gain speed on the way down as the body is accelerated by gravity. This trade-off between kinetic and gravitational potential energy reduces (but does not eliminate) the metabolic cost of walking.

By contrast, when an animal runs, each leg acts like a pogo stick (Figure 10.7B), a process most easily visualized in kangaroos. As a foot contacts the ground, the leg bends, and some of the energy of deformation is stored elastically in the legs' tendons. This stored energy is then used to propel the animal upward and forward, setting the stage for the next foot contact.

These patterns—the inverted pendulum of walking and the pogo stick of running— were first described for humans and their two legs, but (somewhat surprisingly) they apply nicely to antelopes (four legs), cockroaches (six legs), and even crabs (eight legs; Full 1997; Alexander 2003; Biewener 2003).

Simple physics allows us to predict when an animal must switch from walking (in which at least one foot is on the ground at all times) to running (in which all feet are off the ground for at least some part of a stride). In order for a mass to move at a constant rate u along a circular path of radius r, it must constantly be accelerated toward the circle's center (see Chapter 2):

$$acceleration = \frac{u^2}{r}. \qquad (10.15)$$

For example, if you swing a rock on the end of a string, this centripetal (center-directed) acceleration is provided by tension in the string. As a walking animal vaults over a leg, its center of mass moves along a circular path with a radius equal to leg length ℓ, and the required centripetal acceleration is applied by g, the acceleration of gravity (Figure 10.7A). But, unlike the tension in a string, the magnitude of g is fixed, so equation 10.15 implies that there is a critical speed above which gravity can no longer keep the center of mass from flying upward:

$$u_{crit} = \sqrt{g\ell}. \tag{10.16}$$

This is the speed at which an animal must switch from walking to running, although animals often change gaits at a slightly lower speed (Alexander 2003).

In both walking and running, energy is used to accelerate and decelerate the legs, and the importance of this effect varies with size (Supplement 10.5). When body mass is less than a kilogram, the power required to accelerate and decelerate legs is only 2% to 3% of total locomotory power, but it can be 20% to 50% of total power for large animals (100 to 1000 kg).

The complications of legged locomotion make it difficult to predict the speed of running from first principles, so instead we rely on empirical measurements. From cockroaches to lizards to mammals, maximum sprint speed varies with body mass, m_b (Alexander 2003):

$$u_{max} \approx 6.0 m_b^{0.23}, \tag{10.17}$$

from which we can see that, compared to crawling, running is decidedly swift. For a mouse ($m_b = 20$ g), u_{max} is 2.4 m·s^{-1}, 130 times faster than a sprinting abalone of the same size. The highest speed reliably measured for a running animal was that of a cheetah, clocked at 29 m·s^{-1} (64 mi/h; Sharp 1997).

For most animals, these maximal speeds cannot be maintained for long. The metabolic rate needed to power a sprint outstrips the respiratory and circulatory systems' ability to supply oxygen to the muscles, and the animal quickly runs out of available energy. For insects, where trachea can bring air directly to the muscles, oxygen delivery is less of a problem, and sprint speeds can be maintained longer.

When animals need to travel long distances, they do so at much lower speeds. For terrestrial vertebrates, which includes animals with masses from 1 g to 1000 kg (Alexander 2003):

$$u = 1.0 m_b^{0.1}. \tag{10.18}$$

Mass for mass, invertebrates—primarily insects and crustaceans—move roughly a quarter as fast as vertebrates. For body masses from 5 μg to 0.2 kg (Full 1997),

$$u = 0.24 m_b^{0.089}. \tag{10.19}$$

7   WHEELS

As we have seen, legs have disadvantages: during locomotion they require a change in elevation of the body and they must constantly be accelerated back and forth. These disadvantages could be avoided if, instead of legs, animals had wheels. Wheels need not accelerate and declerate as a body moves at constant speed, and they could propel a body without changing the elevation of its center of gravity. In theory, a wheeled organism could travel over a smooth horizontal surface with the expenditure of no more energy than that required to overcome drag. Indeed, some mantis shrimp and caterpillars bend their bodies so that head meets tail, and then turn rapid backward somersaults to escape prey, in effect using their whole body as a wheel (Caldwell 1979; Brackenbury 1997).

However, two practical problems preclude the widespread use of biological wheels (LaBarbera 1983). First, the need for a rotary joint imposes a formidable barrier to the evolution of wheeled appendages. It is unclear how blood vessels and nerves could be routed through a rotary joint, and it would be tricky to engineer the ratcheting apparatus necessary to convert the tensile force provided by muscle into the moment required to power a wheel. The only rotary appendage known in nature is the flagellum of bacteria, a solid structure that requires neither nerves nor circulatory system and is powered by a tiny molecular motor rather than by muscle.

The second and perhaps more important barrier is environmental. Wheels work wonders on roads but are nearly useless on any other type of terrain. Indeed, when faced with the task of climbing a steep wooded slope, even the most ardent mountain biker will pick up her machine and start walking, taking advantage of her legs to step over obstacles. These problems were apparently deciding factors in the course of evolution—animals have legs instead of wheels.

However, legs need not be the jointed structures of a crab, lizard, or elephant. The legs of sea urchins and sea stars are tube feet, hollow muscular cylinders with an adhesive disk on the end. During locomotion, contraction of an internal bulb forces water into the tube foot, causing it to extend, and contraction of muscles on the fore side of the foot steers it forward. The disk then adheres to the substratum and the leg swings back to its starting position, in effect taking a stride

8   SWIMMING

As with other forms of locomotion, swimming requires thrust. Single-celled organisms use cilia or flagella, where changes in the shape of axonemes interact with viscous drag to propel an organism through water (Supplement 10.6). At higher Reynolds numbers, thrust is produced through inertial forces as muscles change the shape of the body. The arrangement of muscles is often complex. For example, a fish's pectoral fin doesn't flap as a simple flat plate. Instead, individual fin rays move in a syncopated fan dance that boosts the efficiency of thrust production (e.g., Gibb et al. 1994; Lauder et al. 2006). To perform these maneuvers, each ray must have its own set of muscles. Body undulations are driven by the rhythmic contraction of muscles attached to both the spinal column and the skin. The next time you dig into a serving of salmon or flounder, pay attention to how the meat separates into flakes at the interface between major muscle segments (myomeres). You can amuse yourself (if not your dinner companions) by trying to make sense of the way in which they interdigitate.

The jet propulsion of squids, jellyfish, and scallops is also powered by muscles, but elastic energy stored in the body wall of squids and jellyfish and in the hinge of scallops powers the subsequent refilling. In squids, intake is assisted by radial muscles, and the combination of elastic tissue and muscle makes for a highly efficient system (Gosline and Shadwick 1983). In jellyfish and scallops, muscles contract in resonance with elastic oscillations of the body, which saves energy (DeMont and Gosline 1988; DeMont 1990).

In addition to drag, swimming animals have to cope with gravity. Some, such as squids, counteract their weight with the vectored application of thrust. Many fish save energy by using a gas-filled swim bladder to make themselves neutrally buoyant, but this tactic involves some risk. If a fish is neutrally buoyant at a certain depth but moves up ever so slightly, pressure decreases, the gas in the swim bladder expands, and the fish becomes positively buoyant. As a consequence, it floats up even farther, which again reduces the pressure. To avoid this potentially lethal positive feedback, fish typically maintain slight negative buoyancy. Nautilus and cuttlefish avoid these problems by having a rigid shell around their buoyant organ.

Speed of swimming varies greatly, depending on an organism's size and mode of thrust production. Using flagella, some marine bacteria swim 400 body lengths per second, equivalent to 1640 mi/h for a 6-ft human (Magariyama et al. 1994; Mitchell et al. 1995). On an absolute scale, however, bacteria are slow, traveling less than $1 \text{ mm} \cdot \text{s}^{-1}$.

Constrained by the same factors that limit the speed of ciliary crawling, ciliary swimming is limited to 0.5 to 1.3 mm $\cdot$ s^{-1} (Sleigh and Blake 1977). Because this speed is set largely by the characteristics of individual cilia, it is (to a first approximation) independent of the size of the organism, a fact we can use to estimate the maximum size a ciliary swimmer (such as an invertebrate larva) can be and still maintain its position in the water column.

In Chapter 8, we equated viscous drag and weight to calculate sinking speed as a function of an organism's size. We can turn that result around to calculate the diameter d that corresponds to the fastest sinking speeds (u_{max}) that can be overcome by ciliary swimming:

$$d = \sqrt{\frac{18 \mu u_{max}}{(\rho_b - \rho_f) g}}. \tag{10.20}$$

For $u_{max} = 1.3 \text{ mm} \cdot \text{s}^{-1}$, $\mu = 10^{-3}$ Pa $\cdot$ s, and $\rho_b = 1080 \text{ kg} \cdot \text{m}^{-3}$ (a typical value), d is 200 μm. Thus, larvae can be up to approximately 200 μm in diameter and maintain altitude using cilia; larger organisms sink. Marine invertebrate larvae generally have diameters in the range of 50 to 500 μm. The smaller of these organisms can potentially use their cilia to maintain altitude, but the large ones will sink unless they adjust their buoyancy or use another mode of locomotion.

For large organisms, the mechanics of swimming are too complex to allow us to predict maximum speed using the sort of simple physical arguments we have used for crawling and aquatic running. Nonetheless, some general conclusions can be drawn. For fish, maximum acceleration—and therefore maximum short-term sprint speed— is limited by the power output of muscles (Wakeling and Johnston 1998), and sprint speed increases with size:[1]

$$u_{max} \approx 0.4 + 3.4 m_b^{0.33}. \tag{10.23}$$

Here body mass (m_b) is measured in kilograms and speed in meters per second. A 20-g fish can sprint at 1.34 m $\cdot$ s^{-1}, 71 times the speed of an equal-sized abalone, but

---

[1] Videler (1993) proposes that maximum sprint speed (m $\cdot$ s^{-1}) is

$$u_{max} = 0.4 + 7.4 \ell, \tag{10.21}$$

and Webb (1975) suggests that for fish in general,

$$m_b = 10 \ell^3, \tag{10.22}$$

where m_b is measured in kilograms. These relationships combine to give equation 10.23.

A.

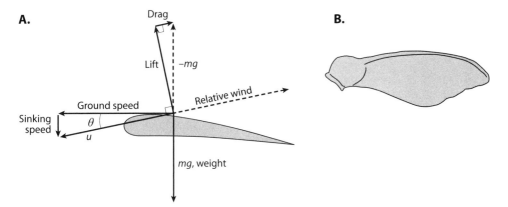

B.

Figure 10.8 **A**. The force balance of a wing during steady gliding. **B**. A maple seed.

only 55% that of a equal-sized mouse. The world's record for sprint swimming speed is held jointly by dolphins and tunas at $11 \ \text{m} \cdot \text{s}^{-1}$ (Alexander 2003).

Sustained swimming speeds are substantially slower. Videler (1993) suggests that for fish larger than about 15 g cruising speed increases only gradually with size:

$$u = 0.47 m_b^{0.17}. \tag{10.24}$$

A 20-g fish swims at $0.36 \ \text{m} \cdot \text{s}^{-1}$, only 15% the speed of an equal-sized mouse.

For a variety of invertebrates with masses ranging from 10 mg to 0.9 kg (Full 1997),

$$u = 0.18 m_b^{0.094}. \tag{10.25}$$

9   GLIDING FLIGHT

There are two types of flying organisms—those that glide and those that flap—and we examine them in turn.

Gliders use gravity to provide the speed needed for lift. Consider an organism moving at constant speed u through still air at angle θ to the horizontal (Figure 10.8A). Because speed is constant, acceleration is zero, implying both that thrust equals drag and upward force equals weight. From the geometry of the situation, we can quantify lift, F_L:

$$F_L = m_b g \cos \theta. \tag{10.26}$$

From the standard model for lift (Chapter 9), we know that $F_L = \frac{1}{2}\rho_f u^2 A_{\text{pl}} C_L$, where A_{pl} is the planform area of the wings and C_L is the lift coefficient. Thus,

$$\frac{1}{2}\rho_f u^2 A_{\text{pl}} C_L = m_b g \cos \theta. \tag{10.27}$$

Solving for u, we can estimate the speed at which animals glide:

$$u = \sqrt{\frac{2 m_b g \cos \theta}{\rho_f A_{\text{pl}} C_L}}. \tag{10.28}$$

The ratio of $m_b g$ to A_{pl} is known as *wing loading,* W, so

$$u = \sqrt{\frac{2W \cos \theta}{\rho_f C_L}}. \tag{10.29}$$

For all reasonably adept gliders, θ is small enough so that $\cos \theta \approx 1$, and a high (but reasonable) estimate of C_L is 1.5 (Alexander 2003). Inserting these estimates into equation 10.29, we arrive at an estimate of minimum gliding speed as a function of wing loading:

$$u_{min} \approx \sqrt{\frac{4W}{3\rho_f}}. \tag{10.30}$$

Values are given in Table 10.2 for a variety of animals.

With their huge wings, butterflies and moths can glide at speeds as low as $1 \text{ m} \cdot \text{s}^{-1}$, but other animals have to glide faster to produce enough lift to support their weight. Gliding albatrosses and swans, for example, fall like bricks if their airspeed dips below 13 or $17 \text{ m} \cdot \text{s}^{-1}$, respectively. There is an important message here: compared to crawling, running, and swimming, gliding is an intrinsically high-speed process.

As equation 10.30 shows, minimum gliding speed depends on W, the ratio of body mass to wing area. For organisms with the same shape but different sizes, $A_{pl} \propto m_b^{2/3}$, so $W \propto m_b^{1/3}$. As a result, we would predict that $u_{min} \propto m_b^{1/6}$. Measured data for the speed of gliding—which is likely to be faster than u_{min}, but should scale in the same fashion—bear this out (Alexander 1998):

$$u = 8.0 m_b^{0.14}. \tag{10.31}$$

Once a gliding organism is up to its minimum equilibrium air speed, it has several options. For example, by reducing drag to increase speed slightly—thereby increasing lift—it can minimize its sinking speed. This tactic has advantages in two circumstances. As for passive drifters, minimizing sinking speed maximizes a glider's time aloft. For a gliding seed, for instance, sinking as slowly as possible maximizes the time available for wind to carry it away from its parent. The seeds of many trees (pines, maples, birches, etc.) have seeds with only one wing (Figure 10.8B). These samaras glide to increase their time aloft, but rather than traveling in a straight line, the wing traces a helical path as it rotates around the seed (Norberg 1973; Vogel 1994).

Sinking slowly also increases the chance that an organism can take advantage of updrafts such as those found in thermals (air rising over a heated portion of the ground) or on the upstream sides of hills. If vertical velocity in the updraft exceeds sinking speed, the organism can *soar*; that is, it can sink through the air but still rise relative to the ground. A wide variety of birds (e.g., hawks, vultures, and eagles) use soaring to travel great distances.

And then there are albatrosses, which take advantage of the ocean's atmospheric boundary layer to stay aloft indefinitely using a technique known as *dynamic soaring* (Jameson 1959; Wood 1972). Starting a few tens of meters above the ocean surface, they glide downwind, gaining air speed as they descend. They then turn into the wind, which decreases their speed relative to the water but increases their speed relative to the air, giving them enough additional lift to gain altitude. Without the boundary layer, the added drag that inevitably accompanies increased lift (induced drag; see Chapter 9) would limit the height to which the bird could rise, and it would eventually

Table 10.2 Wing Loading and Minimum Speeds of Flight for a Variety of Flying Animals. Data from Alexander (2003) and Biewener (2003).

Animal	Mass	W	u_{min}
	(kg)	$(N \cdot m^{-2})$	$(m \cdot s^{-1})$
Pterosaur	15	32	6.5
Andean condor	10	101	11.6
Wandering albatross	8.7	140	13.7
Mute swan	8	230	17.5
Canada goose	1.8	155	14.4
Buzzard	1	33	6.6
Mallard duck	1	113	12.3
Black grouse	1	85	10.6
Herring gull	0.54	51	8.2
Archeopteryx	0.27	55	8.6
Magpie	0.22	35	6.8
Sparrow hawk	0.2	28	6.1
Diving petrel	0.14	64	9.2
Rousettus bat	0.14	25	5.8
Starling	0.075	37	7.0
Budgerigar	0.035	34	6.7
House sparrow	0.028	26	5.9
Swallow	0.024	16	4.6
Greater horseshoe bat	0.023	12.2	4.0
Fruit bat	0.014	12.3	4.0
Little brown bat	0.007	7.5	3.2
Hummingbird	0.005	32	6.5
Sphinx moth	0.0005	1.3	1.3
Bumblebee	0.0002	15.7	4.6
Butterfly	0.0001	0.9	1.1
Dragonfly	0.0001	2	1.6
Fly	0.00002	15.4	4.5

have to flap its wings to return to its initial altitude. But because velocity increases with height in the boundary layer, albatrosses are able to glide back to their starting elevation, where they can start the process anew.

Soaring animals can stay aloft only when air is moving. In still air, gliders inevitably lose altitude, and how far they travel before touching down becomes a matter of minimizing the glide angle, θ, which, in turn, requires animals to minimize drag. There are two components to drag in this system: profile drag—the drag the organism would experience in the absence of lift—and induced drag. These impose contrasting constraints on wing morphology. Traditionally, wings are described by their *aspect ratio*, the ratio of span (the distance between wing tips) to average chord (the distance along the wing in the direction of flow). Profile drag (which for most fliers is primarily friction drag) increases with increasing speed. Conversely, induced drag *decreases* with increasing speed. Both forms of drag decrease with increasing aspect ratio (see Supplement 10.7). As a consequence of these divergent trends, there are different combinations of speed and wing shape that result in the same gliding performance (Alexander 2003).

10   FLAPPING FLIGHT

Gliding works only as long as an organism can trade gravitational potential energy for the speed needed to make lift equal to weight. In still air, gliders eventually run out of altitude. To avoid this drawback, flapping fliers move their wings; as long as the muscle-driven speed of the wings is sufficient to provide the necessary lift, animals can stay aloft.

This concept allows us to estimate the flapping frequency of hovering animals. For example, a hummingbird's wing is approximately 5 cm long. For its wing tip to move at the bird's u_{min} of 6.5 m · s^{-1}, the wing must swing at 130 rad · s^{-1}. Given that the wing can't travel through an arc of more than 180° (π rad), it must change direction at least $130/\pi = 41$ times per second. Each full flap consists of two strokes (one forward and one back), so 41 direction changes per second is a flapping frequency of 20.5 Hz, which is why hummingbirds' wings are just a blur.

There are several factors that render this calculation a rough approximation at best. If a hummingbird's wing tip moves at u_{min}, the rest of the wing moves slower, providing less lift. But at the end of each stroke, the wing rotates about its axis, flipping over so as to provide lift on the return stroke. As the wing flips, it produces a vortex that subsequently increases the wing's lift coefficient beyond that possible in steady flight (values of 2 to 6 have been measured in insects). In essence, in the process of changing direction, the wing stores energy in a vortex, energy that it regains on the return stroke and uses to help keep the bird aloft. These unsteady effects complicate the analysis of hovering flight, but the basic message is nonetheless clear. Hovering animals must flap their wings at high frequencies.

Hovering's frantic flapping places stringent demands on animals' neuromuscular systems (Biewener 2003). Some small midges, for instance, flap their wings at 1000 Hz, far faster than nerves can stimulate muscles. This problem was solved in insects (the fastest flappers) by the evolution of asynchronous muscles, which work in conjunction with the elasticity of the body to form an oscillating system. An initial stimulus from a nerve causes flight muscles to contract, which both flaps the wings down and stores elastic energy in the tendons and cuticle of the thorax. When the muscles relax, the thorax rebounds, pulling the wings up and stretching the muscles. This extension acts as a stimulus for the muscles to contract again. Once started, this resonant system oscillates on its own, requiring only occasional input from the nervous system.

The high-frequency flapping necessary when hovering can be slowed down if animals use some of their lift to propel themselves forward. In this case, the speed of the body through the air adds to the speed of the wing to provide lift. In a downstroke (Figure 10.9A), the wing has a high vertical airspeed, which—combined with the animal's ground speed—both increases the wing's total airspeed and places it at a substantial angle of attack (θ). The resulting lift is angled forward, providing thrust to overcome drag and sufficient upward force to more than offset the animal's weight. Consequently, the bird accelerates forward and upward.

A. Downstroke   **B. Upstroke**

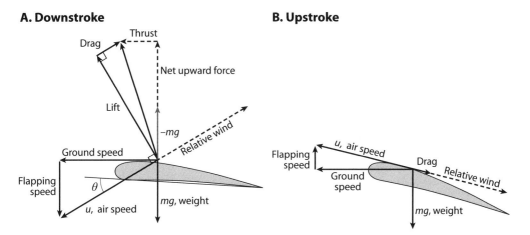

Figure 10.9 The force balance of a flapping wing during (**A**) downstroke and (**B**) upstroke.

In an upstroke, the wing typically has a slower vertical airspeed, reducing the angle of attack to near zero (Figure 10.9B). As a result, lift is negligible and drag, although reduced, is still present. Thus, during the upstroke, the animal decelerates horizontally and accelerates downward. As long as the forward and upward accelerations in the downstroke are sufficient to offset the rearward and downward accelerations in the upstroke, the animal maintains speed and altitude. The exact manner in which this equilibrium is reached varies from animal to animal. For instance, birds and bats can bend their wings to reduce wing area on the upstroke, minimizing the effects of drag. Some insects rotate their wings to increase the angle of attack on the upstroke, gaining additional lift at the cost of additional drag.

The complexities of the flapping stroke make it difficult to predict flying speed from simple principles, so we are again left to rely on measured relationships. The most useful of these involve the *advance ratio*, $\mathcal{J}$, the ratio of an animal's airspeed (u) to the tangential speed of its wings. For a wing of length ℓ, flapping through angle ϕ, the wing tip travels an arc distance of $2\phi\ell$ in each full flapping cycle. Thus, if the wings beat f times per second, wing-tip speed is $2\phi\ell f$, and the advance ratio is (Ellington 1984)

$$\mathcal{J} \approx \frac{u}{2\phi\ell f}. \tag{10.32}$$

For most animals in forward flight, $\phi \approx 1$ rad (Pennycuick 1990; Alexander 2003). Using this value we can estimate flight speed:

$$u \approx 2\mathcal{J}\,\ell f. \tag{10.33}$$

The longer wings are and the faster they flap, the faster an organism flies.

For birds ranging in size from a tree swallow to a bald eagle, $\mathcal{J}$ has an average of 2.7 (Pennycuick 1990), so these animals fly at approximately 2.7 times their wing-tip velocity. For example, a greater egret ($\ell = 0.67$ m, $f = 2.8$ Hz) flies at 10 m·s^{-1}. The northern flicker is a much smaller bird ($\ell = 0.26$ m), but because of its high wing-beat frequency ($f = 9.2$ Hz), it flies faster (13 m·s^{-1}). Bees have even shorter wings ($\ell = 13$ mm) and a smaller advance ratio ($\mathcal{J} = 0.66$; Dudley and Ellington 1990), but these factors are somewhat offset by their high flapping frequency (150 Hz), allowing them to fly at 2.6 m·s^{-1}.

Speed of steady flight increases with body size. Measured across all flying organisms (Alexander 1998),

$$u = 16m_b^{0.14}. \tag{10.34}$$

Mass for mass, insect flight is slower (Full 1997):

$$u = 3.55m_b^{0.049}. \tag{10.35}$$

Maximum flight speeds have not been reliably established. Tucker and others (1998) measured a speed of 58 m · s^{-1} (130 mi/h) for a diving falcon, but since the bird had its wings folded at the time, it was not so much flying as falling. Thompson (1961) recorded a red-breasted merganser maintaining position ahead of his airplane at 40 m · s^{-1} for a short time.

One final note before we move on: the ability to maneuver is governed by the rate at which an organism can change speed, direction, or both. For a flying organism, these accelerations depend on the ratio of lift to mass, and since lift is proportional to wing area, maneuverability should be proportional $1/W$, the inverse of wing loading. As we calculated earlier, for animals of a given shape, wing loading is proportional to $m_b^{1/3}$. Therefore, maneuverability should be proportional to $m_b^{-1/3}$, decreasing with size. This expectation is borne out in Table 10.2. Small, highly maneuverable fliers such as dragon flies, hummingbirds, bats, and swallows have low wing loadings (2 to 32 N · m^{-2}) while large, less maneuverable fliers (geese, swans, ducks, and albatrosses) have much higher Ws (100–230). While it is difficult to calculate maneuverability directly from wing loading, we know that all flapping fliers can produce lift forces sufficient to offset their weight. If that lift were to be directed sideways, the animal would accelerate horizontally at 9.81 m · s^{-2}. In short, fliers are intrisically capable of rapid changes in speed and direction.

11   SPEED AND POWER OF TRANSPORT
A Summary

The findings of our investigation of speed are summarized in Figure 10.10A and B. Constrained by friction with the substratum, transport by crawling is slow—a few millimeters per second at best. Swimming is faster than crawling but slower than running. Flying is by far the fastest way to travel; an insect, bird, or bat can fly 15 to 30 times as far in a day as terrestrial animals can run or fish and squid can swim and thousands of times farther than snails can crawl.

The contrast between swimming and flying can be altered, however, if the medium (air or water) moves. Both a 1-kg buzzard and a 1-kg trout can be brought to a halt when moving upstream against flows they might well face in nature: the bird when flying into a 15-m/s wind and the fish when swimming against a 0.5-m/s current. In other words, there can be times when the difference in speed so evident in Figure 10.10A is misleading. Conversely, when flying with the flow, the buzzard screams along at 30 m · s^{-1} relative to the ground, but the trout manages only 1 m · s^{-1} relative to the streambed. In this case, Figure 10.10A is an understatement. Because the ground doesn't move, these complications don't apply to walking and running.

Power required for locomotion is summarized in Figure 10.10B. Buoyed by water and moving relatively slowly, swimming organisms expend energy at a low rate.

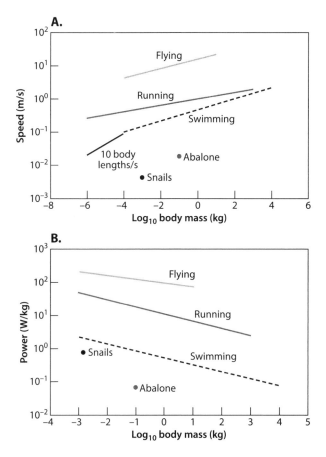

Figure 10.10 A. Speed of transport as a function of body mass. **B.** Power of locomotion as a function of body mass.

The higher speeds of running require more power, and the combination of very high speed and a lack of buoyancy makes flying an extraordinarily energy-intensive process.

12   COST OF TRANSPORT

In many ecological scenarios, it is not speed and power but rather the *cost* of transport that matters. It is worth migrating to greener pastures, for instance, only if the caloric bonus from the change in venue offsets the cost of getting there. Animals often must travel to find mates, but putting too much energy into the search might not leave enough energy to build a gonad. In short, optimal allocation of metabolic reserves often depends on how costly it is to travel.

Cost of horizontal transport is essentially zero for passive drifters, who expend no more energy when moving than when standing still. For all other modes of locomotion, cost is less easily estimated. Unlike speed—which, as we have seen can at least sometimes be predicted on a mechanistic basis—it is difficult to estimate cost from first principles. In large part, this is due to internal inefficiencies involved

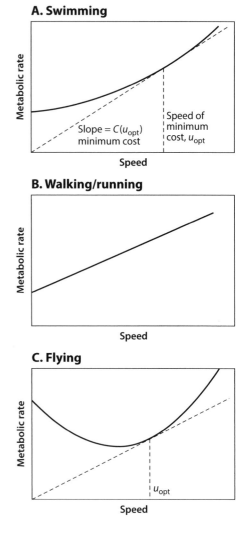

Figure 10.11 Metabolic cost of locomotion as a function of speed for swimming (**A**), walking (**B**), and flying (**C**).

in the process of locomotion. For example, only a fraction of the chemical energy released by ATP hydrolysis appears as mechanical energy of muscle contraction. A conversion efficiency of 25% is often assumed (e.g., Alexander 2003), but the exact value varies from muscle to muscle and, for each muscle, on the rate of contraction and the load against which the muscle pulls. Similarly, the efficiency with which ATP is harvested from food varies from creature to creature and with the type of provender.

Faced with such barriers to predicting the cost of transport, physiologists have traditionally opted to rely on empirical measurements. In a typical experiment, an animal's oxygen consumption rate is measured while it moves at a series of speeds. Consumption of 1 liter of O_2 corresponds to the metabolic release of approximately 20.1 kJ of energy (Schmidt-Nielsen 1997), so oxygen-consumption rate can be translated to power. The resulting plots of mass-specific metabolic rate ($W \cdot kg^{-1}$) as a function of speed ($m \cdot s^{-1}$) have different characteristic shapes for swimming, running, and flying (Figure 10.11).

All organisms metabolize even when standing still. Above this maintenance level, the metabolic rate of swimming organisms rises in proportion to $u^{2.5}$ (Figure 10.11A; Alexander 2003). With few exceptions (e.g., humans and horses) the metabolic rate of legged locomotion increases linearly with speed (Figure 10.11B; Taylor et al. 1982). Flight has yet another pattern (Figure 10.11C). Unlike swimming and walking, where it costs little for an animals just to hang out, hovering is extremely demanding. It is less demanding to fly slowly than to hover, so metabolic rate initially decreases with increasing speed before reaching a minimum; it then increases.

For swimming and flying, the shapes of these graphs allow us to glean a useful metric of locomotory performance. The slope of a line extending from the origin to a point on the curve is the *cost of transport* for that particular speed—the joules required to transport 1 kg of body mass 1 m. The speed where the line is just tangent to the curve is u_{opt}, the velocity where cost is minimal. The cost of transport at u_{opt} provides

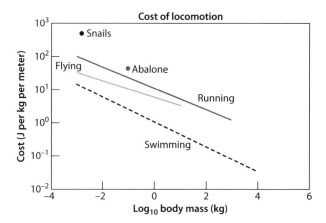

Figure 10.12 Cost of transport as a function of body mass.

a useful metric to compare cost among different swimmers and fliers (Figure 10.12). Because the metabolic rate-speed curve is linear for walking and running, there is no well-defined optimal speed, and in this case we use the cost of transport for the typical speeds at which these animals move (equations 10.18 and 10.19).

There are two important messages to take home from Figure 10.12. First, regardless of locomotory mode, the larger the organism the lower its per-mass cost of transport. It costs 200-fold less to transport a gram of whale shark a meter than it does to transport a gram of goldfish. Second, there are substantial differences among modes of locomotion. It costs swimmers less to transport their mass than any other type of animal. Flight is more costly; walking or running is more costly still.

Because the power required for flight is so high, it might seem odd that flying is less costly than walking or running. But cost is power divided by speed, and the high metabolic rate needed to maintain altitude in air is more than offset by the extraordinary speeds that flight offers (Figure 10.10A).

Few measurements of cost have been made for crawling and burrowing, so we have only a hint of the general pattern of costs for these forms of travel. For small gastropods (e.g., snails and abalone), crawling is even more costly than walking or running. In a functional sense, the high cost of crawling is a tax imposed by the organism's adherence to the substratum during travel. The cost of burrowing has yet to be definitively measured, but Dorgan et al. (2011) suggest that it, too, is substantially higher than the cost for legged locomotion.

The information presented in Figure 10.12 should be treated with caution. There is substantial variation around each of the trends, and this simple overview excludes some notable exceptions. In particular, the line for swimming is based solely on fish. Using a data set that included a wide variety of non-fish swimmers (everything from shrimp to whales) Videler (1993) found that their cost of locomotion was substantially higher than that of fish and very similar to that of flying animals. When examining only marine mammals, Williams (1999) found that their cost of transport was higher still—falling between that of flying and running—a fact she attributed (at least in part) to the high metabolic rate these animals must sustain to maintain their high body temperature.

13   CLIMBING AND JUMPING

Up to now, we have investigated the mechanics of horizontal travel. But organisms move vertically as well. When an organism climbs, energy must be supplied to move its mass upward against the tug of gravity. The change in energy as an object of net weight F_{net} (weight minus buoyancy) moves a vertical distance Δz is

$$\Delta W = F_{net}\Delta z. \tag{10.36}$$

Dividing both sides of this equation by the time required to effect the change in height tells us that climbing power, P, is the product of weight and vertical velocity w:

$$P = F_{net}w. \tag{10.37}$$

Recall from Chapter 2 that net weight depends on the difference between an organism's density (ρ_b) and the density of the surrounding fluid (ρ_f). For an organism of volume V,

$$F_{net} = gV(\rho_b - \rho_f). \tag{10.38}$$

Noting that $V\rho_b$ is body mass (m_b), and substituting equation 10.38 into equation 10.37, we find that the power per body mass ($W \cdot kg^{-1}$) needed to climb at rate w ($m \cdot s^{-1}$) is

$$\frac{P}{m_b} = gw\left(1 - \frac{\rho_f}{\rho_b}\right). \tag{10.39}$$

The faster an organism climbs and the denser its body is relative to the surrounding fluid, the more power required.

How does this conclusion affect animals? There are two cases in which we can use the principles outlined in this chapter to make predictions.

A typical 4-g snail using cilia to crawl has 2 cm^2 of foot area[2] (Miller 1974), and, powered by its cilia, the snail moves at a speed of 2.8 mm $\cdot$ s^{-1} on a horizontal surface. Could the same speed be maintained while climbing in air? From equation 10.37, we predict that 4 g of snail crawling upward at 2.8 mm $\cdot$ s^{-1} expends energy at a rate of 1.1×10^{-4} W. Cilia can produce 4×10^{-10} W m^{-1} of cilium, so a typical 15-μm cilium can contribute 6×10^{-15} W to locomotion. In that case, the snail would need 1.8×10^{10} cilia to meet the power demand of climbing at this speed. Cilia have a diameter of approximately 0.25 μm. If packed cheek to jowl, only 3.2×10^9 would fit in the 2 cm^2 of our snail's foot—less than a fifth of the required number. In reality, 2 cm^2 of snail foot has roughly 2×10^9 cilia, allowing an uphill speed of at most 0.3 mm $\cdot$ s^{-1}. True maximum speed is probably even slower because some power is required to drag cilia through mucus, whether the animal is climbing or not.

In water, ρ_f/ρ_b is larger than in air; as a result, the power required for a 4-g snail to crawl upward at 2.8 mm $\cdot$ s^{-1} is 20-fold smaller, 5.5×10^{-6} W. To produce this power, the foot needs only 9.2×10^8 cilia, half the number actually present. In short, snails can use cilia to climb vertical surfaces at their usual sedate pace while underwater, but when out of water, speed is severely limited by available power. To my knowledge

---

[2] Gram for gram, ciliary crawlers have bigger feet than muscular crawlers, for reasons that will become apparent in a moment.

most (if not all) amphibious and terrestrial snails use muscles rather than cilia to power climbing. As we have seen, muscular crawlers have power to spare.

Among terrestrial mammals, an organism's ability (or at least its willingness) to climb varies dramatically with size. Small rodents scamper up trees with no apparent effort, whereas it takes considerable sweat on the part of a human to climb the same trunk. Equation 10.39 helps to explain this contrast.

A typical mouse has a mass of 20 g and a body length of 10 cm. Traveling horizontally at a rate of 1 body length per second, it expends energy at the high total rate of 26 $W \cdot kg^{-1}$ (Biewener 2003), due in large part to the high metabolic rate needed just to keep its body temperature up. (As we will discuss in Chapter 11, basal metabolic rate ($W \cdot kg^{-1}$) varies as $m_b^{-0.25}$.) Climbing vertically at the same speed requires only an additional 1 $W \cdot kg^{-1}$ (equation 10.39). In other words, the mouse needs to increase its power output by only 3.8% to climb at 1 body length per second.

The situation is different for humans. A woman with a mass of 50 kg and a height of 1.6 m expends energy at 5.6 $W \cdot kg^{-1}$ when walking on a horizontal surface at 1 body length per second. This value is lower than that of the mouse owing primarily to the woman's relatively large mass and, therefore, lower mass-specific metabolic rate. But climbing at one body length per second requires an additional 15.7 $W \cdot kg^{-1}$, an increase of 280%. This large relative increase goes far to explain why adult humans are less inclined than small rodents to climb trees.

Elephants shouldn't even consider it. Data from Alexander (2003) suggest that, with a mass of 3800 kg and a length of approximately 3 m, an elephant traveling at 1 body length per second has a metabolic rate of 2.8 $W \cdot kg^{-1}$. To climb at this rate would require an additional 29 $W \cdot kg^{-1}$, a 10-fold increase the pachyderm probably couldn't produce.

In summary, relative to the cost of horizontal locomotion, the cost of climbing is strongly dependent on size. For small organisms, the climbing cost is negligible; for large organisms, it is prohibitive. This conclusion helps to explain why tropical trees swarm with skittering insects and scampering lizards but are home to few large animals.

Jumping provides an intriguing contrast. Whereas a climb plays out over extended time, a jump unfolds all at once. By explosively extending their legs, fleas, grasshoppers, and antelopes impart an initial upward velocity to their bodies that carries them aloft. Surprisingly, the height to which they jump is nearly independent of size. (Due to the large drag on organisms, jumping in water is a losing proposition, so I'll confine this discussion to jumping in air.)

If an animal of mass m_b starts at a standstill, its muscles must do work equal to $\frac{1}{2}m_b w^2$ to get the body up to vertical speed w at the moment of takeoff. Once the animal is separated from the ground, this kinetic energy is traded for gravitational potential energy, $m_b g z$, where z is the distance through which the animal's center of gravity is elevated. Equating these two energies and solving for maximum jump height (z_{max}), we find that

$$z_{max} = \frac{w^2}{2g}. \tag{10.40}$$

Note that body mass has dropped out of the equation. For instance, fleas can jump up about 0.5 m, and human high jumpers can't do much better, raising their center of mass by only about 0.6 m when jumping from a standing start.[3] Further evidence that the height of a jump doesn't increase with the size of the jumper comes from the fact that the maximum standing high jump on record (2.26 m) was performed by a relative small animal: a bushbaby (*Galago senegalensis*) that launched itself from the floor to the top of a door (Hall-Craggs 1965).

Although the height of jumps is approximately independent of size, the acceleration that an organism must produce to achieve that height varies drastically. To jump to height z_{max} using legs that can extend through distance x, an organism must accelerate its body at rate a (see Supplement 10.8):

$$a = \frac{g z_{max}}{x}. \tag{10.41}$$

The smaller the organism, the smaller x is and the larger the acceleration required. Fleas, which can extend their legs by only about a millimeter, need an acceleration of $4900 \text{ m} \cdot \text{s}^{-2}$ to jump their 0.5 m, which in turn means that the entire extension of the legs must happen in less than a millisecond.[4] Although capable of exerting substantial force, fleas' muscles cannot contract this fast, so they are not capable of providing sufficient power for the jump. In the evolved solution to this problem, fleas' jumping muscles contract slowly, storing energy in rubbery structures in the leg joints' hinges. This stored energy can then be rapidly released, amplifying the power available for the jump.

Plants go fleas one better. The trebuchet-like apparatus in the flower of the bunchberry dogwood accelerates pollen at $24{,}000 \text{ m} \cdot \text{s}^{-2}$ (Edwards et al. 2005).

14   LOCOMOTION IN CONTEXT

When judged on an absolute scale, animals' vast range of locomotory speeds (Figure 10.10A) may seem surprising. How, for instance, can snails survive in a world where lizards, mammals, and birds can move orders of magnitude faster? But absolute speed often isn't the proper measure of locomotory performance; speed needs to be judged in the context of an organism's habitat and physiology.

Take, for example, an intertidal snail crawling at a typical speed of $1.6 \text{ mm} \cdot \text{s}^{-1}$. It takes the animal about 10 min to travel a meter, but its entire vertical range—the portion of the intertidal environment where temperature and hydrodynamic forces are acceptable and proper food is available—is typically only a meter or two. On a steep rock face, the snail, slow as it is, could travel across its entire vertical range in only 10 to 20 min. Alpine fauna (mountain goats, for instance) face analogous environmental and ecological constraints, but most would find it difficult to traverse their entire vertical range in as short a period.

Similar reasoning can be applied to foraging distances. A predatory intertidal snail travels approximately 10 cm between the barnacles it eats, requiring about a

---

[3] For humans, the world-record high jump is 2.45 m, but Javier Sotomayor's center of mass was already well above the ground when he took off, and he used the horizontal velocity of his run-up to help catapult him over the bar.

[4] Actually, this is an overestimate of the time available. In making this calculation, I have ignored the energy lost to drag, so the flea would have to attain an even higher initial speed and a higher acceleration to reach an altitude of 0.5 m.

minute in transit. By contrast, a terrestrial predator playing a similar ecological role—a wolf, for instance—travels perhaps 10 km between meals, a difference in spatial scale of 10^5. To achieve the same time between meals as a snail, wolves would have to to run at 160 m · s^{-1}, half the speed of sound. In reality, their maximum sustained speed is only about 10 m · s^{-1}, which works out to a transit time of 17 min. Thus, when viewed in the context of time between meals, the stately crawl of snails doesn't appear so pedestrian.

Furthermore, time itself must be judged in context. Schmidt-Nielsen (1984) notes that mass-specific metabolic rate $\mathcal{M}$ serves as an index of the pace of an organism's life. The faster a kilogram of tissue respires, the faster it can turn over its molecules, and the more rapidly the organism of which it is a part can grow, reproduce, assimilate food, and eliminate wastes. In many respects, it is this *physiological pace* that determines how fast time is counted off by a biological clock.

We can use this concept to compare time as perceived by snails and wolves. The relative physiological pace of life for these two organisms is the ratio of their resting metabolic rates:

$$\text{relative pace} = \frac{\mathcal{M}_{\text{snail}}}{\mathcal{M}_{\text{wolf}}}. \tag{10.42}$$

At 10°C, an intertidal snail's $\mathcal{M}$ is approximately 0.10 W · kg^{-1} (e.g., Innes and Houlihan 1985), whereas that of a 30-kg wolf (which operates at an internal body temperature of 37°C) is roughly 15-fold higher, 1.5 W · kg^{-1} (Taylor et al. 1982). As a consequence, for every minute the wolf's clock ticks off, the snail's clock counts only 4 s. Thus, the 20 min a wolf would experience as it watched a snail traverse its entire vertical range would appear to the snail as a mere 80 s. In the context of its metabolic pace, the snail perceives itself as practically dashing through life.

The body temperature of snails and other ectotherms can vary, and their pace of life varies with it. But for most animals this variation is unlikely to bring the tempo of ectothermic life up to that of mammals and birds. If we assume a typical increase in metabolic rate of 100% per 10°C, a snail would need a body temperature of nearly 50°C to keep physiological pace with a wolf, a temperature that would likely be fatal. Flying insects are the major exception to this generality. Their flight muscles are designed to work at wolflike temperatures, so the pace of a bee's life is similar to that of a mammalian predator.

15   WINTER MIGRATION

The evolutionary decision to migrate in winter to a distant feeding ground depends on whether the benefits of migration can offset its metabolic costs, and these costs are, in large part, governed by the mechanics of locomotion we have explored in this chapter. Alexander (1998) used these principles to provide insight into this important life-history decision.

The analysis begins with the assumption that an animal's fitness is proportional to the rate at which it can accumulate energy. An organism that can gather more energy than its competitors can grow faster and produce more young. Thus, as a life-history strategy, an animal should migrate only if the trip results in a net gain in energy above what the animal would accrue by staying home. In nature, this trade-off is sensitive to the availability of winter forage at the home site—the better home is, the less likely it is that migration will make things better. But for the sake of simplicity let's assume that the food available at home is just sufficient for maintenance—if the animal stays put, it can survive but cannot grow or reproduce. This sets the bar for migration as low as reasonably possible; if there were less food available, migration would be a necessity, not a choice.

To calculate the costs and benefits of migration, we need to quantify R, the net rate at which energy can be acquired. To allow for convenient comparison among organisms, we normalize R to the animals' standard metabolic rate, the rate at which it expends energy when standing still:

$$R = \frac{\text{rate of energy gain} - \text{standard metabolic rate}}{\text{standard metabolic rate}}. \tag{10.43}$$

As a ratio of rates, R is dimensionless. According to our assumption, $R = 0$ on the home site, but it could be larger on the distant foraging grounds. Our task is to calculate how large R needs to be on the feeding grounds to make migration worthwhile.

We now make two additional assumptions. First, we assume that migration is an annual event to be completed in time t_{win}, the length of winter (nominally 183 days). We also assume that the animal doesn't feed while traveling. These assumptions allow us to calculate the net energy that can be accumlated during migration. The time required to travel one way to the foraging area is t, so the total time required for a round trip is $2t$. Thus, if it migrates, the animal spends $t_{\text{win}} - 2t$ days on the feeding grounds and accumulates total energy

$$\text{energy accumulated on the feeding grounds} = (t_{\text{win}} - 2t)\, R. \tag{10.44}$$

But getting to the feeding grounds requires energy. While migrating, the animal expends energy at C per day, a value that depends on u, the speed of travel. (As with R, cost is normalized to standard metabolic rate.) Energy expended during travel is

$$\text{cost of transport} = 2tC. \tag{10.45}$$

Only if this cost is equal to or less than the energy accumulated on the feeding grounds is migration worthwhile. In other words, the minimum criterion for migration is that

$$(t_{\text{win}} - 2t)\, R_{\min} = 2tC. \tag{10.46}$$

It is useful to take this calculation one step further. Time in transit (t) is equal to the distance traveled to the feeding grounds (x) divided by the speed of locomotion (u). Thus,

$$\left(t_{\text{win}} - \frac{2x}{u}\right) R_{\min} = \left(\frac{2x}{u}\right) C. \tag{10.47}$$

Solving for $R_{\min}$, we find that

$$R_{\min} = \frac{C}{\frac{t_{\text{win}} u}{2x} - 1}. \tag{10.48}$$

The higher the cost of transport and longer the distance, the less likely it is that migration is advantageous. The faster the animal moves, the greater the value of migration.

It is at this point that the mechanics of locomotion come into play. If we assume that swimming and flying animals migrate at u_{opt}, the speed that minimizes their cost,

Table 10.3 Parameter Values for Calculating the Cost of Travel During Migration and the Optimal Speed at which an Animal Should Commute.

Cost, $C(u_{opt})$	k_1	k_2	k_3	k_4
Flapping flight	12	0.10	0	0
Gliding	3	0	0	0
Swimming (all)	3.3	0	0	0
Walking and running	3.2	0.02	1.8	−0.06

u_{opt}	k_5	k_6
Flapping flight	16	0.14
Gliding	8	0.14
Swimming (fish)	0.47	0.17
Swimming (others)	0.85	0.17
Walking and running	1.0	0.10

we can use the graphical approach discussed earlier in this chapter (Figure 10.11) to specify u_{opt} and C at that speed. For animals that walk or run, it isn't possible to calculate u_{opt}, so we use the C at a speed typically used in migration. Knowing u_{opt} and C then allows us to calculate whether migrating a given distance is feasible.

To this end, Alexander derived equations to model the costs and speeds of migrating animals. For cost at optimal (or for walkers, typical) speed, the equations take the general form

$$C = \frac{\text{metabolic rate during locomotion}}{\text{standard metabolic rate}} = k_1 m_b^{k_2} + k_3 m_b^{k_4}. \qquad (10.49)$$

where m_b is body mass in kilograms and k_1 to k_4 take on values given in Table 10.3. For instance, C for a walking animal is

$$C = 3.2 m_b^{0.02} + 1.8 m_b^{-0.06}. \qquad (10.50)$$

For optimal speed, the equations take the familiar form

$$u_{opt} = k_5 m_b^{k_6}. \qquad (10.51)$$

Here u_{opt} is in $m \cdot s^{-1}$ and k_5 and k_6 have the values given in Table 10.3. Optimal speed for a walking animal, for example, is

$$u_{opt} = 1.0 m_b^{0.10}. \qquad (10.52)$$

Alexander's results are summarized in Tables 10.4 to 10.6 for animals of small, medium, and large mass: 3 g, 3 kg, and 3 tonnes (t). As we have noted previously, flying animals have metabolic rates higher than those of swimmers or walkers, but they move much faster (Figure 10.10A) and therefore spend less time commuting (Table 10.5) and more time at the feeding grounds. As a result, the richness of the feeding ground does not need to be nearly as high as it is for other forms of locomotion to make migration worthwhile. For a 3-g bird using flapping flight, for instance, net energy gain on the feeding grounds need be only 12% above standard metabolic rate ($R = 0.12$) to make a 1000 km migration feasible (Table 10.5). By contrast, a 3-g

Table 10.4 Optimal Speed of Travel and the Associated Dimensionless Normalized Metabolic Rate for Migrating Animals.

		3 g	3 kg	3000 kg
Flapping flight	u_{opt} (m/s)	7.1	18.6	
	C	6.7	13.4	
Soaring	u_{opt} (m/s)	3.6	9.3	
	C	3.0	3.0	
Fish swimming	u_{opt} (m/s)	0.2	0.6	1.8
	C	3.3	3.3	3.3
Penguin swimming	u_{opt} (m/s)		1.0	
	C		3.3	
Swimming	u_{opt} (m/s)			3.3
(whales and seals)	C			3.3
Walking	u_{opt} (m/s)	0.6	1.1	2.2
	C	5.4	5.0	4.9

Table 10.5 Travel Times and Minimum Necessary Energy Gains Needed to Make a 1000-km Migration Worthwhile.

		3 g	3 kg	3000 kg
Flapping flight	Time $2t$ (days)	3.3	1.2	
	R_{min}	0.12	0.09	
Soaring	Time $2t$ (days)	6.5	2.5	
	R_{min}	0.11	0.04	
Fish swimming	Time $2t$ (days)	132	41	13
	R_{min}	8.61	0.95	0.25
Penguin swimming	Time $2t$ (days)		23	
	R_{min}		0.47	
Swimming	Time $2t$ (days)			7
(whales and seals)	R_{min}			0.13
Walking or running	Time $2t$ (days)	41	21	10
	R_{min}	1.57	0.63	0.29

walking mammal would need an R of 1.6 (and a 3-g fish nearly 9) to make the trip worthwhile. It would take an exceptionally rich food source to allow an animal to gain energy at ten times its standard metabolic rate.

Migrating 1000 km becomes more practical for larger animals. A 3-kg fish requires an R of 0.95 and a 3-t whale or seal an R of only 0.13.

Table 10.6 Travel Times and Minimum Energy Gains Needed to Make a 10,000-km Migration Feasible.

		3 g	3 kg	3000 kg
Flapping flight	Time $2t$ (days)	33	12	
	R_{min}	1.45	0.97	
Soaring	Time $2t$ (days)	65	25	
	R_{min}	1.66	0.47	
Fish swimming	Time $2t$ (days)	Too slow	Too slow	127
	R_{min}			7.39
Penguin swimming	Time $2t$ (days)		Too slow	
	R_{min}			
Swimming	Time $2t$ (days)			70
(whales and seals)	R_{min}			2.05
Walking or running	Time $2t$ (days)	Too slow	Too slow	104
	R_{min}			6.41

Conditions are more stringent for a 10,000-km migration (Table 10.6). Birds of all sizes can manage the trip, needing an R of 0.47 to 1.66. But small- and medium-size swimmers and walkers move so slowly that the feeding grounds are beyond their reach in the 183 d of winter. Whales, large seals, and very large fish could complete the journey but would require rich forage (an R of 2.05 to 7.39).

This analysis does not take into account several factors that commonly play into migration (quality of forage at the homesite, feeding during travel, trips that last more than a single winter, etc.), but—simple as it is—it successfully explains the observed patterns of migration for a wide variety of animals and highlights the potential of biomechanics and energetics to provide insight into the evolution of life-history traits. For a slightly more realistic model, which takes homesite foraging into account, see Supplement 10.9.

16   CONCEPTS, CONCLUSIONS, AND CAVEATS

Movement is intimately tied to the production of thrust, which is needed to maneuver, overcome the resistance of drag, and offset the pull of gravity. Animals and plants produce thrust through a wide variety of mechanisms—the shedding of momentum, the friction of viscosity, and the reaction of an immovable solid—and knowledge of these mechanisms allows us to draw a series of sweeping conclusions:

- For organisms willing to forego control over where they travel, drifting is easy and inexpensive: one need only stay aloft. But sinking rate varies with size: small organisms inherently sink slowly; to stay aloft, large organisms must adjust their buoyancy (which is impractical in air) or continuously propel themselves upward.

- Evolution has developed ingenious mechanisms to store and release energy during locomotion—vortices, tendons, cuticle, and rubbery joints, for instance—thereby lowering the cost of locomotion.
- Cost of transport is exceptionally low for fish, but so is their sustainable speed, leading to some interesting trade-offs. The low cost of transport makes it possible for these animals to migrate great distances, but these migrations often take multiple years.
- Liberated from water's drag, legged locomotion in air allows for higher speeds than aquatic animals can attain, but the cost is concomitantly high. By contrast, flight speeds are so high that they more than offset the increased power necessary to stay aloft in air, and the low cost of aerial transport makes it possible for a wide variety of insects and birds to forage widely, avoid the ups and downs of uneven terrain, and even migrate.
- Small organisms in air—and any organism in water—can climb with impunity, but climbing is difficult and slow for large terrestrial animals.
- The ability to jump is surprisingly independent of body mass.

Care must be taken in applying these conclusions. For example, I have presented several equations describing how speed varies with body mass. These relationships are valuable as descriptions of general trends, but they cannot be expected to apply exactly to any particular organism. Indeed, there is considerable variation around each of these trends, which must be taken into account if accurate predictions are desired for a given species.

Having focused on two ecologically relevant aspects of locomotion—speed and cost—I have given short shrift to a host of intriguing and important details. To gain a broader and more nuanced understanding of the subject, I urge you to consult the classic texts (Alexander and Goldspink 1977; Videler 1993; Dudley 2000; Biewener 2003; Alexander 2003).

In recent years a new field—movement ecology—has emerged, combining the study of animal behavior, biomechanics, population dynamics, and community ecology to approach questions of why organisms move, how they transport themselves, and what the consequences are (e.g., Nathan et al. 2008; Clobert et al. 2012). In this integrated framework, the physics and physiology we have explored in this chapter provide mechanistic explanations for the issues of speed and cost that govern ecological interactions.

Chapter 11

Thermal Mechanics I
Introduction, Solar Heating, Convection, Metabolism, and Evaporation

Of all the physical characteristics that govern life, temperature is arguably the most important. It determines when and where individuals can survive, how fast they can grow, how fruitfully they can multiply, and often how they interact. Temperature affects organisms in both air and water, and, as a result of ongoing changes in earth's atmosphere, it is increasing.

Despite its fundamental role in biology, the physics of temperature is unfamiliar to many biologists. It is a common assumption, for instance, that the temperature of terrestrial plants and cold-blooded invertebrates (ectotherms) is the same as the air around them. But the flight muscles of bees and the flowers of some plants can be 35°C higher than the surrounding air (Denny 1993). By contrast, some seaweeds and desert herbs can be 5°C to 8°C cooler than air. Ecologists are well aware that average temperature varies with latitude and altitude but are prone to forget that local variation—at scales as small as a few centimeters—can be even larger. These small-scale variations (which depend on organism size and behavior) often have effects that regulate the interaction of individuals (and thereby control community dynamics), and affect the evolution of thermal tolerance. For example, Kaspari and colleagues (2014) found that in a Panamanian rain forest, small-scale variation in ants' body temperature has resulted in species' having widely different thermal tolerances. The variation in tolerance in this single forest is nearly as large as the variation found worldwide. Clearly, understanding the physics of temperature is central to ecological mechanics.

The complexity of this subject is such that it is convenient to split our investigation into two chapters. In this first chapter we review the basic physics of heat, introduce the heat budget, and discuss the four most common mechanisms by which heat can enter or leave an organism: sunlight, convection, metabolism, and evaporation. We explore illustrative examples of how these mechanisms combine to affect body temperature. In the next chapter, we deal with the remaining mechanisms of heat transfer (conduction and infrared radiation), examine the effects of heat storage, and finish up with an exploration of how a heat budget can be used to predict species ranges (an example of environmental niche modeling).

That is where we are headed, but we begin with an overview of the ways in which temperature affects biology.

1   WHY TEMPERATURE MATTERS

Temperature's biological importance begins at the level of molecules. The higher the temperature, the greater the fraction of molecules with the thermal energy needed to complete a particular chemical reaction. As a result, reaction rates increase with increasing temperature. Traditionally, the thermal variability of such processes is expressed as Q_{10}, the relative change in rate accompanying a 10°C increase in temperature. Knowing Q_{10}, we can predict the relative rates at any two temperatures T_1 and T_2:

$$\frac{\text{rate at } T_2}{\text{rate at } T_1} = Q_{10}^{(T_2-T_1)/10} \tag{11.1}$$

For example, most organisms have a metabolic Q_{10} between 2 and 3 (Schmidt-Nielsen 1997; Hochachka and Somero 2002), and this temperature dependence can have cascading effects. In the intertidal zone of rocky shores, for instance, the temperature of plants and animals can change by more than 30°C in a low tide, and, as a consequence, their metabolism can vary by a factor of 8 to 27 in just a few hours. Similarly drastic variations can be imposed on alpine and desert plants.

The biological consequences of temperature variability can be good or bad. The rapid metabolism accompanying high temperatures may allow both plants and animals to grow faster and reproduce sooner. On the other hand, an increased rate of metabolism burns fuel faster and therefore requires increased rates of gas exchange, food intake (for animals), and photosynthesis (for plants). If these temperature-dependent trade-offs differ between producers and consumers, changes in temperature can affect both the top-down and bottom-up regulation of population dynamics (Harley 2013).

Q_{10} effects are just one of the many biological consequences of temperature:

- Metabolic effects of temperature interact with temperature's effects on the environment. Oxygen is required for aerobic metabolism, for example, but at 30°C the oxygen concentration in seawater is only 56% that at 0°C, variation equivalent to the change between air at sea level and at the top of a 16,000-ft mountain peak. Thus, for aquatic animals at the high temperatures where metabolism requires rapid delivery of oxygen, oxygen can be most difficult to deliver.
- The physiological effects of body temperature can also depend as much on the rate of increase as on the peak temperature. Variation in temperature that occurs slowly through an organism's lifetime allows it to acclimatize, whereas the same variation imposed abruptly could be lethal. You have probably encountered a mild version of this phenomenon. It is painful to jump straight into a hot shower. But if you start with warm water and gradually increase the temperature, giving your body time to adjust, temperature that was previously unbearable can be enjoyed.

- Timing also matters. An organism might easily cope with an individual heat shock but be compromised if the same shock is applied repeatedly. On the other hand, given some recovery time, repeated sublethal heat shocks can lead to stress hardening, a form of acclimatization in which the body's tolerance becomes more robust. Response can also vary through ontogeny: a temperature shock might kill juveniles but spare adults.
- In addition to physiological adaptations to temperature, organisms have a broad range of behavioral responses. Both plants and animals can reorient body parts to increase or reduce exposure to the warming effects of sunlight, and, to the same end, some animals can change their color in response to their thermal needs. Plants are limited in their movements, but animals can choose to bask in the sun or amble into the shade as the need arises.
- Ecological interactions can amplify the effects of temperature. Consider a case from intertidal shores. A shift in water temperature from 10°C to 13°C increases the rate at which the sea star *Pisaster ochraceus* eats mussels by 59%, a Q_{10} of 4.8 (Sanford 2002). Because mussels are the dominant competitor for space and *Pisaster* is their primary predator, this minor change in sea-surface temperature—the sort of change associated with an El Niño event—can substantially change the species diversity of the intertidal community.

In summary, body temperature matters to plants and animals but in an exceedingly complex fashion. Only with a detailed history of the variation in temperature—and a thorough understanding of the physical basis for this variation—can we hope to predict the individual and ecological consequences of body temperature.

Therein lies the impetus for the development of *heat-budget models*, the central theme of this and the next chapter. These models' mechanistic predictions provide a critical tool for physiologists and ecologists to explore how organisms cope with the exigencies of the physical environment.

2 HEAT-BUDGET MODELS
A Prospectus

The utility of heat-budget models is manifold. For example, it is often impractical to measure body temperature directly: imagine trying to take the temperature of a fruit fly as it forages or a pollen grain as it wafts on the breeze. Given sufficient time and money, technology might be mustered to do the job, but a heat-budget model can quickly provide accurate answers with much less effort. Even when it is feasible to take an individual's temperature directly, it is often impractical to measure the temperature of multiple organisms simultaneously. Again, heat-budget models can solve the problem; a single, relatively inexpensive weather station can provide all the environmental data required to accurately calculate the body temperature for any number of plants and animals.

Heat-budget models even allow collection of data after the fact. It is impossible to go back in time to directly measure the temperature of a dinosaur, but heat-budget models allow us to make a reasonable estimate.

The ability to forecast temperature is also important. As earth's climate changes in coming decades, heat-budget models allow us to calculate the corresponding effects on body temperatures, providing a means to predict not only the consequences for individual physiology, but also the resulting shifts in species' ranges and interaction strengths (e.g., Kearney et al. 2008).

Heat-budget models can also address a wide variety of what-if? questions. Wind tends to counteract the effects of increased air temperature by augmenting evaporation and convection. If, as a consequence of climate change, air temperatures and wind speed both increase, will future organisms be hotter or colder? What if, in addition to an increase in wind speed, the sky is more often cloudy? It would be difficult to conduct the large-scale field experiments necessary to answer these questions, but they can easily be addressed by heat-budget models.

3   HEAT

Let's begin with kinetic energy. In Chapter 2 we concluded that the kinetic energy of an object is $\frac{1}{2}mu^2$, half the product of the object's mass and the square of its speed. A quantitative example gives this concept some tangibility. To pass the time, you decide to drop a water balloon on a friend. The balloon contains a liter of water ($m = 1$ kg), and, when dropped from your window, it attains a speed of 10 m $\cdot$ s^{-1} before hitting its target. As the balloon drops, each bit of its mass moves at the same speed and in the same direction, and this coherent motion gives the balloon a kinetic energy of 50 joules (J), energy that can do work, in this case the entertaining task of splattering water.

This example treats kinetic energy from the macroscopic perspective of the whole balloon. Let's instead look at individual water molecules. Each has a minuscule mass (3×10^{-26} kg). But, even when the balloon is at rest, molecules rattle around with an average speed of 636 m $\cdot$ s^{-1} (we'll calculate this surprising value later in the chapter), and there are so many of them (3.3×10^{25} per kilogram), that their combined kinetic energy is approximately 200,000 J, 4000 times the energy of the falling balloon taken as a whole.

If the mere 50 J of kinetic energy in a falling kilogram of water can produce a satisfyingly large splash, why don't we notice the effects of the 200,000 J active in the same kilogram at rest? The discrepancy is due to the manner in which individual molecules move compared to the motion of the balloon. At the macroscopic level, water in the balloon moves coherently, allowing it to do work. By contrast, an individual molecule acts like the ball in a pinball machine. It has a high speed, but its direction is constantly and randomly changing as it collides with other molecules. At room temperature, each water molecule collides approximately 60,000 billion times per second, thoroughly randomizing its path, and making it highly unlikely that the 3.3×10^{25} water molecules in a kilogram ever get sufficiently organized to perform much useful work.[1] This is why a liter of water appears to just sit there. It contains a

---

[1] This is an oversimplification. The random motion of molecules in water at room temperature *can* do useful work, but only if the assemblage of molecules is connected to a heat sink at lower temperature. The ability to extract work from the random motion of molecules is the subject of thermodynamics, a field of inquiry that is, unfortunately, beyond our purview.

huge amount of kinetic energy, but because this energy is tied to the random jostling of molecules, its effects are different from those of coherent, macroscopic motion. For this reason we treat the kinetic energy of random motion differently than we treat that of coherent movement, and we give it a different name—the kinetic energy of random motion is *heat*.

4   TEMPERATURE

Heat is the quantity we will track in our exploration of thermal mechanics, but it is an organism's body temperature we desire to predict. How are the two connected?

From equation 2.50, we know that for molecules of mass m,

$$\text{average translational kinetic energy} = \frac{m\overline{u^2}}{2}, \tag{11.2}$$

where $\overline{u^2}$ is average squared speed of a molecule's center of mass. Absolute temperature T is proportional to this average translational energy:

$$T = k_T \frac{m\overline{u^2}}{2}. \tag{11.3}$$

For the SI system, in which temperature is measured in kelvins, $k_T = 4.84 \times 10^{22}$ kelvins per joule $(k \cdot J^{-1})$.

(Unless otherwise specified, all temperatures in this text are in kelvins. Conversion from kelvins to degrees Celsius is simple:

$$°C = K - 273.15. \tag{11.4}$$

For convenience, I will often ignore the pesky 0.15°C—for example, expressing room temperature (20°C) as 293 K rather than 293.15 K.)

We can rearrange equation 11.3 to calculate the average speed of molecules at a given temperature:

$$\sqrt{\overline{u^2}} = \sqrt{\frac{2T}{k_T m}}. \tag{11.5}$$

The term $\sqrt{\overline{u^2}}$—the root-mean-square, or rms, velocity—is a convenient measure of the average speed of molecules. For room temperature water, rms speed is $636 \text{ m} \cdot \text{s}^{-1}$, as advertised.

Note that temperature does not depend of the number of molecules present. For example, in the upper reaches of earth's atmosphere, individual gas molecules are boosted to extraordinary speeds as they are zapped by incoming ultraviolet radiation, resulting in temperatures approaching 1000 K. But at that altitude there might be only a thousand molecules in a liter of atmosphere, so despite molecules' high temperature their combined heat energy $(2 \times 10^{-17} \text{ J})$ is vanishingly small compared to the 200,000 J of a water balloon's cold—but numerous—molecules.

5   SPECIFIC HEAT

Having defined both temperature and thermal kinetic energy, we need a way to translate between them. This is the role of *specific heat capacity*, C_P, the ratio of change in energy (ΔW) to a change in temperature (ΔT) for a given mass of material:

$$C_p = \frac{\Delta W}{m\Delta T}. \tag{11.6}$$

Specific heat varies substantially among materials. For some (in particular, diatomic gases such as H_2, N_2, O_2), much of the energy injected into the material causes molecules to vibrate or rotate and, therefore, does not contribute to an increase in the translational speed of their centers of mass, the speed that counts in the definition of temperature. As a result, these materials have high specific heats; hydrogen has the highest at 14,000 $J \cdot kg^{-1} \cdot K^{-1}$. For other materials (metals especially), virtually all injected energy results in translational motion and they have low specific heats. Gold has the lowest of any material, 128 $J \cdot kg^{-1} \cdot K^{-1}$. The specific heat of air, 1006 $J \cdot kg^{-1} \cdot K^{-1}$, is typical of a gas, but that of water is exceptionally high for a liquid—approximately 4200 $J \cdot kg^{-1} \cdot K^{-1}$—roughly four times that of air.

6   THERMAL DIFFUSION

Heat and temperature are also connected through the process of diffusion. As molecules rattle around due to thermal agitation, they transport heat from areas of high temperature to areas of low temperature according to Fourier's law:[2]

$$\mathcal{F}_{H,z} = -\rho C_p \mathcal{D}_H \frac{dT}{dz}. \tag{11.7}$$

Here $\mathcal{F}_{H,z}$ is heat flux along the z-axis ($W \cdot m^{-2}$), ρ is the material's density ($kg \cdot m^{-3}$), $\mathcal{D}_H$ is the diffusivity of heat ($m^2 \cdot s^{-1}$), and dT/dz is the temperature gradient.

Note the similarity to Fick's equation for molecular diffusion (Chapter 4): heat (like mass) is transported down the relevant gradient at a rate governed by a diffusion coefficient. However, thermal diffusion differs from molecular diffusion in one important respect. Whereas mass is transported as molecules change position, heat can be transported through molecular collisions. In this respect, the diffusion of heat is analogous to the diffusion of momentum we discussed in Chapter 7, where we employed an analogy to billiard balls. A rapidly moving ball can transfer its kinetic energy to a second ball when the two collide. The second ball can then transfer its kinetic energy to another ball, and so forth, transporting kinetic energy (by analogy, heat) across a crowded billiard table without any single ball making more than a fraction of the trip. Because heat can diffuse separately from mass, thermal diffusivity in water and solids ($\approx 10^{-7}$ $m^2 \cdot s^{-1}$) is two orders of magnitude larger than molecular

---

[2] Jean Baptiste Joseph Fourier (1768–1830) was a French mathematician who did his most important work in the employ of Napolean Bonaparte. To calculate the transport of heat through materials, he devised a means of expressing periodic functions as a series of sines and cosines: the Fourier series, which we will encounter in Chapter 20. Fourier was also the first to propose that Earth's temperature is regulated by the greenhouse effect.

Table 11.1 Some Relevant Properties of Water.

Water	Units	0°C	10°C	20°C	30°C
Specific heat capacity	$J \cdot kg^{-1} \cdot K^{-1}$	4218	4192	4182	4179
Thermal conductivity	$W \cdot m^{-1} \cdot K^{-1}$	0.5651	0.5867	0.6011	0.6157
Thermal diffusivity	$10^{-6} \, m^2 \cdot s^{-1}$	0.134	0.140	0.143	0.148
Density (freshwater)	$kg \cdot m^{-3}$	999.87	999.73	998.23	995.68
Density (seawater)	$kg \cdot m^{-3}$	1028.11	1026.95	1024.76	1021.73
Kinematic viscosity	$10^{-6} \, m^2 \cdot s^{-1}$	1.79	1.31	1.01	0.80
Latent heat of vaporization	$10^6 \, J \cdot kg^{-1}$	2.513	2.489	2.465	2.442

diffusivity ($\approx 10^{-9} \, m^2 \cdot s^{-1}$). Due to the relatively large distances molecules travel between collisions in gases (analogous to a sparsely populated billiard table), aerial thermal diffusivity is approximately equal to molecular diffusivity.

The product of density, specific heat, and thermal diffusivity is often combined into a single variable, *thermal conductivity* ($W \cdot m^{-1} \cdot K^{-1}$):

$$K = \rho C_p \mathcal{D}_H. \tag{11.8}$$

Thus, Fourier's law can be written in a more compact form:

$$\mathcal{F}_{H,z} = -K \frac{dT}{dz}. \tag{11.9}$$

This is the relationship we will use when we deal with heat transport by conduction in Chapter 12.

7   LATENT HEAT

Thermal energy as we have discussed it so far is *sensible heat*: heat associated with the kinetic energy of molecules, heat that would raise the temperature of a thermometer (see Supplement 11.1 for a discussion of how thermometers work). There is, however, another form of heat—*latent heat*. The distinction is necessary because, during phase transitions (e.g., solid to liquid, liquid to gas), heat energy can be added to or extracted from materials without changing their temperature. For instance, when heat is added to water at its boiling point (373 K), its primary effect is to break hydrogen bonds. This allows molecules to evaporate, but does not increase their speed and, therefore, doesn't increase their temperature. The latent (hidden) energy in the disrupted bonds becomes sensible when, as water vapor condenses, the bonds reform and heat energy is released. For our purposes the latent heat of interest is water's latent heat of vaporization, Λ_{wat}, a whopping $2.45 \times 10^6 \, J \cdot kg^{-1}$ (Table 11.1).

8   THE HEAT BUDGET

Having familiarized ourselves with the basic teminology of heat and temperature, we now turn to the specific mechanisms by which heat can enter or leave an organism

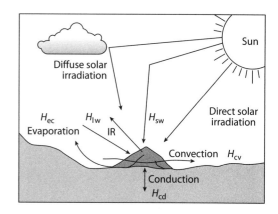

Figure 11.1 Modes of heat flux for an organism.

(Figure 11.1). In each case, we measure the transport of heat in watts, and by convention the sign of transport is positive for heat moving into an organism and negative for heat moving out.

- As sunlight is absorbed by plants, animals, soil, water, or rocks, its short-wave electromagnetic energy can be converted to heat at a rate H_{sw}.
- Objects also exchange heat with their environment through the emission and absorption of long-wave infrared light, leading to a net rate of heat transfer, H_{lw}

These two mechanisms are examples of *radiative heat transfer*, the transport of heat energy via electromagnetic radiation.

- An organism can exchange heat with the solid substratum via *conduction*, as heat diffuses from the hotter object to the cooler at rate H_{cd}.
- Similarly, *convection* can transport heat at rate H_{cv} as air or water moves past an object.

Both conduction and convection involve the transport of heat by diffusion, and thus they resemble the transport of mass and momentum we dealt with when discussing boundary layers in Chapters 6 and 7.

- Heat can be exchanged at rate H_{ec} through *evaporation* or *condensation*.
- And lastly, heat can be produced at rate H_{met} in the organism itself as a by-product of metabolism.

All six of these processes can occur at the same time, some adding heat to an organism, others taking it away. If heat enters faster than it leaves, the net rate of heat accumulation, H_{net}, is positive, heat builds up in the organism, and its temperature increases. Conversely, if more heat leaves than enters, H_{net} is negative, heat drains away, and temperature decreases.

These concepts can be expressed mathematically as the organism's *heat budget*:

$$H_{sw} + H_{lw} + H_{cd} + H_{cv} + H_{ec} + H_{met} = H_{net} \qquad (11.10)$$

Our goal in these two chapters is to turn this simple equation into a practical tool—a response function—for predicting body temperature.

9   SOLAR HEATING

If you have ever sat next to a campfire at night, you have some personal experience with the physics of light production. When the fire is roaring, the surface of the wood

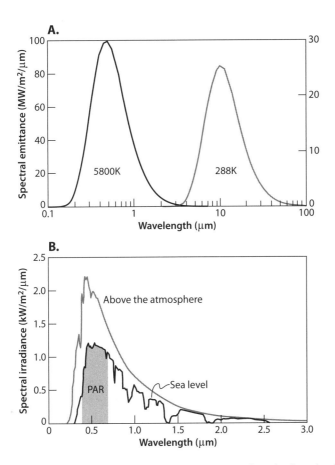

Figure 11.2 A. Sunlight (shortwave radiation) has shorter wavelengths than the long-wave radiation produced by Earth's surface. **B.** Earth's atmosphere absorbs some of the sun's radiation (redrawn from Gates (1980)).

is white hot, but as the blaze subsides and temperature decreases, color shifts from white to orange to red and then fades from sight. The temperature-dependent color of the fire is evidence of two fundamental aspects of physics: heated objects emit light, and the distribution of emitted wavelengths varies with temperature. The higher the temperature, the shorter the wavelength at the distribution's peak, λ_{max}, a relationship quantified by *Wien's displacement formula*:

$$\lambda_{max} = \frac{2.9 \times 10^6}{T}. \tag{11.11}$$

Here λ_{max} is measured in nanometers and T is temperature in kelvins (Incropera and DeWitt 2002).

Two pertinent examples put equation 11.11 into perspective. Heated by nuclear fusion in the star's interior, the visible surface of the sun has a temperature of approximately 5800 K, producing the spectrum of light shown in Figure 11.2A. The peak of the solar spectrum is at a wavelength of approximately 500 nm, a pleasant forest green, which, when combined with the other wavelengths, produces the familiar warm color of sunlight.

By contrast, temperatures of organisms and the environment range from roughly 233 K to 328 K ($-40°C$ to $55°C$). At these low terrestrial temperatures, λ_{max} is in the range of 10,000 nm, far removed from the 500 nm peak of sunlight. As a consequence, 99% of the energy emitted by objects at terrestrial temperatures has wavelengths greater than 4000 nm, whereas 98% of the energy in sunlight occurs at wavelengths less than 4000 nm. This separation allows us to talk unambiguously about sunlight as *short-wavelength radiation* and light emitted by organisms and the environment as *long-wavelength radiation*. It will be handy to refer to this long-wave radiation as *infrared light*, but keep in mind that when I use the term it refers to the far infrared ($\lambda > 4000$ nm), not to wavelengths just beyond the visible range (red = 700 nm).

Sunlight arrives in two forms: direct and diffuse. Light that travels substantially unimpeded from source to absorption is direct. Light that is scattered such that it arrives more or less equally from all directions is diffuse.

As sunlight (either direct or diffuse) impinges on water, the ground, or an organism, it can be absorbed and converted to heat at rate H_{sw}:

$$H_{sw} = \alpha_{sw} \left(A_{pr} \mathcal{I}_{sw,dir} + A_{sky} \mathcal{I}_{sw,dif} \right). \tag{11.12}$$

Here α_{sw} is the *shortwave absorptivity* of the object's surface, A_{pr} is the object's surface area (projected in the direction of the sun), and $\mathcal{I}_{sw,dir}$ is direct solar irradiance. (I'll define irradiance in a moment.) A_{sky} is the total surface area of the object open to diffuse sunlight, and $\mathcal{I}_{sw,dif}$ is diffuse solar irradiance. Through the course of a day, sunlight varies in brightness and strikes objects at different angles. As a consequence, the irradiances and areas in equation 11.12 are likely to change through time. (See Supplement 11.2 for the details.)

A truly bewildering variety of units is used to measure the brightness of light: everything from lumens to sperm candles. The measure we use here, *irradiance*, is based not on how we humans perceive light (as are many measures), but rather on the rate at which light energy can be transferred. A totally black surface absorbs all the light that hits it and converts this energy to heat. The rate at which such a surface accumulates heat energy, expressed per area of surface exposed, is by definition irradiance. Thus, $\mathcal{I}$ has units of watts per square meter.

Above Earth's atmosphere, direct solar irradiance is approximately 1366 $W \cdot m^{-2}$, a value known as the solar constant.[3] Approximately 25% of this incident solar radiation is absorbed or scattered by Earth's atmosphere (Figure 11.2B), leaving roughly 1000 $W \cdot m^{-2}$ of direct irradiance at sea level. This irradiance can be augmented by small-angle scattering. For example, on a partly cloudy day $\mathcal{I}_{sw,dir}$ can momentarily exceed 1100 $W \cdot m^{-2}$ as additional light is redirected slightly downward by water droplets at a cloud's edge.

On an overcast day, sunlight cannot travel directly to the ground. Instead, it is repeatedly scattered by water droplets in clouds and arrives as diffuse illumination, characterized by a lack of shadows. Under a thickly overcast sky, noontime diffuse irradiance may be only a tenth the direct irradiance under a clear sky.

---

[3] The solar constant isn't really constant. Solar irradiance varies slightly with the roughly 11-y period of the sunspot cycle and has varied substantially over Earth's 4.5-billion-year history. Early life evolved in what is jokingly referred to as the Chinese restaurant era because back then there was a dim sun (Bryson 2004).

A.

Sunlight

B.

Sunlight

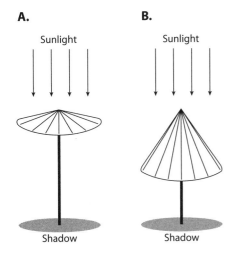

Shadow

Shadow

Figure 11.3 Defining projected area. (See text.)

The tenfold difference in solar irradiance between clear and cloudy days is not readily apparent to us. Our eyes have an automatic gain control analogous to that of a video camera, which cranks up the gain in dim light to maintain apparent illumination and cranks it down in bright light to keep from overloading the system. As a result of our adaptive visual system, estimation of irradiance by eye is nearly futile. Instead, irradiance is measured using a *pyranometer*, a device that directly measures the amount of heat energy absorbed by a totally absorbing surface.[4]

In equation 11.12, it is important to differentiate between projected area A_{pr} and surface area A_{sky}. Consider a black umbrella with the noontime sun shining vertically down on it. By absorbing sunlight, the umbrella casts a shadow on the floor beneath (Figure 11.3). In regard to the energy it absorbs, it doesn't matter whether the umbrella is flat with small surface area (A) or high-spired with more surface area (B). Both parasols cast the same shadow, evidence that they absorb the same amount of light energy. The same is true for any object. The amount of energy it can absorb from direct irradiance depends only on A_{pr}, the size of the shadow it casts on a surface perpendicular to the direction in which light arrives. By contrast, the amount of diffuse light that can be absorbed depends on A_{sky}, the surface area that can "see" the sky. Typically, $A_{sky} > A_{pr}$.

The remaining term in equation 11.12 is shortwave absorptivity, α_{sw}, the dimensionless ratio of light absorbed to light available. A perfectly clear piece of glass and a perfectly reflecting mirror both have $\alpha_{sw} = 0$; a perfectly absorbing surface has $\alpha_{sw} = 1$. If a totally black surface absorbs all light, one might suppose that a white surface would absorb none. But the term white specifies only that a surface reflects all colors equally and is by no means a guarantee that it reflects any of them particularly well. Commercial flat black paints have absorptivities of approximately 0.95. White paints have absorptivities of approximately 0.25 to 0.30. Plants, animals, soils, and rocks have absorptivities that typically fall between those of black and white paints.

Absorptivity is measured with a radiometer, an instrument similar to a pyranometer, where the rate at which heat is absorbed by the surface in question is compared to rate at which heat is absorbed by a totally absorbing surface. The ratio of the two rates is absorptivity.

---

[4] Pyranometers are different from the light meters used in photography and the light sensors used to study photosynthesis, both of which are designed to measure only light of visible wavelengths (photosynthetically active radiation, PAR, for the botanically minded) and therefore underestimate the total light energy available for conversion to heat (see Figure 11.2B).

In summary, the rate at which heat is absorbed from sunlight (equation 11.12) depends on the intensity of light (both direct and diffuse), and the area and absorptivity of the object exposed to it.

To this point, we have assumed that all light energy absorbed by an object is converted to heat. During photosynthesis, however, some light energy is stored in carbohydrates rather than being converted to heat, and H_{sw} as we have calculated it would overestimate the rate of heat absorption. The overestimation is typically slight, however (Nobel 1991), and we will not take it into further account.

10   CONVECTIVE HEAT TRANSPORT

Convective heat transfer is driven by the convective motion of fluids, which comes in two varieties. *Free convection* is driven by the buoyancy of a fluid. For example, air next to a hot rock is heated by conduction, and the resulting decrease in the air's density causes it to rise like a hot-air balloon, transporting heat away from the substratum. Fluid motion due to any mechanism other than buoyancy is known as *forced convection*, wind being a common example.

The rate of convective heat transfer, H_{cv}, can be described by a deceptively simple equation:

$$H_{cv} = h_c A_{cv} \left(T_f - T_b \right).\tag{11.13}$$

Known as *Newton's law of cooling*, equation 11.13 highlights the role of body temperature (T_b) in the exchange of heat energy. If T_b is greater than T_f (the temperature of the surrounding fluid), H_{cv} is negative, indicating that heat flows out of an object. Conversely, if $T_b < T_f$, heat flows in. The greater the temperature difference, the faster heat flows, and for a given temperature difference, the rate of heat transfer (either loss or gain) depends on A_{cv}, the area of the object in contact with the surrounding fluid.

The complexity of equation 11.13 is hidden in h_c, the *convective heat transfer coefficient* ($W \cdot m^{-2} \cdot K^{-1}$), an index of the ability of the system to transport heat. Via the rearrangement of equation 11.13, h_c is defined as

$$h_c = \frac{H_{cv}}{A_{cv}(T_f - T_b)}.\tag{11.14}$$

where h_c depends on the shape and size of the object and the velocity, density, and viscosity of the surrounding fluid. As a consequence, it is difficult to predict accurately from theory and instead must be measured empirically.

We can, however, use boundary-layer physics to understand the general behavior of h_c. Because the boundary layer is an area of relatively sedentary flow, it can act as an insulating blanket: heat that will eventually be carried away from an organism by the fluid's advection must first diffuse across the boundary layer. Because a fluid's speed and an object's shape control boundary-layer thickess, we can expect convective heat transfer to depend on speed and shape.

The rate at which heat diffuses through the boundary layer is governed by Fourier's equation:

$$\frac{H_{cv}}{A_{cv}} = \mathcal{K}\frac{dT}{dz},\tag{11.15}$$

Table 11.2 Some Relevant Properties of Air.

Air	Units	0°C	10°C	20°C	30°C	40°C
Specific heat capacity	$J \cdot kg^{-1} \cdot K^{-1}$	1006	1006	1006	1006	1006
Thermal conductivity	$W \cdot m^{-1} \cdot K^{-1}$	0.0247	0.0254	0.0261	0.0268	0.0276
Thermal diffusivity	$10^{-6} \, m^2 \cdot s^{-1}$	18.9	20.2	21.5	22.8	24.2
Density (at 1 atm)	$kg \cdot m^{-3}$	1.292	1.246	1.204	1.164	1.128
Kinematic viscosity	$10^{-5} \, m^2 \cdot s^{-1}$	1.33	1.42	1.51	1.60	1.70

where z is distance from the boundary, $\mathcal{K}$ is the thermal conductivity of air or water (see Tables 11.1 and 11.2), and the rate of heat flow depends on the steepness of the thermal gradient, dT/dz. (There is no negative sign in front of $\mathcal{K}$ in this case because we define heat influx into the organism as positive.)

For the boundary layer adjacent to an organism, we can use an approximation of the thermal gradient:

$$\frac{dT}{dz} \approx \frac{T_f - T_b}{\delta}, \tag{11.16}$$

where δ is boundary-layer thickness. Inserting this expression into equation 11.15 and rearranging, we arrive at an estimate for H_{cv} that explicitly incorporates boundary-layer thickness:

$$H_{cv} = \frac{\mathcal{K}}{\delta} A_{cv} \left(T_f - T_b \right). \tag{11.17}$$

In a few circumstances we could estimate δ using our knowledge of boundary-layer physics (Chapters 6 and 7), but it is convenient to take a more roundabout approach. The *Nusselt number*, Nu, is defined as the ratio of an object's characteristic length, ℓ_c, to the thickness of its boundary layer:

$$Nu = \frac{\ell_c}{\delta}. \tag{11.18}$$

As a ratio of lengths, Nu is dimensionless. A multitude of experiments over the years have shown that Nu can be modeled as a power function of Reynolds number:

$$Nu = k_1 Re^{k_2}, \tag{11.19}$$

where coefficient k_1 and exponent k_2 depend on an object's shape. Engineers have measured k_1 and k_2 for a wide variety of shapes (Table 11.3).

The Nusselt-Reynolds number relationship is useful to us because it allows us to calculate h_c as a function of object size, flow speed, and the properties of the fluid. Comparing equation 11.17 to equation 11.13, we can conclude that

$$\delta = \frac{\mathcal{K}}{h_c}. \tag{11.20}$$

Inserting this expression for δ into the definition of Nu,

$$Nu = \frac{h_c \ell_c}{\mathcal{K}}. \tag{11.21}$$

Table 11.3 Nussel-Reynolds Number Relationships for Selected Shapes.

Shape	Notes	k_1	k_2
Sphere	Re = 25–10,000	0.37	0.6
Perpendicular to Flow			
Cylinder	Re = 0.4–4	0.886	0.330
	Re = 4–40	0.817	0.385
	Re = 40–4000	0.612	0.466
	Re = 4,000–40,000	0.173	0.618
	Re = 40,000–400,000	0.024	0.805
Vertical plate		0.204	0.731
Parallel to Flow			
Flat plate	Laminar flow	0.593	0.5
Leaf ($W > L$)		1.06	0.5
Leaf ($W < L$)L		1.67	0.5
Cone	$H/D = 0.75$	0.187	0.605
(on substratum)	$H/D = 0.63$	0.208	0.592
	$H/D = 0.50$	0.113	0.657
	$H/D = 0.38$	0.127	0.634
	$H/D = 0.15$	0.098	0.639

Notes: Unless otherwise noted, shapes are in free-stream flow. Data from Gates (1980), Incropera and DeWitt (2002), and Harley et al. (2009). H = height, W = width, L = length, and D = diameter.

Substituting this version of Nu into equation 11.19, we arrive at a practical recipe for estimating h_c:

$$\frac{h_c \ell_c}{\mathcal{K}} = k_1 \mathrm{Re}^{k_2},$$

$$h_c = \frac{\mathcal{K} k_1 \mathrm{Re}^{k_2}}{\ell_c}. \tag{11.22}$$

We can make this formula more explicit by expanding the components of the Reynolds number:

$$h_c = \mathcal{K} k_1 \left(\frac{u}{\nu}\right)^{k_2} \ell_c^{k_2-1}. \tag{11.23}$$

Here u is mainstream fluid velocity and ν is kinematic viscosity. Note that both $\mathcal{K}$ and ν vary with the temperature of the fluid (Table 11.1). In short, given k_1 and k_2, equation 11.23 is a convenient means to estimate h_c for an object of a particular size in a given fluid. In the rest of this chapter, we will make use of this formula as it applies to a sphere, in which case ℓ_c is the sphere's diameter, $k_1 = 0.37$, and $k_2 = 0.6$.

The values in Table 11.3 can be used as rough estimates of h_c for some plants and animals: a cone as a guess for a limpet, for instance; a sphere as a guess for a snail;

Figure 11.4 Littorine snails minimize conductive contact with the rock (photo by Luke Miller).

a cylinder or plate for a cactus. However, if accurate values are required, it is best to measure h_c directly (see Supplement 11.3).

11 A CONVECTION-DOMINATED SYSTEM
Littorine Snails

Having explored the mechanics of solar heating and convective heat transfer, let's put them to use. Littorine snails (periwinkles) are characteristic inhabitants of rocky intertidal shores in many parts of the world (Figure 11.4). Often they are the highest snails on the shore, where their ability to contend with the extreme rigors of their near-terrestrial habitat gives them refuge from marine competitors lower down. When the tide is out, periwinkles glue their shells to the rock and retreat inside, closing the aperture behind them with an impermeable operculum. This strategy conserves water but precludes evaporative cooling. To account for this, some species have a behavioral strategy to keep themselves cool. Before gluing their shell to the rock, they stand it on end so that its spire points away from the substratum (Figure 11.4), a behavior that has several advantages: the area of contact between shell and rock is reduced—minimizing conductive heat input from the rock—and the upright posture facilitates convective cooling by extending the shell higher into the aerial boundary layer, where wind is faster and temperatures are lower (Marshall and Chua 2012). If all else fails, the snails have high thermal tolerances. In California, the common periwinkle (*Littorina keenae*) has a thermal limit of 48°C (118°F; Somero 2002) and the knobbly periwinkle (*Echinolittorina malaccana*), found on the shores of Borneo, can survive body temperatures up to 56.5°C (134°F; Marshall et al. 2010).

Is this combination of behavioral strategies and physiological tolerance effective? In a worst-case scenario, will littorine snails overheat? To answer these questions, we formulate a simple heat-budget model.

Let's assume that for these small snails, body temperature quickly reaches a thermal equilibrium with its environment, a state in which heat energy influx equals heat energy outflow and H_{net} therefore is zero. (In Chapter 12 we'll calculate the time required to come to equilibrium; it's indeed short.) Because conductive contact between shell and rock is negligible, we assume that $H_{cd} = 0$. Similarly, because the snail has sealed its body behind the operculum, $H_{ec} = 0$. The metabolic rate of invertebrates in general is low, and we can guess that this applies doubly to a sealed up littorine that has minimal access to the oxygen necessary to sustain aerobic metabolism. So, let's assume that $H_{met} = 0$. Long-wave radiative heat transfer is a bit trickier, but for the sake of simplicity, let's assume that the snail absorbs as much as it radiates, such that $H_{lw} = 0$.

Having thus assumed away most mechanisms of heat transfer, we are left with a pleasingly simple heat budget:

$$0 = H_{sw} + H_{cv}$$
$$= \alpha_{sw}(A_p \mathcal{I}_{sw,dir} + A_{sky} \mathcal{I}_{sw,dif}) + h_c A_{cv} (T_{air} - T_b). \tag{11.24}$$

We can solve this equation for the difference between body and air temperature:

$$T_b - T_{air} = \frac{\alpha_{sw}(A_p \mathcal{I}_{sw,dir} + A_{sky} \mathcal{I}_{sw,dif})}{h_c A_{cv}}. \tag{11.25}$$

Body temperature is higher than air temperature by an amount that increases with increasing absorptivity, area exposed to sunlight, and solar irradiance. Conversely, the difference between T_b and T_{air} decreases with increasing heat-transfer coefficient and convective area.

Our objective here is to estimate the outcome of a worst-case scenario, and to that end we make several assumptions. Solar irradiance is maximal on clear days, in which case $\mathcal{I}_{sw,dif}$ is negligible. Thus

$$T_b - T_{air} \approx \frac{\alpha_{sw} A_p \mathcal{I}_{sw,dir}}{h_c A_{cv}}. \tag{11.26}$$

In a worst-case scenario, the snail would be totally black ($\alpha_{sw} = 1$) and the sun would be maximally bright ($\mathcal{I}_{sw,dir} = 1000 \text{ W} \cdot \text{m}^{-2}$). Let's assume that the snail's shell is roughly spherical, with radius r. In that case, projected area is

$$A_{pr} = \pi r^2 \tag{11.27}$$

and convective area is the entire surface area of the sphere:

$$A_{cv} = 4\pi r^2. \tag{11.28}$$

Thus, for a spherical snail

$$T_b - T_{air} \approx \frac{250}{h_c} \tag{11.29}$$

Using equation 11.23 for the heat-transfer coefficient (with the appropriate k_1 and k_2 values for a sphere), we find, for instance, that when $T_{air} = 30°C$,

$$T_b - T_{air} = 44.1 \frac{r^{0.4}}{u^{0.6}} \tag{11.30}$$

The slower the wind and the larger the snail, the hotter it's body is relative to the air around it.

Now there are only a few details left to add to the heat budget. Littorine snails are typically small, less than 0.8 cm long, corresponding roughly to a radius of 0.4 cm. Vogel (2009) notes that even on the calmest of days, u cannot be expected to drop below $0.1 \text{ m} \cdot \text{s}^{-1}$ due to the breeze caused by free convection. Using these values, we find that the worst-case scenario for a totally black littorine snail is $19.3°C$ above ambient air temperature. The maximum T_{air} recorded in the intertidal zone in Monterey is $32°C$, which equates to a body temperature of $51.3°C$, well above the snail's lethal limit ($48°C$). Thus, despite their behavioral repertoire and extraordinary thermal tolerance, California littorines are potentially subject to thermal mortality. Similar peril exists for knobbly periwinkles in Brunei where Marshall and colleagues (2010) measured air temperatures of $40°C$ in the vicinity of their snails. In this case, $T_b - T_{air}$ is $19.4°C$, so body temperature would be $59.4°C$, again above the snails' lethal limit ($56.5°C$).

However, equations 11.25 and 11.30 offer several possibilities for how snails could avoid these dire circumstances. First, they could limit their size. If, for instance, snails were half the size we have assumed ($r = 0.2$ cm), their temperature would be only $14.4°C$ above T_{air}, low enough that they would be immune to overheating in both California and Brunei. Alternatively, the snail could produce a shell with a lower absorptivity. Reduction from $\alpha_{sw} = 1$ (which we have assumed so far) to $\alpha_{sw} = 0.7$ (a typical value for a gray snail) would reduce body temperature of a 0.8-cm-long snail from $19.3°C$ to $13.5°C$ above T_{air}, again allowing both Califiornia and Brunei snails to survive.

The snail's third option is to live where it is windy. Even a mild breeze of $1 \text{ m} \cdot \text{s}^{-1}$ (2.2 mi/h) would lower body temperature of a 0.8-cm-long snail from $19.3°C$ to only $4.8°C$ above T_{air}. In this case, air temperature would have to exceed $43.2°C$ ($108°F$) to cook snails in California or $51.7°C$ ($125°F$) in Brunei, neither of which seems likely.

Our simple heat-budget model thus suggests that if high intertidal snails are sufficiently small, have a relatively low absorptivity, and live where it's is even slightly windy, they can save themselves from fatal thermal stress.

These conclusions were verified by McQuaid and Scherman (1988), who experimented with two subspecies of littorine snails in South Africa. *Littorina africana africana* (the lighter colored of the two) maintained slightly cooler body temperatures than did the darker *Littorina africana knysnaensis*. Both morphs had exceptional tolerance for high body temperatures, surviving 2-h exposures to temperatures in excess of $47°C$, but on an unusually hot day, both died when their shells were painted black.

The discussion so far has focused on snails at low tide, in which case body temperature is elevated relative to air. What about snails in water? Comparing equation 11.23 for air and water, we find that the heat-transfer coefficient for a sphere in water is 97-fold higher than that in air. As a consequence, the difference between T_b and water temperature, T_w, is 97-fold smaller than the difference between T_b and T_{air}:

$$T_b - T_w \approx 6.8 \times 10^{-5} \frac{\alpha_{sw} \mathcal{I}_{sw} r^{0.4}}{u^{0.6}}. \tag{11.31}$$

For a black spherical snail and bright, noontime sun, this boils down to

$$T_b - T_w \approx 6.8 \times 10^{-2} \frac{r^{0.4}}{u^{0.6}}. \tag{11.32}$$

A black snail 0.8 cm in diameter in flow of $0.1 \text{ m} \cdot \text{s}^{-1}$ would have a body temperature only about $0.03°C$ above that of the water around it. Even a massive black snail with a diameter of a meter would be only $0.2°C$ above water temperature

This result can be readily generalized. Barring unusual circumstances—very bright light, very low flow, or a very large organism—sunlight cannot heat aquatic plants or animals much above the temperature of the water around them.

12   METABOLISM

When sunlight is absorbed by plants, photosynthesis converts some of its electromagnetic energy into chemical energy and stores that energy in glucose molecules. This stored energy can subsequently be used in metabolism, powering the innumerable chemical reactions of life. In the process, metabolism releases some of the stored solar energy as heat. The rate of metabolic heating varies with the mass of the organism and differs among species, and these factors are traditionally modeled as a power function:

$$H_{met} = Mm^{k_M}. \tag{11.33}$$

where M is the species-specific metabolic heat coefficient ($\text{W} \cdot \text{kg}^{-k_M}$), m is the mass of the organism in question, and k_M is an exponent chosen such that the model matches the data.[5]

It has long been recognized that for most species k_M is less than 1. That is, a gram of tissue from a large organism produces heat at a lower rate than a gram from a small organism (Schmidt-Nielsen 1984). The exact value of k_M has been the matter of heated debate. Measured across animal species, but within a functionally similar taxonomic group (e.g., rodents, ungulates), $k_M \approx \frac{3}{4}$. Within a species, k_M is often closer to $\frac{2}{3}$. (See Hoppeler and Weibel 2005 for an entrée into this extensive literature.)

For our purposes, the precise value of k_M matters less than the magnitude of M, which tailors equation 11.33 to the metabolism of a particular species. A typical resting mammal, for instance, has a metabolic heat coefficient of about $3.4 \text{ W} \cdot \text{kg}^{-k_M}$ (Schmidt-Nielsen 1984). Some birds have an even higher M, about $6.2 \text{ W} \cdot \text{kg}^{-k_M}$. By contrast, invertebrates have metabolic heat coefficients a thousand times smaller ($M = 0.006 \text{ W} \cdot \text{kg}^{-k_M}$). So, while a 1-kg rabbit at rest produces 3.4 W of heat energy, and a 1-kg sea gull perhaps 6.2 W, a 1-kg sea star produces only 0.006 W. To my knowledge, the metabolic heat coefficients of plants have not been measured, but it is safe to assume that they are generally as low as those of invertebrates.

To assess the effect of metabolic heat on body temperature, let's return to our investigation of spherical organisms in thermal equilibrium with convective flow. If metabolism is the only source of heat to an organism and convection the only way heat can escape, $H_{met} = H_{cv}$, and we can calculate the difference between body temperature and the surrounding fluid:

$$T_b - T_f \approx \frac{Mm^{k_M}}{A_{cv}h_c}. \tag{11.34}$$

Let's examine the message of this equation for both aquatic and terrestrial organisms.

---

[5] The strange units of M are a result of the power-law form of equation 11.33.

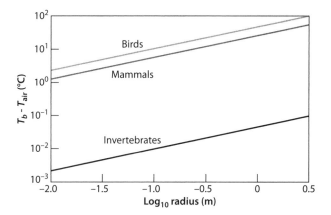

Figure 11.5 Results of our simple metabolic heat-budget model ($k_M = 0.75$).

First, animals in water: for a spherical organism, m is a function of radius and body density ($\frac{4}{3}\pi r^3 \rho_b$) and A_{cv} is a function of radius ($4\pi r^2$). Substituting equation 11.23 for h_c, we can calculate body temperature in water at 10°C:

$$T_b - T_w \approx \frac{M \left(\frac{4}{3}\pi\rho_b\right)^{k_M} r^{(3k_M - 1.6)}}{7000 u^{0.6}}. \tag{11.35}$$

T_b increases with resting metabolic rate and an organism's size, but even for a spherical whale ($r = 3$ m), body temperature would be only about 1°C above water temperature. In order to keep warm, whales, porpoises, seals, and sealions need an insulating layer of blubber, which acts to decrease h_c.

The situation is different in air. Again using equation 11.23 for h_c, we arrive at an equation for the body temperature of a spherical organism in air at 20°C:

$$T_b - T_a \approx \frac{M \left(\frac{4}{3}\pi\rho_b\right)^{k_M} r^{(3k_M - 1.6)}}{71.8 u^{0.6}}. \tag{11.36}$$

This expression is graphed in Figure 11.5 for typical parameter values ($k_M = 0.75$, $u = 1.0$ m · s^{-1}, $\rho_b = 1080$ kg · m^{-3}) and a range of metabolic coefficients. It exposes the fundamental thermal challenges facing terrestrial organisms.

First, unless metabolic rate is high and the organism is large, body temperature in the absence of solar heating approaches that of the surrounding air. This poses no problem for plants and most invertebrates, which function well at ambient temperatures. However, for some animals, bees for instance, convective cooling poses a problem.

To provide enough power for flight, bees' muscles must have a temperature of at least 30°C. On cold days, this requires them to raise their temperature well above that of the surrounding air, a feat they can accomplish only by increasing their metabolic rate or decreasing h_c. They do both (Heinrich 1993). In preparation for flight, bees shiver, using their flight muscles not to beat the wings, but rather to produce heat, which eventually gets them up to temperature. When the flight muscles are active, a bee can have a metabolic rate coefficient twice that of a flying hummingbird. To aid the heating process, bees have evolved a furry coat for their thorax, insulation that effectively reduces their heat-transfer coefficient. The system is so effective at producing and retaining heat that on hot days bees have the potential to overheat.

On those occasions, they use their circulatory system to transfer heat from the furry thorax to the bare abdomen where the blood can be cooled by convection.

Some plants manage similar heroic feats. Voodoo lilies (*Amorphophallus titanum*) and *Philodendron* flowers have tissues that are capable of extraordinary metabolic rates, allowing them to heat their reproductive structures, which then exude an odor of urine, dung, or rotting meat to attract pollinating insects. Skunk cabbage (*Symplocarpus foetidus*) is the king of hot plants. Using starch reserves in its roots as fuel, it heats the growing shoot, allowing it to melt its way through snow to be the first plant to blossom in spring. Its flowers are hot as well, with an odor true to their species' name. Lotus flowers heat themselves, perhaps to provide insects with a thermal reward by reducing the cost of maintaining an elevated temperature in their flight muscles (Seymour and Shultze-Motel 1996).

Small birds and mammals face the same problem as bees—to function, they need to keep their body temperature at 35°C to 40°C—and their solution is the same. They maintain a high metabolic rate and cloak themselves in insulating fur or feathers. To save energy, hummingbirds allow their body temperature to drop at night, and, like bees, they must heat themselves up in the morning before flying.

For large terrestrial animals, metabolic heat has its dark side (Figure 11.5). Even at rest, large mammals can be dangerously hotter than the surrounding air, calling for a reduction in insulation (in elephants and rhinoceroses, for instance). Small mammals and birds can run into the same problem when they are active and their metabolic rate increases by a factor of five to even ten. In these instances, shedding heat by convection alone cannot offset the heat produced during locomotion. Which leads us to our next heat-transfer mechanism.

13   EVAPORATION AND CONDENSATION

As water evaporates, sensible heat is converted to latent heat of water vapor. Conversely, latent heat is converted back to sensible heat as water condenses. In either case, the rate of heat transfer can be described through a simple expression

$$H_{ec} = \Lambda_{wat} \frac{dm}{dt}. \tag{11.37}$$

Here Λ_{wat} is the latent heat of evaporation (measured in $J \cdot kg^{-1}$), and dm/dt is the rate at which water mass is lost or gained ($kg \cdot s^{-1}$). At 20°C, 2.45 *million* joules of heat energy are transferred for each kilogram of water evaporated or condensed. Because Λ_{wat} is so spectacularly large, H_{ec} can have substantial biological importance. To take an extreme example, an average human male has a mass of approximately 70 kg, composed of flesh with a specific heat similar to that of water, $4200 \ J \cdot kg^{-1} \cdot K^{-1}$. If no other heat transfer were allowed, evaporation of only 4.4 L of water (a little over a gallon, and only about 6% of body mass) would suffice to lower his body temperature from 37°C to 0°C.

Just as the complexity of convective heat transfer is hidden in h_c, much of the complexity of evaporative heat transfer is contained in the the superficially simple term dm/dt. Indeed, we can model the rate of evaporative water loss through an expression analogous to Newton's law for convective heat transfer:

$$\frac{dm}{dt} = h_m A_{evap}(\mathcal{C}_{v,b} - \mathcal{C}_{v,air}). \tag{11.38}$$

Here h_m is the *mass-transfer coefficient* $(\text{kg} \cdot \text{m}^{-2} \cdot \text{s}^{-1})$ and A_{evap} is the area of contact between a liquid surface and the air. $C_{v,b}$ is the *mole fraction* of water vapor in air at the liquid's surface and $C_{v,\text{air}}$ is the mole fraction of water vapor in air well away from the surface.[6] (Mole fraction is the the fraction of gas molecules, that are water. For instance, if for every 100 air molecules 5 are water vapor, $C = 0.05$.)

The difference in mole fraction between $C_{v,b}$ and $C_{v,\text{air}}$ is the driving "force" behind evaporation and condensation. If $C_{v,b}$ is higher than $C_{v,\text{air}}$, random motion results in a tendency for more water molecules to move away from the liquid surface than move toward it, and there is a net flux of water molecules into the air—evaporation. Conversely, if $C_{v,\text{air}}$ is higher than $C_{v,b}$, there is a net flux of molecules into the liquid—condensation.

Directly at the air-liquid interface, air is assumed to be saturated with water vapor. That is, air at the surface holds water vapor at a sufficient mole fraction to put it at equilibrium with the liquid. This saturating mole fraction increases rapidly with temperature (Figure 11.6A): a liter of saturated air contains almost five and a half times as many water molecules at $40°C$ as it does at $10°C$. As a consequence, the rate of evaporation is very sensitive to temperature.

At any temperature, the presence of salt in water decreases the saturating mole fraction. This effect is small when salts are dilute, as they are in well-hydrated plants and animals, but can become substantial when salts are concentrated, as in some dehydrated seaweeds.

The mole fraction of water vapor in air far away from an organism also varies. On a humid day, $C_{v,\text{air}}$ can be high, and evaporation is slow. If air is dry, $C_{v,\text{air}}$ is low, and evaporation is rapid.

After water has evaporated, subsequent convective transport of vapor away from the object's surface is analogous to the convective transport of heat. By the same token, its is convection that brings vapor to the object's surface where it can condense. In this respect, transport of water vapor and convective transport of heat are analogous, and the mass-transfer coefficient h_m is proportional to h_c with a proportionality constant that falls out of the math (see Supplement 11.5):

$$h_m \approx \frac{h_c}{1502} \ \text{kg} \cdot \text{J}^{-1}. \tag{11.39}$$

14 SYSTEMS DOMINATED BY CONVECTION AND EVAPORATION

Desert Herbs and Seaweeds

Whenever a terrestrial organism opens itself to gas exchange, evaporation is an unavoidable side effect: animals need to breathe in oxygen but inevitably breathe out water vapor. Similarly, plants need to absorb carbon dioxide for photosynthesis, but for each CO_2 molecule taken in, they lose 300 to 1300 water molecules (Nobel 1991).

---

[6] Equation 11.38 differs from that in most texts in that it uses mole fractions rather than vapor densities to quantify air's water content. As a result, the magnitude and units of h_m differ from those often cited. See Supplement 11.4 for an explanation.

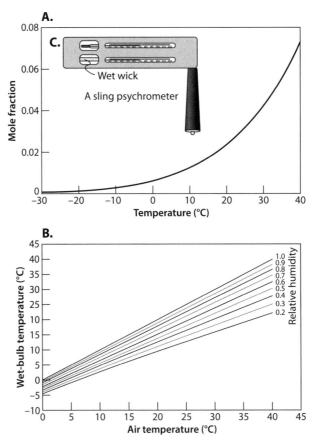

Figure 11.6 A. The saturating mole fraction of water vapor in air increases with increasing temperature. **B.** Wet-bulb depression can be used to measure relative humidity. **C.** A sling psychrometer.

Apart from the important effects evaporation has on organisms' water balance, it can have effects on the organism's temperature that cascade into the plants or animal's ecology.

Consider, for instance, *Datura wrightii*, a perennial herb common in the desert Southwest of North America. As is common in terrestrial plants, *Datura*'s leaves are coated with a waxy cuticle so that evaporative loss is confined to the stomata, pores that can be opened when CO_2 is needed and closed when water is scarce. *Datura* has a deep taproot, which provides it access to water even in the desert, giving it the luxury of keeping its stomata open in the heat of the day. The resulting evaporation (which is substantial due to the desert air's low humidity) cools the leaves 5° to 7° below ambient temperature (Potter et al. 2009). This seeming advantage has a downside, however. Because the eggs of the sphinx moth (*Manduca sexta*) are intolerant of high temperatures, the moths cope by laying their eggs on the underside of cool *Datura* leaves. As the eggs mature into larvae, they eat the leaves. The plant is thus faced with an evolutionary conundrum: open stomata allow for free gas exchange, which keeps leaves cool and facilitates photosynthesis, maximizing growth rate. But cool leaves ultimately increase the intensity of herbivory. Potter and colleagues suggest that the benefit of limiting evaporation—and thereby killing its herbivore's eggs—might outweigh the costs of reduced gas exchange.

Similar trade-offs are possible for the many organisms that have evolved means to control their rate of evaporation. Insects and crustacea, for example, have exoskeletons that reduce the contact area between moist body and dry air, and mammals can control when to sweat. By contrast, evaporation is an unavoidable result of exposure to terrestrial conditions for amphibians and many intertidal organisms, such as anemones, sponges, and all seaweeds. Their inability to control evaporation potentially puts these organisms at risk and provides a simple system in which to employ heat-budget models.

As an instructive example, let's use the sort of turfy seaweeds common in the mid to high intertidal zone. These algae have minimal contact with the rock, so their conductive heat transfer is negligible, and their metabolic rate is presumably too low to matter. For simplicity, let's restrict ourselves to seaweeds in the shade, so that we can ignore the effects of solar heating and infrared radiative heat exchange. As we have done in previous examples, we assume that the seaweed is in thermal equilibrium with its environment. This leaves us with just convection and evaporation in the heat budget:

$$H_{cv} + H_{ec} = 0. \tag{11.40}$$

Unfortunately, what appears to be a simple equation has a serious hidden complication. Put simply, evaporation from the organism affects the temperature, water content, and flow of air in the boundary layer. As a consequence, the two factors in this simple heat budget interact, precluding a simple solution.

As a way around this problem we resort to theory. Through a consideration of the idealized dynamics of evaporating surfaces, it can be shown (e.g., Monteith and Unsworth 2008) that there is a minimum temperature to which a moist surface can fall by evaporating water, a temperature central to the current discussion because it is likely to be a good approximation of the temperature of a shaded amphibean or seaweed. For reasons that will become apparent in a moment, this minimum temperature is known as the *wet-bulb temperature*, T_{wb}.

Whereas equation 11.40 attempts to calculate body temperature by equating rates of evaporative energy loss and convective energy gain, calculation of wet-bulb temperature proceeds by equating the total energy lost from a moist object with the total change in energy of the adjacent air:

$$C_{p,air} (T_{air} - T_{wb}) = \Lambda_{wat} \left[\mathcal{C}_s (T_{wb}) - h_r \mathcal{C}_s (T_{air}) \right]. \tag{11.41}$$

Here, $C_{p,air}$ is the specific heat of air (1006 J·kg^{-1}·K^{-1}), and the left side of the equation is thus a measure of the energy (per mass of air) that would be required to cool air to the wet-bulb temperature. Λ_{wat} is again the latent heat of evaporation (J·kg^{-1}), and $\mathcal{C}_s(T_{wb})$ and $\mathcal{C}_s(T_{air})$ are, respectively, the saturating mole fraction of water vapor in air at the wet-bulb temperature and saturating mole fraction at ambient air temperature. That leaves one term that requires explanation: *relative humidity*, h_r.

There are many ways to quantify the water-vapor content of the atmosphere, but the most familiar metric is h_r, the amount of water vapor in the air relative to the maximum (saturating) amount that air can hold. If a kilogram of air contains 10 g of water vapor but could maximally hold 20 g, the relative humidity is 0.5. Often h_r is multiplied by 100 and expressed as a percent, but, to avoid misunderstanding, we will always express h_r as a simple fraction.

The product of h_r and $C_s(T_{air})$ is the actual mole fraction of water vapor in ambient air, and the difference between $C_s(T_{wb})$ and $h_r C_s(T_{air})$ is the driving 'force' for evaporation. When multiplied by Λ_{wat}, $C_s(T_{wb}) - h_r C_s(T_{air})$ tells us how much sensible heat energy (per mass of air) is converted to latent heat as a moist surface cools. When the moist surface is at T_{wb}, the energy converted to latent heat by evaporation is equal to the energy available to cool the air, and the two sides of equation 11.41 are equal. Thus, equation 11.41 provides a means to calculate T_{wb}. Unfortunately, the solution of equation 11.41 is not just a matter of algebra. Because T_{wb} appears explicitly on the left side of the equation as well as implicitly on the right, it is necessary to solve the equation through a process of trial and error (Supplement 11.6). The results are shown in Figure 11.6B.

It is clear from the form of equation 11.41 that to maintain equality between the two sides of the equation, any change in relative humidity necessitates a change in wet-bulb temperature. Conversely, if you can measure T_{wb} and T_{air}, you can use equation 11.41 to calculate h_r. This is the basis for a device known as a sling psychrometer (Figure 11.6C), two thermometers attached by a swivel to a handle so that they can be rapidly swung through the air. The bulb of one thermometer (the dry bulb) is bare, whereas the bulb of the other (the wet bulb [aha!]) is covered by a damp wick. After the psychrometer has been swung briskly for a minute or so, the temperature of both thermometers is quickly read, the dry bulb providing a measure of T_{air} and the wet bulb a measure of T_{wb}. Figure 11.6B can then be used to estimate h_r.

Now back to seaweeds: moist seaweeds act like a wet-bulb thermometer; in a breeze their body temperature approaches T_{wb}. If we know h_r and T_{air}, we can use equation 11.41 to solve for (or Figure 11.6B to estimate) T_{wb} as a prediction of T_b.

Wet-bulb temperature can be substantially lower than air temperature. For example, on a typical day in a temperate intertidal zone ($h_r = 0.5$), T_{wb} (14°C) is 6°C below T_{air}. Thus, the unavoidable desiccation to which seaweeds are subjected can have the salubrious side affect of keeping them cool.

We humans are well aware of this phenomenon. We are designed to function with a core body temperature of 37°C. A few degrees lower and our muscles seize up, a few degrees higher and we die. There is no particular problem maintaining core temperature when surrounded by cool air; our bodies are capable of producing sufficient metabolic heat to offset the rate of convective heat loss. Unfortunately, we have no way of turning off the metabolic furnace when the air is hot. If air temperature exceeds roughly 35°C (normal skin temperature), the human body cannot lose enough heat by convection alone to keep from overheating. Instead, we rely on the evaporation of sweat to shed the roughly 100 W of heat our bodies produce at rest. As long as the wet-bulb temperature of the atmosphere is less than 35°C, evaporative cooling has the potential to keep us alive. This is how humans manage to live in hellaciously hot desert climates. Deserts are dry, and the low relative humidity results in a wet-bulb temperature sufficiently low to allow human life.

At least for now: Sherwood and Huber (2010) suggest that one of the most troubling effects of global climate change may be an upward shift in wet-bulb temperatures that could render huge areas of the globe uninhabitable for humans. In their calculations it isn't so much a rise in air temperature that creates the problem (although that alone is worrisome), it is an even greater rise in aerial water-vapor content that limits the potential for evaporative cooling. They predict that if global average air temperature rises by 7°C, small areas in the tropics will be physiologically off limits to humans, and if air temperature increases by $11°C$ to $12°C$, most of the areas where humans now live will be uninhabitable for at least part of the year.

But that's the future; intertidal seaweeds are already faced with the life-and-death consequences of evaporation. For them, survival and growth depend on the interplay between body temperature and photosynthesis. Like all organisms, seaweeds die if they overheat. They can avoid overheating by evaporatively cooling, but the accompanying desiccation reduces their growth rate—when dry, they can't photosynthesize. This trade-off interacts with the plant's morphology to determine where on the shore they can survive.

Mastocarpus papillatus is one of the most abundant intertidal algae on the Pacific shores of North America, ranging from Baja California to Alaska. Typically found in the mid- to upper intertidal zone, it is renowned for its morphological variability. The more-or-less planar fronds can be thin or thick, simple in outline or tortuous (*dissected* is the technical term).

Both morphologies are cooled by evaporation. Given a stiff breeze (5 to $7 \text{ m} \cdot \text{s}^{-1}$), they are $5°C$ to $6°C$ below ambient air temperature, near T_{wb}. But beyond that similarity, the different morphologies interact with the environment in different ways. Thin/dissected blades dry quickly, limiting the time available for photosynthesis. By contrast, thick/simple blades stay moist longer—allowing them to photosynthesize longer and grow faster. Thus, under benign conditions, thick/simple blades have a competitive advantage over their thin/dissected relatives. But thick/simple blades have a low heat-transfer coefficient—during long exposures to high air temperatures they dry out, overheat, and die. By contrast, due to their shape, thin/dissected blades have a high convective heat-transfer coefficient, and (like small littorine snails) they never overheat. Thus there are predictable trade-offs between growth and survival.

Based on measurements in a laboratory wind tunnel, Bell (1995) quantified the various components of energy transfer in *Mastocarpus* and used them to construct heat-budget models for both thin/dissected blades and thick/simple blades. Environmental data from local weather stations, wave buoys, and tidal stations then allowed her to calculate how frond temperature and water content varied through time for each blade type. Her results show that up to a height of 1.0 m above mean lower low water, thick blades survive and grow faster than thin ones. But a scant 0.25 m higher on the rock, the evaporative cooling capacity of thick blades cannot sustain them through the increased exposure to hot air and they die, ceding space to thin-bladed plants. In this fashion, the physics of heat transfer leads to a spatial gradient in seaweed morphology.

These results highlight the advantage of using heat-budget models as a means of exploration. Even short-term direct measurements of *Mastocarpus*'s frond temperatures and water content are difficult, and long-term continuous measurements in the field are well nigh impossible. As a result, models are currently the only practical way to estimate the likelihood of lethal thermal stress and the cumulative effects of temperature and desiccation on this seaweed's growth and survival.

15   CONCEPTS AND INTERIM CONCLUSIONS

In this chapter, we reviewed the physics of heat and temperature and have seen how that knowledge can be used to develop the heat budget, a means of defining, parameterizing, and quantifying the pathways by which heat enters and leaves organisms. By measuring or estimating individual terms in the heat budget, we can compile a detailed description of the thermal environment, which we can then use to predict body temperature. A heat-budget model thus provides a mechanistic link between the physical environment and body temperature, a value central to our understanding of the physiology, behavior, and ecology of plants and animals.

We can draw several general conclusions from our exploration so far:

- *In water, body temperature equals water temperature.* The only notable exceptions here are large animals with high metabolic rates.
- *Body temperature seldom equals air temperature.* Time and again in our simple scenarios—snails on sun-baked rocks and seaweeds in a breeze—body temperature is either substantially above or slightly below air temperature. Given that Q_{10} effects can amplify the physiological effects of any shift in body temperature, these deviations from air temperature can have substantial biological importance.
- *Body temperature depends on organism size.* This fact is apparent in several of our scenarios. For example, in situations where convection matters, size enters the equation through the heat-transfer coefficient. The larger the organism, the smaller h_c, and the more body temperature can deviate from air temperature. For littorine snails heated by the sun, this could set a limit to maximum size—large snails can overheat in an environment where small snails do not.
- *Body temperature depends on wind speed.* When air moves slowly, the thick boundary layer around an organism acts as an insulating blanket, allowing body temperature to deviate from air temperature. The faster the wind, the thinner the blanket and the more closely T_b matches T_{air}.

Taken together, these conclusions allow us to make two sweeping generalizations. First, the efficient transfer of heat by convection in water removes temperature as a factor differentiating co-occurring aquatic species. There are, of course, exceptions: a few top predators such as tunas, dolphins, and whales have elevated body temperatures that allow them to increase their rate of prey capture. But otherwise, in a given habitat, all aquatic organisms operate on a level thermal playing field. Not so for terrestrial organisms. Due to the relatively inefficient convective transport of heat in air, different species can have different body temperatures even in the same environment, and these differences have the potential to affect a wide variety of ecological and evolutionary interactions (Harley 2013; Kaspari et al. 2014).

Chapter 12

Thermal Mechanics II
Stored Energy, Conduction, Long-Wave Radiation, and Synthesis

In the last chapter, we began our exploration of the heat-budget equation by examining solar heating, convective and evaporative heat transfer, and metabolic heat production, four factors that commonly control body temperature. However, the scenarios we used to illustrate these mechanisms were chosen to be as simple as possible, requiring us to make several assumptions: that organisms are at thermal equilibrium with their environment, that they don't gain or lose heat through conduction with rocks and soil, and that sunlight is the only source of radiative energy. There are many circumstances in which these assumptions are violated, and to complete our understanding of the heat budget, we need to take these circumstance into account. In this chapter, we will see that the size of an organism affects how rapidly its body temperature responds to changes in heat flow—for large plants and animals, shifts in body temperature lag behind shifts in the environment. Similarly, close contact between an organism and the substratum can induce lags in body temperature as waves of heat are conducted through the rock or soil. And we will see how the exchange of infrared radiation can cause organisms to freeze even when air temperature is relatively warm. We finish with a brief investigation of the concept of the environmental niche and show that it can be used to predict how species ranges will adjust to our changing climate.

1   STORED ENERGY

We begin with a discussion of H_{net}, the difference between the rate at which heat enters an organism and the rate at which it leaves. Recall from Chapter 11 that changes in heat energy and temperature are tied to each other:

$$\Delta W = mC_p \Delta T_b. \tag{12.1}$$

Any change in the energy content of a body (ΔW) is accompanied by a change in body temperature (ΔT_b), with the proportionality between the two set by the product of mass and specific heat, mC_p.

If the change in temperature occurs over time Δt, the average rate of change of energy in this period is

$$\frac{\Delta W}{\Delta t} = mC_p \frac{\Delta T_b}{\Delta t}. \tag{12.2}$$

Taken to the limit as Δt approaches 0, we arrive at an expression for H_{net}, the instantaneous rate at which heat energy is stored or lost:

$$\frac{dW}{dt} = H_{net} = mC_p \frac{dT_b}{dt}. \tag{12.3}$$

Newton's second law of motion (Chapter 2) provides a useful analogy for understanding the role of H_{net} in the overall heat budget. Recall that acceleration—du/dt, the rate of change of speed—is equal to the ratio of force to mass:

$$\frac{du}{dt} = \frac{F}{m}. \tag{12.4}$$

Force provides the impetus for a change in speed, and it is resisted by an object's inertia. Now consider the definition of H_{net} (equation 12.3). Rearranging, we see that the rate of change of body temperature is equal to the ratio of H_{net} to mC_p:

$$\frac{dT_b}{dt} = \frac{H_{net}}{mC_p}. \tag{12.5}$$

Thus, by analogy, the net rate at which an object gains or loses heat (H_{net}) provides the impetus for a change in temperature, but it is resisted by the object's *thermal inertia*, mC_p, also known as *thermal mass*. It is the ratio of H_{net} to thermal inertia, rather than either factor individually, that sets the rate of heating or cooling.

The exceptionally high specific heat of water comes into play here. For a given mass, an organism made primarily from water (as most organisms are) will change its temperature more slowly than it would if made from almost any other substance.

In our previous heat-budget models, we steadfastly assumed that the organism in question was at thermal equilibrium with its environment. We are now in a position to justify or refute this assumption by explicitly calculating the rate of temperature change in an organism. Littorine snails again provide an appropriate example.

We begin by assuming that for these animals conductive and evaporative heat transfer are negligible, as is metabolic heat production, and that the snail has reached a temperature at which heat influx from sunlight is just offset by convective heat efflux to the breeze. We then interrupt this equilibrium by allowing a dense cloud to cover the sun, cutting off H_{sw}. How long will it take for the snail to come to equilibrium with this new thermal environment?

In this scenario, long-wave heat transfer is negligible (we'll deal with the specifics of long-wave heat transfer later in the chapter), and the snail's heat budget boils down to just two terms, convective heat transfer and net heat loss:

$$H_{cv} = H_{net}. \tag{12.6}$$

Inserting the details for these two forms of heat exchange (Newton's law of cooling. Equations 11.13 and 12.3), we obtain the starting point for our calculation:

$$h_c A_{cv} [T_{air} - T_b] = mC_p \frac{dT_b}{dt}. \tag{12.7}$$

If we assume that air temperature is constant, we can solve this differential equation (Supplement 12.1):

$$T_b - T_{air} = [T_b(0) - T_{air}] \, e^{-[h_c A_{cv}/(mC_p)]t} \tag{12.8}$$

where $T_b(0)$ is body temperature at the beginning of the experiment. T_b approaches air temperature exponentially through time at a rate set by the factor $h_c A_{cv}/(mC_p)$.

According to this relationship, it would take infinite time for T_b to reach T_{air}, making it impractical to calculate the exact time required to come to thermal equilibrium. But we can circumvent this problem by borrowing the concept of *response time* from physics and engineering. If $T_{b,0} - T_{air}$ is the total temperature range the organism must traverse to reach equilibrium, thermal response time is the time it takes for T_b to move a fraction $1 - 1/e$ (approximately 63%) of this range. This happens when

$$\frac{h_c A_{cv}}{mC_p} t = 1. \tag{12.9}$$

Thus, the thermal response time of the snail (in seconds) is

$$t_r = \frac{mC_p}{h_c A_{cv}}. \tag{12.10}$$

The larger the thermal inertia of the snail (mC_p), the longer its response time. The larger the heat transfer coefficient or convective area, the shorter the response time.

For a spherical snail, $m = \frac{4}{3}\pi r^3 \rho_b$, where r is the sphere's radius and ρ_b is the snail's density (1080 kg $\cdot$ m^{-3}). We assume that the snail's specific heat is that of water, 4200 J $\cdot$ kg^{-1} $\cdot$ K^{-1}. Convective area is $4\pi r^2$, and we use our previous approximation for h_c (equation 11.23). The response time (in seconds) of a spherical snail in air at 20°C is thus

$$t_r = 2.67 \times 10^5 \frac{r^{1.4}}{u^{0.6}}. \tag{12.11}$$

For a large littorine ($r = 0.4$ cm) in nearly still air ($u = 0.1$ m $\cdot$ s^{-1}), the response time is 467 s (7.8 min). If the wind blows at 1 m $\cdot$ s^{-1}, t_r is reduced to 117 s. In either case, T_b adjusts rapidly to a new thermal environment, as we have assumed. Thermal response time increases with increasing size, however. A sphere with a radius of 5 cm (the size of a sea urchin, for instance) has a response time of 67 min in a wind of 1 m $\cdot$ s^{-1} and more than 4 h if $u = 0.1$ m $\cdot$ s^{-1}.

The substantial response times of large intertidal organisms led Sylvan Pincebourde and his colleagues (Pincebourde et al. 2008) to investigate the thermal biology of the sea star, *Pisaster ochraceus*. *Pisaster* is the keystone predator we briefly encountered in Chapters 1 and 11, and, for an intertidal organism, it is relatively large. The rate at which *Pisaster* eats mussels depends on temperature, with the surprisingly high Q_{10} of 4.8 (Sanford 2002); two *Pisaster*—one at air temperature and one just 2°C warmer—differ in their foraging rate by 37%. Furthermore, *Pisaster* has a relatively low tolerance for elevated body temperature; if T_b exceeds 35°C, it dies (Pincebourde et al. 2009). This thermal limit constrains when and where the sea star can hunt for prey. If the animal ventures out of water on an abnormally hot day, or if it stays out too long on an average day, it risks overheating.

However, the large size and basic morphology of sea stars predisposes them to a novel behavioral adapation. Much of the space inside a sea star's body is empty of innards, and the animal can choose to fill this space with either air or water. Pincebourde et al. (2009) found that *Pisaster* takes advantage of this choice to adjust its thermal inertia. If an individual encounters elevated body temperature on one low tide, it responds by taking on extra water for its next emersion. The increase in thermal inertia substantially increases thermal response time, reducing the risk of overheating and potentially allowing the sea star to forage longer.[1]

For most organisms, this behavioral control of thermal response time would be an interesting footnote, but given its status as a keystone species, the ability of *Pisaster* to regulate thermal inertia takes on added ecological importance. One of the current open questions in ecology concerns the ability of plants and animals to adjust to increased average air temperature. If species must rely on the slow process of evolution to adapt to climate change, they may not be able to respond quickly enough. In this context, the ability of *Pisaster* to behaviorally adjust on a tide-by-tide basis may have an important buffering effect on intertidal community ecology.

2   CONDUCTIVE HEAT TRANSPORT

Conduction is the transport of heat by diffusion. Although conduction occurs in both solids and fluids, its effect in fluids is typically overshadowed by bulk movement of the liquid or gas, leading to convective transport of heat. For present purposes, when we talk about conduction, we focus on the diffusive transport of heat in solids.

Conduction is governed by Fourier's law (equation 11.9):

$$\mathcal{F}_{H,z} = -\mathcal{K}\frac{dT}{dz}, \tag{12.12}$$

where $\mathcal{F}_z$ is heat flux (W · m^{-2}) and $\mathcal{K}$ is thermal conductivity. This relationship can be reworked to solve for H_{cd}, the rate of heat transfer (in watts):

$$H_{cd} = \mathcal{K}A_{cd}\frac{dT}{dz}. \tag{12.13}$$

Here A_{cd} is conductive area, the area over which a plant or animal is in contact with the substratum, and the negative sign has been removed in accordance with our convention that H_{cd} is positive when heat moves into an organism.

As an example of how to apply Fourier's law in a biological context, we consider plants and animals that maintain an intimate contact with their rocky substratum, such as lichens and crustose algae, limpets, acorn barnacles, and chitons. Each is firmly attached to the rock, and heat can be conductively transported between rock and organism across this contact area. For lichens and crustose algae, the body is so thin that it is safe to assume that the entire organism is at the same temperature as the rock's surface. Limpets, barnacles, and chitons are thicker, but haemolymph circulates through their bodies, and we assume that the mixing of heat by circulation

---

[1] It is unclear why *Pisaster* doesn't keep itself full of water all the time.

Table 12.1 Thermal Properties of Rocks and Other Materials.

	H_2O Content Fraction of pores	K $W \cdot m^{-1} \cdot K^{-1}$	ρ $kg \cdot m^{-3}$	C_p $J \cdot kg^{-1} \cdot K^{-1}$	$\mathcal{D}_H$ $m^2 \cdot s^{-1}$
Rock					
Basalt		2.07	2900	860	0.83×10^{-6}
Diabase		1.80	2800	796	0.81×10^{-6}
Dolomite		4.60	2850	900	1.79×10^{-6}
Granite		2.51	2650	886	1.07×10^{-6}
Limestone		2.05	2500	780	1.05×10^{-6}
Marble		2.80	2550	883	1.24×10^{-6}
Quartzite		5.15	2700	872	2.19×10^{-6}
Sandstone		3.14	2500	775	1.62×10^{-6}
Average		3.02	2681	844	1.33×10^{-6}
Soil					
Sandy	0.0	0.30	1600	800	0.24×10^{-6}
(40%) pores	0.2	1.80	1800	1180	0.85×10^{-6}
	0.4	2.20	2000	1480	0.74×10^{-6}
Clay	0.0	0.25	1600	890	0.18×10^{-6}
(40% pores)	0.2	1.18	1800	1250	0.53×10^{-6}
	0.4	1.58	2000	1550	0.51×10^{-6}
Peat	0.0	0.06	260	1920	0.10×10^{-6}
(80% pores)	0.4	0.29	660	3300	0.13×10^{-6}
	0.8	0.50	1060	3650	0.12×10^{-6}
Ice (0°C)		2.22	920	2100	1.15×10^{-6}
Yellow pine		0.15	640	2805	0.08×10^{-6}

Notes: Values for conductivity, density, and specific heat are typical values, from which the corresponding thermal diffusivities have been calculated. Beware, however, that in nature there is substantial variation around each of the typical values.

is sufficiently rapid that body temperature is more or less the same throughout and again equal to rock surface temperature. Thus, in both cases, diffusion of heat through the rock determines the rate of conductive heat transport to or from the organism:

$$H_{cd} = K_r A_{cd} \frac{d T_r}{dz}. \tag{12.14}$$

Here, $d T_r / dz$ is the gradient of temperature in the rock along an axis perpendicular to the interface between rock and organism. If $d T_r / dz$ is positive—that is, if temperature increases with distance into the rock—heat flows out of the rock and into the organism. Here K_r is rock's thermal conductivity. If the need arises it is a relatively simple matter to measure the thermal conductivity of a solid (see Supplement 12.2), but engineers have made these measurements for many materials, and representative values for typical substrata are listed in Table 12.1.

In contrast to $\mathcal{K}_r$, which is constant for a material, the thermal gradient is a dynamic property of the system, and quantifying it is correspondingly complex. As heat diffuses into or out of a barnacle, for instance, the flow of heat changes the thermal gradient in the rock beneath. The shift in thermal gradient in turn affects the rate at which heat diffuses. Because of this interaction between rate and gradient, a conductive system can evolve through time even when external factors are constant. Furthermore, if external factors aren't constant, the rate of conductive heat transport at any particular time can depend on rates in the past, that is, on the *history* of transport. In short, the temperature gradient cannot simply be looked up, it must either be measured directly or calculated on the fly as the system evolves.

Direct measurement is seldom worth the effort. If we have to measure the temperature gradient to predict body temperature, it is likely simpler just to measure body temperature directly. Thus, calculation is usually the way to go. Unfortunately, the math involved—a numerical solution of the differential equation describing the diffusion of heat—is beyond the scope of this text (see Supplement 12.2). Instead, we trust that the calculation can be made, providing us with the appropriate value of dT_r/dz anytime we want it and allowing us to explore the consequences of conductive heat transfer.

Experiments have shown that the body temperature of a limpet, barnacle, or crustose seaweed is indeed very nearly equal to that of the rock (e.g., Wethey 2002; Denny and Harley 2006). With this fact in mind, consider the following hypothetical, but highly relevant, situation. A massive rock (in the jargon of such things, a semi-infinite solid) is allowed to come to thermal equilibrium with seawater at a temperature T_w such that rock temperature (both at the surface and internally) is equal to T_w. We now drain away the water to simulate low tide, turn on the sun so that heat is delivered to the rock's projected area A_{pr} at rate H_{sw} (Figure 12.1A), and follow rock surface temperature through the course of 6 h until the tide comes back in.

This scenario is simple enough that an exact answer is available from theory. Carslaw and Jaeger (1959) calculate that, under these circumstances, temperature at the rock surface ($T_{r,s}$) varies with time t:

$$T_{r,s}(t) = T_{r,s}(0) + \frac{2H_{sw}}{\sqrt{\pi \rho_r C_p \mathcal{K}_r}} \sqrt{t}. \qquad (12.15)$$

Here $T_{r,s}(0)$ is surface temperature at the start of the experiment, in our case the temperature of seawater; ρ_r, C_p, and $\mathcal{K}_r$ are the rock's density, specific heat, and thermal conductivity, respectively. H_{sw}, heat flux into the surface, depends on α_{sw} (the rock's shortwave absorptivity); solar irradiance, $\mathcal{I}_{sw}$; and θ, the angle between sunlight and the plane of the rock's surface:

$$H_{sw} = \alpha_{sw} \mathcal{I}_{sw} \sin \theta. \qquad (12.16)$$

The larger the flux of heat into the rock (the numerator of the fraction in equation 12.15), the faster surface temperature rises. But the denser the rock, the higher its specific heat, or the greater its propensity for conducting heat (in the denominator), the slower surface temperature rises. Finally, surface temperature rises in proportion to the square root of time, a proportionality that should remind you of our discussion of diffusive processes in boundary layers (Chapter 7). The similarity is no coincidence. Because heat diffuses through rock, its conductive transport is directly analogous to that in which mass and momentum diffuse through a boundary layer. Putting all this

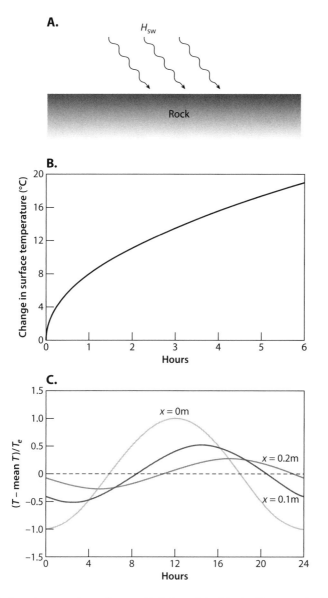

Figure 12.1 Surface temperature of a semi-infinite solid subjected to a constant rate of heating. **A.** The experimental setup. **B.** Results for a typical rock substratum. **C.** The temperature of sandy soil oscillates through the day. The greater the depth, the lower the amplitude of oscillation and the greater the phase shift.

together, we can conclude that surface temperature rises rapidly at the beginning of the experiment and more slowly as time goes on.

We are now in a position to calculate values relevant to limpets, barnacles, and crustose algae. For a typical rock, ρ_r, C_p, and $\mathcal{K}_r$ are 2681 kg·m³, 844 J·kg⁻¹·K⁻¹, and 3 W·m⁻¹·K⁻¹, respectively. The granite of my local shore has an α_{sw} of 0.6, and θ is typically 60°. Given this scenario, rock-surface temperature rises by 7.8°C in 1 h and 19.1°C in 6 h (Figure 12.1B). For many temperate intertidal zones, sea-surface temperatures range from 10°C to 20°C, in which case the scenario we have just painted would, over the course of a 6-h low tide, impose on limpets, barnacles, and seaweeds body temperatures of 29.1°C to 39.1°C. This range kills some species and not others, leading us to suppose that rock temperature can have a substantial ecological influence.

David Wethey used this approach to explore the distribution of two barnacle species on the rocky shores of New England. In the absence of temperature effects, *Semibalanus balanoides* and *Chthamalus fissus* have the same upper limit on the shore (set by the need to be submerged for sufficient time to feed), and the larvae of both species settle at all heights. But *Semibalanus*, the larger of the two, is a better competitor for space on the rock—where they co-occur, *Semibalanus* can squeeze out the smaller *Chthamalus*. By contrast, *Chthamalus* is more tolerant of high temperatures—if the substratum temperature exceeds 26°C, *Chthamalus* survives better than *Semibalanus*, and above 42°C, only *Chthamalus* survives.

These factors set the stage for an inadvertent human experiment (Wethey 2002). To facilitate boat traffic, engineers in the early 1900s dredged a short canal across Cape Cod and spanned the canal with a bridge at each end. Weather at the two bridges is essentially identical (they are only 5 km apart), and barnacle larvae freely traverse the canal, ensuring that the supply of new recruits is the same for the concrete footings of each bridge. One might expect populations on the two bridges to be similar.

However, water temperature at the two bridges is substantially different. Bathed at high tide by the cool waters of Massachusetts Bay, footings of the northern bridge, stay cool at low tide due to the thermal inertia of the substratum. *Semibalanus* takes advantage of the chill and excludes *Chthamalus*. Footings of the southern bridge start each low tide at the warmer temperature of Buzzards Bay. As a result, upper levels of the footings exceed *Semibalanus*'s thermal limits, and *Chthamalus* survives in this thermal refuge.

For many terrestrial organisms, it isn't temperature at the substratum's surface that is important, it is the temperature below that matters—in a prairie dog's burrow or around a tree's roots, for instance. Again, the theory of heat transport provides useful information (Carslaw and Jaeger 1959). In the alternation between day and night, the substratum's surface temperature varies approximately sinusoidally around a mean temperature $\overline{T}$ with an amplitude of T_e:

$$T(0, t) = \overline{T} + T_e \cos\left(\frac{2\pi}{t_p} t\right). \tag{12.17}$$

Here t_p is the period of temperature oscillation, in this example the length of a day, 8.64×10^4 s. At a depth z below the surface, both the amplitude and timing of temperature variation change according to the relationship

$$T(z, t) = \overline{T} + T_e \exp\left(-\sqrt{\frac{\rho_r C_p \pi}{t_p \mathcal{K}_r}} z\right) \cos\left(\frac{2\pi}{t_p} t - \sqrt{\frac{\rho_r C_p \pi}{t_p \mathcal{K}_r}} z\right). \tag{12.18}$$

This relationship is graphed in Figure 12.1C. The amplitude of thermal variation decreases exponentially with depth, and the timing of peak temperature shifts to later in the day.

The depth at which thermal variation is 37% ($1/e$) of its surface amplitude is the *damping depth*, z_d:

$$z_d = \sqrt{\frac{t_p \mathcal{K}_r}{\rho_r C_p \pi}}. \tag{12.19}$$

For rock, which has a high diffusivity (Table 12.1), z_d is 19 cm. The same degree of damping occurs at 15 cm in moist sandy soil and 6 cm in peat. Animals need to get deeper into rock to escape the heat than they do into soil.

Temperature at depth is out of phase with that at the surface by $\sqrt{\frac{\rho_r C_p \pi}{t_p \mathcal{K}_r}} z$ radians, but this shift is unlikely to have important effects. Fluctuations aren't 180° out of phase with the surface until a depth of $\sqrt{\frac{\pi \mathcal{K}_r t_p}{\rho_r C_p}}$, and at this depth the amplitide of variation is only 4% of that at the surface.

Equations 12.17–12.19 tell an important ecological story. By slithering into a crevice in the rock or burrowing into dirt or sand, animals can avoid the drastic thermal variation that effect life at the surface (e.g., Huey et al. 1989). Similarly, as they extend into the ground, plants' roots escape the potentially stressful thermal variation the rest of the plant encounters at the surface. In short, the slow rate of thermal diffusion in soil provides animals and plants access to habitats that would otherwise be uninhabitable.

The same relationships can be applied to seasonal—as opposed to daily—variation in temperature. In this case T_e is the amplitude of seasonal variation; t_p is 3.154×10^7 s, the length of a year; and damping depths are 2 to 3 m. This explains how permafrost can persist in polar habitats even though the soil's surface can get quite hot in summer.

3   LONG-WAVE RADIATIVE HEAT TRANSFER

Just as heat can be gained by absorbing light, it can be lost by its emission. Any object at absolute temperature T_b has the potential to lose heat at rate $\mathcal{R}_{lw,max}$:

$$\mathcal{R}_{lw,max} = \sigma_{SB} T_b^4, \tag{12.20}$$

a relationship known as the Stefan-Boltzmann equation.[2] The ratio of $\mathcal{R}_{lw,max}$ ($W \cdot m^{-2}$) to T_b^4 is set by σ_{SB}, the Stefan-Boltzmann constant ($5.67 \times 10^{-8} \, W \cdot m^{-2} \cdot K^{-4}$).

In practice, however, objects emit only a fraction (ϵ_{lw}, the long-wave emissivity) of their maximum radiance. Thus, the realized rate of light emission is

$$\mathcal{R}_{lw} = \epsilon_{lw} \sigma_{SB} T_b^4 \tag{12.21}$$

$\mathcal{R}_{lw}$ is the infrared *radiance* of the object, the power it emits from each square meter of surface. Polished metals have an ϵ_{lw} near zero, so they emit much less infrared light than equation 12.20 would suggest. However, most objects found in nature readily emit long-wave radiation and have emissivities near 0.95.

Because $\mathcal{R}_{lw}$ varies with the fourth power of an object's surface temperature, a small increase in temperature can result in a large increase in the rate of heat loss.

---

[2] It was derived independently in the late 1800s by Joseph Stefan (1835–93) and Ludwig Boltzmann (1844–1906).

It is very important to note that T_b in equation 12.21 is measured in kelvins (*not* degrees Celsius).

Multiplying both sides of equation 12.21 by A_r, the area from which light is radiated, we arrive at $H_{\text{lw,out}}$, the overall rate at which a body emits long-wave radiation:

$$H_{\text{lw,out}} = \epsilon_{\text{lw}} \sigma_{SB} A_r T_b^4 \tag{12.22}$$

Any portion of the object's surface that can "see" the surroundings contributes to A_r. Thus, in most cases, A_r is approximately equal to the surface area of the object not in direct contact with the substratum.

Just as objects can lose heat by emission of long-wave radiation, they can gain heat by absorbing the infrared light emitted by the environment. The intensity of light emitted by an object's surroundings also obeys the Stefan-Boltzmann law. For the moment, let's assume that everything in an object's environment is at temperature T_e. In that case, the effective infrared irradiance impinging on the object is

$$\mathcal{I}_{\text{lw}} = \epsilon_e \sigma_{SB} T_e^4. \tag{12.23}$$

where ϵ_e is the effective emissivity of the surroundings, a value we will return to later in this section. By analogy to the absorption of sunlight (equation 11.12), we can calculate the rate at which infrared light is absorbed:

$$\begin{aligned} H_{\text{lw,in}} &= \alpha_{\text{lw}} A_r \mathcal{I}_{\text{lw}} \\ &= \alpha_{\text{lw}} A_r \left(\epsilon_e \sigma_{SB} T_e^4 \right) \end{aligned} \tag{12.24}$$

where α_{lw} is the object's long-wave (infrared) absorptivity. Subtracting heat lost (equation 12.22) from heat gained (equation 12.24) gives us an expression for the net heat transferred by infrared radiation:

$$H_{\text{lw}} = \alpha_{\text{lw}} A_r \left(\epsilon_e \sigma_{SB} T_e^4 \right) - \epsilon_{\text{lw}} \sigma_{SB} A_r T_b^4. \tag{12.25}$$

Under certain conditions, the long-wave absorptivity of an object is equal to its long-wave emissivity:

$$\alpha_{\text{lw}} = \epsilon_{\text{lw}} \tag{12.26}$$

a relationship known as Kirchhoff's law.[3] While it does not apply to all objects, Kirchhoff's law is a very good approximation for plants, animals, and their surroundings. This equality allows us to simplify equation 12.25 by substituting ϵ_{lw} for α_{lw}:

$$\begin{aligned} H_{\text{lw}} &= \epsilon_{\text{lw}} A_r \left(\epsilon_e \sigma_{SB} T_e^4 \right) - \epsilon_{\text{lw}} \sigma_{SB} A_r T_b^4 \\ &= \epsilon_{\text{lw}} \sigma_{SB} A_r \left(\epsilon_e T_e^4 - T_b^4 \right). \end{aligned} \tag{12.27}$$

This, then, is the net heat exchanged by long-wave radiation between an object and its environment.

---

[3] One should be exceedingly cautious when applying Kirchhoff's law at visible wavelengths. The equalilty between α and ϵ at long (i.e., infrared) wavelengths is due to the fact that at these wavelengths, neither factor varies much with a change in wavelength. In the terms of the trade, such objects act as *grey bodies* at long wavelengths. In contrast, for most objects α and ϵ both vary substantially as a function of shorter wavelengths, and need not be equal. See Incropera and DeWitt (2002) for a discussion.

Eventually we will want to incorporate H_{lw} into a calculation of body temperature, in which case it will be inconvenient for T_b to be raised to the fourth power. As a practical matter, we use an approximation of equation 12.27—a linearization—in which T_b is raised only to the first power:

$$H_{\text{lw}} \approx \epsilon_{\text{lw}} \sigma_{SB} A_r \left[T_e^4 (\epsilon_e - 1) + 4 T_e^3 (T_e - T_b) \right]. \tag{12.28}$$

A derivation of this expression is given in Supplement 12.3.

To this point, we have assumed that everything in an object's surroundings is at T_e, but that seems more than a little unrealistic. A full accounting of variable T_e is beyond the scope of this chapter (see Incropera and DeWitt 2002), but a simple approximation applies reasonably well in many situations. Often the objects surrounding a plant or animal are at very nearly the same temperature as the organism itself and have very nearly the same emissivity. When this is the case, the net radiative heat exchange between an organism and this fraction of its environment is negligible. Instead, the bulk of the infrared heat exchange occurs between the object and the sky, which can have a different emissivity than the organism and can be at a much different temperature. Thus, if we can quantify the infrared exchange between organism and sky, we can in many cases account for most of the heat transferred by long-wave radiation.

Looking from the ground up, what infrared light does an organism see? In other words, what is the effective long-wave irradiance of the sky? Recall that there is very little long-wave radiation in sunlight (Figure 11.2), so looking up from the ground with infrared eyes, it isn't the sun an organism sees. Instead, it sees two sources of diffuse infrared radiation. First, there is the atmospheric backscattering of light emitted by Earth's surface—the warmer the surface, the more infrared light it emits and the more light scattered back. Second, there is long-wave radiation from the warm, near-surface air itself. Both factors depend on the ability of the atmosphere to absorb and emit infrared radiation, that is, on the absorptivity (emissivity) of the air. To infrared eyes, the sky would appear as a diffuse, uniform source of long-wavelength light with irradiance $\mathcal{I}_{\text{lw,sky}}$.

The nitrogen and oxygen that make up most of our atmosphere have an ϵ_{lw} close to zero, and therefore contribute little to the sky's irradiance. Instead it is the greenhouse gases—primarily water vapor and carbon dioxide—that have appreciable ϵ_{lw} and are responsible for scattering/emitting long-wave radiation toward objects at Earth's surface. On the scale of minutes to days, the concentration of atmospheric CO_2 varies relatively little, and we can treat it as a constant. By contrast, the water-vapor content of the air varies drastically, both spatially and temporally, and we need to take this variation into account.

As a practical matter, we describe $\mathcal{I}_{\text{lw,sky}}$ using an analog of the Stefan-Boltzmann equation:

$$\mathcal{I}_{\text{lw,sky}} = \epsilon_{\text{air}} \sigma_{SB} T_{\text{air}}^4, \tag{12.29}$$

where T_{air} (in kelvins) is air temperature measured a meter or two above an organism. In this expression, all the complexities of the situation have been shoe-horned into the term ϵ_{air}, air's *effective emissivity*. A variety of expressions have been formulated for ϵ_{air} (see Campbell and Norman 1998), but for our purposes, the following expression works well (Figure 12.2; Supplement 12.4):

$$\epsilon_{\text{air}} \approx 7.3 \times 10^{-7} T_{\text{air}}^{2.46} h_r^{0.143}. \tag{12.30}$$

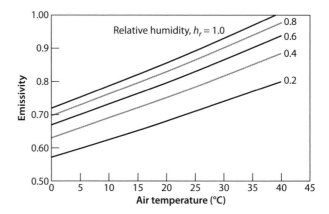

Figure 12.2 Effective emissivity of the sky.

The accuracy of this expression is neither better nor worse than the alternatives, but it has the advantage of containing explicit terms for the factors known to affect long-wave irradiance. The presence of T_{air} tells us that the temperature of near-surface air contributes to the amount of light available to be scattered/emitted by the atmosphere. Similarly, the presence of h_r tells us that water-vapor content is important.

The next step in quantifying net long-wave heat exchange is to estimate the effective fraction of the organism's radiative area that exchanges infrared light with the sky. This fraction—γ, the *view factor*—can be estimated using a camera lens with a 180° field of view. If the camera is substituted for the organism in the environment, it sees everything that the organism's surface would see, and it is then an easy matter to determine what fraction of the visible environment is sky.[4] This fraction is a practical estimate of γ. Substituting ϵ_{air} for ϵ_e in equation 12.27 and incorporating γ, we obtain a reasonable approximation for the net transfer of infrared radiation:

$$H_{lw} \approx \epsilon_{lw} \sigma_{SB} A_r \gamma \left(\epsilon_{air} T_{air}^4 - T_b^4 \right) . \tag{12.31}$$

Again, it is convenient to trade this exact expression for a linear approximation:

$$H_{lw} = \underbrace{\left[\epsilon_{lw} \sigma_{SB} A_r \gamma T_{air}^4 \left(\epsilon_{air} - 1 \right) \right]}_{1} + \underbrace{\left[4 \epsilon_{lw} \sigma_{SB} A_r \gamma T_{air}^3 \left(T_{air} - T_b \right) \right]}_{2} . \tag{12.32}$$

Note that with the exception of T_b all variables and constants on the right side of the equation are already known or easily measurable.

It is worth taking a moment to explore the implications of equation 12.32. Because $\epsilon_{air} \leq 1$, term 1 is negative or zero. The sign of term 2 depends on whether T_b is greater than or less than T_{air}. When terms 1 and 2 are summed, the net result can thus be either a gain or loss of heat, and it is not always intuitive which happens when. For example, one might suppose that if body and air temperature are the same, there

---

[4] To measure γ exactly, one would need to take into account not only the fraction of the "visual field" of the organism subtended by each small solid angle of sky, but also the angle at which light from that area intersects the organism's surface. For a discussion of the complications involved in measuring view factors, consult Campbell and Norman (1998) and Incropera and DeWitt (2002).

should be no net exchange of infrared energy between object and sky. This would be true if ϵ_{air} were 1, but at typical air temperatures near 20°C, ϵ_{air} is 0.7 to 0.8. In this case, even if $T_{air} = T_b$ (so that term 2 is zero), term 1 is negative, and the object loses heat. Conversely, the higher the air temperature, the more closely ϵ_{air} approaches 1 (Figure 12.2) and the more closely term 1 approaches 0. As a result, if T_{air} is high (30°C or above) and a few degrees above T_b, the object can gain heat from the sky.

To give some tangibility to the effects of long-wave heat exchange, we explore a pertinent example. In Chapter 11 we calculated the maximum temperature of littorine snails when heat influx from the sun was offset only by heat lost to convection:

$$H_{sw} + H_{cv} = 0. \tag{12.33}$$

This is a reasonable approximation of daytime conditions, but what happens at night? The simplest scenario in this case would be to switch off the sun, so that $H_{sw} = 0$. Working through the algebra, we would conclude that $T_a = T_b$. In other words, in the absence of heat input from the sun, our simple heat budget predicts that convective heat transfer forces the snail's body temperature to be the same as the temperature of the surrounding air.

This simple conclusion is misleading, however. Recall that in formulating the heat budget for a snail, we assumed that long-wave heat exchange was negligible compared to solar heat influx. At night, however, when solar heating doesn't enter the equation, the relative importance of long-wave heat exchange is amplified, and we need to take it into account. In essence, we can reformulate equation 12.33, substituting H_{lw} for H_{sw}:

$$H_{lw} + H_{cv} = 0. \tag{12.34}$$

Inserting our expressions for H_{lw} and H_{cv},

$$\left[\epsilon_{lw}\sigma_{SB} A_r \gamma T_{air}^4 (\epsilon_{air} - 1)\right] + \left[4\epsilon_{lw}\sigma_{SB} A_r \gamma T_{air}^3 (T_{air} - T_b)\right] + h_c A_{cv} (T_{air} - T_b) = 0, \tag{12.35}$$

and solving for $T_b - T_{air}$, we conclude that[5]

$$T_b - T_{air} = \frac{\epsilon_{lw}\sigma_{SB} A_r \gamma T_{air}^4 (\epsilon_{air} - 1)}{4\epsilon_{lw}\sigma_{SB} A_r \gamma T_{air}^3 + h_c A_{cv}}. \tag{12.36}$$

This equation is not as forbidding as it might seem. All values on the right side are already known or can be measured. The long-wave emissivity of a snail is approximately 0.95 and the Stefan-Boltzmann constant is $5.67 \times 10^{-8} \ \mathrm{W \cdot m^{-2} \cdot K^{-4}}$. If we again model the snail as a sphere, $A_r = A_{cv} = 4\pi r^2$, and the heat-transfer coefficient for our spherical snail is the same at night as it is in the day (equation 11.23). The effective emissivity of air, ϵ_{air}, varies with air temperature and relative humidity (equation 12.30), each of which we can measure. Thus, the only term remaining to be specified is the view factor, γ. Because only the upper half of our spherical snail's surface area has an effective view of the sky, we set $\gamma = 0.5$.

---

[5] Note that the arithmetic here is simple only because T_b is raised to the first power. If we were to use the exact expression for H_{lw} (equation 12.31) rather than its linearized approximation (equation 12.32), we would be faced with an equation in which T_b was raised to the fourth power, a considerable complication.

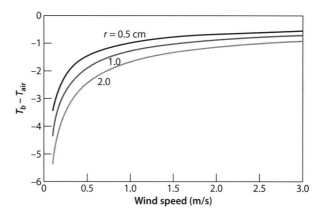

Figure 12.3 Results of our model for snails' nighttime temperature.

We are now in a position to insert specific values. Let's assume a nighttime air temperature of 5°C with a relative humidity of 0.5. In that case, $\epsilon_{air} = 0.68$ and

$$T_b - T_{air} = \frac{-51.6}{2.31 + 5.67 \left(\frac{u^{0.6}}{r^{0.4}} \right)}. \tag{12.37}$$

This expression is graphed in Figure 12.3 as a function of wind speed and for a variety of snail radii.

Three messages emerge. First, if wind speed is slow—minimizing convective heat transfer—body temperature can be 4°C to 6°C below air temperature, flirting with freezing. Increasing wind speed increases convective exchange of heat, and T_b more closely matches T_{air}. This makes it advantageous for littorine snails to glue their shells in an upright position, even on cold nights. By extending into the boundary layer, they encounter higher wind speeds, mitigating their heat loss. Second, the larger the snail, the lower its heat-transfer coefficient, and the more effective long-wave heat exchange is in reducing body temperature. And third, in none of these scenarios is the difference between body temperature and air temperature near the magnitude it reached when the sun is out: the snail is maximally 4°C to 6°C below air temperature at night compared to nearly 20°C above T_{air} in extreme cases during the day.

The relatively small effect of infrared radiative heat loss can still be important, however; heat lost by long-wave emission can cause plants and animals to freeze even when $T_{air} > 0°C$. You likely have some personal experience with this effect. After a night when air temperature has dipped—but not to freezing—you may nonetheless have had to scrape frost off your car's windshield.

The lower the relative humidity, the smaller ϵ_{lw} is (Figure 12.2) and the greater is the divergence between body temperature and air temperature. This is one reason why on clear nights desert plants and animals often have to contend with low body temperatures. By contrast, on foggy nights relative humidity is very nearly 1, and the resulting boost in the effective emissivity of the sky precludes any substantial reduction in T_b.

Citrus growers take advantage of a similar effect. On nights when near-freezing temperatures are predicted, fruit may freeze due to infrared heat loss. As a precaution, farmers burn old car tires in their orchards. It isn't the heat of the fire that matters;

it's the smoke produced. The high long-wave emissivity of soot particles has the same effect as high humidity.

Perhaps surprisingly, radiative heat transfer plays a key role in forest fires. Fires ignited by lightning are often the dominant mode of disturbance in temperate forests, and the size, abundance, and turnover rate of burnt areas is an important factor in determining species diversity. The extent to which a fire will spread—that is, the size of a burnt patch—is governed in part by wind but also by the radiative transfer of heat. Infrared radiation from a burning tree can ignite a neighbor, but recent research shows that the spread of fire by this mechanism depends the morphology of the tree being ignited (Cohen and Finney 2010). When a part of the tree (a leaf, needle, or branch) is heated by a radiant source, the adjacent air is heated by conduction and becomes less dense. The difference in density leads to a buoyancy-driven convective current that transfers heat away from the needle, leaf, or twig.

The efficiency of this transfer depends on boundary-layer thickness. The needles of conifers, for instance, have small diameters, with a correspondingly thin boundary layer and effective heat transfer. As a consequence, it is nearly impossible to ignite needles by radiative heat transfer alone. This new information about the mechanism of ignition has had a substantial impact on foresters' ability to predict the rate at which fires spread and the ultimate size of burnt patches (Finney 1999).

4   THE HEAT BUDGET REVISITED

The examples discussed in this and the preceding chapter demonstrate how the physics of heat transfer can be used to predict body temperature in some simple situations. Before we move on, it will be instructive to outline how these concepts can be combined to predict body temperature in more complex (that is, more realistic) scenarios. To that end, we return to the overall heat budget:

$$H_{sw} + H_{lw} + H_{cd} + H_{cv} + H_{ec} + H_{met} = H_{net}, \qquad (12.38)$$

an expression that embodies the spirit of heat transfer but not enough of the mechanism to allow for calculation. To proceed, we need to fill in the details.

Looking back over our discussion, we can restate equation 12.38 for organisms in air:

$$k_1 + k_2 + k_3 \left(T_{air} - T_b \right) + k_4 + k_5 \left(T_{air} - T_b \right) + k_6 + k_7 = k_8 \frac{dT_b}{dt}, \qquad (12.39)$$

where

$$k_1 = \alpha_{sw} \left(A_p \mathcal{I}_{sw,dir} + A_{sky} \mathcal{I}_{sw,dif} \right) \text{ (solar heat flux)}, \qquad (12.40)$$

$$k_2 = \epsilon_{lw} \sigma_{SB} A_r \gamma T_{air}^4 (\epsilon_{air} - 1) \text{ (a component of IR heat flux)}, \qquad (12.41)$$

$$k_3 = 4\epsilon_{lw} \sigma_{SB} A_r \gamma T_{air}^3 \text{ (a component of IR heat flux)}, \qquad (12.42)$$

$$k_4 = K_r A_{cd} \frac{dT_r}{dx} \text{ (conductive heat flux)}, \qquad (12.43)$$

$$k_5 = h_c A_{cv} \text{ (a component of convective heat flux)}, \qquad (12.44)$$

$$k_6 = \Lambda_{wat} A_{evap} h_m (C_{v,b} - C_{v,air}) \text{ (evaporative heat flux)}, \qquad (12.45)$$

$$k_7 = Mm^{k_M} \text{ (metabolic heat)}, \qquad (12.46)$$

$$k_8 = mC_p \text{ (thermal inertia)}. \qquad (12.47)$$

The details in coefficients k_1 through k_8 may appear daunting en masse, but taken individually they are quite tractable. In particular, note that every term in these coefficients is either known or readily measured.

Dividing both sides of equation 12.39 by k_8, we arrive at a general expression for how body temperature varies through time:

$$\frac{dT_b}{dt} = \frac{k_1 + k_2 + k_3\,(T_{air} - T_b) + k_4 + k_5\,(T_{air} - T_b) + k_6 + k_7}{k_8}. \tag{12.48}$$

This differential equation can be solved in two ways. First, as we have noted, there are many situations in which organisms are effectively at equilibrium with their thermal environment: their body temperature is such that the rate of heat loss is the same as that of heat gain. In this case, $dT_b/dt = 0$ and

$$0 = k_1 + k_2 + k_3\,(T_{air} - T_b) + k_4 + k_5\,(T_{air} - T_b) + k_6 + k_7. \tag{12.49}$$

When this is the case, we can solve for body temperature explicitly:

$$T_b = \frac{k_1 + k_2 + k_3\,T_{air} + k_4 + k_5\,T_{air} + k_6 + k_7}{k_3 + k_5}. \tag{12.50}$$

In short, when an organism is at equilibrium with its environment, we can measure T_{air} and coefficients k_1 to k_7, and use them to calculate body temperature directly.

When the environment changes too fast for the organism to keep up, we need to take dT_b/dt into account. That is, we need to integrate equation 12.48 through time. In practice, this is accomplished using a computer and one of several ingenious algorithms (such as a fourth-order Runge-Kutta; see Supplement 12.5), but the gist of the procedure is easily outlined.

First, we approximate the infinitesimal changes in temperature and time with finite values: ΔT_b is the measurable change in body temperature occurring over Δt, a practical increment in time (usually a few few minutes). Thus, we can approximate equation 12.48 as

$$\frac{\Delta T_b}{\Delta t} \approx \frac{k_1 + k_2 + k_3\,(T_{air} - T_b) + k_4 + k_5\,(T_{air} - T_b) + k_6 + k_7}{k_8}. \tag{12.51}$$

Rearranging, we arrive at a heuristic recipe for solving the equation:

$$\Delta T_b \approx \Delta t \left[\frac{k_1 + k_2 + k_3\,(T_{air} - T_b) + k_4 + k_5\,(T_{air} - T_b) + k_6 + k_7}{k_8} \right]. \tag{12.52}$$

We start at $t = 0$ by measuring all the factors in the square brackets, including an initial value for T_b. Plugging these values into equation 12.52 and multiplying by Δt gives us ΔT_b, how much body temperature changes during the interval Δt. Adding ΔT_b to T_b provides a new value for T_b at $t = \Delta t$, which we can then substitute back into the right-hand side of the equation. By repeating this procedure, we calculate T_b for $t = 2\Delta t$, $3\Delta t$, and so on. T_{air}, and k_1 to k_8 may themselves change through time (e.g., the sun rises and sets), but as long as we can specify their values for each time increment, the integration can proceed. In short, as long as we specify the details of the physical environment, we can predict body temperature.

This ability has unusual importance in these times of rapid climate change. Without the heat budget's mechanistic accounting of thermal physics we would either have to wait until the enviroment changes to predict the effects on individual body temperature, or perform costly—and often impractical—manipulative experiments to simulate the changed environment. It is far easier to perform these experiments using heat-budget models.

5   THE ENVIRONMENTAL NICHE

This leads us to the subject of the environmental niche. The niche concept has been central in ecology since Hutchinson, MacArthur, and Levins popularized it in the 1950s and 1960s (Hutchinson 1957; MacArthur and Levins 1967). Although its definition differs in subtle ways among practitioners, a species' niche is, in essence, a list of the things it needs in order to survive and reproduce: an appropriate temperature, sufficient water, the availability of appropriate food, and so forth. Traditionally, the niche is conceptualized as a volume in an n-dimensional space in which each dimension represents one item on the list. The *fundamental niche* is the volume in which a species can survive and reproduce indefinitely. The *realized niche* is the subvolume within the fundamental niche where the species is actually found. On some axes, the realized niche may bump up against the envelope of the fundamental niche; on other axes, the realized niche may fall well short of its fundamental bounds.

The niche concept has a long and valued history. For instance, it is a basic premise in ecological theory that no two species can coexist at equilibrium while occupying the same niche. If they have similar needs, one species will eventually outcompete the other for a limiting resource, or the species will evolve so that their realized niches do not overlap. Darwin's finches are a classic example of this sort of niche displacement. The bills of the various closely related species have evolved to specialize on different food items, minimizing competition (Weiner 1994).

Recently, the niche concept has gained renewed popularity with the advent of *environmental niche modeling*, a method for characterizing the physical attributes of a species' niche. Advances in digital recording technology and data sharing provide access to climatic information from an expanding system of weather stations, and this information can be correlated with the presence or absence of a given species. Environmental niche models assume that the conditions that co-occur with a species are a quantitative estimate of the species' niche. That is, they assume that one can determine where a species *can* live by measuring the environment where it currently *does* live. Environmental niche models are attractive because they are eminently practical—the necessary data are easily obtained, and the detailed physiology, behavior, and natural history of the species in question need not be known.

As Earth's climate changes, ecosystems worldwide are experiencing compositional shifts as some species' ranges retreat and others extend. Predicting the direction, rate, and extent of these shifts poses a major challenge for ecosystem management, and environmental niche models have often been the method of choice for coping with this problem. A recent example illustrates the process.

The cane toad *Bufo marinus* was introduced into northeastern Australia in 1935, and it has spread extensively across the continent. The toad has a potent toxin in its skin, making it dangerous both to native wildlife and humans, which in turn has sparked an interest in predicting how far its invasion might extend. Are there environmental conditions that could stop the toads before they threaten Perth and Adelaide?

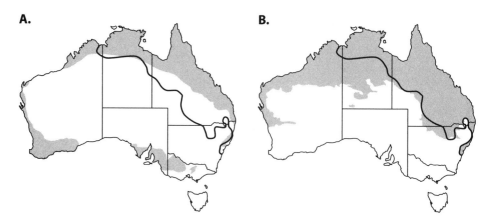

Figure 12.4 Cane toads in Australia. **A.** The current limits to toad distribution (solid line) and the future limits based on a climate envelope model (shaded area). **B.** Future limits based on a mechanistic model (redrawn from Kearney et al. (2008)).

Initial attempts to answer this question used *climate envelope models*, one form of the environmental niche approach. Through a careful investigation of the climatic conditions in the toads' current habitat, these models estimated its environmental niche. As Australia's climate shifts over the next century, areas amenable to toads will expand, suggesting that the toad will eventually invade much of the southern coast, the area containing several of the country's major cities (Figure 12.4A).

However, there are potential problems with these and analogous predictions for other organisms:

- There are many environmental factors that potentially could affect an organism. How can one be sure that the factors currently measured by weather stations are those that really matter? For example, few weather stations measure nighttime cloud cover, which can have an important effect on organisms' minimum temperature. Some trees cannot tolerate freezing, for instance, and nighttime cloud cover could have a profound effect on their environmental niche.
- Environmental factors often interact to determine important biological parameters. For instance, body temperature depends on the interaction among solar irradiance, air temperature, relative humidity, and wind speed. Measurement of these parameters within a species' current range provides evidence that certain combinations of these factors produce viable body temperatures, but they offer no guarantee that these are the only viable combinations. Higher air temperatures than those found in an organism's current range, coupled with higher wind speeds, might allow it to have a comfortable body temperature outside the current climate envelope.
- Due to migration or dispersal from sources within a species' viable range, areas whose environments are outside the fundamental niche may currently be inhabited even though the species will not be able to persist there. By assuming that organisms can live and reproduce

indefinitely wherever they are currently found, climate envelope models mistake these sinks for viable habitat, thereby biasing the climate envelope (e.g., Davis et al. 1998; Hijmans and Graham 2006).

- A similar argument works the other way around. A species may be confined to its current range by biological rather than environmental factors, in which case a climate envelope based on the current range can underestimate the potential for range expansion.

An ecomechanical approach to niche measurement (which includes heat-budget models) can help to avoid these issues. Consider, for instance, the littorine snails we have repeatedly used as an example. These animals can prosper only when their body temperature is within a certain range, and this range delineates the species thermal niche. Our knowledge of thermal mechanics allows us to list all the potential environmental factors that affect snails' body temperature: solar and infrared radiation, air temperature, substratum temperature, wind speed, metabolism, and relative humidity. By combining these factors in a heat-budget model, we can easily ascertain which factors are important. We found, for instance, that metabolic heating is too small to matter.

The littorine heat-budget model also allows us to specify all possible combinations of environmental factors that produce a viable body temperature. Thus, it is an easy matter to determine whether the environment at a particular location could allow for species persistence, even if those conditions differ from the environment where the species is currently found. Similarly, the heat-budget model can identify sinks, areas where the species is currently found but is not viable. Because the heat budget is based on physical principles, what works for a snail can work for any other organism. In short, this mechanistic approach has broad practical applicability.

The utility of the ecomechanical approach is illustrated by Kearney and his colleagues (2008), who used heat-budget models and mechanistic response functions to arrive at an alternative prediction for the cane toad's spread in Australia. In constructing these models, they considered all stages of the toad's life history and the interaction of each stage with the physical environment. In particular, they measured the speed with which adult toads hop as a function of their body temperature. Speed was low at low temperatures, increased with temperature up to 32°C, and then rapidly decreased at higher body temperatures; a typical performance curve (see Chapter 3). Low speed (at either high or low temperature) limits toads' foraging ability as well as their ability to disperse and find mates.

Body temperature is, in turn, a function of a variety of environmental, behavioral, and physiological factors. For example, the body temperature of a toad can be 5°C to 6°C below air temperature due to evaporative cooling, but toads can maintain this effect for only so long before they encounter physiological stress from desiccation and must find water or shelter.

Combining the toad's locomotory capabilities with the heat-budget model for body temperature, Kearney and colleagues (2008) were able to calculate the rate at which adult toads could invade new territory. Due to the animal's ability to evaporatively cool and their rapid hopping at high temperatures, the toad's rate of invasion is predicted to be high in the warm northern part of Australia but is predicted to approach zero as the toads extend their range to the colder south (Figure 12.4B). Both predictions match measured rates of range extension, so it seems likely that Perth and Adleaide are safe.

In sum, adult toads could live on the southern coast if delivered there (which is why climate envelope models include the coast in the toad's potential range), but a mechanistic analysis reveals that toads do not have the locomotory capacity to transport themselves to those locations. Expected change in Australia's climate (increased air temperature, decreased relative humidity, and shifts in rainfall patterns) would allow the toads to extend their range approximately 100 km further to the south, but still north of the major cities.

In the past, mechanistic approaches to niche delineation have been constrained by the lack of appropriate technology. For example, heat-budget models should be validated by measurement of body temperture in animals in their natural environment, but for mobile animals these measurements have been difficult. Recent advances in data loggers can circumvent this problem. Commercially available digital temperature loggers are now small enough to be implanted in individual lizards or toads, allowing researchers to directly measure the body temperature of free ranging individuals.

6   CONCEPTS, CONCLUSIONS, AND CAVEATS

Our findings in this chapter reinforce and extend the conclusions we reached in Chapter 11:

- Body temperature is seldom equal to air temperature. The temporal lags introduced by energy storage in organisms and the conduction of heat in soil and rocks add to the long list of ways in which body temperature is likely to differ from the ambient temperature of the surroundings.
- Body temperature depends on body size. In Chapter 11 we showed that size mattered for equilibrium body temperature. In this chapter, we found that body size governs response time and, thus, the rate at which T_b approaches equilibrium. Large organisms may never have sufficient time to equilibrate with a changing environment.
- The ability to calculate body temperature as a function of environmental conditions can provide a mechanistic link between thermal physiology and population and community ecology, a link demonstrated by the application of environmental niche modeling.
- In addition to its utility in predicting range limits, a mechanistic understanding of the interaction between the physical environment and individual organisms can guide the study of how species' niches will shift through evolution. Mechanistic models of range limitation provide direct evidence of which environmental factor is currently limiting. With this information in hand, it is possible to test for the organism's ability to adapt in response to this constraint (e.g., Kuo and Sanford 2009).

Beyond their utility in ecology and evolution, heat-budget models have the potential to be valuable tools for physiologists. A variety of physiological models (e.g., bioenergetic models and dynamic energy budget models; Kooijman 2010; Nisbet et al. 2012) are poised to use the predictions from heat-budget models to predict organisms' ability to acquire energy through foraging or photosynthesis, the status of their physiological defenses, and their rates of growth and reproduction.

The theory presented in this chapter is intended to give you an appreciation for heat-budget models and their ability to predict body temperature. However, by highlighting a few heuristic examples, I have downplayed the nitty-gritty details that require attention when constructing a realistic heat-budet model: the complicated geometry that governs how solar irradiance changes through a day, the effects of respiration and stomatal conductance on evaporative cooling, the changes in absorptivity and emissivity that often accompany desiccation, the role of insulation (e.g., fur and blubber) and the resulting thermal gradients inside organisms, and many others. If you wish to pursue the subject of heat-budget modeling, you would be well advised to consult the bibles of the field: Angilletta (2009), Gates (1980), Nobel (1991), Campbell and Norman (1998), and Monteith and Unsworth (2008).

Part III

SOLID MECHANICS

Chapter 13

Biological Materials I
Materials Mechanics

Time and again in the last twelve chapters we have discussed the possibility that forces imposed on organisms can deform or damage them, but the gory details have been left vague. What is the largest force an organism can resist? How does an organism's size and shape affect its potential for destruction? How does the strength of a plant or animal depend on the properties of the materials from which it is formed? In turn, how do the properties of those materials depend on the molecules from which they are constructed? Fortunately, engineers and materials scientists have grappled with questions of this sort, albeit in terms of buildings and machines instead of plants and animals. Over the next six chapters we will apply their wisdom in a biological context.

We begin in this chapter with a general introduction to materials science, an overview of basic concepts and terminology. Next, in Chapter 14, we survey the types of molecules from which biological materials are constructed and see how those properties can be used to predict the size and shape of organisms. Along the way, we will discover that the actual strengths of biological materials are far below the strength their molecules can theoretically provide, a disparity we will explain in Chapter 15 using the theory of fracture mechanics. In Chapters 16 and 17 we take the next steps, first investigating the mechanics of adhesion and then tackling beam theory, the branch of engineering that explains how the size and shape of a structure affects its response to an applied force. And lastly, in Chapter 18, we place biological structures in a dynamic environment and study the ways in which their movement affects the forces they experience.

That's where we are headed. But first we must explain why some materials are solid and others aren't.

1   BONDING FORCES

As noted in Chapter 5, materials can be divided into fluids and solids. In a fluid, molecules have at most a transitory attachment to each other. As a consequence, when a force is applied, liquids and gases deform and keep right on deforming for as long as the force is imposed (Figure 13.1A). The larger the force, the greater the rate of deformation; that is, fluids are *viscous* (Figure 13.1B). Because bonds between fluid

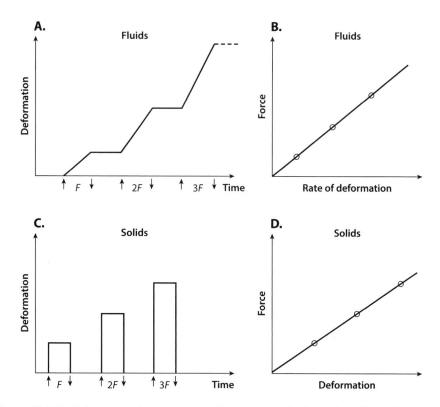

Figure 13.1 A. Deformation through time as forces are applied to a fluid. Upward arrows indicate the imposition of force and the downward arrows indicate its removal. **B.** The results of **A** show that force increases in proportion to deformation rate. **C.** Deformation through time as forces are applied to a solid. **D.** The results of **C** show that force increases in proportion to the amount of deformation.

molecules are transitory, liquids and gases have no fixed shape—once deformed, they stay deformed.

By contrast, in a solid molecules are permanently attached to each other. Application of a force causes the material to deform, but, unlike a fluid, as force is removed a solid recovers its original dimensions (Figure 13.1C, D). In other words, solids are *elastic*.

These differences between fluids and solids can be explained by the nature of the forces that hold atoms together. Consider two generic, electrically neutral atoms in close proximity to each other (Figure 13.2A). Each has a positively charged nucleus embedded in a cloud of negatively charged electrons, and the interaction between adjacent electron clouds binds the atoms to each other. In the absence of an externally applied force, the atoms settle into an equlibrium with their nuclei ℓ_0 apart, the distance at which the energy of the system is minimal (Figure 13.2B). The stronger the bond, the smaller the interatomic spacing. Typically, ℓ_0 is a fraction of a nanometer.

Because energy is minimal at a spacing of ℓ_0, additional energy is required to push the atoms closer together or pull them apart, and we can use this incremental energy to calculate the force required to change the interatomic spacing. Recall that energy is the product of force and distance; thus, force is energy divided by distance. In this

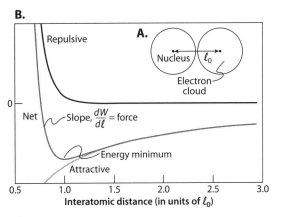

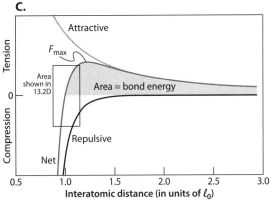

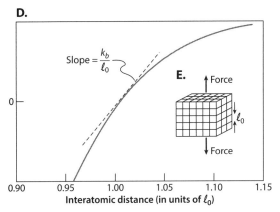

Figure 13.2 A. Definition of inter-atomic spacing. B. Bond energy. C. Bond force. D. Net bond force near the resting separation. E. Atoms arranged in a crystalline array.

case, the force required to change the spacing between atoms is equal to the change in energy per change in interatomic spacing, the slope of the net energy curve.

Force calculated in this fashion is shown in Figure 13.2C. Compressive (i.e., negative) force increases very rapidly as electron clouds are squeezed into each other, so reducing interatomic spacing much below ℓ_0 is a practical impossibility. By contrast, the force required to pull atoms apart initially increases as atoms are separated, but it then reaches a peak and tapers off. The area under this tensile force-distance curve—the work required to fully separate atoms—is the *bond energy* that holds atoms together.

The magnitude of bond energy depends on the type of interaction between adjacent electron clouds, four of which play major roles in biological materials:

1. *The London dispersion force.* Averaged over time, the electron cloud of an isolated atom is symmetrically distributed around the nucleus (Figure 13.3A). At any instant, however (where an instant is on the order of 10^{-16} s), quantum uncertainty can result in an asymmetrical cloud that is more negative on one side and, thus, relatively more positive on the other. This asymmetry—technically, an electric dipole—induces a similar asymmetry in the adjacent atom. As a result, the

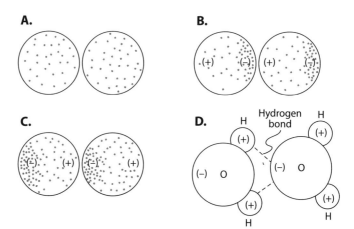

Figure 13.3 On average, the charge density of adjacent neutral atoms is uniform (**A**), but charge density varies through time (**B** and **C**; redrawn from De Podesta (2002)). **D**. The permanent dipoles of water molecules form hydrogen bonds.

positive part of the electron cloud of one atom is near the negative portion of the other (Figure 13.3B). Opposites attract, so for this instant the two atoms are pulled toward each other. An instant later, the charge asymmetries may shift position (Figure 13.3C), but the mutual attraction remains. On average, the two atoms are pulled together.[1] This odd quirk of atomic mechanics was first described by Fritz London, a German-American physicist, and the effect—the London dispersion force—is named in his honor. The London dispersion force is one component of the van der Waals force, a more general term that includes a variety of dipole-dipole interactions. Because of the transitory nature of electron asymmetries, London dispersion forces are relatively weak. It takes only approximately 2 to 17 kJ to break a mole of these bonds (de Podesta 2002). Although Figure 13.3 shows the interaction as occuring between individual atoms, dispersion forces can also act on neutral molecules such as nitrogen (N_2) and oxygen (O_2). However, at room temperature the energy of thermal agitation (approximately $3.7 \, kJ \cdot mol^{-1}$) disrupts the dispersion forces between these molecules, allowing N_2 and O_2 to move freely as a gas.

2. *Hydrogen Bonds*. A stronger bond is formed if the asymmetry in the electron cloud is permanent. This feat is impossible for electrically neutral atoms but is commonplace for molecules, especially those that contain hydrogen and oxygen bonded together. Water—H_2O—is a prime example (Figure 13.3D). Oxygen has a much stronger affinity for electrons than does hydrogen, so when two hydrogen atoms bind to oxygen to form water, the electron cloud at the oxygen end of

---

[1] This is an example of Jensen's inequality. Given their average electron distribution, neutral atoms should not be attracted. However, the average of the instantaneous interaction between electron clouds binds atoms to each other.

the molecule is more concentrated (and therefore more negative) than that near the hydrogen nuclei. The negatively charged end of one water molecule attracts a positively charged end of an adjacent molecule, forming a *hydrogen bond*. The same effect can occur between organic molecules that have OH groups. The energy required to break hydrogen bonds is approximately ten times greater than that of London dispersion forces, about $100 \text{ kJ} \cdot \text{mol}^{-1}$ (de Podesta 2002). The hydrogen bonds in water, for instance, are sufficiently strong to make it a liquid at room temperature.

London dispersion forces and hydrogen bonds are commonly used to stabilize the structure of biological solids, but they are too weak to act as the primary mechanism by which atoms are held in place.

3. Stronger bonds can be formed when electrons shift their allegiance among nuclei. In some cases—bonds between the carbon molecules in proteins and polysaccharides, for instance—an electron is shared between atoms, and energy is required to break the resulting *covalent bond*. Covalent bonds are several times stronger than hydrogen bonds; it takes approximately 347 kJ to disrupt a mole of carbon-carbon bonds, for example (Zumdahl and Zumdahl 2003).

4. Lastly, electrons can jump ship altogether, moving from one atom to another. Having lost a negative charge, the atom left behind acquires a net positive charge, and the atom to which the electron moves becomes negatively charged. Again, opposites attract, and the two ions form an *ionic bond*. Ionic bonds can be even stronger than covalent bonds: it takes 756 kJ to break a mole of the bonds between sodium and chlorine in table salt, for instance.

2 WHY SOLIDS ARE ELASTIC

The nature of the attractive forces between atoms can be used to explain the extensibility and strength of solids. To see how, we return to the force-separation curve of Figure 13.2C, focusing on the portion of the curve near where force is zero and the distance between nuclei is ℓ_0 (Figure 13.2D). If we could somehow grasp the atoms involved and push them closer together, we would find that the compressive force increases approximately linearly with the proportional change in separation ($\Delta \ell$, the small change in distance, divided by ℓ_0, the initial separation). If we pull the atoms apart, tensile force similarly increases as a linear function of proportional change in separation. That is,

$$F = \frac{k_b}{\ell_0} \Delta \ell, \tag{13.1}$$

where k_b/ℓ_0 is the bond's stiffness ($\text{N} \cdot \text{m}^{-1}$). In other words, at the atomic level, force is approximately proportional to deformation, the fundamental characteristic of an elastic solid.

What works for an individual pair of atoms also works for multiple pairs. Consider for instance an orderly array of atoms such as the sodium and chlorine ions in a crystal of table salt (Figure 13.2E). As force is applied to the whole crystal, bonds are stretched between each pair of atoms, and the force-deformation curve for the macroscopic sample looks just like that of its atomic components: force increases linearly in response to proportional deformation.

The linear relationship between force and deformation was first described by Robert Hooke (1635–1702),[2] a contemporary (and competitor) of Isaac Newton's. In Hooke's words, *ut tensio, sic vis*: as the extension, so the force. In honor of his contribution on the subject, materials that have a linear relationship between force and deformation are described as Hookean.

There are limits to this elastic response, however, most evident when bonds are deformed in tension. Returning to the interaction between individual atoms (Figure 13.2C), note that when atomic separation reaches approximately 125% of ℓ_0, bond force reaches a maximum, F_{max}. Further separation requires less and less force, resulting in an instability—in other words, the bond breaks. In essence, for extensions beyond approximately 25% of ℓ_0, the atom being pulled is more strongly bonded to the object doing the pulling than it is to the atom to which it was originally attached. Again, what works for a pair of atoms works for multiple pairs. If the bonds in a material are extended beyond their limits, the material breaks.

3   THEORETICAL STRENGTH OF SOLIDS

Just as the behavior of chemical bonds can explain the limits to deformation, it can be used to estimate the force required to break a material. Consider a crystal of diamond, for example. Made solely of carbon atoms, diamond is an apt model for estimating the strength of organic molecules, which are held together by carbon-carbon covalent bonds.

The density of diamond is 3520 $kg \cdot m^{-3}$, which—at 0.012 $kg \cdot mol^{-1}$ (a mole is 6.02×10^{23} atoms)—amounts to 1.77×10^{29} carbon atoms per cubic meter. If we assume that atoms are uniformly distributed in the crystal lattice, there are $(1.77 \times 10^{29})^{2/3} = 3.15 \times 10^{19}$ atoms in a square meter of diamond surface, i.e., 5.23×10^{-5} $mol \cdot m^{-2}$. Given that it takes 347 kJ of energy to break a mole of carbon-carbon bonds, 18.1 J are required to break all the bonds in a square meter of diamond. This is diamond's *bond energy density*, ρ_{bond}.

Now Figure 13.2C suggests that the area under the entire force deformation curve for a chemical bond is on the order of $F_{max} \times \ell_0$, an estimate we can use to calculate breaking force. Dividing breaking energy (18.1 $J \cdot m^{-2}$) by the ℓ_0 of C–C bonds in diamond (0.154 nm) gives us an estimate of the force per area required to break a diamond:

$$\frac{\rho_{bond}}{\ell_0} = \frac{18.1 \text{ J m}^{-2}}{0.154 \text{ nm}} \approx 118 \text{ GPa.} \tag{13.2}$$

(1 GPa is 10^9 Pa.) Predictions using a more detailed theoretical approach—that accounts for the actual distribution of atoms and the bending of bonds—estimate breaking strengths of 90 to 225 GPa, depending on the fracture plane (Telling et al. 2000), so our rough estimate seems reasonable. This is a remarkable strength.

---

[2] Hooke also occupies a place of honor in biology. Using rudimenatry, handmade microscopes, he noticed that all living things seem to be made of small units that he called "cells."

The bonds in a typical diamond engagement ring (a cross-sectional area of about $6 \times 10^{-6} \, \text{m}^2$) could theoretically suspend a 72-tonne (t) mass without breaking. An average automobile has a mass of about 2 t, so an engagement diamond could theoretically suspend all the cars in a small parking lot.

Because many biological materials are held together by the same C–C bonds found in diamond, one might expect collagen, chitin, and cellulose to have the same sort of strength. In fact, they are orders of magnitude weaker. Among the biological materials that rely on C–C bonds for their strength, the strongest (cellulose and spider silk) have a breaking strength on only 1 GPa.

There are two main reasons for this disparity, both important for how we think about biological materials. First, there are typically far fewer C–C bonds in any cross section of a biological material than there are in diamond. In a protein, for instance, one C–C bond holds two amino acids together, but each amino acid is a relatively bulky molecule. As a result, there can't be nearly as many C–C bonds in the cross section of a protein as there are in diamond. The same restriction applies to polysaccharides such as cellulose and chitin.

The second—and primary—reason biological materials don't measure up to the theoretical potential of diamond concerns an implicit assumption we made in setting up our calculation. Our estimate of diamond's strength assumes that all bonds are equally loaded, allowing each to contribute to resisting the applied force. This theoretical uniformity is seldom achieved in nature. Instead, a few bonds typically bear the initial brunt of the load, and when the local force per area exceeds their limit, they break. Their failure shifts the load to another few bonds, which then break and pass the load on to others. In short, when force is concentrated on just a few bonds, materials can fracture piecemeal under loads far smaller than our calculations would suggest. This is true even of diamond. If real diamonds were always as strong as we have estimated, they would be nearly indestructable. But a jeweler can easily split a diamond by applying a small force along just the right plane.

It is more difficult to perform a similar calculation for an ionic solid such as sodium chloride or calcium carbonate; due to the electrostatic nature of ionic bonds, each atom is both attracted to and repulsed by all other atoms in the crystal lattice rather than just by its neighbors. Estimates using other methods suggest theoretical tensile strengths approximately a tenth that of diamond (Gosline 2016) but still well beyond the measured strength of any biological material.

We will explore the mechanics of fracture in Chapter 15. For the moment, it is sufficient to know that fracture can explain the inability of biological materials to achieve their theoretical strength.

4   POLYMERS AND SOFT ELASTICITY

In our explanation of elasticity so far, we concluded that the nature of chemical bonds limits the extensibility of a material to about 25%. There are, however, many biological materials that extend much farther. Hair can extend by 50%, for instance, spiders' silk and mussels' byssal threads can extend by 100% to 150%, and some rubbery materials can extend 300% or more. How do they do it?

Two tricks are involved. The first is to use covalent bonds to link individual molecules not into three-dimensional crystals, but rather into linear, rope-like structures (Figure 13.4A). In this case, each individual molecule is known as a monomer, and

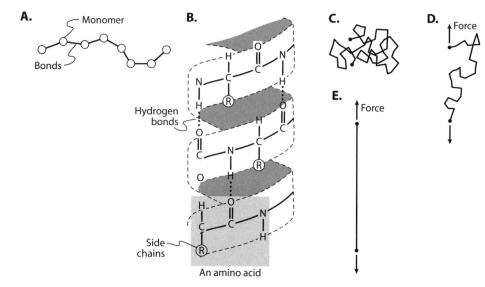

Figure 13.4 **A**. A linear polymer. **B**. The α helix. In the absence of external force, a polymer assumes a compact, random shape (**C**). Application of a force extends the molecule (**D**), but only when it is fully extended can the force pull directly on the bonds between monomers (**E**).

the entire chain is a *linear polymer*. Glucose monomers hooked end to end form cellulose, for example. Glucosamine molecules linked sequentially make chitin, and linked amino acids form proteins such as collagen, keratin, and abductin.

If we were to stretch out each of these minuscule ropes and then pull on their ends, they would extend by at most 25%, just like the diamond discussed earlier. But linear polymers are seldom stretched out in nature. In keratin, for instance (the material from which fur, hooves, and fingernails are made), protein polymers twist around themselves, coiling into an arrangement known as an α helix (Figure 13.4B; Wainwright et al. 1976). The helix is stabilized by hydrogen and disulfide (covalent) bonds. Sufficient tensile force can disrupt these secondary bonds, but their failure doesn't mean that the polymer itself breaks. Instead, it uncoils, allowing the material to extend substantially before the applied force pulls directly on any of the C–C bonds in the polymer's backbone. Thus, the overall extension of the polymer (more than 50%) far exceeds that of any of its individual bonds.

The tactic of molecular rearrangement is taken to an extreme by rubber-forming polymers such as abductin, resilin, and elastin. The bulky side groups of the amino acids in these protein chains, and the kinks introduced by proline monomers,[3] prevent the polymer from having an ordered shape. Instead, through the action of thermal agitation, the chain takes on an ever-changing, random configuration (Figure 13.4C). Just as particles randomly walking the x-axis tend to stay near the origin (Chapter 4), the ends of a randomly arranged polymer chain tend to stay close to each other. As a result, the distance between ends can increase drastically before the chain is fully

---

[3] The angle at which proline bonds to its neighbors is different from other amino acids, which creates problems for orderly packing.

extended, and only then can an applied force pull directly on the polymer's covalent bonds (Figure 13.4D and E). Because, in the absence of applied force, they assume a compact configuration, rubber-forming protein polymers can be extended to several times their resting length before they break.

5   RUBBER ELASTICITY

The random configuration of a rubber-forming protein explains its impressive extensibility but leaves open an important question: why is it elastic? In order to take on a random shape, the polymer must be free to reconfigure, but if the polymer can freely rearrange, why is a force necessary to separate its ends?

The answer lies in the second law of thermodynamics, which dictates that disorder in a system (its *entropy*) tends to increase. We have seen how this tendency explains the diffusion of molecules (Chapter 4), momentum (Chapter 6), and heat (Chapter 11). In similar fashion, the tendency toward disorder explains the force required to extend a randomly arranged polymer.

When the ends of a flexible polymer are close together, there are many different ways in which the chain can be arranged. Indeed, there is a maximum number of arrangements—and the conformational entropy of the chain is therefore at its highest—when the average separation between ends is a small fraction of the overall length of the polymer chain (Figure 13.4C). If there are n links in the chain, each of lengh ℓ,

$$\text{average end-to-end distance} = \sqrt{n}\ell. \tag{13.3}$$

For example, if the chain is 100 links long, its entropy is greatest when the ends are only about 10 link lengths apart.

Because the distance between ends is small, the chain's extensibility is large:

$$\text{extensibility} = \frac{\text{total chain length}}{\text{average end-to-end distance}} = \frac{n\ell}{\sqrt{n}\ell} = \sqrt{n}. \tag{13.4}$$

Our hypothetical 100-link polymer could stretch to ten times its resting length before becoming fully extended.

If we separate the ends of the chain beyond their average distance, we reduce the number of possible arrangements, thereby reducing the polymer's entropy (Figure 13.4D). Taken to the extreme, if the ends of the chain are separated by a distance equal to the polymer's total length, there is only one possible arrangement and entropy is zero—the chain is maximally ordered (Figure 13.4E). The more order we impose on the chain, the more force required (see Wainwright et al. 1976), and as soon as we remove the force, thermal agitation returns the ends of the chain to their minimum entropy separation. The properties of a single chain can be scaled up by crosslinking multiple chains into a three dimensional network. Thus, the resulting material—a rubber—is elastic.

The details of the entropy-based restoring force can be counterintuitive. For instance, the force required to hold rubber at a fixed deformation increases in proportion to absolute temperature. As a result, a stretched piece of rubber shrinks when heated, just the opposite of most materials. But the physics and math behind these details are a bit complex, so we will not pursue them here. The interested reader should consult Denny and Gaines (2000) for the basic mathematics, Gosline (2016)

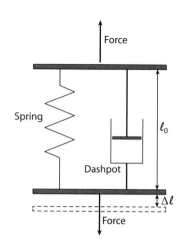

Figure 13.5 A Voigt element. (See text.)

for a more thorough discussion, and Treloar (1975) for the full-blown theory. For our purposes, it is sufficient to know that entropy's influence in a rubbery material results in a force-deformation curve similar to that of other solids—force increases with deformation until the material breaks. The main difference in this case is that rubber can stretch much farther than other materials.

6   VISCOELASTICITY

The extensibility of polymeric materials raises an important issue. As a keratin helix uncoils or a rubber polymer is stretched, individual monomers slide past each other. As they do, weak bonds—dispersion forces and hydrogen bonds—can be formed between adjacent polymer chains, and these bonds resist being sheared in the same way that those of water molecules do. Deformation of polymers can thus entail the same sort of viscous resistance found in a fluid. Unlike the stretching of ionic or covalent bonds and the change in entropy of a disordered polymer—both of which create force proportional to the *amount* of deformation—viscous resistance depends on the *rate* of deformation. In other words, when the rearrangement of polymeric molecules involves viscous interactions, the total force required to deform the material has both elastic and viscous components.

To understand how this combination works, engineers invoke a hypothetical mechanical model—a spring and a dashpot acting in parallel (Figure 13.5). The spring acts as ideal elastic element in which force is proportional to deformation. The dashpot—a plunger immersed in a viscous fluid—acts as an ideal viscous element: the force required to deform it is proportional solely to the rate of deformation. (In this purely heuristic model, we assume that the bars connecting the spring and dashpot stay parallel, ensuring that the dashpot is deformed only as far as the spring is stretched.)

Given this arrangement, the total force required to stretch the model (technically known as a Voigt element) is the sum of two components:

$$F = \underbrace{\frac{k_e}{\ell_0} \Delta\ell}_{\text{elastic}} + \underbrace{k_v \frac{\Delta\left(\frac{\Delta\ell}{\ell_0}\right)}{\Delta t}}_{\text{viscous}}. \qquad (13.5)$$

Here k_e/ℓ_0 is the elastic stiffness of the spring ($\text{N} \cdot \text{m}^{-1}$); k_v is the viscous resistance of the dashpot ($\text{N} \cdot \text{s}$); $\Delta\ell$ is deformation; ℓ_0 is the resting length of the spring and dashpot; and Δt is change in time.

If the viscosity of the system is low or one deforms the system very slowly (that is, if $\Delta\left(\frac{\Delta\ell}{\ell_0}\right)/\Delta t$ is small), the dashpot provides negligible resistance, and the Voigt element behaves like a spring. If viscosity is substantial, deformation is fast, or both, the dashpot adds to the spring's resistance.

Solids that exhibit this sort of rate dependent behavior—and that includes virtually all polymeric biological materials—are said to be *viscoelastic*.

7 LOAD

Our discussion so far has been couched in terms of force and deformation, but these quantities have been left vague. How much force is required to produce a given deformation? How, exactly, do we quantify deformation in the first place? To answer these questions we need the appropriate vocabulary.

An external force applied to a material is called a *load*, and loads come in three basic varieties. Two forces acting along the same line but directed away from each other form a *tensile load*, which tends to extend a sample (Figure 13.6A). Conversely, two coaxial forces directed toward each other form a *compressive load*, squishing the material between them (Figure 13.6B). Both tensile and compressive loads are applied perpendicular to the surface of the material, and together they comprise *normal loads*.

By contrast, *shear loads* are applied when opposing forces are imposed along parallel, but different, axes (Figure 13.6C). Shear loads act parallel to a material's surface. (The shear loads we deal with in solids are the same as those we encountered when dealing with boundary layers (Chapter 6).)

Now consider the equilibrium between loads and the material to which they are applied. When subjected to a tensile load, an elastic solid deforms until the resistive force of its bonds (or, for a rubber, of its entropy) just equals that of the applied load. At that point, there is no net force acting on the sample, and the system doesn't move. The same logic applies to a sample in compression.

The situation is different, however, for a sample loaded in shear. Because they are not coaxial, each of the two forces applied to the sample in Figure 13.6C has a lever arm acting relative to the sample's center (Figure 13.6D). Consequently, each force applies a moment to the sample. (Two equal moments acting in concert like this are called a *couple*.) If these two forces were the only load applied to the sample, it would rotate, in this case clockwise. To keep the sample from rotating, we must apply an equal and opposite couple, as shown in Figure 13.6E. In short, for a material to be at equilibrium in shear, two sets of forces must be applied, each acting at right angles to the other.

These forces can be viewed from a different perspective. It can be shown (Supplement 13.1) that applying counteracting couples to a material is equivalent to applying tensile and compressive loads at 45° to the sample's surface (Figure 13.6F). In other words, when a sample is at equilibrium in shear, it is stretched in one direction and compressed at right angles to that direction.

8 STRESS

We know intuitively that the relationship between force and deformation depends on the size of a sample. It takes more force to break a tree trunk than it does to snap

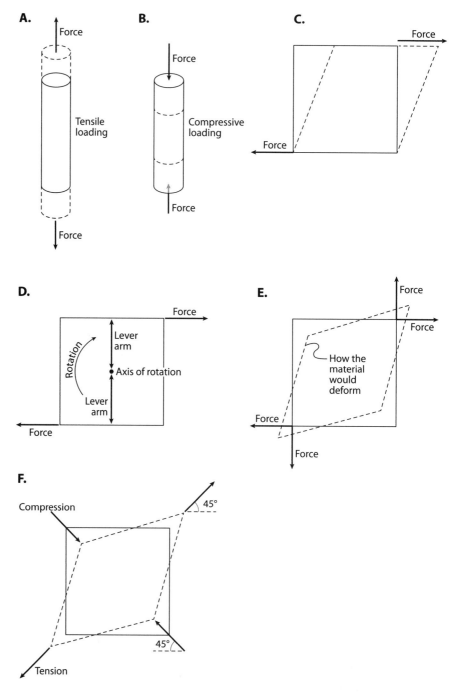

Figure 13.6 A. Tensile loading. **B.** Compressive loading. **C.** Shear loading. **D.** A single set of shearing forces applied to an object causes it to rotate. **E.** To resist rotation, a second set of shearing forces must be applied. The dashed outline is the shape that results. **F.** Overall shearing force can be thought of as orthogonal tensile and compressive forces acting at 45° to the shearing forces.

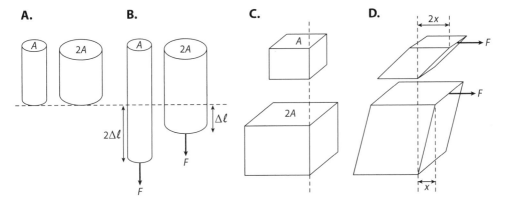

Figure 13.7 Defining stress. **A** and **B**. Tensile stress. **C** and **D**. Shearing stress. (See text.)

a toothpick, even though both are made of wood. Because size matters, measures of raw force and deformation can describe the mechanical properties only of particular objects. How can we scale force and deformation to describe the intrinsic properties of the material from which an object is made, properties that are independent of an object's size and shape?

Let's begin with force. Consider the two samples of material shown in Figure 13.7A. They are the same length, but one has twice the cross-sectional area of the other. When subjected to the same tensile load, the thicker sample extends less than the thinner one (Figure 13.7B). If, however, we were to double the load on the thicker strip, it would extend just as far as the thin one. That is, as long as we apply the same force per cross-sectional area—the same tensile *stress*—we get the same deformation. The same idea applies if we load the samples in compression. Tensile and compressive stress are given the symbol σ, and stress has units of $N \cdot m^{-2}$, that is, pascals (Pa).

The basic concept of stress (the intensity of force) applies in shear as well (Figure 13.7C and D). The larger the cross-sectional area of a sample, the greater the force needed to achieve a given deformation. In this case, however, the area in question is parallel to applied force. As with shear stress in fluids, shear stress in solids is given the symbol τ.

When a solid is deformed in tension or compression, it is common for its cross-sectional area to change. Stretch the sample and cross-sectional area decreases; compress the sample and area increases. This raises an important question—what area do we use when defining tensile or compressive stress: the area of the unloaded material (the *nominal area*, A_n) or the area that results when the load is applied (the *instantaneous area*, A_i)?

Instantaneous area provides a better picture of what is happening in the material, and for that reason it is often preferred. In this case, one obtains *true stress:*[4]

$$\sigma_t = \frac{F}{A_i}. \tag{13.6}$$

---

[4] Note the new use of the symbol σ here. Stress has no relation to the standard deviation, the context in which we have previously used this symbol.

Although preferred in theory, instantaneous area can be difficult to measure in practice. In these cases, it is more convenient to use nominal area, from which one obtains *nominal stress* (often referred to as *engineer's stress*):

$$\sigma = \frac{F}{A_n}. \tag{13.7}$$

For small deformations, nominal area is approximately equal to instantaneous area, and nominal and true stress are equivalent. Most structural materials used by human engineers (e.g., steel, concrete, and glass) deform very little during their intended use, and it is, therefore, common for engineers to use nominal stress.

When deformed in shear, we have seen that a solid is simultaneously subjected to both a tensile load (which tends to decrease cross-sectional area) and a compressive load (which tends to increase it; Figure 13.6F). The two tendencies cancel each other, so nominal shear stress is true shear stress.

9   STRAIN

Now consider two samples with the same cross-sectional area, but one is twice as long as the other (Figure 13.8A). If we apply a tensile load to the short sample, it stretches by $\Delta\ell$. What happens when we apply the same load to the longer sample? To answer this question, think of the longer sample as two short samples connected in series. The applied load acts equally on each component, so each stretches by $\Delta\ell$, and the overall sample stretches by $2\Delta\ell$ (Figure 13.8B). Thus, although made of the same material and subjected to the same stress, our two samples extend different amounts. Deformation depends on size.

Note, however, that each sample deforms the same amount *relative to its initial length*. In other words, change in length per original length—what engineers call *strain*—is independent of a sample's size and, therefore, is a useful way to quantify deformation.[5] As the ratio of lengths, strain is dimensionless.

This concept can be implemented in several ways. For tension and compression, we can calculate *nominal strain* (also known as *engineer's strain*) by comparing deformation, $\Delta\ell$, to the initial length of the sample, ℓ_0 (Figure 13.8C):

$$\varepsilon = \frac{\Delta\ell}{\ell_0}. \tag{13.8}$$

Alternatively, we can repeatedly calculate the ratio of deformation to length as a sample is extended or compressed, comparing each increment in deformation to ℓ, the sample's length immediately preceeding that increment. Toting up all these infinitesimal strains from initial length ℓ_0 to final length ℓ_f provides an alternative description of the overall strain. Expressed mathematically, this *true strain* is

$$\varepsilon_t = \int_{\ell_0}^{\ell_f} \frac{d\ell}{\ell}$$

$$= \ln\left(\frac{\ell_f}{\ell_0}\right). \tag{13.9}$$

---

[5] The terms stress and strain have made their way into the vernacular, allowing one to speak of the stress and strain of daily life, for instance. It is important to keep in mind that in material science, stress and strain are different, and each has a precise meaning.

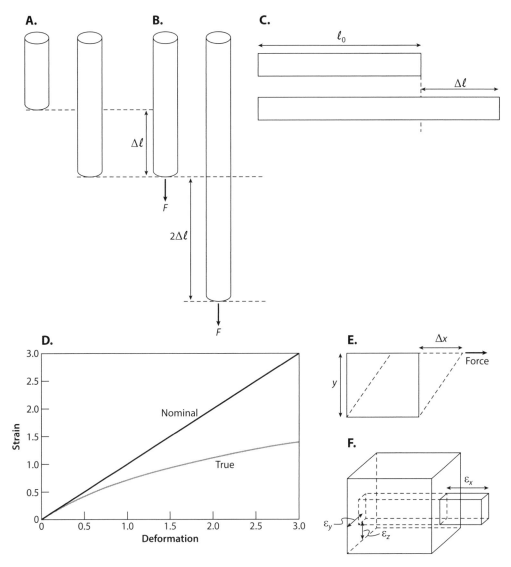

Figure 13.8 Defining strain. **A** and **B**. Tensile strain. **C**. Resting length and change in length. **D**. Comparing nominal and true strain. **E**. Shear strain. **F**. Poisson's ratios. (See text.)

Just as true stress provides a more detailed description of stress by taking into account a sample's instantaneous cross section, true strain provides a more detailed description of strain by taking into account the sample's instantaneous length. For small deformations ($\varepsilon < 0.2$), $\varepsilon_t \approx \varepsilon$; at larger deformations, $\varepsilon_t < \varepsilon$ (Figure 13.8D).

Which definition of strain should you use: true or nominal? There are a few instances in which ε_t is preferred (see Supplement 13.2 for an example), but true strain has a major disadvantage—it is nonintuitive. If I were to tell you that I had stretched a sample to a true strain of 0.405, it would take a bit of head scratching to realize that I was, in fact, describing a 50% extension. By contrast, if I describe the stretch as a nominal strain of 0.5, the percent extension is obvious. Because it is more intuitive, I will use nominal strain unless otherwise noted.

For a sheared solid, we use the same definition of strain we used in Chapter 6 for shear in fluids (Figure 13.8E). For a sample of thickness y, deformation by Δx produces a strain of

$$\gamma = \frac{\Delta x}{y}.$$ (13.10)

Because y doesn't vary as the sample deforms, γ is a true strain.

The strains we have defined so far are measured in the direction of the applied force. However, when a material is extended or compressed in one direction, its dimensions change in other directions as well. Stretch a rubber band, for instance, and its width and depth decrease. To fully specify the deformation of a material, one thus needs to measure strain along three mutually perpendicular axes. For example, if we designate ε_x as strain in the direction of applied force, we can measure ε_y and ε_z at right angles to the force (Figure 13.8F). Positive strain along the x-axis results in negative strains along the y- and z-axes.

The magnitudes of ε_y and ε_z, taken relative to ε_x, define the *Poisson's ratios* for a material:

$$v_{yx} = -\frac{\varepsilon_y}{\varepsilon_x},$$ (13.11)

$$v_{zx} = -\frac{\varepsilon_z}{\varepsilon_x}.$$ (13.12)

The negative signs are included so that Poisson's ratios are positive.

Poisson's ratios differ among materials, ranging from 0.2 to 0.5. Typically, $v_{yx} = v_{zx} = v$, in which case Poisson's ratio can be related to ΔV, the volume change accompanying tensile or compressive strain. If V_0 is the unstrained volume,

$$\frac{\Delta V}{V_0} = \varepsilon_x (1 - 2v).$$ (13.13)

Thus, for a material with a Poisson's ratio less than 0.5, volume increases as the material is strained, as we would expect from our understanding of chemical bonds.

By contrast, a Poisson's ratio of 0.5 implies that volume doesn't change as a material is deformed. This is approximately true for mucins (which have a high water content) and rubbers (whose mobile molecules allow them to change shape without changing volume). A Poisson's ratio greater than 0.5 is possible for an isotropic material only if volume decreases as a material is stretched. While feasible in theory, this odd property has yet to be found in biological materials.

Note that equation 13.13 applies only if $v_{yx} = v_{zx}$. Fabrics are an obvious exception. Stretch a swatch of fabric along an axis diagonal to the threads, and it will contract along the perpendicular axis while retaining the same thickness, a condition known as plane strain. In this case, $v_{zx} = 0$, and $v_{yx} = 1$. The skin of many animals behaves this way.

10 ELASTIC MODULUS

Recall from equation 13.1 that the force required to deform an individual bond between atoms is

$$F = \frac{k_b}{\ell_0} \Delta \ell.$$

But $\Delta \ell / \ell$ is ε, so

$$F = k_b \varepsilon, \tag{13.14}$$

where k_b is the bond's stiffness.

As before, we expect that the properties of a material will resemble the properties of the individual bonds within the material. Thus, when we bring equation 13.14 up to the macroscopic scale, we expect stress to be proportional to strain (Figure 13.9A):

$$\sigma = E \varepsilon. \tag{13.15}$$

Here the constant of proportionality, E, is the *elastic modulus*, the material's stiffness. Because ε is dimensionless, E has the same units as stress; that is, force per area, pascals.

Solving equation 13.15 for E, we see that

$$E = \frac{\sigma}{\varepsilon}. \tag{13.16}$$

In other words, E at a given strain is the slope of the line connecting that point on the stress-strain curve to the origin. All is well with this simple relationship as long as a material is precisely Hookean; in that case, E is constant. However, for many biological materials, the curve of stress versus strain is nonlinear, and E varies with ε (Figure 13.9B).

E is a measure of a material's overall stiffness, the stress required to bring it to a given strain. At times it is more informative to know the *incremental* stiffness—how much additional stress is needed to obtain a small additional bit of strain. This value is the *tangent modulus*, the slope of the stress-strain curve at some particular strain (Figure 13.9B):

$$E_{tan}(\varepsilon) = \frac{d\sigma}{d\varepsilon}. \tag{13.17}$$

Regardless of the overall shape of the stress-strain curve, E_{tan} at infinitesimal strain (a value know as *Young's modulus*[6]) is approximately equal to E. In many human engineering projects, the materials involved are only minimally strained, and their stiffness is adequately described by their Young's modulus. Consequently, tables of moduli in engineering books often list Young's modulus without any explicit warning that it may apply only at small strains.

For many materials, modulus in compression is the same as modulus in tension, but this equality isn't universal. Kelp stipes, for instance, are stiffer in tension than in compression (Gaylord and Denny 1997). In Chapter 14 we will find that moduli vary drastically among biological materials.

---

[6] Named in honor of Thomas Young (1773–1829), a pioneer in the exploration of materials mechanics.

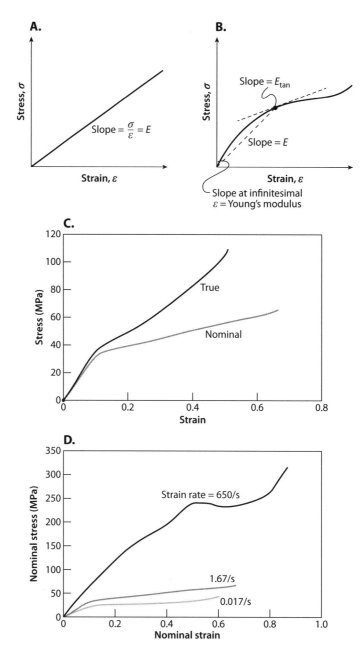

Figure 13.9 Defining modulus. **A.** A Hookean material. **B.** A non-Hookean material. **C.** Stress-strain curves for byssal threads plotted using true and nominal stress and strain. **D.** For viscoelastic materials, modulus depends on strain rate; data for byssal threads.

When reporting the modulus of a material, it is often a matter of personal preference whether to use nominal or true stress and strain. It is, however, important to realize that the moduli calculated are different. To illustrate this point consider the stress-strain behavior of a byssal thread, the proteinaceous material that attaches mussels to rock (Figure 13.9C). The same data were used to plot both curves shown here, but in one case nominal stress and strain were used and in the other, true stress and true strain. Because true stress is greater than nominal stress and true strain is smaller that nominal strain (Figure 13.8D), E calculated using σ_t and ε_t is larger than that calculated using using σ and ε. It would be wrong to employ the high modulus calculated using true stress and strain in subsequent calculations involving nominal stress and strain.

Note that these complications disappear when strains are small; in that case true stress and strain equal nominal stress and strain, and all is right with the world. In this respect, engineers should count their blessings that they commonly deal with small strains. Biologists often aren't so lucky.

Just as tensile modulus E is the ratio of tensile stress to tensile strain, *shear modulus*, G, is the ratio of shear stress to shear strain:

$$G = \frac{\tau}{\gamma}. \qquad (13.18)$$

Similarly, the *tangent shear modulus* is

$$G_{\tan}(\gamma) = \frac{d\tau}{d\gamma}. \qquad (13.19)$$

Materials are stiffer in tension and compression than they are in shear. To be precise, if $v_{yx} = v_{zx} = v$,

$$G = \frac{E}{2(1+v)}. \qquad (13.20)$$

Shear modulus is never more than half the tensile modulus, and for an isovolumetric material (one whose Poisson's ratio is 0.5), G is a third of E. This relationship assumes the material is *isotropic*; that is, its properties are the same in all directions. Many biological materials are anisotropic (the properties of wood, for instance, are different with and across the grain), and in these cases equation 13.20 can be misleading. For a discussion of the dangers of using equation 13.20, see Vogel (2003).

If a material is viscoelastic, its modulus depends on the rate at which it is deformed. Recall that in our simple Voigt model for a viscoelastic material, the force required to extend a polymer is

$$F = \frac{k_b}{\ell_0} \Delta \ell + k_v \frac{\Delta\left(\frac{\Delta\ell}{\ell_0}\right)}{\Delta t}$$

$$= k_b \varepsilon + k_v \frac{d\varepsilon}{dt}, \qquad (13.21)$$

where k_b and k_v are values related to the material at a microscopic scale. Scaling up to the whole material, we can infer that for a viscoelastic material the relationship between stress and strain takes a similar form:

$$\sigma = E_S \varepsilon + \mu \frac{d\varepsilon}{dt}. \qquad (13.22)$$

Here E_S is the *storage modulus*, the contribution of elasticity to overall stiffness, and μ is the material's dynamic viscosity. Dividing stress by strain to calculate E (the overall

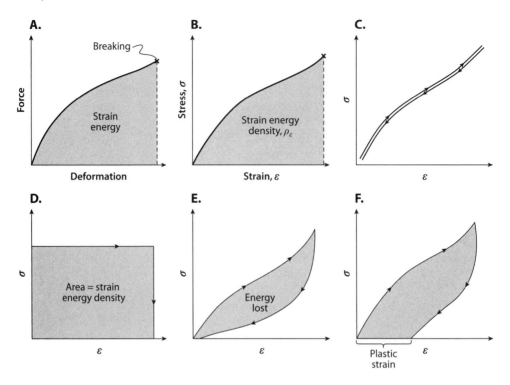

Figure 13.10 Defining strain energy (**A**) and strain energy density (**B**). In a purely elastic material, the stress-strain curve of retraction is the same as that of extension and energy is stored (**C**). Energy is not stored in a purely viscous fluid (**D**). Viscoelastic materials have properties intermediate between those of pure solids and pure fluids (**E** and **F**).

modulus), we find that

$$E = E_S + E_L \frac{d\varepsilon}{dt}, \tag{13.23}$$

where $E_L (= \mu/\varepsilon)$ is the *loss modulus*. If the material is strained very slowly (so that $d\varepsilon/dt \approx 0$), $E \approx E_S$, but when strain rate is substantial, $E > E_S$. This effect is shown in Figure 13.9D for mussels' byssal threads: the faster a thread is stretched, the higher its modulus.[7] For viscoelastic materials it is inappropriate to report a modulus without also reporting the strain rate at which that modulus was measured.

11   STRAIN ENERGY

Consider the force-deformation curve for a sample of a particular size (Figure 13.10A). Realizing that deformation is a measure of distance, we can conclude that the area under this curve (the integral of force times distance) is the

---

[7] The astute reader will have noticed something odd about equation 13.23: it predicts that—for a viscoelastic material subjected to a constant rate of strain—E decreases with increasing strain because $E_L (= \mu/\varepsilon)$ decreases with increasing strain. This is an artefact of the simple model we have used to predict the effects of viscosity. More realistic models avoid this quirk (see Supplement 13.3), but their added complexity adds little to the current message—stiffness of a viscoelastic material increases with increasing strain rate.

strain energy required to extend the sample. In this case, because we have extended the sample to its breaking point, the shaded area is the sample's breaking strain energy.

Of course, if we had a larger sample, more force would be required to deform it, and thus more energy would be required to break it. To describe the breaking energy of the material—rather than of a particular sample—we can again use stress and strain to normalize our measures of force and deformation (Figure 13.10B). Stress has the units of force per area, and nominal strain the units of length per length, so the integral of stress and strain has the units of

$$\frac{F}{\ell^2} \times \frac{\ell}{\ell} = \frac{\text{energy}}{\text{volume}}. \tag{13.24}$$

This quantity is known as *strain energy density*, ρ_ε.[8] As we will see, ρ_ε varies greatly among materials.

The viscous nature of viscoelastic materials affects their strain energy. The faster a viscoelastic material is deformed, the more force is required to reach a certain deformation and the larger the strain energy. Unlike the energy used to extend an elastic solid, however, the energy used to deform the viscous component of the material is not stored. Instead, it ultimately ends up as heat. Consider, for instance, Figure 13.10C, the stress-strain curve for a purely elastic solid that has been stretched to a certain strain and then returned to its initial state. Stress is required to strain the sample, and for this material each stress uniquely corresponds to a particular deformation. Thus, as stress is removed, the sample follows the same path back to its undeformed dimensions. As a result, the area under the unloading curve is precisely equal to the area under the loading curve—all the strain energy put into the material is stored and can be regained. This ability to store strain energy is a corollary to our definition of an elastic solid, and it is the reason why E_S is called the storage modulus.

By contrast, Figure 13.10D shows the stress-strain behavior of a pure fluid deformed at a constant strain rate. Here, a given stress doesn't uniquely correspond to a particular strain. Instead, as long as stress is applied, the material deforms at a constant rate. Because energy is the product of force and deformation, strain energy constantly acrues, but when the stress is removed, the rate of deformation instantly goes to zero; the material stops in its tracks and does not return to its original shape. As a consequence, the strain energy that went into deforming the material cannot be recovered. This is why E_L is called the loss modulus.

Figure 13.10E and F shows the cyclic stress-strain behavior of viscoelastic materials. Even if the material returns to its original dimensions (Figure 13.10E), strain energy has been lost to viscous processes. In Figure 13.10F, the material does not return to it original dimensions; instead, there is some residual, *plastic* strain. Here, too, energy is lost to viscosity. In either case, the energy lost—expressed as a fraction of the total energy expended—is the material's *hysteresis*. Conversely, the energy recovered, expressed as a fraction of the total, is the material's *resilience*.

The viscosity of liquids increases as temperature is reduced (Chapter 5). Similarly, the viscous character of viscoelastic materials is magnified at low temperatures.

---

[8] Note that to obtain ρ_ε in standard units of joules per cubic meter, it is advisable to use nominal stress and strain. True stress and strain can be used if the volume of the material is constant (see Supplement 13.4). However, if volume changes or other combinations of stress and strain are used—nominal stress and true strain, for instance—spurious values result.

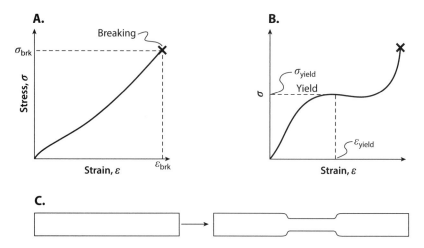

Figure 13.11 A. Failure in a brittle material. **B.** Yielding can be a form of failure. **C.** Some soft plastics neck down as they yield.

Resilience decreases and the modulus becomes very strain-rate dependent. These facts were demonstrated in spectacular and tragic form in 1986 when a solid-fuel booster of space shuttle *Challenger* exploded soon after takeoff. The spacecraft was launched in unusually cold weather, which caused the rubber O-rings sealing the boosters to lose their resilience and malfunction.

12   FAILURE

As we have seen, there are limits to how far materials can be strained before the chemical bonds that hold them together begin to give way and the material fails. Failure can take several forms. In the case shown in Figure 13.11A, stress rises with increasing strain until the material abruptly breaks, as indicated by the cross; σ_{brk}, the stress at which the material breaks, is its *strength*, and ε_{brk}, the strain at which it breaks, is its *extensibility*. Materials such as glass, coral skeleton, mollusc shell, and many rubbers fail in this fashion, and these materials are described as being *brittle*.

For other materials, failure takes a different form (Figure 13.11B). As strain increases, stress initially rises. But the material then *yields*—further extension requires no additional stress; indeed, stress may actually decrease. Only at much greater strain does the material break. The sharp inflection in stress defines the material's *yield stress* and *yield strain*. The soft plastic used to tether cans in a six pack of beer or soda has this type of behavior, which you can demonstrate for yourself. As you pull harder on the plastic it gradually extends, but then, all of a sudden, the material yields. The hydrogen bonds holding the plastic's polymer chains in place give way; the chains realign, becoming more parallel with the applied force; and the material necks down (Figure 13.11C). Because of the reduction in area, less force is required to extend the material, and its strain rapidly increases.

Yield can be a type of failure. If, in order to perform a certain function, a material is required to exhibit a certain stiffness, yielding can compromise that function even though the material stays in one piece. A leg bone that yields wouldn't be of much use to a running gazelle.

In other cases, yield can be functional. The ability to yield increases the area under the stress-strain curve, greatly increasing the energy required to break a material. For this reason, materials that yield are often described as *tough*. This term is intended to draw a contrast with materials that require little energy to break, a character often associated with nonyielding materials, that is, with materials that are brittle.

There is considerable room for confusion here. Rubber doesn't yield—and it is, therefore, brittle—but due to its large breaking strain, it nonetheless has a high breaking strain energy density, so in one sense it is tough. Bone does yield—so could legitimately be described as tough—but it still breaks at low strain and has a lower breaking strain energy density than rubber. When encountering tough and brittle in the literature, take care to determine how the terms are being used. We will deal with an alternative definition of tough in Chapter 15 when we deal with fracture mechanics.

13 CONCEPTS, CONCLUSIONS, AND CAVEATS

Whether a material is fluid or solid depends on the nature of the bonds that hold the material's molecules together. In general, the bonds in a liquid (dispersion and hydrogen bonds) are substantially weaker than the bonds in organic solids (covalent and ionic bonds).

Molecules bound up in crystals have little freedom to move. As a result, force imposed on a crystal is transmitted directly to the bonds between molecules, which resist being stretched. Thus, crystals are elastic but relatively inextensible. By contrast, the bonds in a linear polymer provide substantial flexibility, allowing the molecule to fold into a compact configuration that must be extended before force can be transmitted to the bonds themselves. As a result, linear polymers are often much more extensible than crystalline solids.

Rubbers take this tactic to an extreme. Their polymer chains are so flexible that thermal agitation can fold them into compact shapes, conformations that both maximize entropy and allow for extraordinary extensibility. Stretching the chain decreases its entropy, so for rubbers it is not the stretching of bonds that makes them elastic, it is instead nature's tendency to maximize disorder.

The concepts of stress, strain, modulus, and strain energy allow us to quantify the properties of materials, and we have briefly discussed the ways in which materials can fail. In the next chapter we will put these concepts to use as we survey the materials from which plants and animals are constructed.

It is important to note that this chapter is only an introduction to materials mechanics. In the interest of keeping things simple I have given short shrift to the nature of viscoelasticity and the methods by which it is measured. All biological solids have some viscous component, however, so if you want to pursue biomaterials mechanics in earnest, you will need to build on the introdution provided here by consulting texts on viscoelasticity. Gosline (2016) provides an overview, and Ferry (1980) is the bible.

Chapter 14

Biological Materials II
The Spectrum of Biological Materials

Evolution has provided organisms with an incredible smorgasbord of materials from which to choose, a variety far too extensive for us to cover in depth. Instead, this chapter takes a broad-brush approach to the spectrum of biological solids, beginning with an overview of the basic classes of natural materials: slimes, rubbers, fibers, and ceramics. For each we will explore the properties most often of interest in ecomechanics: stiffness, strength, and toughness. We then proceed to a brief explanation of how these component materials are combined to make structural composites such as bone, wood, and insect cuticle. At the end of the chapter, we will use our understanding of material properties to analyze the design of giant kelps and the pattern of disturbance in mussel beds.

1 STIFFNESS

When and where a particular material is useful is governed in large part by its stiffness. To grow tall against the pull of gravity, trees need wood that is stiff enough to support the weight of the crown. To run fast enough to escape a cheetah, the legs of an impala need bones that are stiff enough to act as an effective skeleton. On the other hand, to allow the impala's legs to move, its skin must be compliant. In response to biology's broad range of functions, evolution has produced materials with an equally broad range of stiffnesses (Table 14.1).

Slimes. Included in the slimes are all materials with a shear modulus less than 0.1 MPa, an arbitrary but useful criterion. The vast majority of slimes are mucins of one form or another: the slippery mucus that coats fish, the pedal mucus secreted by snails, and the mucins that form the matrix of materials such as seaweeds and sea-anemone body wall. All of these materials are formed primarily of water—mucus is typically more than 95% water and in some cases more than 99%.

Slimes make use of water primarily as a means to fill space. For all practical purposes water is incompressible, and in the habitats where slimy organisms are found, it is readily available. In slimes, water is converted into a viscoelastic solid by the addition

Table 14.1   Physical and Mechanical Properties of Biological and Synthetic Materials. Source: Data from Hale 2001, Wainwright et al. 1976, Denny 1993, Gordon 1976, Meyers et al. 2008, Gosline et al. 1999, Barber et al. 2015, and Cannell and Morgan 1987.

	E	σ_{brk}	ε_{brk}	ρ_ε	ρ
Slime	GPa	MPa		MJ·m^{-3}	kg·m^{-3}
Pedal mucus	2×10^{-7}	0.001	5	0.005	1035
Rubbers					
Resilin	0.002	3	1.9	1.9	1200
Abductin	0.004	6	1.5	4.5	1200
Elastin	0.002	2	1.5	1.5	1200
Fibers					
Spider silk	10	1100	0.27	160	1300
Collagen	1.5	150	0.12	7.5	1200
Cellulose	100	500	0.02	50	1200
Chitin	45	580	0.013	4	1200
Ceramics					
Calcium carbonate	137	75	0.0005	0.02	2700
Silica	25	200	0.008	0.8	2500
Composites					
Limpet teeth	120	4900	0.06	150	2000
Coral skeleton	60	40	0.0003	0.006	2000
Mussel shell	31	56	0.0018	0.05	2700
Bone	20	160	0.03	4	2300
Wood	6	100	0.01	0.5	500–1300
Keratin	4	200	0.5	50	1200
Arthropod cuticle	0.6	25	0.05	0.6	1200–1900
Algal stipe	0.008	2.1	0.3	0.3	1060
Synthetic Materials					
Steel	200	3000	0.015	20	7900
Glass	100	100	<0.001	<0.05	2500
Cement	4	4	<0.001	<0.02	2800
Fiberglass	70	300–1000	0.01	1.5–5	1750
Kevlar 49	130	3600	0.027	50	1440

of high molecular weight mucopolysaccharides (long polysaccharide chains attached to a short protein core) or glycoproteins (short saccharide chains linked to a relatively long protein core). Negatively charged carboxyl groups (COO$^-$) on the monomers cause these polymers to repel themselves and "puff up," incorporating water into the chain's interstices.

The process of puffing up helps to explain how many invertebrates and some seaweeds can produce extraordinary amounts of slime. For example, when bothered, the sea star *Pteraster* produces a volume of mucus several times that of its body. How could that huge volume have been packed into the animal? The answer, apparently, is that the pH within the mucus producing cells is sufficiently low that the acidic groups on the mucus polymers are not dissociated. Lacking these repellant positive charges, mucus molecules can be stored compactly (Verdugo et al. 1987). Only when released into the surrounding seawater do the acidic groups dissociate, causing the material to expand drastically.

Puffed-up mucopolysaccharides and glycoproteins occupy a volume so large that even at a concentration of 1%, the blobs they form more than fill the available space, forcing the polymers to intertwine and interact. This temporarily cross-linked network forms a gel (Smith 2002, 2006), and the interaction of the expanded polymers gives the gel some semblance of stiffness. For example, the pedal mucus of terrestrial slugs has a shear modulus of approximately 200 Pa (Ewoldt et al. 2007).

In addition to being used on their own, mucins find wide utility as matrix materials in soft composite solids. A prime example is jellyfish jelly, where mucin fills the spaces in a loose reticulum of collagen fibers.

Rubbers. The rubbery materials found in animals are all proteins—randomly coiled polypeptide chains cross-linked to form a three-dimensional network. Biological rubbers generally have elastic moduli in the range of 1 to 10 MPa. It's a strange quirk of nature that there are no rubbery materials in plants. Latex (the base material of natural rubber) is the viscous sap of rubber trees, but until the sap is artificially cross-linked, it isn't very elastic.

Although stiffer than slimes, rubbers aren't stiff enough to serve well as primary structural elements. Instead, their main use is in energy storage devices, with which we will deal shortly.

Fibers. In each of nature's fibrous materials, some portion of each polymer chain (proteins for collagen and silk; polysaccharides for cellulose and chitin) are tightly bound into ordered regions. While these regions do not have the latticelike order of inorganic crystals such as diamond, their defined structure nonetheless qualifies them as crystals, and this crystalline nature has the effect of making these materials very stiff. Their elastic moduli range from 1 GPa (collagen) to 100 GPa (cellulose), three to five orders of magnitude stiffer than rubbers.

Fibers are often used as stand-alone tensile elements. Collagen fibers, for example, are commonly used as tendons and ligaments—arranged so that they act in tension— and the silks in a spider's web form the tensile elements of an aerial net. We will see in a moment that fibers are also used as one component of composite materials.

Ceramics. The stiffest materials found in organisms are those formed from inorganic ceramics: calcium carbonate (calcite or aragonite) or silica (a hydrated form of silicon dioxide). Due to its highly structured lattice (which causes tensile forces to be placed directly on the bonds holding atoms together), calcite is very stiff ($E = 137$ GPa; Wainwright et al. 1976), and aragonite is likely to be similar. Silica has a slightly lower modulus (25 GPa; Meyers et al. 2008). For these ceramics, stiffness comes with a significant penalty, however—they are more than twice as dense as organic solids, so structures made from ceramics are heavy.

There are a few biological structures made from more or less pure ceramics. For example, the ossicles found in sea stars and sea urchins, the skeletons of reef-forming corals, and the eggshells of birds are made of nearly pure calcium carbonate, and the spicules of some sponges are pure silica. More commonly, however, ceramics are used as a component of composite materials.

2   STRENGTH

Next to stiffness, strength is perhaps the most important attribute of biological materials—few biological structures function properly if they break when loaded. Typical failure stresses of biological materials are listed in Table 14.1, a compilation notable in two respects. First, as I mentioned in Chapter 13, the strength of real materials never approaches the theoretical value we estimate for chemical bonds—on the order of 100 GPa. The current record holder for biological materials is the tooth of the limpet *Patella vulgata*, a composite of chitin and the ferric mineral goethite, with an average breaking stress of only 4.9 GPa (Barber et al. 2015). (We will account for this discrepancy in Chapter 15 when we explore fracture mechanics.) Second, the range of strengths is relatively narrow. While the stiffnesses of biological materials varies over more than nine orders of magnitude, strength covers only about six.

Slimes. The strength of slimes is difficult to measure because they behave primarily as fluids—if subjected to a continuously applied force they tend to flow rather than break. The strength cited in Table 14.1 is for pedal mucus when sheared beneath a snail's foot, where it yields at a stress of 1000 Pa.

Rubbers. The theory of rubber elasticity is so elegant that most experimental work on biological rubbers has been directed towards investigating the molecular basis of the material's modulus, and few researchers have bothered to tug on a sample until it snaps. Consequently, the values cited here are the result of educated guesswork. Biological rubbers can extend to strains of 1.5 to 1.9 before breaking, which, when combined with the measured modulus and an assumed linear stress-strain curve, allows us to estimate the breaking stress of rubbers at 3 to 6 MPa.

Fibers. Fibers are 10 to a 100 times stronger than rubbers (150–1100 MPa). Indeed, the strength of spider silk is comparable to that of steel and Kevlar, the material from which bulletproof vests are made.

Ceramics. You may be surprised that the strength of ceramics (on the order of only 100 MPa) is on par with collagen and two- to fivefold less than cellulose, chitin, and silk. As we will see, this relative weakness is the price these materials pay for being very stiff.

3   TOUGHNESS

Biological materials are often called upon to absorb (and sometimes store) energy. When a strand of spider silk is impacted by a flying insect, it must be capable of absorbing the insect's kinetic energy or the prey will escape. When a kangaroo hops, the tendons in its legs must be capable of storing the energy needed to propel the next bound. One might suppose that strength and the capacity to absorb energy would go hand in hand. But breaking energy depends on both strength and extensibility, and, among the component biological materials for which toughness is likely to matter (rubbers and fibers), those with high breaking stress generally have low breaking extensions (Figure 14.1). It all tends to even out such that the range of toughness is surprisingly small.

Slimes. As noted previously, the strength of slimes is an ill-defined characteristic (they yield rather than break), but if yield is viewed as a form of failure, slimes' toughness is approximately $0.005 \text{ MJ} \cdot \text{m}^{-3}$, a small value on the scale of such things.

Rubbers. Rubbers' failure strain energy densities are nearly a thousand times

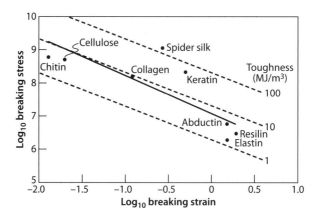

Figure 14.1 Among fibers and rubbers, breaking stress typically decreases as breaking strain increases. The solid line is the regression through the data points.

greater than those of slimes: 1.5 to 4.5 $MJ \cdot m^{-3}$. Indeed, they are among the highest of any biological materials, befitting the use of rubbers to store energy.

A classic example is provided by bivalves (clams, mussels, and scallops), who encase their bodies in a rigid shell. When the animal is bothered, the two valves are clamped tightly together by contraction of the large adductor muscle (which, among its other attributes, is quite tasty). When clammed up in this fashion, no living portion of the animal is open to the environment or available to predators. However, because the muscles are inside the shell, they can't provide the force required to open it—except in unusual circumstances (see Kier and Smith 1985), muscles can't push. Instead, the opening force is provided by abductin in the shell's hinge. As the shell closes, strain energy is stored in the rubber, which can then be used to reopen the shell.

Fibers. As noted earlier, the high strength of fibers compensates for their relatively low extensibility, giving them toughness comparable to that of rubber. The exception here is spider silk with a toughness of 160 $MJ \cdot m^{-3}$, the highest of any biological material. As with its stiffness and strength, silk's toughness is ideally suited to the job of catching flying insects, a utility that is enhanced by the fact that silk also has a high hysteresis (Denny 1976). Energy absorbed as an insect hits a web is dissipated as heat rather than being returned to the insect, allowing the web to act as a net rather than a trampoline.

Ceramics. The trade-off between extensibility and strength that works for fibers doesn't work for ceramics. Although fairly strong, calcium carbonate breaks at a strain of only 0.0005, and as a result its toughness is comparable to that of slime. Silica is a bit more extensible but still has a low toughness.

4   COMPOSITE MATERIALS

Slimes, fibers, and ceramics provide organisms with a wide variety of properties from which to choose, and we have briefly explored cases in which these materials function on their own. Often, however, these components are combined to make composite materials where the utility of the whole exceeds the sum of its parts. Fibers, in particular, have found exceptionally wide utility in biological composites.

Fibers are admirably strong and extensible, but to be useful in the construction of materials that shear or compress, they must be packaged in a way that inhibits their tendency to buckle. This is typically accomplished by winding fibers in the shape of the desired structural element and then embedding them in some sort of matrix—the matrix reduces fibers' likelihood of buckling, and fibers add strength to the matrix.

Bones provide an excellent example. The need to function effectively in locomotion imposes stringent design criteria on these skeletal elements: they must be stiff, strong, *and* tough. Collagen (a fiber) can provide the toughness but not the high modulus. On the other hand, calcium phosphate (a ceramic) can provide the stiffness but neither the strength nor the extensibility needed to be tough. Together, though, collagen and calcium phosphate form bone, a composite that meets skeletons' needs.

The interplay between fibers and matrix plays out differently in other materials. In sea-anemone body wall, the gel-like matrix is a millionfold less stiff than the collagen fibers it embeds, and the stiffness of the material is consequently much less than that of the fibers, allowing anemones to reinflate themselve after they have been disturbed (Supplements 14.1 and 14.2). By contrast, in arthropod cuticle, chitin fibers are embedded in a stiff matrix of cross-linked proteins, and the composite has a stiffness approaching that of chitin itself, an advantage when the material is used to make an exoskelton. Limpet teeth are stiffer still. Here the fibers are made of the mineral goethite ($FeO(OH)$) with chitin as the matrix. The resulting composite has a glasslike stiffness of 120 GPa, useful for a structure that scrapes diatoms from rocks.

The design of keratin—the material from which hair, fur, hooves, horn, scales, feathers, and beaks are made—emphasizes toughness rather than stiffness. Although it can form fibers, keratin is actually a composite material, combining α-helical protein chains with a stiff, cross-linked protein matrix. (Unlike all other materials we have discussed so far, keratin is produced inside cells, so keratinous structures are made of dead cells.) Keratin is less stiff than bone or wood ($E = 4$ GPa vs. 6 to 20 GPa), but the ability of α-helices to uncoil increases keratin's extensibility, giving it substantial toughness. The ability to absorb energy without breaking is a decided advantage for a material used in hooves, horns, and beaks that are subject to large impact loads.

As the reinforcing fiber in plant cell walls, cellulose is probably the most abundant polymer on Earth. Cellulose itself is the stiffest (and among the strongest) of organic fibers, but the stiffness and strength of its composites are highly sensitive to the properties of the embedding matrix. In the cell walls of seaweeds, the matrix consists of relatively low-modulus polysaccharides: agar, alginic acids, and other compounds. The modulus of the resulting material is low (on the order of 10 MPa) and its strength is only a tenth that of other fiber-reinforced composites. By contrast, cellulose fibers in woody plants are tied together by a highly cross-linked matrix of hemicellulose and lignin, and the resulting material (wood) has great strength (100 MPa) and a stiffness (approximately 6 GPa) only about tenfold less than cellulose itself.

In summary, fiber reinforced composites are incredibly adaptable, and they are consequently the most common structural materials in nature.

The utility of fiber-reinforced composites has not been lost on human engineers. Surfboards, auto bodies, boats, and bath tubs are commonly made from fiberglass, a composite of glass fibers embedded in an acrylic or epoxy matrix. Tennis rackets, golf clubs, and aircraft fuselages employ composites reinforced with carbon fibers. Because of their technological prevalence, fiber reinforced composites have been intensively studied, and the theory underlying their performance is well worked out (e.g., Bunsell

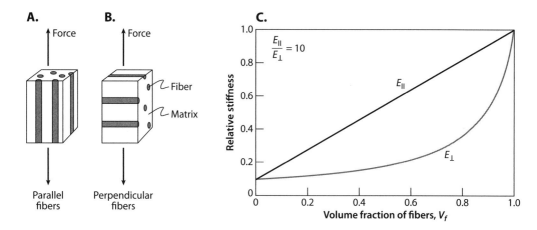

Figure 14.2 In a composite material, fibers can run either parallel to (**A**) or perpendicular to (**B**) a force's line of action, and this orientation affects the material's stiffness (**C**).

and Renard 2005; Daniel and Ishai 2005). The principles governing their strength will be covered in the next chapter; here we review the the theory of their stiffness (with derivations provided in Supplement 14.3).

The overall modulus of a fiber-reinforced composite depends on three factors:

1. the moduli of the fibers and matrix;
2. the relative amounts of each material in the composite; and
3. the orientation of the fibers relative to the applied load.

To see how these factors interact, we begin with a composite in which the fibers are aligned parallel to a tensile load (Figure 14. 2A). Fibers and matrix have moduli E_f and E_m, respectively ($E_f > E_m$), and occupy volume fractions V_f and V_m ($V_f + V_m = 1$). The overall modulus of the composite is

$$E_{\parallel} = V_f E_f + V_m E_m. \tag{14.1}$$

In short, when loaded parallel to the fibers, a composite's stiffness is the sum of its components' stiffnesses, each weighted according to its volume fraction. Overall modulus increases linearly with fibers' volume fraction (Figure 14.2C).

Alternatively, fibers and matrix can be aligned perpendicular to a tensile load (Figure 14.2B). In this case, the material's overall stiffness is

$$E_{\perp} = \frac{1}{\dfrac{V_f}{E_f} + \dfrac{V_m}{E_m}}. \tag{14.2}$$

Again fibers' effect depends on their volume fraction, but the increase in stiffness isn't linear. Modulus initially increases gradually with increasing volume fraction of fibers, but the increase becomes more rapid at high V_f. For any composite, stiffness is lower if fibers are perpendicular to the load than if they are parallel.

Composite materials reinforced by fibers running in only one direction are severely anisotropic—their stiffness is maximal when force is applied parallel to the fibers and minimal when the load is applied perpendicular to the fibers. If such a material is used in a situation where the direction of force is constant, anisotropy can be advantageous—fibers can be aligned in the appropriate orientation, minimizing the number needed and thereby minimizing the metabolic cost of producing the material. However, if the direction of the load varies or isn't known, anisotropy can be disastrous. A material capable of resisting a load in one direction can fail if the same load is imposed from another.

This problem can be avoided by laying down fibers in multiple orientations. Arthropods are the masters of this tactic. Chitin fibers in the cuticle of insects and crustaceans are produced in layers. Within a layer, all fibers run in the same direction, but that direction shifts from one layer to the next, usually by 5° to 10°. Once enough layers have been produced, there are fibers running in all directions, and the material's stiffness becomes isotropic.

Of course there is a cost to this jack-of-all-trades strategy. Whatever the direction of the applied force, some fibers are at right angles to it and contribute little to stiffness. Unless the modulus of the matrix is a substantial fraction of fibers' stiffness, the overall modulus of a composite with uniformly distributed fiber directions is only 37.5% of the maximum stiffness that would be obtained if all fibers were parallel to the load (see Supplement 14.4).

Stomatopods embody the full range of possibilities for fiber-reinforced composites. These benthic crustacea are found widely in nearshore waters, where they are renowned both for their unusual method of prey capture and their social interactions. Their feeding appendages superficially resemble those of praying mantises (hence stomatopods' common name: mantis shrimp), and they use a unique system of "latches, linkages, and lever arms" to strike out at prey (Patek et al. 2007). In some species, the distal segment of the appendage is used as a spear to stab prey; in others, the distal and near-distal segments combine to form a hammer used to break fish skulls and snail shells. In both cases, the strike is incredibly swift. Over the course of 2.7 ms, the business end of the appendage is accelerated to 23 $m \cdot s^{-1}$ (Patek et al. 2004; Patek and Caldwell 2005). The hammer can deliver a blow of 1500 N, thousands of times the stomatopod's weight, and the rebound from the blow is so fast that the water between hammer and prey cavitates. The implosion of the cavitation bubble delivers a second blow to the prey, amplifying the effect of the strike.

In addition to their function in prey capture, stomatopod feeding appendages are used in the process of divvying up territory among individuals. Given the potency of their weapons, mantis shrimp could easily kill each other. Instead, they chose to gauge their position in the local pecking order by conducting ritual fights where each animal hammers on the posterior of its opponent. The force of the blow indirectly tells each individual whether it is stronger of weaker than its opponent, allowing the social hierarchy to be maintained without killing the participants (Taylor and Patek 2010).

The chitin of the shrimp's exoskeleton is finely tuned to these functions. For example, the superficial layers of the hammer are hardened by the inclusion of calcium phosphate crystals aligned parallel to the surface (Weaver et al. 2012), and this resilient exterior is backed up by several layers of fiber reinforcement that safely spread the force of impact, allowing the appendage to survive thousands of blows. The acceleration of the feeding appendage (up to 104,000 $m \cdot s^{-2}$) is far too fast to be provided by muscle. Instead, muscles contract slowly, storing energy by bending various stiff elements of the appendage. The "loaded" appendage is then held in place by a latch (Patek et al. 2007; Zack et al. 2009). When the animal pulls its trigger, the stored energy is abruptly released.

In contrast to the resilient cuticle of the feeding appendage, the cuticle of the telson—the fanlike segments at the animal's posterior—is designed to absorb the energy of an attack. It is less calcified than the hammer, and its fibers are aligned differently (Taylor and Patek 2010). The larger the animal, the greater the fraction of energy the telson dissipates, and Taylor and Patek hypothesize that this size-dependent material property provides information that helps individuals to evaluate their social standing.

As notable as cuticle is for its mechanical properties, it is just as remarkable for the manner in which it is produced. Arthropods grow by periodically molting, requiring a new skeleton to be formed inside the old one. When the old, hard skeleton is discarded, the new one is pumped up to an increased size. This requires the new skeleton to be initially compliant. Once it has reached its new size, however, the skeleton must be hardened. Although the molecular basis of this hardening process is still uncertain, it undoubtedly involves cross-links between matrix molecules. As long as cross-links are few and far between, the matrix has a low modulus and the cuticle is compliant. The material can then be hardened by chemically crosslinking matrix molecules. Additional stiffness can be had by adding calcium carbonate or calcium phosphate crystals.

Table 14.1 summarizes the full panoply of biological materials: from compliant to stiff, pitifully weak to incredibly strong, and brittle to tough. The full story of how these properties interact with biology would fill several texts, and we will encounter a few examples as we continue to investigate ecological mechanics. I would be remiss, however, if I didn't give you a taste for how the knowledge of materials and their mechanics can be used in an ecological and evolutionary context.

5   DESIGNING A GIANT KELP

A glance back at Table 14.1 reveals that algal stipes (the "stems" of seaweeds) are outliers among composite materials. They are a hundred times more compliant and ten times weaker than any other composite, and although their extensibility is large (30%), it is not sufficient to give them a notably high toughness. Given that stiffer, stronger, more energy absorbant materials are possible, why would such a wimpy material have been retained in the seaweeds? As a partial answer to this question, we analyze the mechanics of the world's largest algae.

Giant kelps—such as the bull kelp, *Nereocystis*—are hundreds of times larger than typical seaweeds, but despite living on wave-beaten coasts, they have a surprisingly gracile form: a holdfast tethering the plant to the ocean floor, a ropelike stipe of length ℓ and cross-sectional area A, and a floating mass of fronds that takes advantage of its position near the water's surface to absorb light and photosynthesize (Figure 14.3). The seafloor is a crowded place, and the kelp's holdfast must compete for space with sessile invertebrates and other seaweeds. Consequently, as the stipe and fronds grow, the holdfast cannot expand its grasp on the seafloor to keep pace. The availability of space thus limits the force the holdfast can resist, and this often makes the holdfast the weak link in the kelp's chain. Given this basic architecture, what material properties are best suited to the kelp's survival?

We begin by estimating the maximal load a kelp must withstand. As a wave passes by, the frond mass goes with the flow, obtaining a maximum speed u, which gives the

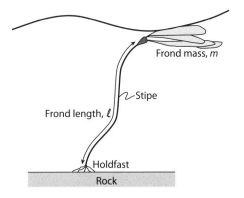

Frond mass, m

Stipe

Frond length, ℓ

Holdfast

Rock

Figure 14.3 The bull kelp, *Nereocystis luetkeana*.

frond mass (m) kinetic energy:

$$\text{kinetic energy} = \frac{mu^2}{2}. \quad (14.3)$$

However, the mass can travel only so far before it reaches the end of its tether, at which point the stipe pulls on the fronds, slowing them down and eventually bringing them to a halt. Of course, as the stipe pulls on the fronds, the fronds pull on the stipe, which is consequently stretched by the tensile load. If all goes well, when the fronds have been brought to a halt, the stipe has absorbed strain energy equal to the original kinetic energy of the frond mass, and the maximum force generated in the process ($F_{\max}$) has not dislodged the holdfast. Given these assumptions, our task is to specify how the stipe material's stiffness contributes to the plant's survival by minimizing $F_{\max}$.

We start with the definition of elastic modulus:

$$E = \frac{\sigma}{\varepsilon}.$$

Rearranging this expression, we see that maximum strain in the stipe is the ratio of breaking stress to modulus:

$$\varepsilon_{\max} = \frac{\sigma_{\text{brk}}}{E}. \quad (14.4)$$

Now, the strain-energy density absorbed by the stipe when loaded to σ_{brk} is the area under the stress-strain curve. If we assume that the material is Hookean:

$$\text{strain-energy density} = \frac{\sigma_{\text{brk}}\varepsilon_{\max}}{2}.$$

Substituting σ_{brk}/E for $\varepsilon_{\max}$, we see that

$$\text{strain-energy density} = \frac{\sigma_{\text{brk}}^2}{2E}. \quad (14.5)$$

Multiplying strain energy density by the volume of the stipe ($A\ell$), we estimate the maximum strain energy the kelp can absorb:

$$\text{strain energy} = \frac{\sigma_{\text{brk}}^2 A\ell}{2E}. \quad (14.6)$$

From the way I have proposed the problem, we know that this strain energy must be at least equal to the frond mass's kinetic energy. Therefore,

$$\frac{\sigma_{\text{brk}}^2 A\ell}{2E} = \frac{mu^2}{2}. \quad (14.7)$$

Solving for σ_{brk} we find that

$$\sigma_{\text{brk}} = \sqrt{\frac{mu^2 E}{A\ell}}. \quad (14.8)$$

Finally, by recalling that force is the product of stress and area (in this case, the cross-sectional area of the stipe), we can calculate the maximum force imposed on the holdfast:

$$F_{max} = \sigma_{brk} A$$
$$= \sqrt{\frac{mu^2 E A}{\ell}}. \qquad (14.9)$$

It is now clear why it is advantageous for the stipe to be made from a low-modulus material: the lower E is, the smaller the force pulling on the holdfast. Thus, given our assumption that the holdfast is the system's weakest link, the lower the modulus, the lower the risk of failure.

To give this exercise tangibility, let's plug in actual values. A typical large *Nereocystis* has a length of 10 m and a frond mass of 20 kg (10 kg of fronds with an additional 10 kg of entrained water). The cross-sectional area of the stipe is approximately 10^{-4} m^2, and the modulus of the stipe material is approximately 5 MPa. A 2-m-high wave would give the frond a velocity of about 1 m · s^{-1}. Putting all this together, we predict that the maximum force acting on the holdfast is only 32 N, which should be easily managed by a well-attached holdfast.[1] By contrast, if the stipe had the modulus of wood (6 GPa), the force imposed on the holdfast would be 35 times as large (1120 N, 256 lb), and the kelp would likely be dislodged.

As an added bonus, equation 14.9 helps to explain why the stipe is skinny; the smaller its cross-sectional area, the smaller the force on the holdfast. equation 14.8.sets the limit, though. The smaller A is, the larger the stress in the stipe, so the stipe can't get too skinny lest it (rather than the holdfast) become the weak link.

6   MUSSELS, PATCH DYNAMICS, AND DISTURBANCE

In many ecological communities, succession never has time to reach a climax before the community is disturbed and succession must start all over again. For these communities, the rate at which disturbances are imposed, and the size and distribution of the patches they create, can affect species diversity and community resilience. Nonequilibrium dynamics of this sort have been a major research area in community ecology for the last several decades, and the study of *patch dynamics* has focused attention on the role of the physical environment in structuring communities in both terrestrial and aquatic systems (e.g., Connell 1978; Pickett and White 1985; Levin et al. 1993). This physical/biological/ecological interaction is fertile ground for ecological mechanics.

The mussel beds of rocky shores offer an example. Not only do mussels exemplify the potential for mechanistic prediction of the rate of disturbance, they also provide the most comprehensive example to date of the potential of ecomechanics to integrate

---

[1] This force can also be easily handled by the stipe. When applied to the 10^{-4} m^2 cross section of the stipe, the resulting stress is 0.32 MPa—well below the 2.1 MPa the material can withstand— and the corresponding strain is 6.4%, only a fifth of what the material can support.

across scales—from molecules to communities—with the mechanics of materials at the core of the analysis.

Outside the tropics, mussels are the 600-lb gorillas of wave-washed rocky shores. The heat and desiccation that accompany low tides excludes their major predator (sea stars) from the midshore, so mussels there are to a large extent safe from predation. Within this refuge, mussels have the potential to outcompete other species for space on the rock. Their rapid growth, prodigious reproductive output, and tenacious grip allow them to form tightly packed beds that leave little room for seaweeds, barnacles, or anemones. Thus, where mussels dominate, the species richness of space-occupying species is low (Paine 1974). This tight coupling between mussel beds and community composition has made them a model system for observations and experiments in patch dynamics (e.g., Paine and Levin 1981).

Mussel dominance is never complete, however. At high tides mussel beds are subject to breaking waves and their potentially lethal hydrodynamic forces. Only if mussels can withstand the impact of the surf can they maintain their hegemony. Through an extensive and ongoing set of measurements, ecological mechanics is beginning to account for both mussels' tenacity and the forces to which they are exposed, allowing us to make ever more accurate predictions of when and where mussels will be dislodged.

Each mussel tethers itself to the rock using a flexible system of threads known as the byssus, with each byssal thread composed of four distinct components: glue that adheres to the rock, a plaque that anchors the thread to the glue, a stiff distal portion of the thread itself, and a more compliant proximal portion. Each component has its own chemical composition, which is reflected in its mechanical properties (Carrington et al. 2015).

The glue that sticks byssal threads to rocks is perhaps the best studied of all biological adhesives. There are five proteins in the glue, the most notable being two that have high mole fractions (20% to 30%) of the modified amino acid 3,4-dihydroxyphenyl-L-alanine, more commonly known as Dopa. Dopa readily forms cross-links both to other proteins and to the substratum, making it one of the stickiest substances on record. In the presence of oxygen, Dopa is readily oxidized to dopaquinone, which is much less adhesive, but mussels avoid this problem by secreting yet another protein that is rich in reactive thiol (SH) groups. The sacrificial oxidation of thiols rescues the Dopa, enhancing adhesion (Yu et al. 2011).

The adhesive plaque is the next link in the byssal chain. Made from four different proteins, its material expands into a foam as it sets. As we will see in Chapter 15, the gaps in this foam can act as crack stoppers, allowing the glue to resist the repeated application of force.

In turn, the plaque transitions to the byssal thread itself, which is constructed primarily from a consortium of three protein components. Each consists of a central stretch of collagen-like protein flanked on either end by proteins of a different sort. One of these components (preCol NG) is found throughout the thread; its flanking regions are formed from a protein similar to that found in plant cell walls. The second component (preCol D) has flanking regions that resemble spider dragline silk; preCol D is found primarily in the distal portion of the thread. The third component (preCol P) has flanking regions that resemble elastin, and this component is found primarily in the thread's proximal portion. Interspersed among these primary structural components is yet another protein (thread matrix protein), and the whole thread is coated by a proteinaceous cuticle that protects the thread from abrasion.

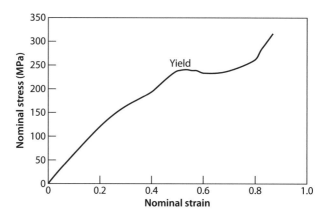

Figure 14.4 The stress-strain curve for a mussel byssal thread, showing yield.

Because it has both rubber- and fiberlike molecules, the stiffness of the thread's proximal portion (approximately 0.01 GPa) is intermediate between these two materials. Because of its silklike flanking regions, the distal portion of the thread is stiffer (0.2 GPa), and the two portions work together to give the thread the properties it needs to act effectively in the byssus. When tension is applied to a thread, it extends, but at a strain of approximately 0.5, it yields before again becoming stiff (Figure 14.4). The ability to yield assures that the load is shared equitably among threads in the byssus. Any thread that takes more than its load yields, and the consequent extension allows other threads to take up the slack. The result is an exceptionally effective tethering system that allows mussels to resist all but the most extreme forces imposed by waves (Bell and Gosline 1996).

With this information about mussels' mechanical capabilities, the ecomechanical analysis can proceed. Empirical measurements, wave theory, and the statistics of extremes (Chapter 22) can be used to predict when and where hydrodynamic forces will exceed mussels' tenacity (Denny 1995), and these predictions have been varified by field measurements (Carrington 2002). Information about when and where mussels will be dislodged can then be used as an input into the theory of patch dynamics to predict species richness and (as we will see in Chapter 23) provide a starting point for an analysis of self organization and criticality in mussel beds.

The elegant design of byssal threads may be compromised by ocean acidification. O'Donnell and others (2013) found that plaque strength decreases with a reduction in pH of the sort predicted for the next century. Taking this effect into account, we can (in theory) predict either the increased outlay mussels will incur to produce enough additional threads to maintain their tenacity, or the concomitant shift in patch formation and community dynamics as weakened threads lead to greater rates of dislodgment.

7   CONCEPTS, CONCLUSIONS, AND CAVEATS

The stiffness of biological materials is exceptionally variable and is tightly correlated with each material's function. Low modulus mucins take up space and provide adhesion and lubrication. Moderate-modulus rubbers store strain energy. High-modulus fibers combined with appropriate matrix materials provide design flexibility—everything from extensible kelp stipes to stiff skeletons. The ability to control the orientation of fibers in fiber-reinforced composites allows plants and animals to fine-tune the properties of their structural materials.

The material properties reported here must be taken with a grain of salt, however. They are typical values, but there can be substantial variation around them. Slimes and fibers are viscoelastic, for instance, so their stiffness depends on the rate at which they are strained. The stiffness of silk and keratin decreases drastically when the material is wet. Futhermore, the strengths reported here also suffer from reporting bias—with few exceptions they refer to pristine, flawless samples. But flaws are inevitable in nature. Scratches, cuts, and the vagaries of growth and development all can affect the properties of biological materials in the real world, and one must take care to keep these effects in mind when extrapolating from lab to field. We will begin this accounting in the next chapter as we deal with fracture.

We end with an added note of complexity: materials' properties can also change through time in response to changes in the environment. To cite a worrisome example, the ability of marine organisms to form crystalline composites may be jeopardized in the near future as the ocean absorbs ever increasing amounts of carbon dioxide. As CO_2 dissolves in water, it lowers the pH, with a resulting tendency to dissolve calcium carbonate and calcium phosphate.

Chapter 15

Fracture Mechanics and Fatigue

One of the notable messages of Chapters 13 and 14 was that biological materials never match the potential strength of the chemical bonds that hold them together, typically falling one to two orders of magnitude short. In this chapter, we account for this universal weakness. Along the way we will investigate how the minute imperfections intrinsic to materials have an impact vastly out of proportion to their size and how structures across a broad spectrum of sizes are designed to cope with these exigencies. We will explore the process of fatigue, which—even in the absence of biological senescence—can limit the lifetime of plants and animals, thereby affecting their life-history strategies. And we will see how fracture and fatigue can act as important feedback mechanisms linking organisms to their surroundings.

1 STRESS CONCENTRATIONS

First, let's explain why materials are weaker than they might be. The story begins with the calculations of C. E. Inglis, a mechanical engineer working in Britain at the beginning of the twentieth century. Inglis was interested in how holes affect the strength of steel plates (Inglis 1913), an important topic at a time when most large man-made structures—from ships to trains to skyscrapers—were held together with rivets. Installing a rivet involves drilling a hole.

Without going into the detailed mathematics of the situation, we can understand Inglis's results through consideration of Figure 15.1A. Here we have a plate loaded uniformly by tension applied at its ends. At its center is a circular hole, and the pattern in which force is transmitted from one end of the plate to the other is illustrated by *stress trajectories*, lines following the path of force as it is passed from one molecule to the next. Near the ends of the plate, stress trajectories are uniformly spaced, but because force cannot be transmitted across the hole, they bunch together in its vicinity, forming stress concentrations. Inglis calculated that stress near a circular hole is *three times* that elsewhere in the plate.

That's bad enough, but now imagine that we drill a second, smaller hole in the middle of the stress concentration (Figure 15.1B). The new hole amplifies the stress in its vicinity, tripling the already-trebled value. If we were to drill an even smaller

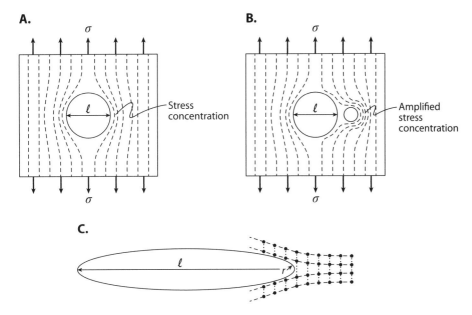

A.

B.

C.

Figure 15.1 A circular hole in a plate held in tension (**A**) creates a stress concentration that can be amplified by a second, smaller hole (**B**). A sharp-ended crack can be modeled as having length ℓ and a tip radius r (**C**).

hole in this new stress concentration, we could triple the stress yet again. Continuing in this vein, a series of ever smaller holes could be drilled, forming a sharp-ended crack with an immense stress concentration at its tip (Figure 15.1C). For a crack of length ℓ with a radius of curvature r at its tip (the radius of the smallest hole), Inglis found that the ratio of stress at the tip (σ_{tip}) to that in the bulk of the material (σ) is

$$\frac{\sigma_{tip}}{\sigma} = 1 + 2\sqrt{\frac{\ell}{2r}}. \tag{15.1}$$

When r is small relative to crack length, σ_{tip} can be much larger than σ.

This is a disturbing thought. It implies that for even a small stress applied at the ends of a plate, stress near the crack's tip could easily exceed even the theoretical breaking stress of the plate's material. Inglis's calculations seemed to suggest that one could walk up to the Golden Gate bridge with a diamond stylus, scratch a few small cracks into the supporting cables, and cause the entire structure to collapse. In fact, you wouldn't have to bother scratching the cables—all structures have naturally occurring cracks that can amplify stress in the same fashion. How, then, do structures stay intact?

There is nothing wrong with Inglis's calculations—sharp ended cracks can indeed produce tremendous stress amplifications—but our analysis to this point is incomplete. The second half of the story not only explains why the strength of all materials falls short of theoretical values, but also has much to say about the design of composite biological materials.

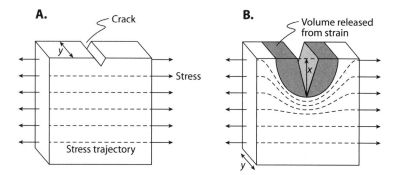

Figure 15.2 As a crack extends into the side of a sample (**A** to **B**), strain energy is released from a hemicylindrical volume.

2   ENERGY AND FRACTURE

In our discussion of chemical bonds, we noted that energy is required to separate one atom or molecule from another. Simply applying a force isn't sufficient; the force must be capable of acting over a distance in order to supply the necessary energy. It turns out that the need for energy—rather than just force—is what allows structures to survive.

Consider a material sample of thickness y stretched to a fixed length. Let's assume that the material is elastic, in which case the work it took to extend the sample is stored as strain energy. We now introduce a sharp-ended crack into the sample's side (Figure 15.2A). The resulting stress amplification is sufficiently large that the stress acting on the intact bonds near the tip exceeds the material's breaking stress. But to break these bonds, energy must be supplied. As long as the overall length of the sample doesn't change, this energy can't be provided by the force acting on the sample's ends. Instead, the energy needed to break bonds must come from within the sample itself, that is, from its stored strain energy.

Let's suppose that, as the crack extends into the material, strain energy is released from a hemicylindrical volume around the crack (Figure 15.2B) and that this energy can be used to break bonds at the crack's tip. If the crack is x long and the sample thickness is y, the hemicylinder's volume is $\frac{1}{2}\pi x^2 y$. If the strain-energy density stored in the material is ρ_ε J·m^{-3}, the total energy released by a crack of length x is

$$\text{total strain energy released } = \frac{1}{2}\pi x^2 y \rho_\varepsilon. \tag{15.2}$$

As the crack extends, the incremental energy released is

$$\text{incremental energy available} = \frac{d\left(\frac{1}{2}\pi x^2 y \rho_\varepsilon\right)}{dx}.$$
$$= \pi x y \rho_\varepsilon \tag{15.3}$$

Note that the energy released by each small, local extension of the crack depends on *overall* crack length, x. The longer the existing crack, the more energy released by the next small increment.

To be useful in extending a crack, strain energy must be delivered to the crack's tip, and the material's shear stiffness provides the mechanism for delivery. As a crack

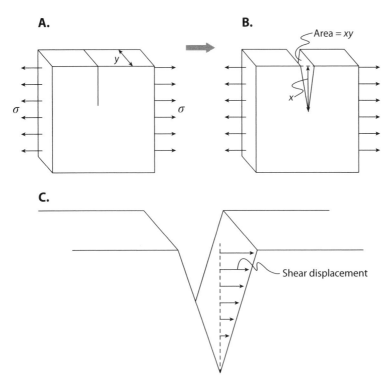

Figure 15.3 As a crack propagates into a sample (**A**), material retracts and is sheared (**B**). **C**. Detail of **B** showing shear.

extends, the material behind its tip is loaded in shear (Figure 15.3A–C), and the resulting deformation transmits strain energy to the crack tip.

Next, let's suppose that for a crack of length x in a material y thick, the energy needed to form a crack is proportional to xy, the area over which bonds are broken:

$$\text{energy required} = xy\rho_{\text{frac}}, \qquad (15.4)$$

where ρ_{frac} is the *fracture energy density*[1] ($\text{J} \cdot \text{m}^{-2}$). (Note that density in this case denotes energy per *area* whereas the density in strain energy density is energy per *volume*.) The incremental rate at which energy is consumed in extending the crack is

$$\text{incremental energy required} = \frac{d\,(xy\rho_{\text{frac}})}{dx}$$

$$= y\rho_{\text{frac}}. \qquad (15.5)$$

Available and required incremental energies are graphed in Figure 15.4 as a function of crack length. At first, the energy available from the incremental release of strain energy is less than the energy required to extend the crack. As a result, even though force per area at the crack tip exceeds breaking stress, the crack cannot

---

[1] In the literature, fracture energy density is usually symbolized by G or T (see Mach, Nelson, and Denny 2007), symbols which have other meanings in this text. To avoid confusion, I use the nonstandard symbol ρ_{frac}.

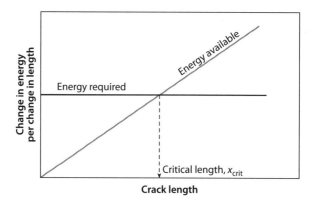

Figure 15.4 Incremental energy available to extend a crack increases with crack length while incremental energy required for crack extension is constant.

extend. But, once x exceeds a critical length x_{crit}, energy released is greater than energy required, and the crack propagates spontaneously. Setting incremental energy required equal to that released and solving for x, we see that:

$$x_{crit} = \frac{\rho_{frac}}{\pi \rho_\varepsilon} \tag{15.6}$$

The greater the fracture energy density of the material, the longer the crack must be to become self-sustaining. The larger the strain-energy density, the shorter the critical crack.

We can take this analysis one step farther. For a Hookean material, strain-energy density is the area under the stress-strain curve:

$$\rho_\varepsilon = \frac{\sigma \varepsilon}{2}. \tag{15.7}$$

But strain is

$$\varepsilon = \frac{\sigma}{E}. \tag{15.8}$$

Thus,

$$\rho_\varepsilon = \frac{\sigma^2}{2E}. \tag{15.9}$$

It may seem odd that the stiffer the material, the lower its strain energy density, but this is again due to the fact that energy is the product of force and distance. For a given force, the stiffer the material, the smaller its deformation, and the lower its strain energy density.

Substituting equation 15.9 into equation 15.6, we find that

$$x_{crit} = \frac{2E\rho_{frac}}{\pi \sigma^2}. \tag{15.10}$$

The greater the stress applied to the material, the shorter the critical crack. For a given stress, the stiffer the material and the higher its fracture energy density, the longer the critical crack.

3   BREAKING STRESS

With that thought in mind, let's turn equation 15.10 around to tell us what stress is required to make a crack grow spontaneously. In other words, let's use our knowledge of fracture mechanics to calculate a material's strength. If the largest crack in a sample has length x_{max},

$$\sigma_{brk} = \sqrt{\frac{2E\rho_{frac}}{\pi x_{max}}}. \tag{15.11}$$

The longer the crack a material contains, the lower the stress at which it breaks.

Consider diamond, for instance. It has a theoretical breaking strength as high as 225 GPa and a modulus of approximately 10^{12} Pa (Telling et al. 2000). We earlier calculated that it has bond energy density of 18.2 J·m^{-2}, which for the moment we use as a substitute for ρ_{frac}. If a diamond has a crack a mere 23 nm long—the length of only a few hundred atoms—its breaking strength is reduced by a factor of 10. A crack 2.3 μm long reduces its strength by a factor of 100. All materials have small flaws, so this exercise goes a long way toward explaining why real biological materials are weaker than their chemical bonds would potentially allow.

In Chapters 13 and 14 we treated breaking stress as an intrinsic property of a material, but equation 15.11 tells us that material strength is actually a structural property—it depends on crack size. Because the size of flaws is likely to vary from sample to sample, the apparent strength of the material is likely to vary as well.

Glass provides a classic example. Large panes are notoriously fragile, but as anyone who has spent time working with them can tell you, they are also fickle. Some panes are forgiving—you can drop them and they just bounce. Others fall apart under the slightest provocation. The variability is due to the random presence of small cracks—some panes have them, others don't.

The random presence of cracks leads to an odd conclusion. The bigger the pane, the more likely it is to have a crack, and—on average—the weaker it should be. Conversely, the smaller the sample, the higher the average strength. This helps to explain the utility of glass in fiberglass. If thin glass fibers were as fragile as large panes, they would be useless as a reinforcing material. Instead, because the fibers used in fiberglass are so thin (a few micrometers in diameter), they are relatively strong. For example, Griffith (1921) found that glass has a modulus of 62 GPa and a fracture energy density of 43 J·m^{-2}. An intact fiber 10 μm in diameter cannot contain a crack longer than 10 μm, so by setting x_{max} to 10 μm we can estimate the minimum breaking stress for a fiber this size:

$$\sigma_{brk} = \sqrt{\frac{2 \times 62 \times 10^9 \times 43}{\pi \times 10^{-5}}} = 412 \text{ MPa}. \tag{15.12}$$

Thus, 10-μm glass fibers are approximately as strong as the strongest biological fibers (silk and cellulose, 500 to 1000 MPa, see Table 14.1), hence their utility in fiberglass. Finer fibers would be stronger still since the theoretical strength of glass is 11 GPa (Griffith 1921[2]).

---

[2] If you get a chance, read Griffith (1921); it is an astounding piece of insight. Not only does Griffith derive the basic explanation for the energetic constraints on crack propagation, but he also uses an innovative experimental technique to directly measure the bond energy of glass.

Limpet teeth take advantage of this principle. The mineral fibers in their goethite-chitin composite are only 20 nm in diameter, thin enough that their strength approaches the theoretical limit (Barber et al. 2015).

Because bigger samples of a material are more likely to contain larger flaws—and thereby be weaker—it is possible that the random distribution of cracks could limit the size of organisms. We will return to this thought later in the chapter.

4   FRACTURE TOUGHNESS

Given that breaking strength is sensitive to crack length—and therefore is not an intrinsic property of a material—it behooves us to look for a better descriptor of material strength. Equation 15.11 again provides the answer. Breaking stress depends on $\sqrt{E\rho_{\text{frac}}}$, and because both E and ρ_{frac} are intrinsic properties of a material, we can think of $\sqrt{E\rho_{\text{frac}}}$ as itself a material property—the ability to resist crack formation. $\sqrt{E\rho_{\text{frac}}}$ is known as *fracture toughness*; it has units of $\text{Pa} \cdot \text{m}^{1/2}$, and is symbolized by K_C (Mach, Nelson, and Denny 2007). The stiffer the material and the greater the energy required to extend a crack, the greater its fracture toughness.

Note that fracture toughness depends on both modulus and fracture-energy density. If having a high modulus were the only factor contributing to fracture resistance, materials such as coral skeleton ($E = 60\,\text{GPa}$) and mussel shell ($E = 31\,\text{GPa}$) should be relatively fractureproof. Instead, they are much more prone to fracture than is keratin, which has a modulus 10 times smaller ($E = 4\,\text{GPa}$). The difference is in the fracture energy density, ρ_{frac}, and therein lies an important story.

In Chapter 13 we calculated bond energy density, ρ_{bond}, the energy required to break the chemical bonds in a square meter of material. One might assume—as we just did in our calculation of crack length in diamond—that this is the same as ρ_{frac}, the fracture-energy density, but in fact the two can be quite different: ρ_{bond} is the energy that must be delivered to the crack's tip, the net energy required to propagate a crack. However, transfer of strain energy to the crack tip is never 100% efficient. As a crack grows, some energy is dissipated by viscous processes as the material deforms, some is lost to plastic deformation, and only the remainder is available to break bonds. As a consequence, fracture-energy density—the gross release of energy—is always greater than bond energy density—the net. Sometimes, ρ_{frac} can be much larger than ρ_{bond}.

The difference between ρ_{frac} and ρ_{bond} is accentuated by a subtle assumption involved in the definition of bond energy. As a crack extends a distance x, we assume that bonds are broken over area xy (Figure 15.5A). For many materials, the path of fracture isn't so clean and planar (Figure 15.5B). Instead, as it advances, the crack wanders up and down along a complex three-dimensional path, and as a result the area over which bonds are broken is much larger than xy. The larger number of bonds requires a larger input of energy, boosting ρ_{frac} and, by extension, fracture toughness.

Fracture energy density can be increased by other mechanisms as well. Consider a crack propagating through a plant. As the crack arrives at an intact cell, the anisotropic material of the cell wall is placed in tension, and the cellulose fibrils realign to become more parallel to the applied load. But in the process, fibrils shear the matrix in which they are embedded, causing the inner layers of the wall to crumple like a twisted soda can. This crumpling absorbs a lot of energy, increasing ρ_{frac} (Farquhar and Zhao 2006).

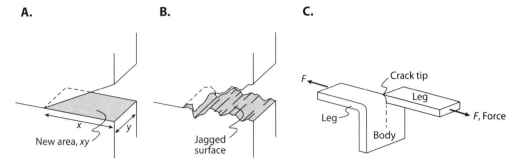

Figure 15.5 Crack area is nominally xy (**A**), but in many materials, actual area of the new surface is much larger (**B**). **C.** A trouser tear test to measure fracture-energy density.

Table 15.1 Fracture-Related Properties of Some Materials.

Material	ρ_{frac} (J·m^{-2})	E (GPa)	K_c ($10^6 \times$ Pa $\sqrt{m}$)
Horse-hoof keratin	5200–13,500	4	4.6–7.4
Bone	400–2900	40	4.0–11
Grasses	850–21,650	0.2	0.4–2.1
Marine algae	256–2420	0.01	0.05–0.16

Because strain energy can be drained from material in a variety of ways, it is difficult to predict ρ_{frac} from first principles. Instead, fracture energy is typically measured empirically using one of two methods (Mach, Nelson, and Denny 2007). In the first, a crack of length x is introduced into a sample and the stress σ_{brk} required for fracture is then measured. Because the sample breaks, $x = x_{crit}$, and

$$K_c = \sigma_{brk} \sqrt{\frac{\pi x_{crit}}{2}}. \tag{15.13}$$

The modulus of the material is then measured in a separate experiment. One can then rearrange equation 15.11 to compute ρ_{frac}:

$$\rho_{frac} = \frac{K_c^2}{E}. \tag{15.14}$$

Alternatively, one conducts a trouser-tear test. A slice is made in a rectangular test piece, forming two "legs." The legs are then pulled apart, and the force F required to tear the trouser is measured (Figure 15.5C). If y is the thickness of the unstressed sample,

$$\rho_{frac} = \frac{2F}{y}. \tag{15.15}$$

Fracture toughness has been measured for a few biological materials (Table 15.1), data which reinforce the conclusion reached earlier—it isn't modulus or fracture energy density alone that determines toughness, it is their product. A stem of grass is 200-fold less stiff than bone but, due to its remarkable fracture-energy density, it is only about 10-fold less tough. On the other hand, seaweeds are 2 orders of

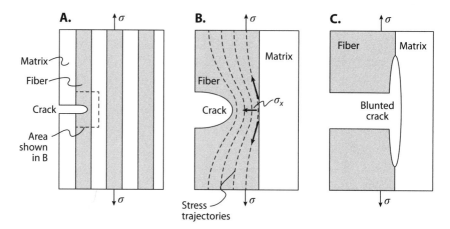

Figure 15.6 As a flaw extends from the matrix into a fiber (**A**), reorientation of stress trajectories (**B**) causes the matrix to separate from the fiber, blunting the crack (**C**).

magnitude less resistant to fracture than is bone, although their ρ_{frac} is comparable. In this case, seaweeds' fragility is due to their low modulus, which causes them to effectively convert stress to stored strain energy, energy that can be used to fuel crack propagation. Compared to the higher-modulus sea grasses found in the same habitat, seaweeds are easily 10 times as prone to fracture, a difference you can demonstrate for yourself. The simplest way to break the stipe of a giant kelp is to nick it with your thumbnail; slight tension then causes the stipe to fracture. The same procedure is futile for a blade of seagrass.

5   CRACK STOPPERS

As we have seen, three criteria must be met to propagate a crack:

1. Local stress at the crack tip must exceed the material's breaking stress,
2. Sufficient energy must be made available to meet the demands of propagation.
3. There must be a mechanism to deliver this energy to the crack tip.

In our analysis of crack propagation we have taken Inglis's warning to heart and have implicitly assumed that criterion 1 is met, but it is time to cast a critical eye on this assumption.

In our discussion so far we have assumed that a crack propagates through a single, uniform material. Interesting things can happen when a second material is brought into the system, as in the case of fiber reinforced composites. Consider the simple composite shown in Figure 15.6A where a crack is propagating through a fiber, headed toward the adjacent matrix. At the crack tip (Figure 15.6B), there is the type of stress concentration discussed earlier in the chapter.

Note, however, the direction of tension in the stress trajectories near the crack tip. In addition to applying a force parallel to the fiber's axis, they also have a component pulling on the material at right angles to the fiber. When the crack reaches the matrix, this lateral force can cause fiber and matrix to separate (delaminate), as shown in Figure 15.6C. As the crack propagates through the fiber and reaches the delamination, that delamination becomes an extension of the crack, but an extension with its long axis parallel to the applied force. Where before there was a sharp end to a long crack, producing the stress amplification needed to fulfill criterion 1, there is now a short crack (short, that is, across the width of the sample) with a blunt end. According to equation 15.1, the new crack has negligible stress amplification, and the local stress may no longer be greater than the material's strength. In short, when the crack runs into the matrix, it is blunted. This tactic—known as a *Cook-Gordon crack stopper* after the the two scientists who discovered and explained it—tends to make fiber-reinforced composites exceptionally tough.

The Cook-Gordon strategy works at both the micro- and macroscopic levels. The vascular tissue in a blade of grass, for example, serves to stop cracks from propagating across the blade. It is much easier to tear the blade along the length of the fibers, that is, along the length of the blade. The same basic strategy is found in many tropical seaweeds (Padilla 1993). In both grasses and seaweeds, cracks are often initiated by grazers. Structural redirection of a propagating crack thus allows plants to minimize tissue loss resulting from herbivory.

The Cook-Gordon mechanism also contributes to the fracture resistance of mollusk shell. Shell material is constructed much like an intricate brick wall, with calcium carbonate crystals forming the bricks and a protein matrix as the mortar. The fracture toughness of $CaCO_2$ is quite low, so a crack can easily propagate through a "brick," but the mortar's toughness is high, and the crack can be blunted. At the very least, the organic matrix (which amounts to less than 0.1% of the material's weight) causes a crack to take a circuitous route, increasing fracture energy density, and decreasing the likelihood of catastrophic failure.

Sea urchins have taken this one step farther. The urchin's skeleton is one single crystal, but a protein matrix is included *within* the crystal's calcium-carbonate structure (Berman et al. 1990). The protein tends to direct cracks against the "grain" of the inorganic crystal, thereby increasing fracture energy density.

So far, we have treated fracture as a purely destructive process; something to be avoided at all cost. There are at least two cases, however, in which fracture provides a distinct advantage: worms' burrows and birds' eggs. Across a wide range of particle sizes and water contents, mud can be brittle, a fact utilized by worms to reduce the force needed to extend their burrows (Dorgan et al. 2005, 2007). Similarly, when a chick is ready to hatch, it must break its way out of the shell, so it is necessary to have a shell material that is susceptible to fracture. In both cases, the weakness incurred by cracks is put to good biological use.

6   FATIGUE

To this point, we have considered what happens when a single cycle of stress is applied to a material containing a crack. If the crack is less than critical length, the material survives; if the crack exceeds x_{crit}, the material breaks. However, nature seldom operates in such binary fashion. Plants and animals are subjected to repeated stresses, and this repetition can lead to unexpected results.

To get a feel for the discussion that follows, try the following experiment. Take a metal paper clip, unfold it, and then repeatedly bend it at one of its kinks. After a few cycles, the metal fatigues and breaks. The farther you bend it in each cycle—that is, the more stress you apply—the fewer cycles required for failure.

There is nothing in our discussion of materials so far that would predict this phenomenon. From our current understanding, deforming a material to the same strain in each cycle should take the same force. If the first force isn't sufficient to break the sample, why should repeated application of the same force lead to failure? More importantly, does the same phenomenon happen in biological materials? If so, they might fail at loads smaller than what we would currently predict, with all sorts of ramifications.

The analysis of fatigue in biological materials has not been well developed, so the discussion that follows is a mixture of empirical measurement and informed speculation.

6.1   Crack Growth

Central to the analysis of fatigue is the concept of stable crack growth. It is supposed—although seldom demonstrated—that an intact material contains microscopic cracks. When a load is applied, stress is never precisely uniform. Local variations in cross-linking, fiber orientation, and so on, can lead to local variations in stress, and in areas of higher stress cracks may begin to propagate. If the overall stress on the material is well below breaking stress, these small cracks usually do not propagate catastrophically. Instead, they grow until they hit an area of lower-than-average stress, whereupon they stop. Alternatively, they might extend into a more compliant portion of the material, where viscoelastic deformation at the crack tip increases the tip's radius, reducing the stress amplification, and again bringing the crack to a halt. In short, one application of a sublethal load causes a small increment in the length of some cracks.

If the load is released and then reapplied, the spatial distribution of stress might be slightly different, or cracks that were blunted might have returned to sharpness. In either case, flaws can again extend some incremental amount, growing from one cycle to the next. Eventually, after enough cycles, one crack reaches critical length, and the sample breaks. This, then, is the basic idea of fatigue: stable extension of intrinsic flaws eventually leads to catastrophic crack propagation.

6.2   Documenting Fatigue

This process has been observed in detail in bone and mollusk shell (Kuhn-Spearing et al. 1996; Taylor et al. 2007). These studies took advantage of the propensity of some dyes to attach to the exposed faces of calcareous crystals (e.g., calcium carbonate, calcium phosphate). Samples of bone or shell were soaked in the dye, sectioned, and examined under a microscope, allowing the researchers to visualize the initial size of cracks. Other samples were subjected to repeated loads before being dyed and sectioned, allowing the researchers to document crack growth.

Unfortunately, this method is currently limited to calcareous crystalline composities. In other materials, fatigue has been analyzed by empirical measurements of the consequences of crack growth. A variety of methods have been used (see Mach, Hale, et al. 2007 and Gosline 2015 for reviews), but one is particularly straightforward: an intact sample is cyclically loaded to a known and constant stress until it breaks, and

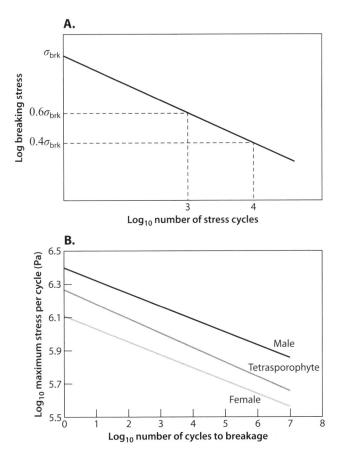

Figure 15.7 **A**. Breaking stress decreases with the number of load cycles. **B**. Fatigue in *Mazzaella flaccida* (redrawn from Mach (2009)).

the number of cycles to failure is recorded. The experiment is then repeated for a different stress, leading eventually to a graph such as that shown in Figure 15.7A. This relationship can then be used to predict the number of cycles of a given stress required to break the material. For example, if a stress 40% of one-time breaking strength requires 10^4 cycles to break the material, each cycle takes the material 10^{-4} of the way toward failure. It might take only 10^3 cycles at 60% of σ_{brk} to reach failure, so each cycle at this stress level advances the sample 10^{-3} of the way. If we know the stress imposed by each loading cycle, we can sum up the material's incremental progress toward breaking and predict its survivorship.

Mach et al. (2011) used this approach to estimate the survivorship of a common intertidal seaweed (*Mazzaella flaccida*) when exposed to ocean waves. Maximum water velocities were measured wave by wave at sites where *M. flaccida* is commonly found. These time series were used to calculate the probability of breakage, and the probabilities were then compared to the survival of actual plants. The predictions closely matched the measurements, indicating that fatigue plays an important role in the survival of this species.

7 FRACTURE AND FATIGUE IN EVOLUTION

Fatigue failure has implications for both evolution and ecology. As a red alga, *M. flaccida* has a complicated life history. An individual thallus might be a male, a female, or a tetrasporophyte, and fatigue resistance differs among these life-history stages (Figure 15.7B). Males are more resistant than tetrasporophytes, and both are more resistant than females.

The relative fragility of females is due in part to the fact that their reproductive structures act as crack initiators. When gametes are released, a hole is formed in the blade, which concentrates stress and increases the rate of fatigue. For equal-size blades subjected to the same flow-induced forces, the fatigue lifetime of females is thus less than that of males, with possible consequences. Females could extend their survival by limiting the size of their blades since smaller blades incur less force. But smaller blades are likely to produce fewer gametes, so there is a trade-off between survival and reproductive output.

Fracture and fatigue can also provide seaweeds with adaptive feedback from the hydrodynamic environment. During the summer, when waves are calm, rockweeds such as *Fucus gardneri* can grow large, putting the plants in danger of dislodgment when storm waves arrive in the fall. Seaweeds have no mechanism to actively decrease their size, but *Fucus* has evolved a structural design that allows it to shrink passively by taking advantage of fracture. The branching points in the plant's distal parts are weaker than the rest of the plant, so when *Fucus* is subjected to large waves, fracture at these branching points prunes the plant to a size appropriate for current wave conditions (Blanchette 1997). *Fucus*'s design is thus reminiscent of the perforations in toilet paper—local weakness is used to initiate graceful failure.

Fatigue can also affect predation. Clams, for example, build robust shells capable of withstanding the loads imposed by a crab's claws. Crabs have learned, however, that repeated application of a sublethal stress can eventually fatigue a shell (Boulding and Labarbera 1986). A small crab can garner a large meal by squeezing a few hundred times on a shell that easily rebuffs a single attack.

When given a choice, crabs prefer small clams, which they can break with a single crunch. But when competiton for food is intense, small clams can be scarce. In this case, there is strong selection pressure for any adaptation that allows an individual crab access to larger clams. Larger claws or stronger muscles would do the trick, but acquiring these structural adaptations involves more cost than does the simple shift in behavior that tells a crab to squeeze repeatedly. Thus, the fatigue characteristics of mollusk shell may have played a role in the behavioral evolution of crabs.

One of the best examples of a natural crack stopper is found in erect coralline algae, which have managed to conquer fatigue. These unusual seaweeds calcify their cell walls—which makes them rigid—but the plants maintain their overall flexibility by having joints known as genicula (Figure 15.8A). In members of the family Corallinacea, each joint is a single tier of cells, each cell acting as a cable linking the calcified rocky segments at its ends. The cell wall material of these genicular cells has both a relatively high modulus (27 MPa) and a high breaking strain (1.2), setting it apart from other algae (with typical moduli of 10 MPa and breaking strains of 0.3). As a result, the overall flexibility of a coralline frond is similar to that of other seaweeds even though only the genicula—which make up approximately 15% of the frond—actually deform.

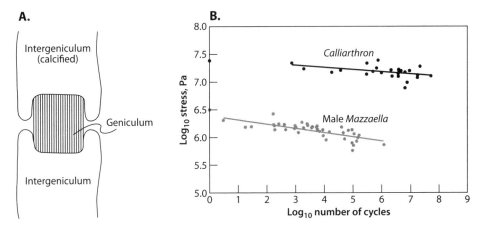

Figure 15.8 The genicula of an articulate coralline alga (**A**) are much more resistant to fatigue than are the blades of *M. flaccida* (**B**). (Redrawn from Denny et al. (2013).)

Because each genicular cell acts as an independent cable, the joint material as a whole has a low shear stiffness. As a result, if one cell breaks, its subsequent deformation can't pull on adjacent cells, and the release of its strain energy cannot be used to propagate a crack. Thus, geniculae should be highly resistant to fatigue.

My colleagues and I tested this hypothesis in *Calliarthron cheilosporioides*, a large erect coralline common on the shore of California (Figure 9.9A), with the results shown in Figure 15.8B (Denny et al. 2013). Not only does it take nearly a hundred times as much stress to break *C. cheilosporioides* as it does to break the most robust form of its neighbor, *M. flaccida*, the coralline can survive a thousand times as many cycles. When loaded to 50% of its one-time breaking stress, *C. cheilosporioides* can withstand more than fifty million cycles before breaking. This is many times the number of stressful cycles the seaweed would ever see in its lifetime—6 y or less— so it is essentially immune to fatigue failure, a factor that likely contributes to it ability to dominate the competition for space in its wave-swept habitat.

8   CONCEPTS, CONCLUSIONS, AND CAVEATS

Flaws concentrate stress, making materials weaker than their chemical bonds allow. The fact that materials have any strength at all is due to the mechanics of crack propagation. Even though stress often exceeds critical values locally, a crack can propagate spontaneously only if sufficient strain energy is available and there is a mechanism for delivering that energy to the crack's tip.

Repeated stress can cause intrinsic flaws to grow, eventually leading to failure by fatigue. For at least some plants and animals, their ability to resist fatigue determines their survivorship in stressful environments.

Despite its obvious importance to the survival of plants and animals, fracture mechanics and its application to fatigue in living organisms have received distressingly little attention. The values listed in Table 15.1 are virtually all that are known for fracture of biological materials. There is a desperate need for both a broad survey of the fracture mechanics of biological materials and focused research on the mechanisms of crack initiation and growth.

Of course, this plea is easier made than answered. There are two major hurdles facing the study of fracture and fatigue in biological materials. First, the theory described in this chapter is only a first approximation of the fracture mechanics of real materials. It assumes, for instance, that a material is isotropic and has a linear stress-strain curve. In the jargon of the field, the simple theory outlined here is known as *linear elastic fracture mechanics*, often referred to as LEFM. We have touched on one consequence of nonlinear material properties—energy dissipation by plastic flow and viscosity—but for a broader understanding of biological fracture and fatigue, one must use theories with less restrictive assumptions. These theories exist (see Mach, Nelson, and Denny (2007) or Gosline (2015) for an introduction), but they are only beginning to be applied to biology.

The second hurdle is the fact that plants and animals are alive. As cracks form in a living material, it is quite possible that they can be repaired. If so, the number of stress cycles imposed on a material may be less important to its fatigue than the time allowed for repair between cycles. If stresses come in quick succession, a material might break. If the same forces were applied with intervals sufficient to allow growing flaws to heal, the material might survive. Denny et al. (1989) observed another example of seaweeds' ability to actively cope with the presence of flaws. When a sharp ended crack was introduced into the edge of a *M. flaccida* blade, cells at the crack tip died in what appears to be a controlled fashion, rounding the crack tip. This strategy is familiar to many motorists. A small crack in the windshield of a car can grow, eventually threatening the integrity of the whole structure. Drilling a hole at the tip of the crack can stop it in its tracks.

Our understanding of repair is most advanced in bone (Taylor et al. 2007). Small groups of cells—osteoclasts, which break down bone, and osteoblasts which rebuild it—constantly move through a bone, remodeling it as they go. Renovation includes both the repair of cracks and readjustment of the bone's mineral content in response to imposed loads. In this fashion, a bone continually remakes itself, allowing it to fulfill its structural function with minimum weight and minimal metabolic expenditure (Martin 2003). Similar remodeling is not possible in many invertebrate shells—which, unlike bone, are external, dead structures—but analogous processes are likely in soft living materials such as seaweed stipes and blades.

The literature on fatigue is notoriously difficult to assimilate. Terminology varies from one paper to the next and is seldom intuitive. If you want to delve further into the subject, it is advisable to consult a primer such as those offered by Farquhar and Zhao (2006) and Mach, Nelson, and Denny (2007).

Chapter 16

Adhesion and Adhesion Resistance

In the last two chapters, we investigated how materials break. We now turn our attention to adhesion, the means by which they can be reattached. The ability to stick to surfaces opens up new niches for plants and animals. We will see, for instance, how lizards and insects manage to run up walls and hang from the ceiling and how limpets, sea stars, seaweeds, and barnacles adhere to wet rocks in the face of crashing waves. In fact, there are so many ways in which objects can stick to each other that nature has had to evolve specific mechanisms to avoid or manipulate adhesion. We will explore how lotus leaves and butterfly wings use these principles to shed water and how desert beetles use a combination of water-loving and water-repelling surfaces to harvest fog.

1 ATTACHMENT AND ADHESION

Structures joined together form a joint where they meet, and joints can be constructed in two basic ways. In the first, *mechanical attachment* binds objects to each other by some sort of fastener. The rivets that attach girders in bridges and skyscrapers are the classic examples, but there are many others. The hooks of one-half of a Velcro pad act as fasteners by snaring the "hair" of the other half. Root hairs become fasteners when they insinuate themselves around soil particles, thereby attaching plants to dirt, and a twined vine forms a whole-body fastener as it wraps itself around a tree. In each case, the fastener is arranged such that somewhere in the structure, the force attempting to pull objects apart is resisted by solid surfaces loaded in compression (Figure 16.1A). As a result, fasteners must be bent or broken to pull the joint apart. In this sense, attachment is an application of beam theory, which we will discuss in Chapter 17. We will not deal with fasteners here, but you may wish to consult Nachtigall (1974), who reviews the vast diversity of fasteners found in plants and animals: everything from plugs and sockets and hook and eyes, to clamps, grippers, anchors, and combs.

The second mechanism that can hold objects together is *adhesion*. In this case, objects—the *adherands*—are held together by materials acting in tension (Figure 16.1B). A seaweed glued to a rock resists a tensile force attempting to dislodge it; the suction cup of a toy dart resists the tensile force attempting to separate it

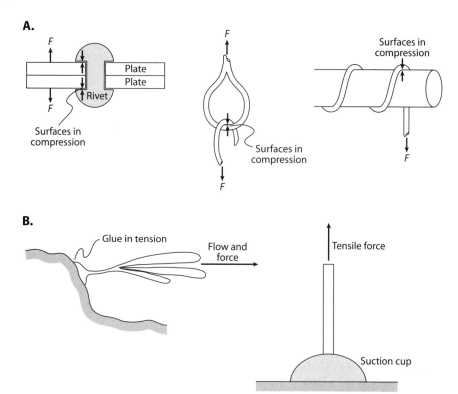

Figure 16.1 A. A sampling of fasteners. **B.** Adhesive joints.

from your refrigerator door. In these cases, the surface properties of the adherands *do* matter. A dart won't stick to a porous surface, and if the rock were made from Teflon®, the seaweed wouldn't stick nearly as well.

There are four basic types of adhesives: pressure difference adhesives, viscous adhesives, glues, and adhesives that rely on van der Waals forces.

2   CLOSE PHYSICAL CONTACT

Van der Waals forces are a quantum mechanical phenomena that cause even uncharged atoms to be attracted to each other—bring any two atoms close enough together and they will bond. (We discussed the London dispersion force, one type of van der Waals force, in Chapter 13.) In this respect, van der Waals forces have the same ubiquitous reliability that gravity does—they are always there. Why then can't van der Waals forces serve as a universal, all purpose adhesive?

The primary problem becomes apparent when we try to reassemble an object from its broken parts. A shattered pane of glass, for example, can't be made whole just by piecing the fragments back together. The reason lies in the fact that van der Waals forces operate effectively only over *very* short distances. Their strength decreases as the inverse seventh power of distance, limiting their useful range to approximately 1 nanometer (nm) a distance equivalent to 3 water molecules stacked end to end. Objects must approach exceedingly close for van der Waals forces to be effective.

However, if two pieces of material can be made to fit together within these tolerances, they adhere quite well. In theory, the tenacity of van der Waals bonds—the tensile stress they can resist—can be as high as 20 MPa, a strength similar to that of insect cuticle (Federle 2006). But the intimate contact necessary for effective adhesion is difficult to come by. When glass is broken, the process of fracture imparts some permanent deformation to each new surface such that (on the nanometer scale relevant to van der Waals forces) two pieces that appear to the naked eye to fit together are actually mismatched. At the micro- and nanoscale, their landscapes are not exactly molded to each other, and the two touch only where a high spot on one—an *asperity*—hits a high spot on the other (Figure 16.2A). Van der Waals forces between these restricted areas are not sufficient to make the object whole.

This small-scale mismatch helps to explain why the frictional shear force required to slide one smooth object over another depends on the compressive force pushing them together (Chapter 2). The greater the compressive force, the greater the compressive stress imposed on the surface asperities and the more they deform. The more they deform, the greater the area of close contact and the more effective van der Waals adhesion is. The frictional force required to initiate sliding is (at least in part) due to the force required to break these molecular-scale welds.

Close contact is facilitated when one or both of the adherands has a sufficiently low modulus. For instance, when soft wax is pressed onto a solid surface, it initially makes contact at only a few spots. However, because the entire compressive force acts over the small area of these asperities, local stress is high and the wax deforms, bringing more area within the van der Waals grasp. The fact that wax has a viscous as well as an elastic component to its stiffness aids in this process. Under sufficient force—and given sufficient time—the wax deforms to match up more precisely with the surface onto which it is pressed, and it adheres by van der Waals forces. Materials that adhere in this fashion are described as *tacky*.

Dahlquist (1969) measured the adhesive properties of materials as a function of their stiffness and found that adhesion by close physical contact (tack) is effective only for materials with compressive moduli less than approximately 0.1 MPa. Thus, mucous secretions (with moduli of 0.001 MPa or less) can be tacky, while rubbers (with moduli of 1 MPa and above) cannot.

There is a catch associated with the adhesion of most tacky materials. Their ability to adhere is enhanced by their compliance, but the low modulus that makes for good tack is commonly associated with a correspondingly low breaking stress. For example, many mucins are admirably sticky (at least in part from van der Waals forces), but they can't form strong glues because they themselves are weak. The yield stress for gastropod pedal mucus is only approximately 1000 Pa, for instance, so a snail's circular foot with a diameter of 2 cm could withstand a force of only 0.31 N before its mucus gave way. We saw in Chapter 10 that the ability of pedal mucus to yield allows slugs and snails to walk on glue, but it also means that they are easily dislodged from the surface to which they adhere. The same caveat applies to manufactured tacky adhesives. Double-sided sticky tape (which is coated with a tacky adhesive) sticks well to many surfaces, but you wouldn't want to fly in an airplane or drive in a car stuck together only with tape.

Some lizards and spiders have found a way around the catch of tackiness; geckos have been most thoroughly studied, and I use them as an example. The adhesive surfaces on the feet of these lizards are organized into a hierarchy of structures.

A.

An asperity

B.

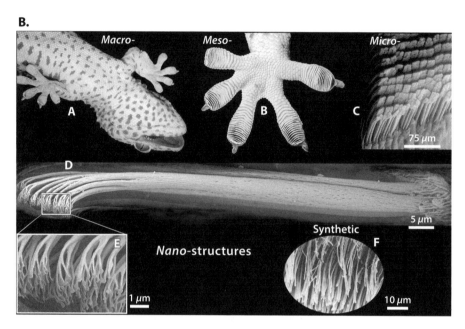

Macro-   Meso-   Micro-

Nano-structures

Synthetic

Figure 16.2 A. At the microscale, local asperities prevent adherands from making close contact. This problem has been solved by geckos (**B**). Structural hierarchy of the gecko adhesive system: (A) ventral view of a tokay gecko (*Gekko gecko*) climbing a vertical glass surface; (B) ventral view of the foot of a tokay gecko showing a mesoscale array of setae-bearing scansors (adhesive lamellae; images A and B by Mark Moffett); (C) microscale array of setae are arranged in a nearly gridlike pattern on the ventral surface of each scansor (in this scanning electron micrograph, each diamond shaped structure is the branched end of a group of four setae clustered together in a tetrad); (D) micrograph of a single gecko seta assembled from a montage of five Cryo-SEM images (image by Stas Gorb and K. Autumn). Note individual keratin fibrils comprising the setal shaft; (E) nanoscale array of hundreds of spatulae tips of a single gecko seta; (F) synthetic spatulae fabricated from polyimide at UC Berkeley in the lab of Ronald Fearing using nanomolding (Campolo et al. 2003). Reprinted from K. Autumn. "Properties, Principles, and Parameters of the Gecko Adhesive System." In *Biological Adhesives*, edited by A. M. Smith and J. A. Callow, 225–26. Springer-Verlag, Berlin, 2006. With kind permission from Springer Science+Business Media LLC.

At the end of each toe there are a series of flexible lamellae (Figure 16.2B), each covered by a dense thicket of hair-like setae made of keratin. A single seta is only 1 to 2 μm in diameter but nonetheless bears at its tip 100 to 1000 smaller flattened hairs known as spatulae. The shaft of each spatula is 2 to 5 μm long and 0.1 to 0.2 μm in diameter, and its tip spreads out to a flattened wafer 0.2 to 0.3 μm wide and a mere 0.01 μm thick.

These spatular tips are so tiny that they fit nicely between asperities in the microscale topography of just about any surface and can thus make effective van der Waals bonds (Autumn et al. 2002). To achieve this close contact, the gecko needs to do just two things: push the lamellae onto the surface and pull the toe in towards its body (Autumn et al. 2000). The resulting combination of compression and shear bends the cantilever-like spatulae so that their tips lie flat on the surface, allowing van der Waals forces to act.

Once attached, a spatula can resist forces in both shear and tension, an important factor for a lizard that encounters both forces when climbing vertical surfaces and hanging from branches. The overall adhesive capability of geckos increases with the number of spatulae attached. Each spatulae adheres with a force of approximately 40 nN and each seta, with a force of 20 to 40 μN; if all 6.5 million setae of a 50-g gecko adhered, they could support the weight of a small child (13 to 26 kg; Autumn 2006).

As beautifully designed as this hierarchical system is, it has its limitations. The gecko has no way of controlling the placement of each individual lamella, so there is a large element of chance in whether a lamella's spatulae will bend in just the right way to make close contact. Even on a smooth glass surface, only about 3% of spatulae achieve a good bond. That is more than enough, however. Only 0.4% need bond to support the gecko's weight.

Keratin—from which geckos' setae and spatulae are constructed—has a modulus of approximately 4 GPa, far too high to be a tacky material. However, because both setae and spatulae are long and thin, they readily bend (see Chapter 17). As a result, a toe's surface behaves like a much more compliant material than it would if it were made from a solid block of keratin. The effective compressive stiffness of the toes' surface is approximately 0.1 MPa, within Dahlquist's limit for tacky materials (Autumn 2006).

The hierarchical structure of gecko toes is a nifty adhesive mechanism, but it would be useless if that adhesion could not be controlled. In order to run and climb, geckos need their feet to stick when planted but release on command so that the foot can be repositioned. The same modular structure that makes adhesion possible also allows for controlled release. When it wants to take a step, the gecko peels its foot loose, releasing the compression on the lamellae and reorienting the spatulae one row at a time. This piecemeal prying minimizes the force required.

Which brings us back to our original question: If van der Waals forces allow for such a strong, controllable adhesive, why don't all organisms employ them? Indeed, the legs and feet of some spiders are covered with fine chitinous hairs that act like the spatulae of geckos, and the adhesive capabilities of spiders approach those of geckos (Bhushan 2007). Why don't other animals follow their lead? There are two apparent reasons. First, it takes real talent to make the precisely designed structures required for gecko-type adhesion. Over the last decade, human engineers have tried mightily to produce adhesives that use this principle but with very limited success. The difficulty of purposefully creating a practical van der Waals adhesive suggests that the evolutionary road to this kind of adhesion might not be an easy one.

More importantly, however, gecko-type adhesion is less effective on wet surfaces. Water molecules are polar and therefore induce stronger van der Waals bonds than those formed by spatulae. As a result, water outcompetes spatulae for space on the substratum. Spatulae can form van der Waals bonds with the attached water, but if the water layer is more than a few molecules thick, the bond between spatula and substratum is weak.

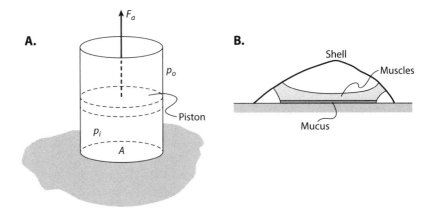

Figure 16.3 When force pulls upward on a piston, the resulting inside-outside pressure differential provides adhesion (**A**). The foot of a limpet acts like a piston to create pressure-difference adhesion (**B**).

3   PRESSURE-DIFFERENCE ADHESION

There are several adhesives that avoid the limitations of van der Waals forces. Consider, for instance, a cylinder of cross-sectional area A resting on a solid surface (Figure 16.3A). Inside the cylinder is a piston, and between the piston and the surface is a fluid (either air or water). The apparatus is surrounded by air at pressure p_o, and pressure in the internal fluid, p_i, is initially the same as that in the air. In the absence of a pressure difference between cylinder and surroundings, only gravity keeps the apparatus attached to the surface.

Now consider what happens when we apply a tensile force between the top of the cylinder and the piston, a force tending to move the piston away from the surface. For the moment, let's assume that the fluid beneath the piston is air. As the piston moves upward, it increases the gas's volume, decreasing its pressure and creating a pressure difference, Δp, between the cylinder's exterior and interior:

$$\Delta p = p_o - p_i. \tag{16.1}$$

Acting over A, Δp creates a force that pushes the apparatus onto the surface:

$$F_a = \Delta p\, A \tag{16.2}$$

F_a acts as an adhesive force because an oppositely directed force of at least this magnitude must be applied to dislodge the cylinder.

This type of adhesion does not depend on the surface properties of the substratum. A thin coating of water wouldn't bother it at all. As long as the surface is solid and doesn't contain any holes through which air can flow (which would equalize p_i and p_o), this pressure-difference apparatus sticks just fine.

Note that the adhesive tenacity of this system (the force per area it can resist) is

$$\sigma_a = \frac{F_a}{A} = \Delta p. \tag{16.3}$$

In other words, the tenacity of a pressure-difference adhesive is equal to the pressure difference it maintains.

This is the principle behind suction cups. The piston force of a rubber suction cup (the type found in toy darts) is provided by the elasticity of the rubber. When the cup is forced against a surface, the air beneath it is squeezed out and the lip of the cup forms a seal. (Licking the dart helps to form a good seal.) The cup then attempts to return to its original shape, pulling up on the gas beneath and decreasing its pressure. The resulting pressure difference holds the cup in place.

Maximum difference in pressure—and thereby maximum tenacity—is set by the nature of the fluid within the apparatus. Because air is a gas, the lowest pressure it can have is 0 Pa—a vacuum—and in this case maximum tenacity is set solely by the ambient pressure outside the apparatus. Atmospheric pressure at sea level is 0.1 MPa; therefore, the maximum adhesive strength of a gas-filled suction cup in air is 0.1 MPa. Tenacity decreases with altitude as air pressure decreases; a suction cup at the top of Mount Everest has only a third the adhesive strength of one at sea level.

The limits on maximum tenacity change if the enclosed fluid is water. Unlike gas molecules, which have minimal attraction to each other, water molecules are held together by hydrogen bonds. As a consequence, a piston pulling up on water can induce tension—a negative pressure—in the fluid and, thereby, a larger pressure difference between p_o and p_i. Thus, using water as the enclosed fluid potentially increases the strength of the adhesive. A wide variety of marine animals use water-filled suction cups (octopuses, squids, and cling fish, to name just a few; see Nachtigall (1974) for a complete list), but limpets—small gastropods with conical shells—are the poster children of pressure-difference adhesives, and we take them as a heuristic example.

Limpets use their pressure-difference adhesive to stick to intertidal rocks (Figure 16.3B). The limpet's foot acts as the piston, pulled on by muscles that run from the foot's ventral surface upward to the shell. There is a thin layer of mucus below the foot, which serves both as the low-pressure fluid and a caulk to seal the foot's edge.

When a limpet sits undisturbed on the rock, its pedal muscles are relaxed, pressure difference is minimal, and the animal is easily dislodged. (The residual adhesive force is due to viscous adhesion and/or glue, which we will discuss later in the chapter.) At any indication of trouble, however, the limpet hunkers down—the pedal muscles contract, pulling up on the foot and creating a negative pressure in the mucus below. The resulting pressure difference increases the limpet's tenacity, and, once clamped down in this fashion, the animals often cannot be dislodged without breaking the shell or ripping the muscles loose from their tendons. Smith (1991b) measured limpet adhesive tenacities as high as 0.23 MPa at sea level—more than double atmospheric pressure, a clear indication that pedal mucus can sustain negative pressures.

Although pressure-difference adhesives provide a convenient, controllable means of resisting tensile force, they have several potential drawbacks. First, their tenacity can be limited by the tendency for aqueous materials to *cavitate*. As noted above, water can withstand substantial negative pressure as long as there is no free space into which it can evaporate. Scrupulously cleaned water can withstand a pressure of -27.4 MPa (Briggs 1950), and the water in plants' xylem tubes can withstand -8 MPa. But the presence of a small bubble—or even a small, pointed bit of solid on which a bubble can nucleate—provides space into which fluid can boil. Once this liquid-gas interface is formed, more water can evaporate into it, the bubble grows spontaneously, and the adhesive bond is broken. Working with octopus suckers and unfiltered seawater, Smith (1991a) found that cavitation generally occurred at pressures of -0.1 to -0.2 MPa, so total tenacity is 0.2 to 0.3 MPa.

The second potential problem is that the force due to pressure difference acts normal to the substratum and therefore provides no inherent resistance to shear. For instance, a toy dart stuck firmly to your refrigerator door can nonetheless slide sideways. Lack of shear resistance could be a problem for a limpet trying to maintain its position in the face of crashing waves.

There are two mechanisms by which a pressure-difference apparatus might resist shear: the viscosity of the internal fluid and friction with the substratum. Viscosity is unlikely to help much. The mucus layer under a limpet's foot is approximately $10 \ \mu m$ thick. If the foot is circular with a diameter of 2 cm and the mucus has the viscosity of water, a shear force of only 4×10^{-4} N is needed to slide the limpet sideways at (what is for it) the dangerous speed of $1 \ cm \cdot s^{-1}$. Converting from water to mucus makes for only a modest increase in resistance. As noted earlier, a force of only 0.31 N would be required to initiate sliding.

Despite this problem, the actual shear resistance of limpets on natural surfaces can be as large—or even larger—than their resistance to tensile forces owing to friction between the edge of the shell and the rock (Denny and Blanchette 2000). As muscles pull up on the foot, they force the shell down onto the substratum, increasing static friction. Often the shell edge is butted up against macroscopic asperities in the substratum, and the animal is dislodged only when the shell chips or breaks. Thus, on a rugose substratum, a limpet-style pressure-difference adhesive can be an effective, controllable means of holding an organism in place.

There is still one potential problem, however. In order to produce a pressure difference, animals have to contract muscles, which requires energy. In other words, when they hunker down there is an ongoing metabolic cost to their adhesion. An alternative means of creating a pressure difference—*capillarity*—requires no energy expenditure on the part of the animal, and thus has some advantages. Before describing capillarity, however, we digress briefly to discuss the concepts of surface tension and surface energy.

4   SURFACE TENSION AND SURFACE ENERGY

Consider a drop of water. A molecule in its interior is attracted to all the molecules around it, and because this attraction acts equally in all directions, the net force on the molecule is zero. A molecule at the drop's surface, however, is more strongly attracted to the water in the drop than to the air outside, resulting in a net force pulling the surface molecule in. For the molecule to have gotten to the surface from the interior, it must have been moved there against this net force, requiring that work be done. This work is γ_s, the *surface energy* of the water-air interface.[1] For clean water at room temperature, γ_s is 73 mJ m^{-2}. (Surface energy decreases slightly with increasing temperature.)

To get a feel for how surface energy manifest itself, imagine a film of water enclosed within a U-shaped frame (Figure 16.4A). A movable wire with length y spans the top of the U, and, by applying a force F to the wire, we move it out a distance x. The energy required to move the wire—the product of force and distance—is thus Fx.

---

[1] Note that γ_s, surface energy, has nothing to do with γ, shear strain. It just happens that they both are traditionally given the same symbol.

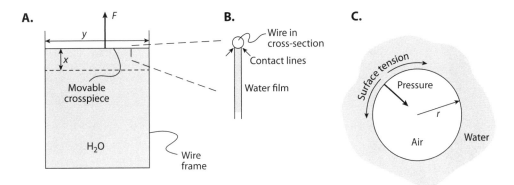

Figure 16.4 A. An apparatus for measuring surface tension. (See text.) **B.** The movable crosspiece in cross section. **C.** Surface tension pulling inward on a bubble of air increases the internal pressure.

In moving the wire a distance x, we have increased the area within the frame by yx. But the water film has two sides, so we have actually increased the area of air-water interface by $2yx$. Dividing energy by area, we see that

$$\gamma_s = \frac{Fx}{2yx} = \frac{F}{2y}. \tag{16.4}$$

Now let's look at the apparatus in cross section (Figure 16.4B). Because the water film has two sides and each has an interface with the movable wire, the total length of air-water interface in contact with the movable wire is $\ell = 2y$. Substituting this relationship into equation 16.4 gives

$$\gamma_s = \frac{F}{\ell}. \tag{16.5}$$

In other words, because it takes energy to create a new area of air-water interface $(J \cdot m^{-2})$, there is a *surface tension*, a force per length $(N \cdot m^{-1})$, that tugs on any object in contact with the air-water interface. The surface tension of water in contact with air is $73 \ mN \cdot m^{-1}$.

The concept of surface tension explains why water drops suspended in air are spherical. Just as the tension in an inflated balloon tends to mimimize the balloon's surface area, the tug of surface tension has the effect of minimizing a drop's surface area, and, for a given volume, minimum surface area is obtained when the drop is spherical.

For future reference, note that surface tension increases the pressure inside the drop in the same way that rubber increases the pressure inside an inflated balloon (Figure 16.4C). The smaller the drop's radius, r, the more curved its surface, and the more effectively surface tension can increase internal pressure. According to Laplace's law (Vogel 1994):

$$\Delta p = p_o - p_i$$

$$= -\frac{2\gamma_s}{r}. \tag{16.6}$$

(The negative sign tells us that pressure inside is greater than pressure outside.)

Table 16.1 Surface Energies for Some Solids and Liquids. Data from Wake (1982) and Pocius (2002).

Solids	Critical Surface Energy (mJ · m^{-2})
Polytetrafluoroethylene (Teflon®)	18.5
Polyethylene	31
Wool	42.5
Starch	39
Tooth enamel	38–40
Cellulose	45
Polyglycine	45–51
Glass	170
NaCl	300

Liquids	Surface Energy (mJ · m^{-2})
1% gelatin	8.3
Ethanol	22.8
Benzene	28.9
Phenol	40.9
Water	72.8

Solids also have surface energies. However, because their molecules are held fixed in place, it doesn't work to think of the surface energy of a solid as the energy required to move molecules from the bulk to the surface. Instead, one should think of a solid's surface energy as we did in our discussion of material strength (Chapter 13)—it is the minimum energy required to create new surface when we break a solid in two. (As we noted in Chapter 15, this minimum energy can be considerably less than the actual energy required to create a new surface, which includes energy lost to viscous and plastic processes.) The surface energy of solids varies considerably. Solids such as wax and Teflon® have low surface energies (a few tens of mJ · m^{-2}), and solids such as glass have high surface energies (a few hundred mJ · m^{-2}; Table 16.1). At 73 mJ · m^{-2}, water is in between.

5   INTERACTION AT THE INTERFACE

When a liquid drop comes in contact with a solid, the result depends on their relative surface energies. For reasons that will become clear in a moment, if a high-surface-energy drop were to spread over a low-surface-energy solid, the overall surface energy of the system would increase. Nature tends to minimize energy, so spreading is discouraged, and the drop remains beaded up. By contrast, if the solid has a higher surface energy than the liquid, total surface energy is reduced by any increase in contact between liquid and solid. In this case, the drop spreads spontaneously and is said to wet the solid surface.

The physics behind these phenomena concern the forces at the interfaces be-tween materials. When a liquid drop contacts a solid surface, three interfaces are

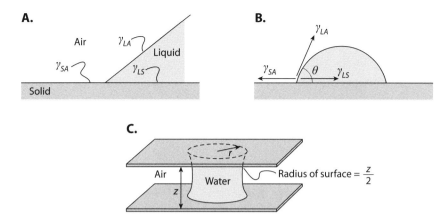

A.

Air γ_{LA}

Liquid

γ_{SA} γ_{LS}

Solid

B.

γ_{LA}

γ_{SA} θ γ_{LS}

C.

Air   Water   Radius of surface $= \dfrac{z}{2}$

z

Figure 16.5 A. The interface between air, liquid, and a solid surface. **B.** Surface tensions acting at the interface. **C.** Surface tension acting on water sandwiched between two solid plates reduces the pressure in the water, leading to capillary adhesion.

involved: between liquid and air, between solid and air, and between solid and liquid (Figure 16.5A). Each interface has a surface energy: γ_{LA}, γ_{SA}, γ_{SL}, respectively.

Once solid and liquid are in contact, energy W_{SL} is required to separate them. In other words, when a solid and a liquid are in contact, they adhere. The greater the energy required to separate them, the greater the adhesive tenacity.

The energy it would take to pull the liquid away from the solid is the difference between their combined surface energies when they are apart and the surface energy when they are together. In other words,

$$W_{SL} = (\gamma_{LA} + \gamma_{SA}) - \gamma_{SL} \tag{16.7}$$

If we knew these various surface energies, we could calculate the work required to overcome the adhesive bond of liquid to solid.

That is a substantial if, however. While it is easy to measure the surface energy of the liquid (γ_{LA}), it is difficult to directly quantify either γ_{SA} or γ_{SL}. We can avoid this problem by switching our perspective from surface energy to surface tension, a shift we accomplish by considering the shape of a liquid drop on a solid surface (Figure 16.5B). Let's assume that the drop is stationary and small enough so that gravitational forces are negligible relative to surface tension. In that case, the combined tensions acting at the intersection of the three interfaces (liquid-air, solid-air, and liquid-solid) must cancel each other out. (If they didn't, the interface would move.) Taking the geometry of the situation into account, we see that

$$\gamma_{SA} = \gamma_{SL} + \gamma_{LA} \cos \theta_{\text{stat}}, \tag{16.8}$$

where θ_{stat} is the interior angle between the solid and the liquid. (The subscript *stat* denotes that this is the static contact angle, the angle measured when the drop is stationary. We will see later in the chapter what happens to θ if the drops moves.) Combining equations 16.7 and 16.8 allows us to quantify W_{SL} without having to measure either γ_{SA} or γ_{SL}:

$$W_{SL} = \gamma_{LA} \left(1 + \cos \theta_{\text{stat}}\right). \tag{16.9}$$

It is worth taking a moment to contemplate the important take-home messages of this relationship. The higher the surface energy of the liquid (γ_{LA}), the greater the work required to separate the liquid from the solid and the greater the effectiveness of adhesion. W_{SL} also depends on the contact angle between liquid and solid. If the liquid wets the solid, $\theta_{stat} = 0$, and W_{SL} is maximal at $2\gamma_{LA}$. If the liquid beads up, θ_{stat} is large, and W_{SL} can approach (but never reach) zero.

This last conclusion fits with what we know about van der Waals bonds. A liquid makes intimate contact with a solid and is, therefore, subject to van der Waals bonding. As a result, it will always take some energy to separate the two. This adhesive energy is minimal, however, if θ_{stat} is large.

We will make use of this information later in the chapter when we deal with glues. For the moment, let's assume that the liquid we are dealing with is water, and that it wets the surfaces it contacts.

6   CAPILLARY ADHESION

Surface tension enables a second type of pressure-difference adhesion, known as *capillary adhesion*. Consider a system in which water is sandwiched between a horizontal substratum and the surface of a biological adherand, each made of a wettable solid (Figure 16.5C). The water responds to this situation in two ways. As surface tension attempts to minimize the water's contact with air, the water forms a disk sandwiched between the two surfaces. Given a certain volume of water and a fixed distance z between surfaces, the disk has radius r. Acting in conjunction with surface tension, the curvature associated with this radius increases the water's internal pressure by

$$\Delta p_1 = -\frac{\gamma_s}{r}.\tag{16.10}$$

Again, the negative sign signifies that p_i is greater than p_o. (The increase here is half that for a sphere of the same radius (equation 16.6) because it is due to curvature in only a horizontal plane.)

But this is only part of the story. Because both the animal's surface and the substratum have higher surface energies than water, the water sandwiched between them tries to spread, and spreading stretches the perimeter of the watery disk, such that in a vertical cross section the interface curves outwards. The radius of this curve is approximately $z/2$, half the distance separating the solids, and surface tension acting in conjunction with this curvature pulls outward on the water, lowering its pressure:

$$\Delta p_2 = \frac{2\gamma_s}{z}.\tag{16.11}$$

These effects are additive, with the result that the pressure difference between the air and the water is

$$\Delta p = \Delta p_1 + \Delta p_2$$

$$= \gamma_s \left(\frac{2}{z} - \frac{1}{r} \right).\tag{16.12}$$

Δp acts over the area of the aqueous disk (πr^2), so that the overall result—*capillary adhesion*—is

$$F_a = \pi \gamma_s \left(\frac{2r^2}{z} - r \right). \tag{16.13}$$

When z is much smaller than r (as it typically is), $r \ll 2r^2/z$, and

$$F_a \approx \frac{2\pi r^2 \gamma_s}{z}. \tag{16.14}$$

The adhesive force F_a acts over area πr^2, so the force per area (the tenacity) of capillary adhesion is

$$\sigma_a \approx \frac{2\gamma_s}{z}. \tag{16.15}$$

Thus, the strength of capillary adhesion is large if the separation between disk and substratum is small. For example, if disk and substratum are separated by a layer of water 10 μm thick, the pressure difference is 0.015 MPa, 15% of the adhesive capability that could be attained by a well-designed suction cup using air as its enclosed fluid and 6% the tenacity of a limpet using mucus. A thinner layer of water would allow for higher tenacity.

In a real apparatus of this sort, z would likely be set by the matchup of asperities between disk and substratum. The more compliant the epithelium, the better the matchup, the smaller z can be, and the greater the adhesive force. The toe pads of crickets—which use this kind of adhesion—have a notably low compressive modulus (0.027 MPa; Gorb et al. 2000).

For terrestrial organisms, capillary adhesion is easily managed. Tree frogs, for instance, have a pad on each toe that acts like the apparatus described above (Hanna et al. 1991). As with geckos, release of adhesion is accomplished by peeling the pads from the surface. A wide variety of insects (flies, beetles, and earwigs) also use capillary adhesion. Tiny bristles on the insects' feet exude small drops of liquid at their tips, and these drops form capillary bonds when they contact a surface (Federle 2006). Others, such as cockroaches, bees, bugs, grasshoppers, and the crickets mentioned before have wet adhesive pads rather than hairs.

Capillary adhesion requires high-surface-energy solids to sandwich the adhesive fluid. As a result, tree frogs, grasshoppers, and insects can adhere nicely to glass but not to Teflon®.

There are two inherent problems with capillary adhesion. First, it works only in air. If the apparatus of Figure 16.5C were submerged in water, the tension of the air-water interface would be abolished, and with it the pressure difference holding the disk in place. Furthermore, like other pressure-difference adhesives, capillary adhesion has little inherent resistance to shear; indeed, water can act as an effective lubricant.

7   VISCOUS ADHESION

Although there can be no capillary adhesion in water, an adhesive organ near a solid substratum can stick nonetheless. Consider the apparatus shown in Figure 16.6. It is similar to that of Figure 16.5C except that the air has been replaced by water, and the adhesive organ is a disk of radius r. To pull the disk away from the substratum, water

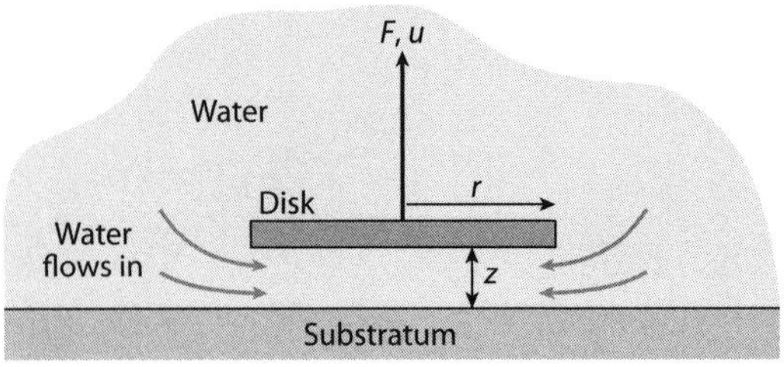

Figure 16.6 The terms used to define Stefan adhesion. (See text.)

must be drawn into the gap between them. As water flows in, it is sheared—the faster the rate of separation, the higher the rate of shear. Because water is viscous, it resists being sheared, and a force is thus required to separate the adherands.

The analysis of this system was first carried out by Joseph Stefan[2] in 1874, and viscous adhesion is often called Stefan adhesion in his honor. When μ is the dynamic viscosity of water and z is the separation between disk and substratum ($z \ll r$), adhesive force is (Wake 1982)

$$F_a = 1.5\mu\frac{\pi r^4}{z^3}\frac{dz}{dt}. \tag{16.16}$$

Stefan adhesion also works in air. For adherands separated by water but immersed in air, the adhesive force is half as large:

$$F_a = 0.75\mu\frac{\pi r^4}{z^3}\frac{dz}{dt}. \tag{16.17}$$

In both cases, force increases in direct proportion to the rate of separation.

Note, however, that F_a is proportional to πr^4, so when we divide force by area to calculate tenacity, we find that σ_a is proportional to r^2:

$$\sigma_{a,\text{wat}} = 1.5\mu\frac{r^2}{z^3}\frac{dz}{dt}, \tag{16.18}$$

$$\sigma_{a,\text{air}} = 0.75\mu\frac{r^2}{z^3}\frac{dz}{dt}. \tag{16.19}$$

Like capillarity, the thinner the liquid layer in a viscous adhesive, the greater the tenacity. But viscous adhesion is also sensitive to the radius of the adherand—the larger the radius, the greater the adhesive force. Furthermore, the requirements of viscous adhesion—water sandwiched between two surfaces—are so simple that it is difficult to see how organisms *avoid* using this mechanism.

To give these ideas some tangibility, consider a submerged disk 2 cm in diameter separated from the substratum by a 10-μm-thick layer of water. The disk would have to be separated from the substratum at a velocity of only 0.7 mm·s^{-1} to exhibit a tenacity of 0.1 MPa, the tenacity of a suction cup.

Although Stefan adhesion can resist large loads for short periods, we must not forget that it owes its strength to viscosity. A small force applied for a considerable

---

[2] Stefan (1835–93) was a Slovenian physicist better known for the derivation (with his student Ludwig Boltzmann) of the law governing how the rate of electromagnetic radiation varies with temperature (see Chapter 12).

time will slowly separate adherands. As the separation grows, the rate of separation increases, and the system eventually fails. This mechanism is therefore well suited to locomotion, during which a foot needs to adhere but only briefly. Stefan adhesion accounts for roughly half the adhesive ability of tree frogs (Hanna and Barnes 1991), likely helps grasshoppers to stick, and probably accounts for much of the adhesion of limpets when they are not hunkered down. (We will deal with their long-term adhesion in the next section.) Part of the adhesive ability of snails and slugs is due to the high viscosity of their pedal mucus.

Unlike the pressure-difference system of limpets—which creates low pressure under the foot by forcing the shell down onto the rock—Stefan adhesion has no tendency to create friction with the substratum, leaving an organism susceptible to displacement by shear. In the absence of friction, the shear resistance of a Stefan adhesive is due solely to viscosity.

A related form of adhesion is used by many plants, which produce sticky secretions that trap insects (Adlassnig et al. 2010). In these cases, the fluid (a polysaccharide mucilage or terpenoid resin) isn't necessarily sandwiched in a thin layer between plant and prey—so adhesive tenacity is low—but the sheer volume of the sticky goo makes it difficult for an insect to disengage. It exhausts itself fighting against viscosity, and its rotting corpse provides the plant with valuable nutrients.

8 GLUE

The lack of shear resistance inherent in pressure-difference and Stefan adhesives can be remedied by solidifying the fluid that separates adherands; that is, by turning the liquid into a glue. If the glue has a high shear strength, the adhesive joint is capable of resisting large, statically applied forces. This advantage comes with a catch, however—it takes time for a glue to solidify. Thus, there is a trade-off: pressure difference and viscous adhesives provide instant tenacity but low shear resistance; glues resist shear but take time to set.

The adhesive strength of a glue depends on two factors. First, as with any solid, glue has a tensile strength. If stress in a joint exceeds the glue's breaking stress, the joint gives way. Tenacity can also be limited by the adhesion of the glue to the surface of the adherands. It is here that surface energy comes into play. The solidification of the fluid does not affect its intrinsic ability to adhere to solids. If glue in its fluid state wets the adherands, the solidified glue forms a strong bond; if it doesn't, the bond is weak.

All our everyday glues make use of these principles. The more venerable adhesives such as library paste, model airplane glue, and Elmer's glue are low-surface-energy liquids that solidify when a solvent evaporates. Glues such as the cyanoacrylates ("superglue") and epoxies solidify as a catalyst chemically crosslinks the liquid. Perhaps the simplest glue is ice; anyone who has inadvertently left a pair of wet mittens outside on a freezing day can attest to the tenacity of solid-state water.

Nature has devised a wide variety of glues. Barnacles glue themselves to rocks and other substrata using a protein glue that is extruded in liquid form and then crosslinked in place (Kamino 2006). When a limpet sits still for an extended period, it introduces a special protein into its pedal mucus, which appears to crosslink the mucus's mucopolysaccharides (Smith et al. 1999). Brown algae glue themselves in place with polyphenolic compounds that are probably cross-linked in the same way chitin is cross-linked in the exoskeletons of arthropods (Potin and Leblanc 2006; Vreeland et al. 1998).

Glues' solidity can lead to problems. Unlike liquids—which flow when stressed—solids can form sharp-ended cracks with the concomitant stress concentrations. As a result, the strength of a glued joint may be substantially less than the potential strength of the glue itself. Furthermore, as with other solids, glued joints are subject to fatigue.

In this respect, it can be advantageous to subdivide the area of a glued joint. A single flaw in a large area can propagate across the entire joint, causing it to fail. The same flaw in a subdivision can lead to local failure, but as long as there is no mechanism for transmitting stored elastic energy from one segment to another, the joint can persist. This subdivision can even take place within the glue itself. Mussels provide an example. As we have seen (Chapter 14), these bivalves tether themselves to intertidal rocks with an array of byssal threads, and each thread is glued to the rock by a small adhesive disk. As it sets, the glue beneath a byssal plaque forms a closed-cell foam, and Stewart and colleagues (2004) suggest that the concomitant subdivision of the solid glue acts as a crack-stopping mechanism.

9   ADHESIVES COMPARED

Having explored the principles of adhesion, we are now in a position to compare adhesives' properties. Because organisms come in different sizes, a fair comparison involves normalizing each adhesive's abilities to the area of an adhesive organ. In other words, we compare the tensile stress σ_a that each adhesive can withstand.

Geckos are the most adept at implementing close-contact adhesion. Although individual setae have high tenacity, at most 3% of setae make intimate contact with the substratum. As a consequence, the effective tensile strength of these lizards is approximately 0.02 MPa (Autumn 2006). A suction cup using air as its enclosed fluid would be about five times as strong (0.1 MPa), and limpets (which use mucus) are 2.3 times stronger still, 0.23 MPa.

Glues can be considerably stronger. Acorn barnacles have shear tenacities of approximately 0.5 MPa (Denny 1995). At times it is the rock substratum that gives way instead of the barnacle's glue, indicating that this adhesive system has reached its practical limit.

For all three of these adhesive systems (close contact, pressure difference, and glue), strength (breaking force per area) is—at least in theory—independent of the area of the adhesive organ: adhesive force increases in direct proportion to the organ's area. There are several assumptions behind this conclusion, however. Consider, for instance, the gecko adhesive system. Doubling the area of a gecko-type adhesive organ would presumably involve doubling the number of spatulae rather than doubling the area of each spatula. The system works only because its individual adhesive elements are small enough to make intimate contact with the substratum.

In assessing the capabilities of a limpet-style pressure difference adhesive, I have assumed that there are no fissures in the seal at the edge of the foot and no air bubbles or cavitation nuclei in the pedal mucus. Either sort of flaw would weaken the adhesive. The chance of encountering an edge flaw would presumably increase with the perimeter of the foot, while that of encountering an interior bubble or nucleating site would increase with foot area. In either case, one might suspect that breaking stress would decrease with increasing area of the adhesive organ. There is some evidence of this effect. Force scales as area to the -0.18 power in the limpet *Lottia scutum* loaded in tension (Smith 1991b). As noted previously, solid glues are subject to the same concerns.

For capillary and Stefan adhesion, size matters. First, capillary adhesion. Suppose that z, the distance separating an adhesive disk from the substratum is a constant fraction k_z of the disk's radius, r. This might be true, for instance, when an organism attempts to adhere to a rugose surface—a small disk could be able to nestle in closer to the substratum than a large disk. Given a constant ratio of z to r, the pressure difference acting on an adhesive disk—and hence the capillary adhesive strength—is

$$\sigma_a = \frac{\gamma_s}{r}\left(\frac{2}{k_z} - 1\right). \tag{16.20}$$

The smaller the adhesive organ—that is, the smaller r is—the more stress it can resist.

This raises an interesting possibility. Rather than having a single capillary adhesive disk of area A, even when flaws are not taken into account it could be advantageous to have multiple small disks whose total area adds up to A. Because each small disk is stronger, their combination should be able to withstand greater force. To analyze this effect, let's take a fixed adhesive area, A, and divide it into n individual disks. The area of each disk is πr^2, so[3]

$$\pi r^2 \approx \frac{A}{n}, \tag{16.21}$$

$$r \approx \sqrt{\frac{A}{\pi n}}. \tag{16.22}$$

Capillary tenacity is thus

$$\sigma_a = \gamma_s \left(\frac{2}{k_z} - 1\right) \sqrt{\frac{\pi n}{A}}. \tag{16.23}$$

For a given overall area, the tenacity of the adhesive increases as the square root of the number of individual disks. Thus, it makes sense to divide a capillary adhesive into small parts. Many of the insects that use capillary adhesion (flies, beetles, and earwigs) adhere to this principle—each foot is covered with a myriad of tiny adhesive hairs (Federle 2006).

Stefan adhesion provides a stark constrast. Substituting equation 16.22 into equation 16.18, we find that for a Stefan adhesive system,

$$\sigma_a = 1.5\mu \left(\frac{1}{z^3}\right)\left(\frac{A}{\pi n}\right)\left(\frac{dz}{dt}\right). \tag{16.24}$$

Dividing a Stefan system into smaller parts—that is, increasing n—*reduces* its strength.

10   ADHESION IS EASY

So far, we have evaluated adhesives primarily by comparing them to each other. It is high time that we compare them to something of biological relevance—the largest

---

[3] This calculation ignores the fact that circular disks would not tesselate efficiently to cover the entire area, but a more rigorous calculation would not change the overall message.

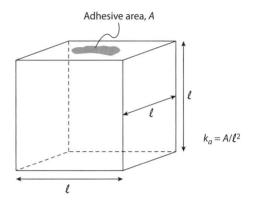

Figure 16.7 The cubic organism adheres with an area A that is a fraction k_a of one of its faces. (See text.)

stress imposed by the environment. In air, the largest force an organism encounters is often its own weight, so for terrestrial plants and animals we explore the ability of adhesion to support a organism against the pull of gravity. In water, weight is seldom a problem; instead, fluid dynamic drag is commonly the largest force an organism experiences. For aquatic adhesive systems we explore the ability of adhesion to resist drag.

Starting with the terrestrial, consider a cubic organism made from a material with density ρ_b (Figure 16.7). If the length of the cube's edge is ℓ, its volume is ℓ^3, and its weight in air is

$$F_g = \rho_b g \ell^3. \tag{16.25}$$

A fraction k_a of one face of the cube adheres to the substratum with tenacity σ_a. Thus, the tensile force the organism can resist is

$$F_a = \sigma_a k_a \ell^2. \tag{16.26}$$

By setting F_a equal to F_g, we can calculate what fraction of the available adhesive area must be used to just support the animal:

$$\sigma_a k_a \ell^2 = \rho_b g \ell^3,$$
$$k_a = \frac{\rho_b g \ell}{\sigma_a}. \tag{16.27}$$

For a typical ρ_b (1050 to 1080 kg · m^{-3}), this expression becomes

$$k_a \approx 10^4 \frac{\ell}{\sigma_a}. \tag{16.28}$$

For example, a cubic organism 1 mm on a side (the rough equivalent of a fruit fly) adhering with air-filled suction cups ($\sigma_a = 0.1$ MPa), could support its weight by devoting just 0.01% of one face to adhesion ($k_a = 0.0001$). A beetle-sized cube 1 cm on a side needs to devote 0.1% of a face to adhesion, an organism 10 cm on a side (a mass of 1 kg, a large squirrel for instance) needs 1%, and an animal 1 m on a side (with a mass equivalent to that of a cow) needs 10%. In theory, a 1000-t cubic dinosaur could hang from the ceiling by devoting one whole face to suction cups!

And this is for one of the weaker adhesives. Using mucus as the internal fluid in a pressure difference adhesive, σ_a can be more than twice as large. The precise tenacities of capillary and viscous adhesives are less easily nailed down—depending as they do on the number of substructures, the thickness of the adhesive layer, and the allowed rate of separation—but each can easily be as tenacious as suction. Solid glues can be 10 to even 100 times as tenacious still.

In short, effective adhesion is easily available to terrestrial organisms. The larger the plant or animal, the more difficult it becomes, but unless the organism is very large, adhesion should not be an insurmountable problem.

The calculations for aquatic organisms proceed along similar lines, the only difference being that the applied load—drag—is proportional to area rather than volume:

$$F_D = \frac{1}{2}\rho_w u^2 C_D \ell^2. \tag{16.29}$$

Here ρ_w is the density of water (1000 kg $\cdot$ m^{-3}), u is water velocity, and C_D is the drag coefficient (which for convenience we can assume is approximately 1).

Setting F_a equal to F_D, we again calculate what fraction of the available adhesive area must be used to keep the animal stuck in place:

$$\sigma_a k_a \ell^2 = 500 u^2 \ell^2, \tag{16.30}$$

$$k_a = 500\frac{u^2}{\sigma_a}. \tag{16.31}$$

In this case, k_a is independent of the size of the organism. Assuming, as we did for terrestrial organisms, a minimum adhesive tenacity of 0.1 MPa (roughly that of our suction cup from before),

$$k_a \approx 0.005 u^2. \tag{16.32}$$

The higher the velocity, the larger the fraction of available area that must be devoted to adhesion.

For all but the most extreme velocities, however, that fraction is small. For a velocity of 10 cm $\cdot$ s^{-1}, a mere 0.005% of available area must be devoted to adhesion. For $u = 1\,\text{m} \cdot \text{s}^{-1}$, 0.5% of available area is needed, and even at 10 m $\cdot$ s^{-1}, an animal can get by with just 50% of its available area devoted to adhesion. Not until water velocity reaches 14 m $\cdot$ s^{-1} would all the available area need to be used, and animals could withstand even higher velocities by using stronger adhesives or by evolving more streamlined shapes.

The ability to adhere easily has cascading consequences. For instance Carol Blanchette and I (Denny and Blanchette 2000) found that the relatively high adhesive tenacity of limpets has obviated the need to streamline their shells. There is a well-defined shape that minimizes the hydrodynamic force on limpets (Denny 2000), but few limpets have it. Instead, by devoting such a large area to the foot and using mucus as the enclosed fluid, limpets have the permission of the environment to evolve shell shapes that are less than optimal for avoiding drag and lift but are well suited for other needs such as territorial defense.

11   ADHESION IN REVERSE

The ready availability of adhesion raises a different set of concerns. Put an organism in almost any environment, and things begin to stick to it, which can cause problems. In particular, rain drops have a tendency to stick to surfaces, a phenomenon that keeps windshield-wiper manufacturers in business. But what for us is an annoyance can actually be dangerous for plants and animals. Water droplets from rain or fog stick to flying insects, for instance, weighing them down. Similarly, rain drops can stick to

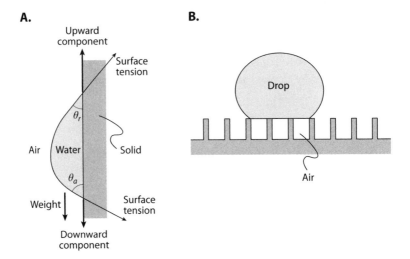

Figure 16.8 A. The internal angle of a moving drop differs between the leading edge (θ_a) and the trailing edge (θ_r). **B**. The Cassie-Baxter state, by which small-scale topography can make a surface hydrophobic. (See text.)

leaves, adding to their gravitational burden. In response to these problems, nature has evolved some ingenious surfaces that resist adhesion.

To see how they work, let's revisit the adhesion of a liquid drop to a surface. As we have seen, the drop adheres because energy is minimized by the formation of the liquid-solid interface. But this is adhesion only in the sense that the drop resists being pulled away from the surface. By contrast, rolling or sliding the drop along the surface doesn't change its area of contact and, therefore, doesn't change the overall surface energy of the system. As a result, it would seem that—although adherent—the drop would have zero resistance to moving along the surface. From what we have learned so far, a raindrop hitting your windshield should just slide off. There is something we have missed.

To this point, I have described the contact angle θ_{stat} between a drop and a surface as a static phenomenon (Figure 16.5). However, if the drop moves and there is any roughness on the surface, θ changes. At its leading edge, θ increases as liquid has to be coaxed to advance. This advancing contact angle, θ_a, is greater than θ_{stat}. At its trailing edge, θ decreases as liquid has to be coaxed to retreat, so the retreating contact angle, θ_r is less than θ_{stat}. The difference between advancing and retreating contact angles— *contact angle hysteresis*, $\Delta\theta = \theta_a - \theta_r$—explains the adhesion of rain drops to your windshield.

Consider, for instance, a drop on a vertical pane of glass (Figure 16.8A). The depiction here is similar to that of Figure 16.5B, but with a twist. In Figure 16.5B, I drew vectors denoting the forces that surface tension applied to the solid-liquid-air interface. Here, the vectors denote the accompanying reaction forces, the forces that surface tension applies to the drop.

There are two forces pulling the drop downward: its weight and the force due to surface tension acting at the leading-edge liquid-air interface. These forces are resisted by the upward tug of surface tension at the trailing edge. Because of contact-angle hysteresis, the upward pull of surface tension is greater than its downward pull. For small drops this net force is enough to offset the drop's weight, and the drop sticks. Both surface tension and weight increase with the size of the drop, but weight increases with the drop's volume while surface tension increases only with its periphery. Eventually a size is reached at which the force of gravity overtakes the pull of surface tension, and the drop breaks loose—this critical diameter is about 2 mm for water on glass.

Experiments have shown that the larger the static contact angle, the smaller the contact angle hysteresis. Reducing hysteresis allows smaller drops to slide. In short, anything a plant or animal can do to increase θ_{stat} can ameliorate the problems of sticky drops.

Static contact angle is maximized if the liquid is physically prevented from contacting the solid. A classic example is the *Leidenfrost effect* seen when a drop of water lands on a hot skillet or a drop of liquid nitrogen splashes onto a smooth floor. Beneath the drop, the liquid rapidly evaporates, and the resulting cushion of gas keeps the drop from ever making contact with the substratum. Its contact angle is very nearly 180°, and the drop readily slides across the surface.

Adhesion-resistant surfaces accomplish the same thing, not by evaporation, but rather by trapping pockets of air between small low-surface-energy projections (Figure 16.8B). The liquid makes contact with the tips of the projections, but they form only a small fraction of the surface's area. Thus, unless the liquid can displace the trapped air, its contact with the surface is small, its static contact angle is large, and its adhesion to this type of *superhydrophobic surface* is drastically reduced. This state of affairs is known as the *fakir condition*—a reference to Indian fakirs who demonstrated their spiritual prowess by lying comfortably on a bed of nails—or the *Cassie-Baxter condition* after A.B.D. Cassie and S. Baxter, who described the phenomenon in influential papers published in 1944 and 1945.

A wide variety of plants and animals have evolved superhydrophobic surfaces. The classic example is the leaf of the lotus plant, which has a static contact angle greater than 160° and a hysteresis of less than 10°. Raindrops easily roll off the leaf, taking any dust with them (Barthlott and Neinhuis 1997). Thus, the lotus leaf is not only self-bailing but self-cleaning as well. More than 200 plants species are known to have superhydrophobic surfaces (Quéré and Reyssat 2008).

Water drops roll off lotus leaves equally well in all directions; butterflies have taken the process one step further. It is advantageous for these flying insects if drops fall off the lateral edges of the wings rather than off the leading and trailing edges (which are important in creating lift) or onto the body (where they would continue to add to weight). This directionality is accomplished by minute shinglelike scales on the wings' surface. The scales form a Cassie-Baxter surface ($\theta_{stat} = 152°$), but their asymmetry makes it easiest for drops to roll away from the body's midline (Zheng et al. 2007).

Superhydrophobic surfaces allow other insects and some spiders to walk on water (Bush and Hu 2006). If water were at all attracted to the legs of these animals, surface tension would suck them in. Instead, the legs of water striders and other water-walking arthropods are coated with a layer of minute hairs that create a Cassie-Baxter surface. The resulting large contact angle allows surface tension to support the animal at the water's surface.

Stenocara, a beetle found in the Namibian desert, plays a more nuanced game. Rainfall is exceedingly scarce in the Namib, but fog is common. These beetles survive by capturing fog droplets, which requires a uniquely designed surface (Parker and Lawrence 2001). The beetle's large abdomen sports a haphazard arrangement of small hydrophilic knobs scattered across an otherwise hydrophobic (although not superhydrophobic) cuticle. As the beetle stands head into the wind, fog droplets impact the abdomen, where they roll across the hydrophobic surface until they arrive at a hydrophilic knob. Through time, a drop of water accumulates on each knob, eventually growing large enough to break loose and roll upwind into the beetle's mouth. Without the hydrophilic knobs, small droplets would simply roll off the back of the abdomen. In this case, it is a combination of adhesion and adhesion resistance that keeps these beetles hydrated.

12 CONCEPTS, CONCLUSIONS, AND CAVEATS

There is a wide variety of mechanisms by which organisms adhere to objects in their environment: van der Waals forces; pressure difference, capillary and viscous adhesion, and solid glues. Most adhesives are sufficiently strong that organisms need to devote only a small fraction of their surface area to adhesion to support their weight or resist fluid-dynamic forces, and the ready availability of adhesive tenacity opens niches that would otherwise be off-limits. Some adhesive mechanisms (such as van der Waals forces) require careful design, while others (such as viscous adhesion) have requirements so simple that they are difficult to avoid. Indeed, adhesion is so common that many plants and animals have evolved surfaces specifically designed to resist it.

In this chapter I have provided an overview of biologically relevant types of adhesion, but this review is far from comprehensive. For example, charged particles can adhere by electrostatic attraction. This effect is unlikely to be important in water (especially seawater), where the ions inherently present in the liquid shield charged particles from each other. But electrostatic attraction may have some consequences in air—charged dust particles tend to stick to a wide variety of surfaces—but exploration of this effect is left to another venue. For a full discussion of adhesion, one should consult Wake (1982), Bhushan (1999), or Pocius (2002) and keep one's eye on the literature. The development of new techniques for measuring the characteristics of molecules at interfaces (Waite and Broomell 2102) promises to open new vistas in adhesives research.

Chapter 17

Statics
Bending and Twisting

This is a chapter about shape. In previous chapters we investigated a few examples of how materials' strength affects plants and animals, but these examples were exceptionally simple. Typically, the materials in question were loaded solely in tension, and, for an object loaded in tension, its cross-sectional shape doesn't matter. A rope that is oval in cross section is just as good as one that is round, so when dealing with tension we are free to ignore the precise shape of a kelp stipe, strand of spider silk, or a byssal thread. In most cases of biological relevance, however, loads are more complex. Materials are not only stretched but compressed and sheared as well. Indeed, different parts of the same structure might have to respond to all three types of load at the same time. For the relatively complex loads imposed on a coral colony, leg bone, or tree trunk, shape matters a great deal, and we need to take it into account. The mathematics that uses a structure's shape to predict its deformation is known as *beam theory*, a well-developed branch of engineering to which this chapter is a brief introduction.

Through an exploration of beam theory, we will learn how the shape and size of trees, cacti, grasses, and corals affect their survival and will predict shapes that are optimal for different types of loads. We will see how mechanics automatically adjusts sea anemones' shape to facilitate feeding, and we will have a first look at why kelps are smaller than redwoods and how daffodils twist in the wind.

1 BENDING

Recall from Chapter 2 that a moment, M, is the product of force and length, where length—the lever arm—is the distance between the force's line of action and the axis of rotation (Figure 17.1A). Application of a net moment causes an object to rotate. To this point, we have dealt with moments in their capacity to make objects rotate as a whole—gametes or larvae in a boundary layer, for instance. We now explore what happens when moments are applied in a fashion such that different parts of an object rotate different amounts—that is, when moments cause plants and animals to bend.

We start with some terminology. Any relatively long, evenly shaped piece of material is a *bar*. Used as a structural element and loaded along its axis, a bar is a

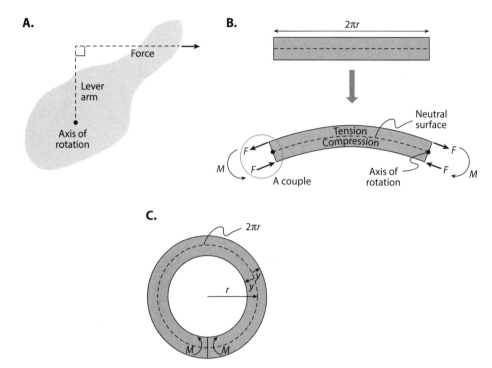

Figure 17.1 A. Moment is the product of force and lever-arm length. **B.** Applied moments stretch the upper half of a beam and compress the lower half, leaving a neutral surface in between. **C.** Bending the beam into a complete circle allows us to calculate strain. (See text.)

column, familiar as one of the structures holding up the Parthenon or the portico of an antebellum southern mansion. If, instead, a bar is subjected to a lateral force, it is a *beam*. Beams are everywhere in human technology. The floor joists and rafters in your house are beams. Bridges are built from beams, as is the chassis of your car. Natural beams are also common: trees, bones, seaweeds, lobsters' legs and bees' wings all are beams, and we need to know how they work.

Consider the generic beam shown in Figure 17.1B. It is initially straight, and—for reasons that will become clear in a moment—we specify that it has length $2\pi r$. To bend the beam, we apply a bending moment to each end, one rotating clockwise, the other counterclockwise. More specifically, each bending moment is applied by a couple: two forces acting in opposite directions, both with the same length of lever arm. Because each couple consists of equal and opposite forces, no net force is applied to the beam, and it is in translational equilibrium. This type of loading—which is rare in nature but useful for heuristic purposes—is known as *pure bending*.

We now shift our focus from the moments applied at each end to the individual forces that contribute to those moments. In the upper half of the beam, the force acting to the right is resisted by one acting to the left, putting the top of the beam in tension. In the lower half, the two forces load the beam in compression. It is this combination of tension and compression that causes the beam to bend.

Now, if the upper portion of the beam is stretched and the bottom compressed, there must be points in between that are left undeformed. These points form the *neutral surface*, the surface within the beam that isn't strained.

As a thought experiment, let's increase the applied moments until our beam bends completely around and its ends meet (Figure 17.1C). Because the neutral surface is neither stretched nor compressed, it maintains its original length, $2\pi r$. As a result, when its ends meet, the beam's neutral surface forms a circle of radius r. Or, to put it another way, r is the beam's *radius of curvature*.

At radial distance y from the neutral surface, the circumference of the now circular beam is $2\pi(r + y)$. (Note that y is positive for locations outside the neutral surface and negative for locations inside.) Consequently, in the process of bending, the length of the beam at y changes by

$$\text{change in length at } y = 2\pi(r + y) - 2\pi r$$
$$= 2\pi y. \tag{17.1}$$

Expressing change in length relative to original length, we calculate the strain at y:

$$\varepsilon = \frac{2\pi y}{2\pi r}$$

$$= \frac{y}{r}. \tag{17.2}$$

In other words, strain in the beam's material increases linearly with distance from the neutral surface (both positive and negative), and is inversely proportional to the beam's radius of curvature.

Having calculated strain, we are now in a position to calculate stress. If we assume that the beam's material is Hookean, stress is determined by the product of modulus and strain (Chapter 13):

$$\sigma(y) = E\varepsilon. \tag{17.3}$$

But $\varepsilon = y/r$ (equation 17.2), so in the beam,

$$\sigma(y) = \frac{Ey}{r}. \tag{17.4}$$

Now, the smaller its radius of curvature, the more tightly a beam is curved. Thus *curvature* (ξ) and radius of curvature (r) are inversely related:

$$\xi = \frac{1}{r}. \tag{17.5}$$

Substituting this relationship into equation 17.4, we see that

$$\sigma(y) = \xi E y. \tag{17.6}$$

For a given distance from the neutral surface, the more curved the beam and the stiffer its material, the greater the stress incurred. In short, by bending a beam to a certain curvature, stress is imposed on the beam's material.

But how large a bending moment do we need to apply to get this curvature? It depends on the beam's shape. Consider the scenario shown in Figure 17.2A, a cross section through a beam bent to curvature ξ. At each distance y from the neutral surface, there is an infinitesimally thin layer of material parallel to the neutral surface,

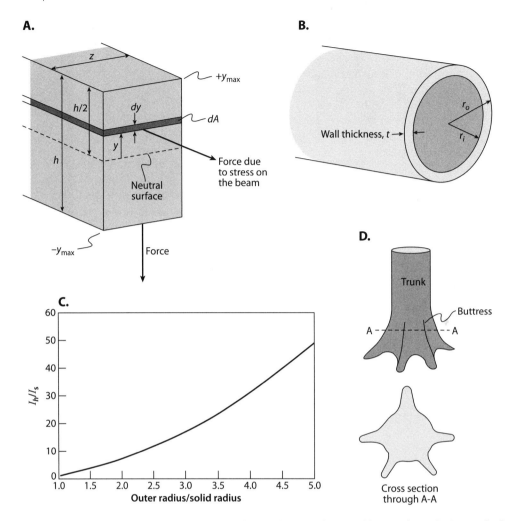

Figure 17.2 A. Cross section through a rectangular beam showing the variables used to calculate I. **B.** Cross section through a hollow circular beam. **C.** The ratio of I for a hollow beam to that of a solid beam with the same volume. **D.** Many trees use buttresses to increase the I of their trunks.

and, looking end-on at the cross section, we see that each infinitesimal layer has cross-sectional area dA.

Now, equation 17.6 tells us that for a layer at y, stress is $\xi E y$, and thus the force acting on the layer's cross section—the product of stress and area, dA—is

$$dF = \xi E y \, dA. \tag{17.7}$$

This force acts with lever arm y to apply moment dM around the neutral surface:

$$dM = (\xi E y \, dA) \, y$$

$$= \xi E y^2 \, dA. \tag{17.8}$$

To calculate the total bending moment acting on the beam, we integrate dM from the bottom of the beam $(-y_{max})$ to its top (y_{max}):

$$M = \int_{-y_{max}}^{y_{max}} \xi E y^2 \, dA$$

$$= \xi E \int_{-y_{max}}^{y_{max}} y^2 \, dA. \tag{17.9}$$

The integral in equation 17.9 describes how the beam's cross-sectional shape—that is, the disposition of its area—affects the moment required to give a certain curvature. As such, it plays a leading role in beam theory and is given its own name—the *second moment of area*, I:

$$I = \int_{-y_{max}}^{y_{max}} y^2 \, dA. \tag{17.10}$$

The moment in this term refers to the fact that area is multiplied by distance. It is the second moment because distance (y) is squared.

Substituting I into equation 17.9, we arrive at a pleasingly simple expression:

$$M = \xi E I. \tag{17.11}$$

In other words, the moment required to bend a beam to a given curvature depends not only on the stiffness of the beam's material (E), but also on the way in which that material is distributed (I), that is, on the beam's shape. Because these two terms work together to determine a beam's resistance to bending, their product, $E I$, is known as *flexural stiffness*.

It will be useful to take this analysis one step further. From equation 17.6 we know that $E = \sigma(y)/\xi y$. Combining equations 17.6 and 17.11, we see that

$$\sigma(y) = \frac{My}{I}. \tag{17.12}$$

Thus, stress in a beam increases with increasing bending moment and distance from the neutral surface and decreases with increasing second moment of area. This relationship is the key to calculating the structural consequences of bending. It tells us how the shape of the beam—indexed by I—interacts with an applied moment to impose stress on the beam's material.

2   SPECIFYING I

However, to calculate σ, we still need to specify the second moment of area, which involves applying equation 17.10 to the cross section in question. Consider again a rectangular cross-section (Figure 17.2A). The total height of the rectangle is h, and its width is z. To calculate I, we need to multiply each element of area by the square of its distance from the neutral surface. But to do that we need to know where the neutral surface is. We locate the neutral surface by applying two rules: (1) it is oriented perpendicular to the plane in which the bending moment is applied, and (2) if a material has the same stiffness in tension and compression (which most materials do), it passes through the cross section's center of area. In this case, the neutral surface is located as shown, with the top of the section $h/2$ above it and the bottom of the section $h/2$ below.

Next, one specifies the amount of area $(d A)$ at each distance from the neutral surface. For this rectangular cross section, $d A = z d y$ for all y.

Given these preliminaries, we now can calculate I:

$$I = \int_{-h/2}^{h/2} y^2 \, d A$$

$$= z \int_{-h/2}^{h/2} y^2 \, d y$$

$$= \frac{z y^3}{3} \Big|_{-h/2}^{h/2}$$

$$= \frac{z h^3}{12}. \tag{17.13}$$

That is, for a beam with a rectangular cross section, the second moment of area is $\frac{1}{12}$ the product of width and the cube of height. Height contributes more to I than does width because height affects y, and y is squared in calculating I. For this, and all other shapes, I has units of m⁴.

A similar calculation can be made for a circular cross section of radius r (Supplement 17.1), with the result that

$$I = \frac{\pi r^4}{4}. \tag{17.14}$$

In biology, beams are often hollow: the legs of insects and crustacea, the wing bones of birds, the stems (technically, culms) of bamboo. To calculate I for hollow cross sections of this sort, one first calculates the second moment as if the section were solid and then subtracts the I for the part of the beam that is missing. For instance, for a hollow circular cross section with inner and outer radii r_i and r_o, respectively (Figure 17.2B):

$$I = \frac{\pi r_o^4}{4} - \frac{\pi r_i^4}{4}$$

$$= \frac{\pi}{4} \left(r_o^4 - r_i^4 \right). \tag{17.15}$$

For a tube whose wall thickness $t = r_o - r_i$ is much less than r_o, equation 17.15 has a simple approximation (Supplement 17.2):

$$I \approx \pi r_o^3 t. \tag{17.16}$$

Hollow beams present intriguing possibilities. Material removed to hollow out a beam comes from the part of the structure nearest the neutral surface, where, due to the factor of y^2 in the formula for I, it contributes little to the beam's bending stiffness. As a result, a hollow beam has nearly the same stiffness as a solid beam with the same outside dimensions but with less expenditure of materials. For plants and animals, the savings in materials corresponds to savings in the metabolic energy required to construct the beam, allowing an organism to put that energy to some other use. Reduction of a beam's mass also reduces its weight. The long bones in birds' wings are a classic example. They are hollow and filled with air,[1] so for the same stiffness they weigh less than a solid beam, making it easier for the bird to get off the ground.

For other animals, the goal is to get stiffness on the cheap. Given a fixed volume of material, how can they make the stiffest structure? For a given volume of chitin, for instance, how can insects and spiders make the stiffest legs? Instead of making a solid beam with length ℓ and a small radius r_s (with a volume $V = \pi r_s^2 \ell$), the organism could make a stiffer hollow beam with a larger radius, r_h. In this case, the ratio of I for the hollow beam to that of a solid beam with the same volume is (Figure 17.2C; Supplement 17.3):

$$\frac{I_h}{I_s} = 2 \left(\frac{r_h}{r_s} \right)^2 - 1. \qquad (17.17)$$

For a fixed volume of material, the larger the radius of a hollow beam, the greater its I and, therefore, the greater its flexural stiffness.

Followed to its logical limit, this reasoning suggests that to get maximum stiffness from a given amount of material, animals and plants should make their beams with radii as large as possible. There is a problem, however. To maintain constant volume, the larger the radius, the thinner the wall of the beam, and thin walls can be dangerous.

Consider what happens when you bend a thin-walled tubular structure such as a plastic drinking straw. Initially, it curves nicely, maintaining a circular cross section. But when stress in the tube's wall reaches a critical level, σ_{crit}, the tube's cross section starts to flatten at some point along its length, becoming wider and thinner. This local deformation reduces the second moment of area, increasing stress (equation 17.12), which causes the local cross section to become even more flattened. The process then feeds back on itself, resulting in *local buckling*—a kink—and the beam fails.

At what stress does a beam buckle? For this kind of thin-walled tubular beam, the critical stress at which a kink forms is (Wainwright et al. 1976)

$$\sigma_{crit} \approx \frac{E t}{4 r_h} \qquad (17.18)$$

For a circular cylinder of length ℓ, material volume V, and a wall thickness t much smaller than r_h (Supplement 17.4),

$$\frac{t}{r_h} \approx \frac{1}{2} \left(\frac{r_s}{r_h} \right)^2. \qquad (17.19)$$

Thus,

$$\sigma_{crit} \approx \frac{E}{8} \left(\frac{r_s}{r_h} \right)^2. \qquad (17.20)$$

For a tube of fixed volume and length, the larger its outer radius, the smaller the critical buckling stress and, thus, the greater the risk of kinking. As a result, although an increase in radius can increase the stiffness of a hollow beam, the utility of this strategy is limited by the tendency to buckle. We will return to buckling later in the chapter when we deal with anemones.

---

[1] In addition to their structural role, the hollow interior of wing bones form part of the bird's respiratory system (Schmidt-Nielsen 1997).

The preceding calculations assume that the neutral surface passes through a cross section's center of area. However, this is true only if a beam's material has the same stiffness in compression and tension. If $E_c \neq E_t$, the neutral surface shifts toward the side of the cross section with the higher modulus. For example, Brian Gaylord found that kelp stipes are three- to fivefold stiffer in tension than in compression (Gaylord and Denny 1997). As a result, their neutral surface is shifted toward the side of the beam in tension.

This shift may be adaptive. First, their relatively low compressive modulus causes kelps such as *Pterygophora*, *Laminaria*, and *Eisenia* to bend more than they otherwise would, allowing them to effectively go with the flow, which reduces their velocity relative to the surrounding water. Bending also potentially brings the fronds down into the benthic boundary layer, where flow is slower. Both effects reduce drag, thereby reducing the risk of exceeding the stipe's breaking strength.

In addition, the difference in moduli reduces the likelihood of fracture. Materials are most likely to fracture when loaded in tension. Because the tensile stiffness of kelp stipe is greater than its compressive stiffness, a relatively small fraction of material is in tension when the beam is bent. If waves arrive from unpredictable directions and cracks are distributed randomly, reducing the volume of stipe in tension reduces the chance that a critical crack will propagate catastrophically.

One final note before we leave the subject of I: many beams of ecomechanical interest have complex cross-sectional shapes. For instance, as a part of the arms race to acquire light, canopy trees in tropical rain forests commonly have extensive buttresses radiating out from the lower part of the trunk (Figure 17.2D), thereby increasing their second moment of area. By placing material as far as possible from the trunk's neutral surface, buttresses give maximum stiffness at minimal expense, providing stability as the tree grows to keep up with its neighbors. The fronds of some giant kelps develop longitudinal corrugations, which increase the frond's I and, thereby, its flexural stiffness, an apparent adaptation to reduce flapping and the drag it causes (Rominger and Nepf 2014). The Is of such cross sections are not found in standard engineering tables, but they can easily be calculated (Supplement 17.5).

3 CANTILEVER BENDING

Our analysis so far has dealt with pure bending in which curvature is the same throughout a beam. As with Hookean elasticity, an ideal fluid, or a principled politician, one seldom encounters pure bending in nature. Instead, another, more complicated, type of bending is especially common. Consider the beam shown in Figure 17.3A. It is fixed at one end and free at the other and is known as a *cantilever*. When we apply a lateral force to the end of the cantilever, it imposes a bending moment, which tends to rotate the beam. However, because its base is rigidly fixed, the cantilever can't simply pivot. Instead, its top stretches, its bottom compresses, and it bends.

Unlike a beam in pure bending, the bending moment in a cantilever varies along the beam's length. At the beam's base, the applied load acts over the beam's entire length, and the bending moment is maximal. At the beam's free end, the applied force has no lever arm, and the moment is zero. As a result of this varying moment, material in a cantilever is maximally stressed at the base and minimally stressed at its tip (equation 17.12).

A.

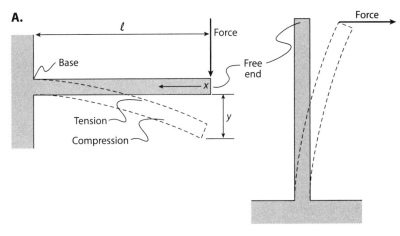

B.

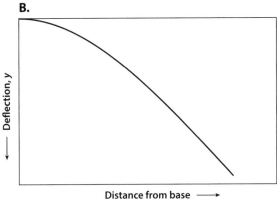

Figure 17.3 A cantilever: a beam with a fixed base and a free end (**A**). When loaded at its free end, a cantilever deforms with a characteristic shape (**B**).

This variable stress causes a cantilever to bend in a particular shape (Figure 17.3B), which we can describe by quantifying the beam's deflection at each point along its length. For a beam of length ℓ and constant cross section, deflection y at distance x from the beam's free end is (Supplement 17.6)

$$y \approx \frac{F}{EI}\left(\frac{x^3}{6} - \frac{\ell^2 x}{2} + \frac{\ell^3}{3}\right).$$ (17.21)

Directly at the free end, where $x = 0$,

$$y_{tip} \approx \frac{F\ell^3}{3EI}.$$ (17.22)

As one would expect, the stiffer the beam's material (E) and the larger its second moment of area (I), the smaller the overall deflection. Deflection is very sensitive to beam length: all other things being equal, a beam twice as long deflects eight times as far.

Note that equations 17.21 and 17.22 are approximations. They are reasonably exact only when y_{tip} is less than a tenth of beam length (see Supplement 17.6). For larger deflections, the shape of a beam is described by a set of exact equations known as the *elastica*. Application of the elastica is a bit mind bending and laborious, but at times it is both necessary and useful. For example, Brian Gaylord and I (Gaylord and Denny 1997) used the elastica in our exploration of wave-induced deformations in subtidal kelps (mentioned earlier), allowing us to document how the these seaweeds reduce stress in their stipes by changing shape as they grow. Readers interested in large-deflection formulas for beams should consult engineering texts such as Gere and Timoshenko (1984).

So far, we have predicted how a cantilever bends based on prior knowledge of its material's stiffness. Contrawise, cantilever bending provides a handy tool for measuring a material's modulus. One obtains a beamlike sample of the material, measures the I of its cross section, and mounts it as a cantilever. A known force is then applied to the beam's tip by hanging a weight from it, and the resulting deflection is measured. E is then obtained through a rearrangement of equation 17.22:

$$E \approx \frac{F\ell^3}{3I y_{\text{tip}}}. \tag{17.23}$$

This method is accurate as long as y_{tip} is less than 10% of ℓ and the modulus in tension is the same as that in compression. If $E_t \neq E_c$, the modulus obtained from equation 17.23 is E_b, the *effective modulus in bending*, which is intermediate between E_t and E_c.

4   STRESS IN CORALS

We are now in a position to apply beam theory in a biological context. Consider a cylindrical coral colony such as the staghorn coral, *Acropora cervicornis* (Figure 17.4A). As water flows past the colony, drag applies a bending moment at the colony's base, potentially dislodging it. However, unlike the end loading we have considered so far, the hydrodynamic load imposed on a staghorn coral is distributed along its length, and we need to take this new type of load into account.

Let's assume that the radius of the coral cantilever is r. For a small portion of the beam with height dz, the area projected into flow is

$$dA = 2r\,dz. \tag{17.24}$$

Drag on this area is

$$dF = \frac{1}{2}\rho_f u^2 C_D dA$$

$$= \rho_f u^2 C_D r\,dz, \tag{17.25}$$

where ρ_f is the density of seawater, u is water velocity, and C_D is the beam's drag coefficient.

The moment applied to the colony's base by this force is the product of drag and its lever arm z, where z is distance from the cantilever's base:

$$dM = \left(\rho_f u^2 C_D r\,dz\right) z. \tag{17.26}$$

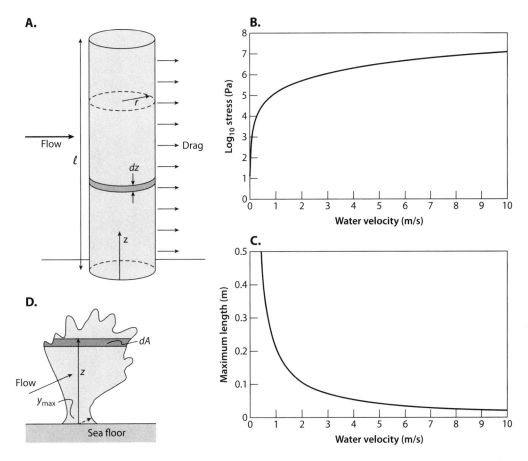

Figure 17.4 A. A coral, which we approximate as circular cylinder in flow, showing the variables needed to calculate the distributed bending moment. **B.** Stress at the base of a coral increases with increasing water velocity. **C.** Maximum length of a coral decreases with increasing water velocity. **D.** Calculating bending moment for a coral of arbitrary profile (y_{max} is measured in the direction of flow). (See text.)

Integrating this expression across the entire length of the colony, ℓ, we arrive at the total bending moment:

$$M = \rho_f u^2 C_D r \int_0^\ell z\, dz$$

$$= \frac{1}{2}\rho_f u^2 C_D r \ell^2. \tag{17.27}$$

Let's use this bending moment to calculate the stress imposed on the colony's base. From equation 17.12 we know that the maximum stress in any cross section of the colony is Mr/I, and I for a circular cross section is $\pi r^4/4$ (equation 17.14). Thus,

$$\sigma_{max} = \frac{\left(\frac{1}{2}\rho_f u^2 C_D r \ell^2\right) r}{\left(\frac{\pi r^4}{4}\right)}$$

$$= \frac{2\rho_f u^2 C_D}{\pi}\left(\frac{\ell}{r}\right)^2. \tag{17.28}$$

The higher the aspect ratio of the coral—the ratio of its length to its radius—the higher the stress at the base. This poses a problem for a long, skinny coral such as *A. cervicornis*, where ℓ can be 20 times r.

Equation 17.28 is graphed in Figure 17.4B for $\ell/r = 20$. The tensile strength of an *A. cervicornis* skeleton is approximately 5×10^6 Pa (Madin 2005), a stress that isn't reached until $u = 6$ m $\cdot$ s^{-1}. Thus, one might suppose that *A. cervicornis* could resist all but the most rapid wave-induced water velocities.

However, the strength of the substratum to which the coral is attached is considerably weaker, only approximately 3×10^5 Pa. The same stress that acts on the base of the coral is imposed on the substratum. As a result, the colony breaks at a velocity of only 1.6 m $\cdot$ s^{-1}. Indeed, *A. cervicornis* is often dislodged even in mild storms (Tunnicliffe 1981).

It would be a mistake, however, to assume that breakage is entirely a bad thing. Corals such as *A. cervicornis* play the role of weeds in reef communities. Between storms they proliferate rapidly, overgrowing their neighbors with little regard for the risk of dislodgment. In the next storm, they pay the price and suffer disproportionately. But the broken pieces of these colonies are capable of resurrecting themselves. If a piece by chance gets wedged into the reef, it can cement itself in place and start growing again (Tunnicliffe 1981). In this sense, breakage serves as a means of dispersal.

For corals that have not opted for this weedy lifestyle, there is an obvious strategy to avoid breakage: reduce the colony's aspect ratio. For example, a short, squat colony with a length equal to its radius exerts only 0.25% as much stress on the substratum as that of a staghorn coral of equal radius and can therefore resist water velocities 20 times higher (approximately 32 m $\cdot$ s^{-1}), rendering it more or less immune to breakage from bending stress. Many massive corals employ this tactic.

So far we have used equation 17.28 to predict the water velocity required to break a coral, assuming that the coral maintains a fixed aspect ratio as it grows. Alternatively, we can assume that the coral has a fixed radius and use equation 17.28 to predict its maximum length. Holding r constant, setting σ_{max} to the substratum's breaking stress, and solving equation 17.28 for ℓ, we find

$$\ell_{max} = \sqrt{\frac{\sigma_{brk}\pi}{2\rho_f u^2 C_D}} r. \qquad (17.29)$$

The larger the radius and the stronger the substratum, the higher the coral can grow. The faster the flow and the larger the drag coefficient, the shorter the coral must be. For a cylindrical coral with a radius of 1 cm, a C_D of 1, and a substratum breaking stress of 3×10^5 Pa, maximum length is quite short (Figure 17.4C): just 21 cm at 1 m $\cdot$ s^{-1} and 11 cm at 2 m $\cdot$ s^{-1}.

Our analysis has dealt with the simple case of a cylindrical coral in uniform flow. Most corals have more complex shapes, and velocity in the benthic boundary layer is likely to vary with distance from the substratum, but the same logic applies. Consider the silhouette of a generic coral (Figure 17.4D). In this case, the area projected into flow varies in a complex fashion with height above the substratum, but we can nonetheless define the bending moment imposed on its base (presumably the maximum moment imposed on the coral):

$$M_{max} = \frac{\rho_f}{2} \int_0^{z_{max}} zu^2(z)C_D(z)\,dA(z). \qquad (17.30)$$

Here $u(z)$ is velocity at distance z from the sea floor (set by mainstream flow and the nature of the boundary layer), $dA(z)$ is the element of area at z (set by the shape of the colony), and $C_D(z)$ is the drag coefficient of that element.

Having quantified the bending moment, we then calculate stress as we did before:

$$\sigma_{max} = \frac{M_{max} y_{max}}{I}, \tag{17.31}$$

where y_{max} is maximal lateral distance from the neutral surface at the colony's base:

$$\sigma_{max} = \frac{\rho_f y_{max}}{2I} \int_0^{z_{max}} z u^2(z) C_D(z)\, dA(z). \tag{17.32}$$

Madin and Connelly (2006) simplified this relationship by assuming that boundary layer thickness in the oscillatory flow experienced by corals is small compared to the length of a colony, such that $u(z) = u_{max}$ everywhere along the coral's length. They also assumed that $C_D = 1$ for all segements of the colony. Given these assumptions,

$$\sigma_{max} = \frac{\rho_f y_{max} u_{max}^2}{2I} \int_0^{z_{max}} z\, dA(z). \tag{17.33}$$

Setting σ_{max} equal to the substratum's breaking stress and dividing both sides of the equation by dynamic pressure ($\frac{1}{2}\rho u_{max}^2$) provides a useful perspective on coral mechanics:

$$\frac{2\sigma_{brk}}{\rho u_{max}^2} \leq \frac{y_{max} C_D}{I} \int_0^{z_{max}} z\, dA(z). \tag{17.34}$$

The left side of this expression—the ratio of strength to dynamic pressure—is an index of the coral's fortitude—its ability to remain intact in face of flow—and Madin and Connelly dubbed it the *dislodgment mechanical threshold, DMT*. The right side of the expression contains all the factors that relate to the shape of the colony, the *colony shape factor, CSF*. If $CSF > DMT$, the coral fails by breaking at its base.

Madin and Connelly measured CSF for a variety of coral species, allowing them to predict the dislodgment of these species as a function of the strength of the substratum and maximum water velocity, each of which varies both through time and from place to place on the reef. These calculations were the basis for their predictions discussed in Chapter 1. As storm intensity increases (due to climate change) and substratum strength decreases (due to ocean acidification), species with a high CSF are predicted to decrease in relative abundance, while those with a low CSF are predicted to increase in relative abundance (Figure 1.1B).

5   SHEAR

Our calculations for the risk of breakage in corals have assumed that the maximum stress imposed is due to bending. Before we blithely move on, its is best if we cast a critical eye on this assumption.

Consider again a cantilever with a force applied to its free end. Not only does this force apply a bending moment (with consequences shown in Figure 17.3A), it also applies a shear load. In response, the beam shears, and an equilibrium is reached between the external force and shear stress in the beam's material (Figure 17.5A). Recall from Chapter 13, however, that this equilibrium requires a second set of

A.   **B.**   **C.**

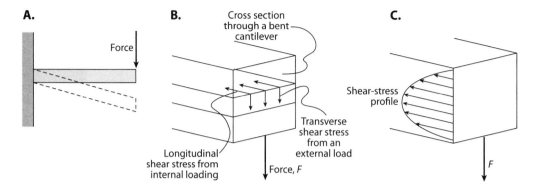

Figure 17.5 A. An applied load shears a cantilever. **B.** Transverse shear stress is accompanied by longitudinal shear stress. **C.** As a result, shear stress varies across the beam.

shearing forces (Figure 17.5B). Therefore, as the external load shears the beam transversely, an internal load shears the beam longitudinally.

In the interior of a beam, this longitudinal shear is resisted by the elasticity of the beam's material. That is, each lamina in the beam is tugged on by material lateral to it. However, at the outer surface of the beam there is no adjacent material to supply a tug. This leads to the disconcerting conclusion that longitudinal shear stress disappears at the beam's lateral surfaces. Because equilibrium shear stresses always come in matched perpendicular pairs, this implies that transverse shear stress also goes to zero at the lateral surfaces.

The strange fact that there can be no shear stress on the outer surface of a beam does not negate the need to offset the applied shearing load F: for a beam with cross-sectional area A, the *average* shear stress acting on the beam's cross section must be F/A. Instead, our logic implies that shear stress varies across the section from zero at the lateral edges to a maximum in the middle. For a rectangular cross section, shear stress varies parabolically (Figure 17.5C; Gere and Timoshenko 1984), with a maximum in the middle:

$$\tau_{max} = \frac{3F}{2A}. \tag{17.35}$$

For a circular cross section,

$$\tau_{max} = \frac{4F}{3A}. \tag{17.36}$$

Values for other shapes can be found in Roark and Young (1975).

We are now in a position to compare shear stress to the tensile and compressive stresses in a loaded beam. Consider perhaps the most general biological example, a cantilever with a circular cross section. The beam has length ℓ, radius r, and is loaded at its free end by force F. Combining equations 17.12 and 17.14, we find that maximal bending stress is

$$\sigma_{max} = \frac{F\ell r}{\left(\frac{\pi r^4}{4}\right)} = \frac{4F\ell}{\pi r^3}. \tag{17.37}$$

From equation 17.36 we know that

$$\tau_{max} = \frac{4F}{3\pi r^2}.$$ (17.38)

Thus, the ratio of maximum bending stress to maximum shear stress is proportional to the beam's aspect ratio, ℓ/r:

$$\frac{\sigma_{max}}{\tau_{max}} = 3\frac{\ell}{r}.$$ (17.39)

Shear stress is equal to bending stress only if length is a third of radius: a very short, very squat beam. For more typical biological cantilevers—where length is perhaps ten times radius—shear stress is much smaller than tensile and compressive stresses, as we have tacitly assumed.

It is instructive to take this analysis one step farther. For Hookean materials, stress is proportional to strain:

$$\tau = G\gamma,$$ (17.40)

$$\sigma = E\varepsilon,$$ (17.41)

and for an isotropic, isovolumetric material, $G = E/3$. Thus, for beams made from these materials,

$$\frac{\varepsilon}{\gamma} = \frac{\ell}{r}.$$ (17.42)

The ratio of strain in tension or compression to that in shear is equal to the beam's aspect ratio. Short, squat beams shear as much as they stretch and compress. Long skinny beams stretch and compress much more than they shear. As we will see in a moment, some sea anemones have evolved in response to these facts.

But before we turn to biology, I need to make one last point about the role of shear in cantilever bending. Although shear stress is commonly less than tensile or compressive stress, it nonetheless plays an important role. As an example, hold a paperback book by its spine with its cover horizontal. Because of the lack of shear resistance between the pages, the book's free end flops down. (The slip between its pages is visual evidence of the longitudinal shear that accompanies the transverse shear imposed by the book's weight.) In effect, each page acts as its own separate cantilever with the result that the book is much more flexible than it would be if the pages were glued together. The relative stiffness of the book-beam varies as $1/n^2$, where n is the number of pages (see Supplement 17.7)—because of its lack of shear resistance, a book with 100 pages is only 1/10,000 as stiff as a block of wood of the same dimensions.

The absence of substantial shear resistance affects the stiffness of large cacti such as the cardons and saguaros. From the outside, these cantilever-like plants appear to have the same general structure as tree trunks except for the presence of vertical pleats, which allow the cactus to expand laterally when water is available, increasing its storage capacity. But to accommodate this expansion, the trunk is typically divided internally into 8 to 14 separate spars, one per pleat. Because the spars are only loosely joined to each other in shear, the cactus bends much more easily than a tree trunk of the same dimensions.

Reduction of shear resistance is a potential problem for tubelike beams as well. Being hollow, they don't have any material at the very location where, in a solid beam,

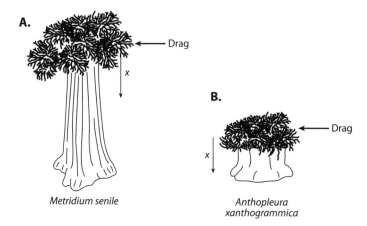

Figure 17.6 Sea anemones with different body plans and different life-history strategies: **A.** *Metridium senile* and **B.** *Anthopleura xanthogrammica* (redrawn from Denny 1988).

shear resistance would be maximal. In apparent response to this issue, the hollow bones in birds' wings have interior struts oriented at approximately 45° angles to the bone's axis. The struts also inhibit local buckling.

6   ANEMONES AS BEAMS

Now we turn to the curious case of anemones. In a classic paper published in 1977, Mimi Koehl applied beam theory to an exploration of the functional anatomy of two species of sea anemone: *Metridium senile*, a plumose species found in protected waters, and *Anthopleura xanthogrammica*, the great green anemone common on wave beaten intertidal shores (Figure 17.6A and B). Both species are cylindrical cantilevers with hollow cross sections, beams that we now know how to analyze.

Both anemones make a living in part by capturing food particles brought to them by moving water, but the nature of the food and the method of capture is distinctly different. *M. senile* has a large, fluffy crown of tentacles it uses as a filter to sieve small particles from the flow. *A. xanthogrammica* acts more like an open garbage can; it sits with its oral disk and short tentacles horizontal, waiting for waves to dislodge a mussel or snail. When an animal falls on the disk, its is engulfed by the tentacles and ingested. These different modes of feeding are facilitated by structural adaptations.

Because *A. xanthogrammica*'s length is less than its radius, we can conclude from equation 17.42 that when a wave strikes, *A. xanthogrammica* deforms more in shear than in tension or compression. As a result, its oral disk stays nearly horizontal, ready to catch any falling prey.[2] By contrast, *M. senile* is much longer than it is wide, indicating that it experiences more bending stress than shear stress. The anemone's thin-walled columnar body is widest at the base, tapering toward the crown of tentacles. As a result of this taper, *I* increases from crown to base, more than

---

[2] *Anthopleura* also has symbiotic algae, which photosynthesize and contribute some of the resulting carbohydrates to the animal (along with a striking green color). Keeping the oral disk horizontal thus also benefits the anemone by maximizing the light that strikes it.

offsetting the corresponding increase in bending moment. Because bending stress is proportional to the ratio of M to I (equation 17.12), stress in *M. senile* is greatest just below the crown. In fact, even for the slow flows encountered by these animals (0.2 m $\cdot$ s^{-1}), bending stress exceeds critical buckling stress, and the anemone kinks below its crown.

But this local buckling is adaptive. As the column buckles, the crown of tentacles reorients, presenting more area to food-laden flow. In this fashion, the anemone's cantilever-like body serves to lift the tentacles up out of the benthic boundary layer into more rapid flow, and local buckling then automatically helps to orient the crown into an optimal feeding configuration.

7 MAXIMUM HEIGHT

In the course of evolution, competition for light has fueled a structural arms race among plants. A tree or seaweed that lifts its photosynthetic apparatus incrementally nearer to the sun can attain a competitive advantage by shading its neighbors. To put it another way, plants have ulterior motives. The results of this arms race are easily observed. Terrestrial forests are dominated by tall trees and kelp forests are dominated by canopy forming species. Even intertidal seaweeds such as the sea palm *Postelsia palmaeformis* have joined the race, using their palm-tree-like form to compete for light.

The morphological outcomes of these races have varied, however. With their ropelike stipes, giant kelps bear little resemblance to trees. Sea palms do look like trees but reach maximal heights of less than a meter, a far cry from majestic redwoods, which can grow to heights of 100 m or more. Why the differences?

One aspect of the answer involves the concept of stability: a system is stable if, after being disturbed, it returns to its undisturbed state. For a plant with its photosynthetic apparatus supported by a vertical stem or stipe, stability implies that, if the system is pushed to the side, it can return to vertical. Having worked our way through the mechanics of cantilevers, we are now in a position to analyze stability in greater detail, thereby to predict the maximum height of plants and their concomitant competitive advantage.

Consider the generic plant shown in Figure 17.7A, a vertical cantilever beam with a crown of leaves or fronds mounted at its free end. We assume that the plant's roots or holdfast fix it rigidly to the ground, and, for simplicity of calculation, we assume that the mass of the cantilever is negligible relative to that of the crown.

If the beam is precisely vertical with the crown balanced on its end, there is no tendency for the cantilever to topple. However, if the crown mass is nudged even a slight distance to one side, its weight interacts with the resulting lever arm to exert a bending moment (Figure 17.7B). If this moment is not counteracted by the beam's elasticity, a small initial displacement leads to a larger displacement, which increases the lever arm. A larger lever arm increases the bending moment, which increases the displacement, and so forth. In short, the potential exists for the cantilever to become unstable and topple over. Under what conditions does instability become a problem?

First, let's specify the bending moment exerted by the crown's mass. If the crown has volume V and density ρ, and is immersed in a fluid with density ρ_f, its weight is

$$F_g = (\rho - \rho_f)\, Vg. \tag{17.43}$$

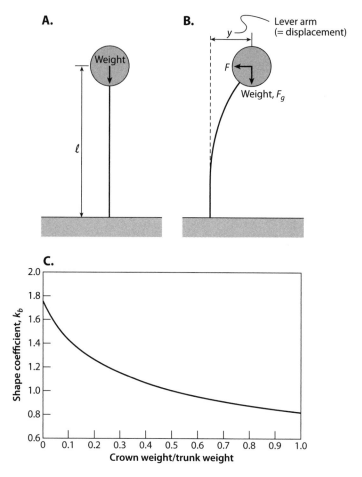

Figure 17.7 Calculating the maximum stable height of a vertical cantilever. **A.** A tree, simplified as a mass balanced on the cantilever. **B.** Deflection of the mass imposes a bending moment. **C.** For a tree, the shape coefficient k_b depends on the ratio of crown weight to trunk weight.

Here, as always, g is the acceleration of gravity. The bending moment exerted by the crown's weight is

$$M_c = (\rho - \rho_f) \, Vg \, y_{tip},$$ (17.44)

where y_{tip} is the crown's horizontal displacement.

This bending moment is resisted by the elastic deformation of the beam's material. From equation 17.22 we know that a force F is required to deflect the tip:

$$F = \frac{3EI\,y_{tip}}{\ell^3}.$$ (17.45)

Here ℓ is the length of the cantilever, and EI is its flexural stiffness. This force acts over the length of the beam to exert a moment M_b tending to bring the beam back to vertical:

$$M_b = F\ell = \frac{3EI\,y_{tip}}{\ell^2}.$$ (17.46)

The system is stable as long as—for a small deflection—the moment exerted by the crown's weight is less than the restorative moment from the beam's elasticity. That is, the cantilever will return to vertical if

$$\frac{d M_c}{d y_{\text{tip}}} < \frac{d M_b}{d y_{\text{tip}}}. \tag{17.47}$$

Now,

$$\frac{d M_c}{d y_{\text{tip}}} = \left(\rho - \rho_f\right) Vg$$

$$\frac{d M_b}{d y_{\text{tip}}} = \frac{3 E I}{\ell^2}$$

Inserting these values into equation 17.47 and solving for ℓ, we see that

$$\ell < \sqrt{\frac{3 E I}{\left(\rho - \rho_f\right) Vg}}. \tag{17.48}$$

As long as the beam's length is less than the expression on the right side of this inequality, the system is stable. The stiffer the beam's material and the larger its second moment of area, the higher the cantilever can be and retain its stability. The denser the crown material and the larger its volume, the shorter the beam must be.[3]

Equation 17.48 also tells us why sea palms are shorter than redwoods. The modulus of wood is approximately 6 GPa, whereas that of seaweeds is approximately 0.01 GPa. The density of both stipe material and the wet wood of a tree is roughly $1000 \text{ kg} \cdot \text{m}^{-3}$, and that of air is $1.2 \text{ kg} \cdot \text{m}^{-3}$. Thus, for organisms with equal I and crown volume, a cantilever made of wood can remain stable to 24 times the height of a beam made from stipe material.

Our analysis so far fails to take into account several important aspects of real plants. For example, it ignores the role of the cantilever's mass, which in a tree can be a substantial fraction of overall mass. Similarly, it suggests that a tree could grow to a great height by drastically reducing its crown volume. But, with a reduced crown volume, a tree might not have the photosynthetic capacity to grow at all. A more detailed analysis that takes the shape of the organism into account—thereby fixing the relative size of the crown and letting us calculate the mass of the beam—allows us to predict of how tall a plant can stably grow while remaining the same shape (Supplement 17.8). The results of this analysis reveal that, for a beam of circular cross section with basal radius r,

$$\ell_{\text{max}} = k_b \left[\frac{E r^2}{\left(\rho - \rho_f\right) g}\right]^{1/3}. \tag{17.49}$$

[3] Note that if the crown's density is less than that of the surrounding fluid, the expression within the radical is negative, and the inequality is indeterminate. This indicates that, under these conditions, the beam is always stable regardless of its length, a scenario that applies to giant kelps. The pneumatocysts of a giant kelp's fronds make its crown less dense than seawater. As a result, any deflection of the crown acts in concert with the stipe's stiffness, exerting a moment tending to pull the stipe vertical.

Here k_b is a coefficient that depends on the ratio of crown mass to the mass of the cantilever and on the taper of the beam (Figure 17.7C; King and Loucks 1978).

To put stability analysis in a biological perspective, let's compare the maximum stable height of a cantilever-like coral to the maximum height allowed by drag (equation 17.29). The density of coral skeleton is approximately 2000 kg $\cdot$ m^{-3} and its modulus is approximately 10 GPa. The density of seawater is 1025 kg $\cdot$ m^{-3}. For a vertical, cylindrical colony with a radius of 1 cm ($k_b = 1.26$), maximum stable length is 5.85 m. This is far in excess of the hydrodynamically imposed breaking length we calculated earlier in the chapter—0.2 m in 1-m $\cdot$ s^{-1} flow. Clearly hydrodynamic force, rather than stability, is a coral's major worry. A similar conclusion applies to trees. The bending moment from wind dragging on the crown is much larger than that imposed by gravity acting on the displaced crown mass, compelling trees to stay shorter than our stability analysis would predict.

By contrast, sea palms have taken the arms race for light to the limit (Holbrook et al. 1991). In an apparent attempt to keep up with the Joneses, *Postelsia* in dense thickets grow to heights where they cannot stand on their own, and, unless propped up by their neighbors, they topple. Solitary sea palms grow differently, suggesting that this growth is indeed due to competition for light. Off on their own, and therefore not subject to competition for sunlight, solitary individuals stay relatively short and are always stable.

8   OPTIMAL SHAPE

In 1858 Oliver Wendell Holmes penned an insightful essay on the theory of optimal design. Perhaps because it was disguised as a poem ("The Deacon's Masterpiece"), the essay has never gotten its due in the engineering literature, but it is a classic nonetheless. The focus of the essay is the construction and subsequent use of a small carriage—a one-hoss shay—designed such that each part was equal in strength to every other. By constructing the carriage in this fashion, the Deacon used a minimal amount of material—having wasted none in making any part too strong for its function—and the shay served admirably for 100 years until, it "wore out" and disintegrated:

> All at once, and nothing first,/Just as bubbles do when they burst./End of the wonderful one-hoss shay./Logic is logic. That's all I'll say.

The one-hoss-shay principle allows us to calculate the optimal shape for cantilever beams (such as corals) where fluid-dynamic force is likely to be the selective factor. Given a material with a particular strength, we can mathematically construct a beam shaped such that, when under load, no part of the beam's surface experiences more stress than any other and all parts are loaded to just shy of their breaking stress. Because the beam's material is used to maximum effect, the beam is built with the minimum volume. Building a coral from as little material as functionally necessary would presumably save the colony metabolic energy, allowing it to grow faster or produce more young.

Consider, for instance, a cantilever-like coral with a circular cross section loaded at its end by a localized force (Figure 17.8A), a situation corals might encounter when impacted by wave-borne debris. Our task is to specify how the radius of the beam must change along its length so that maximum stress along the entire length of the

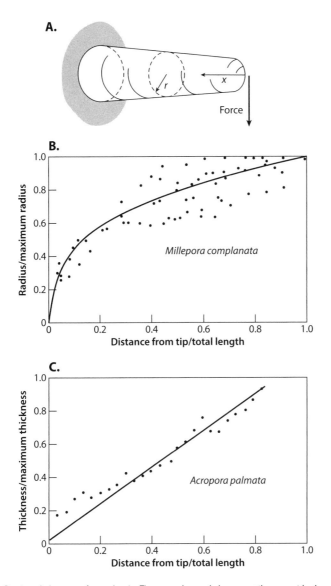

Figure 17.8 Optimal shape of corals. **A.** The coral model: a cantilever with the dimensions shown. **B.** The shape of fire coral approximates the optimum predicted for a cantilever loaded at its end. **C.** The shape for *Acropora palmata* approximates that predicted for a cantilever loaded by drag.

coral is constant at the material's breaking stress, σ_{brk}. The resulting beam requires the minimum amount of material to resist impact.

From equation 17.12, we know that

$$\sigma_{brk} = \frac{F\,xr}{I}, \tag{17.50}$$

where x is distance from the beam's free end and r is the beam's radius at x. For a circular cross section, $I = \pi r^4/4$, so

$$\sigma_{\text{brk}} = \frac{F\,xr}{\left(\frac{\pi r^4}{4}\right)} = \frac{4Fx}{\pi r^3}. \tag{17.51}$$

Solving for r, we have our answer:

$$r = \left(\frac{4F}{\pi\,\sigma_{\text{brk}}}\right)^{1/3} x^{1/3}. \tag{17.52}$$

To resist the load while using the minimal amount of material, the beam's radius should increase in proportion to the cube root of distance from its free end (Figure 17.8B). The larger the force, the larger r should be; the stronger the material, the smaller r can be. These calculations have some ecological credibility: *Millepora complanata*, a fire coral found in the shallow waters of Carribean reefs, is often subjected to the sort of point loading modeled in this example, and its dimensions indeed scale according to the cube-root rule we have just derived (Denny 1988).

Our assumption that the largest force a coral colony faces is a point load at its tip allows for relatively straightforward calculation of optimal shape but makes for limited application. Most corals are more likely stressed by drag or their own weight. Calculation of optimal shape in these cases is a bit more involved (Supplement 17.9), but the results are worth investigating even if we skip their derivation.

Drag is proportional to the area projected into flow and is, therefore, distributed along the length of a cantilever. To resist drag with minimal allotment of material, an organism should increase its radius in direct proportion to distance from the free end:

$$r \propto x. \tag{17.53}$$

Colonies of *Acropora palmata* grow and branch horizontally like the branches of a tree, extending proud of their neighbors, and are thus subject to drag. If drag is the largest force these colonies face, the thickness of an *A. palmata* colony should increase in direct proportion to distance from the free end, and this indeed seems to be the case (Figure 17.8C).

A horizontal coral beam is also subject to its own weight, which is proportional not to its area, but to its volume. If weight is the largest force the coral feels, radius should increase in proportion to the square of distance from the free end:

$$r \propto x^2. \tag{17.54}$$

I know of no coral that follows this pattern, perhaps an indication that self-weight is not an important factor for these aquatic organisms.

Lest we get carried away with this line of reasoning, there is a caveat. First, the examples I have cited here (*M. complanata* and *A. palmata*) are more the exception than the rule—species cherry-picked to make a point. Many, if not most, corals depart from the optimal shapes predicted here. The shape that is truly optimal is that which best serves all an organism's needs—feeding, capturing sunlight, competing with the neighbors, and so on—and the shape that best fulfills these combined tasks need not be optimal for any single task.

In summary, it can be a useful exercise to predict the optimal shape of biological beams. When organisms such as corals match these predictions, it allows us to

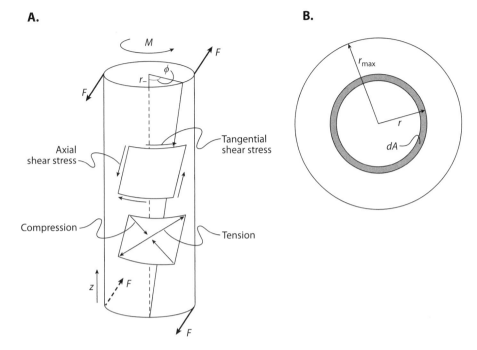

Figure 17.9 Torsion. **A.** A cylinder loaded in torsion, showing the variables used to quantify shear stress and strain. **B.** A cross section through a circular beam showing the variables used to calculate the polar second moment of area, J.

hypothesize that the stresses in question have been important evolutionary drivers. When organisms deviate from predicted optima, the pattern of the deviation can provide information as to what alternative factors are involved.

9 TORSION

Beams bend, but they also twist, and analysis of twisting is neatly analogous to that of bending. Consider, for instance, the cylindrical beam shown in Figure 17.9A. A couple is applied to each end of the beam, with moments acting in opposite directions around the beam's axis (one tending to rotate the beam's cross section clockwise, the other counterclockwise). In response, the material in the beam is sheared and the beam twists. This type of deformation is known as pure torsion.

If we fix one end of the beam, we can measure the rotation, ϕ (in radians), of the other end. As this end rotates, material a radial distance r from the cylinder's axis moves a tangential distance $r\phi$. This tangential distance, expressed relative to the beam's length, ℓ, is the shear strain in the beam at a distance r from its axis:

$$\gamma(r) = \frac{r\phi}{\ell}. \tag{17.55}$$

(To understand why tangential deflection per beam length is shear strain, it may help to revisit Figure 13.8E, in which γ is defined.) Knowing shear strain, we can now calculate shear stress, τ, at each point in the cylinder. For a material with shear modulus G,

$$\tau(r) = G\gamma(r) = \frac{Gr\phi}{\ell}. \tag{17.56}$$

Recall again that shear stresses always come in orthogonal pairs. Thus, the tangential shear stress we have just described must be matched by an axial stress of equal magnitude (Figure 17.9A). If a beam's shear strength is less in the axial than in the tangential direction (as it is in a tree branch, for example), the beam cracks longitudinally. Recall, also, that shear stress is equivalent to tensile and compressive stresses each applied at a 45° angle to shear. If a material is weaker in tension than in shear (human bone, for instance), a twisted beam breaks at a 45° angle. The spiral fractures suffered by skiers' leg bones are prime examples.

Having calculated the shear stress for every point in a twisted cylinder, we are now in a position to calculate the moment required to twist the beam. Consider the cross section shown in Figure 17.9B. At distance r from the beam's axis, there is area dA, which, due to the local shear stress, exerts a tangential force:

$$dF = \tau \, dA \tag{17.57}$$

$$= \frac{Gr\phi \, dA}{\ell}. \tag{17.58}$$

This force applies a moment:

$$dM_t = r \, dF$$

$$= \frac{Gr^2\phi \, dA}{\ell}. \tag{17.59}$$

Integrating across r from zero to r_{max} (the cylinder's radius), we arrive at our answer. The total moment required to twist the cylinder through angle ϕ is

$$M_t = \frac{G\phi}{\ell} \int_0^{r_{max}} r^2 \, dA. \tag{17.60}$$

The integral in this equation—the *polar second moment of area*, J—is the torsional equivalent of I:

$$J = \int_0^{r_{max}} r^2 \, dA. \tag{17.61}$$

Thus,

$$M_t = \frac{\phi G J}{\ell}. \tag{17.62}$$

The longer the beam, the smaller the moment required to twist it through a given angle. For a beam of a given length, the moment required to twist the beam depends not only on the beam's stiffness (G) but also on the way in which that material is distributed (J). Because G and J work together to determine a beam's resistance to twisting, their product, GJ, is known as *torsional stiffness*, the rotational analogue to flexural stiffness, EI.

Working through the math (Supplement 17.10), we see that for a circular cross section

$$J = \frac{\pi r_{max}^4}{2}.$$ (17.63)

In other words, for a beam of circular cross section,

$$\frac{I}{J} = 0.5.$$ (17.64)

In nature, biological cantilevers are commonly bent and twisted simultaneously. The crowns on trees are seldom perfectly symmetrical, for instance, so that wind-induced drag imposes both bending and twisting moments. The relative extent of twist to bend depends on the ratio of flexural to torsional stiffness:

$$\frac{\text{twist}}{\text{bend}} = \frac{\text{flexural stiffness}}{\text{torsional stiffness}} = \frac{EI}{GJ}.$$ (17.65)

Suppose, for example, that flexural stiffness is twice torsional stiffness. For a given applied moment, the beam will twist twice as much as it bends.

Recall from Chapter 13 that for an isovolumetric, istropic material, $G = E/3$. We have just established that, for a circular cylinder, $I/J = 0.5$. Thus, for a beam with a circular cross section made from an isovolumetric, isotropic material,

$$\frac{EI}{GJ} = 1.5.$$ (17.66)

The beam will twist more than it bends. Vogel (2009) reviews the available measurements of E and G for biological materials, and notes that shear modulus is commonly much less than tensile modulus. As a consequence, structures such as tree trunks and algal stipes have large EI/GJ ratios, so they preferentially twist rather than bend.

Daffodils provide an aesthetic example (Etnier and Vogel 2000). The EI/GJ ratio of their stems is 13, sufficiently high that in even a moderate breeze, the drag-induced torque on their flowers is sufficient to twist the stem so that the flowers pivot like wind vanes into stable, down-wind postures. For many species of trees, the petioles of their leaves have similarly high EI/GJ ratios, and Vogel speculates that this is an adaptation allowing the leaves to reconfigure into a streamline shape when subjected to wind.

10 CONCEPTS, CONCLUSIONS, AND CAVEATS

When a beam bends, some material is stretched and some compressed, and the amount of deformation increases with distance from the neutral surface. Because of this dependence on distance, the shape of a beam's cross section—indexed by the second moment of area, I—becomes important. The more of a beam's material that is located far from the neutral surface, the stiffer the structure. Analogous conclusions apply when a beam is twisted. The farther material is from the center of twist, the more it is sheared. The effect of cross-section shape in this case is indexed by the second moment of area, J.

Cantilevers are common in nature, and their mechanics provide a useful perspective on the interaction between structure and function. We can predict the optimal shape of a beam designed to resist a given load, and some corals hew to these predictions.

Several caveats are attached to these conclusions. First, the equations we have used for the deformation of cantilevers are accurate only when deflection is small. This doesn't pose a problem for our analysis of corals because they break before they deflect appreciably. It may be a concern when exploring bending in trees, and it is certainly a concern when investigating seaweeds. If large deflections are suspected of a structure, it is best to use the exact equations of the elastica rather than the approximations presented here.

Second, our analyses have assumed that the base of a cantilever is rigidly fixed. This is approximately true for corals, but substantial deviations from this assumption are common for other biological cantilevers. There is considerable give in tree roots, for instance, and at the base of feathers. A precise analysis of these and similar structures must take these basal deformations into account. This accounting can be difficult. For example, despite considerable effort, the mechanical behavior of tree roots and their contribution to the overall mechanics of trees have yet to be nailed down. As a result, an important and basic question—what limits the height of trees?—remains open (see Vogel (2009) for a discussion).

Lastly, our analysis of bending has tacitly assumed that a deformed beam is in static equilibrium with its load. The operative word here is static. In a steady wind, a tree has plenty of time to deform to an equilibrium deflection. But what happens when a turbulent gust imposes a rapidly varying velocity? As the tree deforms in response to the varying load, its mass accelerates, and the resulting inertial forces can have important consequences. Similar complications apply to kelps buffeted by crashing waves. In short, to apply beam theory accurately in the real world, we often need to account for dynamic as well as static forces. This accounting is the subject of the next chapter.

Chapter 18

Dynamics
The Mechanics of Oscillation

Many mechanical systems—both environmental and biological—have characteristic oscillations, and these oscillations have ecological consequences across the full range of spatial and temporal scales. The wings of flying animals flap up and down, the legs of walking animals swing back and forth, and the dynamics of these motions affects the cost of transport. The ability of trees and grasses to sway in the wind and kelps to bend in wavy flow helps to explain their survival in stressful environments. Sensory hairs take advantage of oscillatory physics to detect faints sounds, and the rise and fall of ocean tides is a major environmental influence in nearshore marine communities. To understand these examples—and to recognize similar phenomena in nature—we need to explore the fundamentals of oscillating mechanical systems.

1   HARMONIC OSCILLATION

In order to oscillate, a mechanical system must have two basic elements: an inertial component that is free to move and a restorative element that opposes that motion. The classic example is a mass (an inertial element) suspended from a spring (a restorative element, Figure 18.1A). If we pull the mass down and then let it go, it oscillates sinusoidally up and down in what is known as *harmonic motion*.

The essence of this oscillation is twofold. First, there is a trade-off between force and momentum. At the extremes of its travel, the mass is stationary, so it has no momentum. But at the extremes of displacement, the spring is maximally stretched or compressed, so it applies maximal force to the mass. At the midpoint of its oscillation, the spring is neither stretched nor compressed, so no force is applied, but the mass moves rapidly, giving it maximum momentum. Back and forth the mass goes, trading force for momentum, and vice versa.

Alternatively, oscillation can be viewed from the perspective of energetics. When maximally deflected, the stationary mass has no kinetic energy, but the deformed spring has maximal elastic potential energy. At the midpoint of the oscillation, the spring isn't deformed, so it has no potential energy, but the mass is moving fast, maximizing its kinetic energy. Oscillation thus involves the trade-off between potential and kinetic energy, and as long as no energy is lost from the system, oscillation continues.

Figure 18.1 A. A mass suspended by a spring can oscillate vertically. **B.** A mass on the end of a cantilever can oscillate horizontally. **C.** In harmonic motion, velocity leads displacement by $\pi/2$ rad and acceleration lags displacement by π rad.

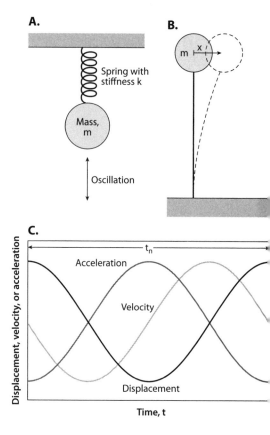

These sorts of trade-offs are inherent in a wide variety of systems. The restoring force need not be a spring; it could be the pull of gravity or the pressure in a gas. The inertial component need not be the translational momentum of a mass as it moves, it could instead be the rotational inertia of an object as it spins. In fact, analogues of the spring-mass oscillator can be found outside mechanical systems. The size of a population can oscillate around its carrying capacity with the reproductive potential of the population providing the "inertia" and food availability providing the restoring "force." The widespread occurrence of harmonic oscillations provides the impetus to investigate how they work.

Rather than deal with hypothetical springs and masses, let's base our discussion on a system already familiar to us from the last chapter. Consider a hypothetical tree in which crown mass m is supported by a trunk that acts as a cantilever beam (Figure 18.1B). The tree has the following, very idealized, properties:

- The trunk's mass is negligible relative to that of the crown, allowing us to ignore it.
- The crown is free to move side to side with no resistance from the air. As a result, no energy is lost to drag as the mass moves.
- If we bend the trunk by pulling the mass to one side (along the x-axis), the beam acts like a linear spring, resisting with a force proportional to displacement x:

$$F = -kx. \tag{18.1}$$

In this context, k is the beam's *structural stiffness*.

Having thus defined the properties of the system, we now perform an experiment. We displace the mass to x_0 and hold it stationary. We then let it go, starting a clock at the moment of release, and record the ensuing results (Figure 18.1C). The crown mass oscillates sinusoidally back and forth as we expect, but we are now in a position to analyze this motion in greater detail.

From Figure 18.1C we see that the crown completes one full cycle in t_n seconds, its *natural period*. The frequency of oscillation—the *natural frequency*, f_n (in Hz)—is the inverse of t_n:

$$f_n = \frac{1}{t_n}. \tag{18.2}$$

With f_n thus defined, we can describe the tree's swaying mathematically:

$$x(t) = x_0 \cos(2\pi f_n t). \tag{18.3}$$

Here x_0 is the oscillation's *amplitude*, the magnitude of its maximum deflection, and t is time from release. The term $2\pi f_n t$ (radians) is the oscillation's *phase*, the angle that determines the value of the cosine function. As phase increases through time, the cosine cycles between 1 and -1, repeating itself every time the phase is an integer multiple of 2π.

Equation 18.3 is an accurate description of the tree's oscillation, but it does not tell us why the system oscillates with this frequency. As a first step to that end, let's calculate the mass's velocity, u. From calculus we know that

$$\frac{d \cos(x)}{dt} = -\sin(x)\frac{dx}{dt} \tag{18.4}$$

and

$$\frac{d(2\pi f_n t)}{dt} = 2\pi f_n. \tag{18.5}$$

Thus, the derivative of equation 18.3 is

$$u(t) = \frac{dx}{dt} = -2\pi f_n x_0 \sin(2\pi f_n t). \tag{18.6}$$

In other words, because the mass's displacement oscillates as a cosine function, its velocity is shifted by $90°$, cycling as a sine function (Figure 18.1C).

Next, we calculate the mass's acceleration by taking the time derivative of velocity. From calculus we know that

$$\frac{d \sin(x)}{dt} = \cos(x)\frac{dx}{dt}. \tag{18.7}$$

Thus,

$$a_x(t) = \frac{du}{dt} = -4\pi^2 f_n^2 x_0 \cos(2\pi f_n t). \tag{18.8}$$

Like the mass's displacement, its acceleration varies as a cosine function, but the negative sign tells us that when the mass is displaced to the left, it accelerates to the right, and vice versa (Figure 18.1C).

We now return to basics. From Newton's second law of motion, we know that for mass m to accelerate at rate a_x, a force F must be applied. Thus, for our oscillating mass,

$$F(t) = ma_x(t) = -4\pi^2 f_n^2 m x_0 \cos(2\pi f_n t). \tag{18.9}$$

(The negative sign tells us that the restoring force acts in the opposite direction from displacement.) Where does this force come from? For our tree, F is provided by the cantilever—the system's spring—and from equation 18.1 we know that the magnitude

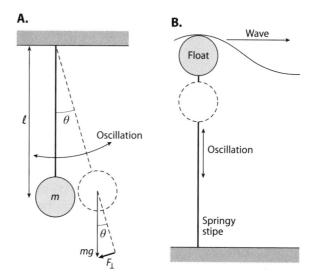

Figure 18.2 A. A pendulum is a harmonic oscillator. **B.** *Nereocystis*, a giant kelp, acts like a buoyant mass on a spring.

of the applied force is k times the beam's displacement (equation 18.3). Thus,

$$F(t) = -kx(t) = -kx_0 \cos (2\pi f_n t) . \qquad (18.10)$$

We now have two equally valid equations for force (equations 18.9 and 18.10). Setting them equal to each other, and solving for f_n, gives us the answer we seek:

$$f_n = \frac{1}{2\pi} \sqrt{\frac{k}{m}} . \qquad (18.11)$$

The frequency at which a spring-mass system oscillates is set by the ratio of stiffness to mass—the greater the stiffness, the more frequent the oscillations; the larger the mass, the slower the oscillations.

With this result in hand, our investigation of beam theory in Chapter 17 allows us to calculate the natural frequency for our tree. The stiffness of a cantilever beam—the ratio between applied force and tip deflection—is

$$k = \frac{3EI}{\ell^3} , \qquad (18.12)$$

where E is the modulus of the beam's material, I is the second moment of area of the beam's cross section, and ℓ is the beam's length. Wood has a modulus of approximately 6 GPa (Cannell and Morgan 1987), so for a trunk 20 cm in diameter and 10 m long, $k = 1,414 \, \text{N} \cdot \text{m}^{-1}$. With a crown mass of 250 kg, a tree with this trunk would have a natural frequency of 0.38 Hz—a period of approximately 2.6 s—the sort of graceful sway we expect.

Let's take a moment to expand on the concept encapsulated in equation 18.11. In essence, this equation tells us that the natural frequency of a spring-mass system is set by the ratio of the system's stiffness to its inertia. In the example we have examined so far, a beam provides the stiffness and mass provides the inertia, but other elements can fill these roles. Consider, for instance, a pendulum with mass m suspended by a massless arm of length ℓ (Figure 18.2A). If the mass is deflected to one side, its weight

mg has a component $F_\perp$ acting at right angles to the arm. From a consideration of geometry, we see that

$$F_\perp = mg \sin \theta, \qquad (18.13)$$

where θ is the angle from vertical through which the arm is deflected. This force, acting over length ℓ, provides a moment M tending to return the arm to vertical.

$$M = mg\ell \sin \theta. \qquad (18.14)$$

The larger the angle through which the arm is deflected, the greater this restoring moment. We can thus define the *rotational stiffness* of the system as the ratio of change in restoring moment to change in angle of deflection:

$$k_{rot} = \frac{d\,(mg\ell \sin \theta)}{d\theta} \qquad (18.15)$$

$$= mg\ell \cos \theta. \qquad (18.16)$$

For small values of θ, $\cos \theta \approx 1$, so

$$k_{rot} \approx mg\ell. \qquad (18.17)$$

We now need an element of inertia. In Chapter 2 (equation 2.30) we learned that the rotational analog of mass is the rotational moment of inertia, J_m, which serves as the inertial element in our pendulum.[1] For this simple system with one discrete mass,

$$J_m = m\ell^2. \qquad (18.20)$$

With expressions now in hand for the rotational analogues of stiffness and inertia, we can calculate the natural frequency of the pendulum:

$$f_n = \frac{1}{2\pi} \sqrt{\frac{k_{rot}}{J_m}}$$

$$= \frac{1}{2\pi} \sqrt{\frac{mg\ell}{m\ell^2}}$$

$$= \frac{1}{2\pi} \sqrt{\frac{g}{\ell}}. \qquad (18.21)$$

Odd as it may seem, frequency is independent of mass—it depends only on the pendulum's length and the acceleration of gravity. It is this characteristic that makes pendulums useful in clocks. For a grandfather clock to tick and then tock at one

---

[1] The rotational moment of inertia, J_m, is closely related to the polar second moment of area, J, we encountered in Chapter 17:

$$J_m = \int_0^{\ell_{max}} y^2 dm, \qquad (18.18)$$

$$J = \int_0^{\ell_{max}} y^2 da. \qquad (18.19)$$

second intervals (a period of 2 s between ticks, $f_n = 0.5$), its pendulum must have a length of one meter, but the weight of the pendulum bob is immaterial.[2]

A playground swing is another classic example of a pendulum. The chains holding the swing's seat are roughly 2 m long, and working through the math, we find that a swing's natural frequency is approximately 0.36 Hz, high enough to be thrilling but low enough to allow for productive interaction with the child involved.

The interaction I have in mind is pumping, in which the rider swings his or her legs forward as the swing moves forward and back as the swing moves back. If one pumps at the swing's natural period, energy is added to the system with each cycle and the amplitude of oscillation increases, augmenting the exhilaration of the ride. The pumping of a swing is one example of the general phenomenon of *resonance*; when an oscillating system is periodically pushed or pulled and the frequency of the force matches the system's natural frequency, large amplitude oscillations can result.

The bull kelp *N. luetkeana* provides a biological example (Figure 18.2B). The fronds of this giant kelp can have a mass of 10 kg and entrain with them another 10 kg of water, for a total of 20 kg. The ropelike stipe has a tensile stiffness of roughly 100 $N \cdot m^{-1}$. Thus, using equation 18.11,

$$f_n = \frac{1}{2\pi}\sqrt{\frac{100}{20}} \approx 0.36 \text{ Hz}. \tag{18.22}$$

In other words, if the frond mass is tugged away from the seafloor and then released, it bobs up and down every 2.8 s. This period is close to that of choppy waves kicked up by local winds. If, by chance, the period of the waves matches that of the kelp, waves can potentially pump the kelp's oscillation, dangerously increasing its amplitude.

But in nature, this doesn't happen. For kelps and other biological oscillators, resonance is usually held in check. To see how, we investigate the frequency-dependent nature of damped oscillations.

2   DAMPED HARMONIC MOTION

We begin by returning to the push-pull relationship between the acceleration of a mass and the deflection of a spring. If we assume that ma and kx are the only forces acting as the system oscillates,

$$ma = -kx. \tag{18.23}$$

Or, to put it another way,

$$ma + kx = 0. \tag{18.24}$$

The left side of this equation is a list of the forces acting in the system, albeit a very short list in this idealized scenario. The zero on the right side of the equation signifies the lack of any external force. As we have seen, this equation embodies the trade-off between kinetic and potential energy that keeps an oscillation going. Noting that

---

[2] To be exact, $\ell = 0.994$ m. In the initial formation of the SI system of measurement, the length of a pendulum with a natural fraquency of 0.5 Hz was considered as a possible definition for the meter.

$a = d^2x/dt^2$, we can rewrite this relationship in the form of a standard differential equation:

$$m\frac{d^2x}{dt^2} + kx = 0. \tag{18.25}$$

In any real system, however, at least one additional force intrudes. As a tree's leaves or a kelp's fronds move through fluid (air or water, respectively), viscous drag opposes their motion. This *viscous damping force* does not depend on how far the system is displaced or how rapidly it accelerates but rather on how fast it moves. We can, therefore, characterize it as

$$\text{viscous damping force} = cu = c\frac{dx}{dt}, \tag{18.26}$$

where c is the *damping coefficient* ($\text{N} \cdot \text{s} \cdot \text{m}^{-1}$). The effect of viscous damping can be taken into account by adding it to the list of forces in our differential equation for oscillation:

$$ma + cu + kx = 0, \tag{18.27}$$

$$m\frac{d^2x}{dt^2} + c\frac{dx}{dt} + kx = 0. \tag{18.28}$$

The solution to this new equation requires a bit more math than is appropriate here (see Supplement 18.1), so we jump straight to the bottom line. If we again start with an initial displacement of x_0 and an initial velocity of zero, subsequent displacement of a damped system varies according to the following relationship:

$$x(t) = x_0 \exp\left(-\frac{c}{2m}t\right)\left[\frac{c}{4\pi mf_d}\sin(2\pi f_d t) + \cos(2\pi f_d t)\right], \tag{18.29}$$

where f_d is the frequency of the damped oscillation:

$$f_d = \sqrt{\frac{k}{m} - \left(\frac{c}{2m}\right)^2}. \tag{18.30}$$

There are three relevant messages to take away from equation 18.29 (Figure 18.3A):

1. As energy is lost to viscosity, the system's amplitude decays exponentially. For a given mass, the larger the damping coefficient, the faster amplitude decays.
2. Damped oscillation has frequency f_d rather than f_n. The larger c is, the lower the damped frequency (equation 18.30).
3. As c approaches $2\sqrt{mk}$, the term inside the radical in equation 18.30 approaches zero.[3] As a result, f_d tends toward zero, and period tends

---

[3] When $c = 2\sqrt{mk}$, equation 18.30 is indeterminate, so equation 18.29 can't be used. However, by allowing c to approach $2\sqrt{mk}$, we can, in the limit, find an alternative expression. When $c = 2\sqrt{mk}$,

$$x(t) = 2\pi x_0 \exp\left(-2\pi f_n t\right)(f_n t). \tag{18.31}$$

If $c > 2\sqrt{mk}$, the system doesn't oscillate at all, and a different equation is required to describe its motion.

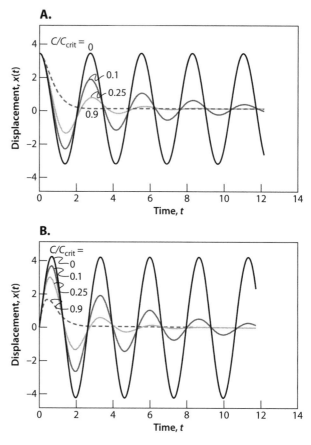

Figure 18.3 Viscosity drains energy from an oscillating system, causing its amplitude to decay. **A.** A mass released from an initial deflection, with no initial velocity. **B.** A mass starting from zero deflection but with an initial velocity.

toward infinity. In other words, as c approaches $2\sqrt{mk}$, the system doesn't so much oscillate as gradually dissipate. Because of this effect, $2\sqrt{mk}$ is an important benchmark for c, known as the *critical damping coefficient*, c_{crit}.

In formulating equation 18.29, we have assumed that our experiment started with the mass stationary at displacement x_0. There are alternative starting conditions. For example, we could start the experiment with the mass at $x = 0$ but give it initial velocity u_0. In this case,

$$x(t) = \exp\left(-\frac{c}{2m}t\right)\left(\frac{u_0}{2\pi f_d}\right)\sin(2\pi f_d t). \tag{18.32}$$

The behavior of the system is much the same (Figure 18.3B), the primary difference being that the oscillation is shifted in time: displacement is zero at $t = 0$ and reaches its first peak one quarter of the damped period $(1/f_d)$ later. We will make use of this relationship later in the chapter when we deal with impulsive loading.

3   FORCED HARMONIC MOTION

All this is valuable information, but in nature we are seldom presented with a scenario in which an initial deformation is abruptly released or an initial velocity is suddenly applied, as the descriptions above require. Instead, we are often more concerned with how a spring-mass system behaves when an external force is applied periodically. Trees bend under the influence of periodic wind gusts and kelps sway in response to periodic loading by ocean waves. Insects' wings beat when driven by the periodic contraction of muscles, and sensory hairs vibrate in reponse to the periodic push and pull of sound waves.

To explore systems of this sort, we examine a case in which a spring-mass system is subjected to F_{ext}, a sinusoidally varying force:

$$F_{ext} = F_0 \cos(2\pi\varphi t). \tag{18.33}$$

Here F_0 is the amplitude of the external force and φ is the frequency with which it cycles. It is important to note that φ is completely independent from either f_n, the system's natural frequency, or f_d, its damped frequency—we can set it to anything we please.

We begin by returning to the differential equation listing the forces acting on a spring-mass system:

$$ma + cu + kx = 0. \tag{18.34}$$

Recall that the zero on the right side of this equation signifies the lack of an external force. To take F_{ext} into account, we replace the zero with equation 18.33:

$$ma + cu + kx = F_0 \cos(2\pi\varphi t), \tag{18.35}$$

$$m\frac{d^2x}{dt^2} + c\frac{dx}{dt} + kx = F_0 \cos(2\pi\varphi t). \tag{18.36}$$

The solution to this equation is again less than straightforward (see Supplement 18.2), and we proceed right to the punch line:

$$x(t) = x_{max} \cos(2\pi\varphi t - \phi_f), \tag{18.37}$$

where x_{max}, the amplitude of the driven oscillation is

$$x_{max} = \frac{F_0}{\sqrt{\left(k - 4\pi^2 m\varphi^2\right)^2 + (2\pi c\varphi)^2}} \tag{18.38}$$

and

$$\phi_f = \arctan\left(\frac{2\pi c\varphi}{k - 4\pi^2 m\varphi^2}\right). \tag{18.39}$$

When driven by a periodic force, the system oscillates at a frequency equal to that of the driver (φ), although there can be a phase shift between the driving force and the system's displacement. The more damped the system, the smaller the amplitude of oscillation and the larger the phase shift.

These effects are most easily visualized by multiplying both sides of equation 18.38 by k/F_0. This yields an expression that compares the maximum internal force acting

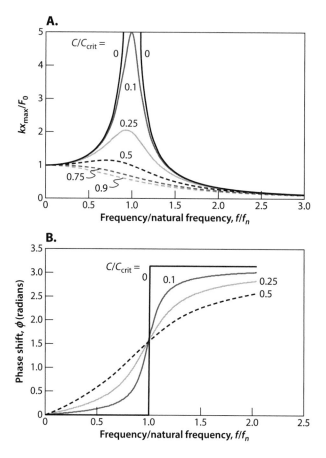

Figure 18.4 A. The force imposed on an oscillating cantilever (here expressed relative to the externally applied load) is amplified near the natural frequency. **B.** Phase shift varies with the frequency of the driving force.

on the spring, kx_{max}, to the amplitude of external force, F_0 (Figure 18.4A):

$$\frac{kx_{max}}{F_0} = \frac{k}{\sqrt{\left(k - 4\pi^2 m\varphi^2\right)^2 + (2\pi c\varphi)^2}}. \qquad (18.40)$$

For simplicity, the abscissa is drawn as the ratio of driving frequency to the system's natural frequency, f_n.

There are several important messages in this graph. At driving frequencies well above the system's natural frequency, large applied forces result in negligible deflection of the spring and, therefore, negligible internal force on the spring's material. In our tree example, the crown mass and cantilever-like trunk have a natural frequency of 0.38 Hz. Turbulent eddies in the wind might attempt to drive the crown mass at a frequency of 10 Hz, but these high-frequency forces would expend themselves working against the crown's inertia, and the trunk would hardly bend at all. Because bending is minimal, wood in the trunk is not appreciably stressed by high-frequency gusts, even if the gusts are strong.

By contrast, at driving frequencies well below the natural frequency, the system is essentially at equilibrium with the applied force. Because F_{ext} varies slowly, the mass is not required to accelerate rapidly, and it is only the beam's displacement that resists the external driver. In a steady wind, our tree would come into equilibrium with drag, and its wood would be stressed.

And then there are driving frequencies near the natural frequency. In this case, the system becomes very dynamic—it resonates. In fact, in the absence of damping, if the system is driven at its natural frequency, displacement amplitude increases without limit. This is the pumping effect we noted earlier for a child on a playground swing and a bull kelp in waves. Unlimited displacement implies unlimited force on the system's spring, which can end only in destruction.

Extreme forces can be avoided, however, if the system is damped. The larger the damping coefficient, the lower the resonant peak in force, until, when the system is critically damped, the peak disappears. Viscous damping thus provides a means to control resonant oscillation.

There are times, however, when resonance can be advantageous. Insects, for instance, need to sense the sounds of approaching predators. If a sensory hair is tuned to the frequecies produced by a predator, resonance can amplify the hair's oscillation, allowing for detection of very weak signals (Humphrey and Barth 2007; Casas and Dangles 2010). Crickets, for example, can detect a sound that imparts as little as 2.5×10^{-20} J to a sensory hair. In fact, sensory hairs are so sensitive that they are on the verge of being able to detect their own thermal jostling. As we learned in Chapter 12, all objects have a kinetic energy proportional to their absolute temperature. At 20°C, a sensory hair has a thermal kinetic energy of $k_T T = 4 \times 10^{-21}$ J, only slightly below the threshold of detectability.

The incredible sensitivity of insect's sensory hairs can lead to cascading ecological effects. Some wasps lay their eggs in living caterpillars, and, in response, caterpillars stop feeding when they hear flight noises. But caterpillars can't discriminate between wasps and honey bees. As a result, among the plants subject to grazing by caterpillars, those that are pollinated by bees are eaten less and presumably grow faster and produce more seeds (Tautz and Rostas 2008).

Resonance can also play a role in locomotion. Jellyfish move by rhythmically contracting their bell to produce propulsive thrust. By pulsing its bell at its resonant frequency, the jellyfish *Polyorchis penicillatus* increases the amplitude of oscillation by 40%, resulting in an energetic saving of 24% to 37% (DeMont and Gosline 1988). Similarly, a wide variety of running animals reduce their metabolic costs by striding at the resonant frequency of their springlike legs (Heglund and Taylor 1988).

By contrast, resonance in other structures could be catastrophic. As noted previously, wave-driven resonant oscillation of a bull kelp could break its stipe, and wind-driven resonant swaying could topple a tree. However, because both plants are highly damped, these lethal displacements seldom occur. As leaves and fronds move through the fluid around them, they dissipate sufficient energy to keep resonant oscillation in check.

In addition to its effects on the amplitide of displacement, resonance affects the timing of displacement. The phase difference ϕ_f between driving force and displacement is graphed in Figure 18.4B. For driving frequencies well below the natural frequency, ϕ_f is small, indicating that the spring moves in time with the force applied. However as frequency approaches and then exceeds f_n, an odd thing happens. Displacement gets increasingly out of phase with the driving force until at high frequencies, the mass moves 180° (π rad) out of phase with the force driving it. The abruptness of this phase switch depends on damping. If $c = 0$, the switch is maximally abrupt—at driving frequencies less than f_n, $\phi_f = 0$; at $f > f_n$, ϕ_f is 180°. If the system is damped, the phase gradually shifts from 0° to 180° as frequency increases through f_n.

The phase shift of a driven spring-mass system can be easily demonstrated with a a small mass (100 g or so) suspended from a rubber band that you hold with your hand. If you move your hand up and down with a low frequency, the mass moves in concert with the driving force: up when your hand moves up, down when your hand moves down. If you gradually shift to higher and higher frequencies, you will find that the mass abruptly switches to an out-of-phase motion when the period of your hand exceeds the natural period of the system. The mass moves down when your hand moves up, and vice versa.

The shift in phase of a spring-mass system can have practical consequences. Ocean tides are caused by the interaction of gravitational and centrifugal forces as the moon orbits Earth and Earth orbits the sun. The net effect of these forces is to drive an harmonic oscillation of the ocean surface. When the natural period of an ocean basin is longer than that of its celestial drivers (approximately 12.5 h), the tides are in phase with the heavens; that is, when the moon is overhead, its gravitational attraction for the ocean produces a high tide. In the tropical Pacific, however, the natural period of the ocean basin is approximately 54 h, much longer than the period of the tidal driving force. As a result, tides in Panama (9°N) oscillate out of phase with the moon. I found this very convenient when working on intertidal barnacles near Panama City. It is hot in Panama, and conducting field work during a noontime low tide is a sweaty business. No problem! Due to the phase shift, the tides were at their lowest when the full moon was high overhead, allowing me to work comfortably in the moonlight. Northern latitudes aren't so lucky. At high latitudes, the east-west distance across the ocean is less, resulting in a shorter natural period and a smaller phase shift.

4   ABRUPTLY APPLIED LOADS

In the analysis we just completed, we assumed that the oscillating system was subjected to a load that varied sinusoidally through time. But natural loads can be aperiodic, and we need to take their temporal quirks into account. We do so by returning to equation 18.32, in which we described how a spring-mass system responds when given an initial velocity, u_0:

$$x(t) = \frac{u_0}{2\pi f_d} \exp\left(-\frac{c}{2m}t\right) \sin(2\pi f_d t). \qquad (18.41)$$

The greater the initial velocity, the greater the displacement at time t.

But how did this initial velocity come about? In order for the system's mass to have velocity, it must have accelerated for some period of time. To take this into account, let's rewrite equation 18.41 in terms of acceleration and time. Recalling that acceleration is force per mass, we can express a change du in the system's velocity as

$$du = \frac{F(\epsilon)}{m}d\epsilon, \qquad (18.42)$$

where $d\epsilon$ is the infinitesimal period over which $F(\epsilon)$, the force at time ϵ, is applied. (The reason for introducing a new symbol for time will become apparent in a moment.) The quantity $F(\epsilon)d\epsilon$ is the *impulse* applied to the system.

The system's response to this velocity increment depends on how much time has elapsed since the velocity was imposed. In other words, if a velocity is applied at time ϵ and we are interested in the system's displacement at some later time, t, the interval in between is $t - \epsilon$ seconds long. Thus, at time t the response to the velocity increment applied at time ϵ is

$$dx(t) = \frac{1}{2\pi f_d} \exp\left[-\frac{c}{2m}(t - \epsilon)\right] \sin\left[2\pi f_d\,(t - \epsilon)\right] \frac{F(\epsilon)}{m} d\epsilon. \tag{18.43}$$

This is the response to the force applied at a specific ϵ. Other forces could have been applied at other times. To calculate total displacement at time t, we sum up—that is, integrate—the responses to all the forces applied previously. In other words, we integrate equation 18.43 for ϵ between 0 and t. Noting that $2\pi f_d$ and m are constant, we can pull them out of this integration, leaving us with the conclusion that

$$x(t) = \frac{1}{2\pi f_d m} \int_0^t \exp\left[-\frac{c}{2m}(t - \epsilon)\right] \sin\left[2\pi f_d\,(t - \epsilon)\right] F(\epsilon)\,d\epsilon. \tag{18.44}$$

This is the answer we seek. If we know the time course of the imposed force between times 0 and t—that is, if we know $F(\epsilon)$—we can predict the displacement at time t.

Equation 18.44 is quite general; we can provide it with any time course for F, and it will crank out the response. Two specific examples, however, can give us a feel for how the equation works.

Earlier in our analysis of spring-mass systems, we considered what happens when the mass at the end of a cantilever (e.g., a crown of leaves supported by a tree trunk) is held deflected by a constant force and then released. Let's turn that scenario around. What happens when a constant force is abruptly applied to an initially stationary mass on an undeflected beam? Although this loading regime is uncommon in nature, analyzing it is an informative first step toward understanding the effects of more realistic dynamics loads. To formulate this question in mathematical terms, take a beam whose tip mass is at $x = 0$ until, at $t = 0$, we apply force F, which is subsequently maintained. In this scenario, $F(\epsilon) = F$, and equation 18.44 becomes

$$x(t) = \frac{F}{2\pi f_d m} \int_0^t \exp\left[-\frac{c}{2m}(t - \epsilon)\right] \sin\left[2\pi f_d\,(t - \epsilon)\right] d\epsilon. \tag{18.45}$$

This result is graphed in Figure 18.5A for several damping coefficients.

There are two things to notice here. First, even though the force is applied instantaneously, it takes time for the system to respond—the deflection of the system doesn't peak until half of a damped period after force is applied. In other words, the greater the system's inertia or the lower its stiffness, the longer the response time.

Second, because the initial effect of the external force is to increase the mass's speed, the resulting inertia can cause the mass to overshoot its equilibrium displacement. As a consequence, the internal force imposed on the spring can be greater than the external force applied to the system. Indeed, if the system is undamped, a sudden force can impose an internal force twice what the system would experience if force were applied statically. The more damped the system, the less the overshoot in force.

Both these factors often come into play in nature. A spring-mass system such as a tree, kelp, or sensory hair can be stationary until a force is suddenly applied for a

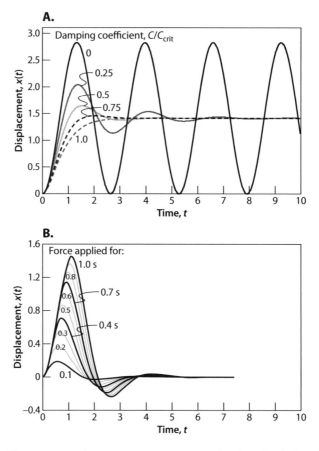

Figure 18.5 A. The response of a spring-mass system to a load applied abruptly and then maintained. **B.** Response to a load applied for varying amounts of time and then removed.

brief time. This is similar to the situation we just analyzed, but in this case the force may be transient. How the system responds depends on how long the external force is applied (Figure 18.5B). If force is imposed for only a small fraction of the damped period, the mass never has a chance to get up to speed before its source of acceleration is removed. As a consequence, maximum deflection is small, and the force felt by the spring is correspondingly small. For example, a gust of wind might imposed a momentary drag force of 2000 N to our hypothetical tree. If this drag were applied as a static force, the crown mass would deflect by 1.4 m. However, if the gust lasts only a tenth of a second, the crown deflects just 0.34 m. (This calculation assumes that the tree's oscillation is undamped. If damped to half the critical value, the tree sways only 0.18 m.) This inertial effect is also important in intertidal environments where organism are subjected to large—but brief—loads from breaking waves. Stiff organisms, such as acorn barnacles, respond quickly to these loads and experience substantial internal forces. By contrast, flexible seaweeds have relatively long response times, which allows them to avoid these peak forces.

5   NONLINEAR SPRINGS AND CHAOS

Our analysis so far is based on the assumption that the restoring force is linearly proportional to displacement. This is seldom precisely true. For example, the gravitational restoring force in a pendulum is very nearly a linear function of angle for small angles but deviates substantially at large angles. In effect, the larger the angle, the lower the system's stiffness (equation 18.15), so the period of oscillation increases as the amplitude of the swing increases.

Nereocystis provides an extreme example of a nonlinear restoring force. When pulled in tension, the ropelike stipe acts as a linear spring, and we have used this behavior to predict the kelp's natural frequency. There is a potential complication, however. Like all ropes, the stipe can pull but it can't push. As a result, the force resisting displacement of the frond mass goes through a discontinuity when the stipe is just at its resting length—when the stipe is stretched, it pulls back; allowed to go slack, it offers no restoring force.

This gross nonlinearity can lead to strange results. If the buoyancy of the kelp's crown is sufficient to keep the stipe in tension at all times, the plant behaves as the sort of predictable harmonic oscillator we have encountered. Reduce the buoyancy just a bit, however, and motion becomes unpredictable. In this case, if, during the passage of a wave, downward velocity releases tension on the stipe, it takes awhile for the crown mass to float back up and reengage the tensile spring. The ensuing motion is very sensitive to the exact time at which tension is restored. The crown mass bounces around apparently at random, and the resulting forces imposed on the stipe can be quite large (Denny et al. 1997).

The sort of sensitivity engendered by the kelp's nonlinear spring is a trademark of an intriguing class of dynamics known as *deterministic chaos*. It is well beyond the scope of this chapter to delve into chaos. Instead, this brief introduction is intended simply as a warning that nonlinear springs can lead to complex oscillations. For an introduction to chaos, see Moon (1992).

Even when nonlinear springs do not trigger chaotic motion, they can affect the forces imposed on an oscillating system. With a group of my colleagues, I examined a variety of biological systems that have springs characterized by low stiffness at small displacements and high stiffness at large displacements (e.g., mussels and giant kelps; Denny et al. 1998). When the mass of such a system is subjected to either a periodic or transient load, the low initial stiffness allows the mass to acquire substantial momentum as it is displaced. The system then pays a price when, at some high displacement, stiffness increases. In essence, the mass comes to the end of its tether and is abruptly jerked to a halt. The rapid accelerations associated with increased stiffness can result in potentially lethal forces.

6   CONCEPTS, CONCLUSIONS, AND CAVEATS

Spring-mass oscillators and their analogues are common in nature, and plants, animals, indeed entire communities, need to cope with their effects. The stiffness and inertia of a system set its natural frequency of oscillation. If the driving frequency is higher than the system's natural frequency, the system's response is muted by inertia. On the other hand, if the driving frequency is lower than the system's natural frequency, the system moves in concert with the driving force. If the natural frequency matches that of the driver, the system resonates. In the absence of damping, resonant displacement is amplified, a boon for a cricket attempting to sense faint sounds or a jellyfish pursuing prey. In other cases, however, damping may be necessary to avoid destructive displacements.

The analysis presented here is intended to serve as a general introduction to the vast topic of oscillation. You should keep in mind, however, that many important aspects of oscillatory systems have not been covered. For instance, in our analysis of forced harmonic motion, we assumed that damping was independent of the driving force. There are many situations in which this is not true, our example of a tree in wind being one of them. Drag (a driving force) is applied as wind moves past leaves. But the same drag also damps the tree's sway. Because relative motion between tree and air provides both the driving and damping forces, the mechanics of the system are complicated.

We have also assumed that the spring in our spring-mass systems responds instantaneously; that is, we have tacitly assumed that deflection of the spring's free end is felt instantaneously as a force at its fixed end. This is true if the period over which a deflection is applied is long compared to the time it takes for the spring to adjust, but that isn't always the case. If you have ever played with a Slinky (a toy consisting of a long, coiled spring), you may have seen this effect in action. When stretched out between you and a friend, a quick jerk on one end of the toy travels as a wave, taking a second or more to arrive at the other end. In cases such as this, the instantaneous force felt at one end need not be directly proportional to displacement at the other, again complicating the mechanics. Gaylord and colleagues (2001) explored this effect in kelps, and concluded that traveling waves of strain in impulsively loaded kelp stipes can substantially amplify the stresses imposed at a stipe's base.

We have also ignored interactions among oscillators. The sensory hairs on a cricket's abdomen do not act in isolation—movement by one can affect the forces on others some distance away. A tree by itself will sway differently than a tree in a forest.

The complexities offered by oscillating systems are endless, interesting, and often important, and I urge you to dig into the primary literature.

Part IV

ECOLOGICAL MECHANICS

Chapter 19

Ecological Variation and Its Consequences

... variation itself is nature's only irreducible essence. Variation is the hard reality, not a set of imperfect measures for a central tendency.
Stephen Jay Gould (1985)

Nature is variable: the physical environment fluctuates in both predictable and unpredictable ways, plants and animals fatigue and sustain damage, and interactions among organisms shift through time and space. On occasion, these variations are sufficiently extreme to pose a substantial threat: an unseasonal freeze can kill all the song birds on an island, for instance; corals can be broken by storm waves; and urchin stampedes can decimate a kelp forest. (We will deal with the biology of these sorts of extremes in Chapter 22.) Most of the time, however, life's variations fall within organisms' normal working range, and to understand the resulting biological dynamics we must take these more typical fluctuations into account.

We do so in the course of three chapters. In this, the first, we explore *scale transition theory*, a statistical method that allows one to predict how physiological, environmental, and ecological variability affect the average response of a system (Chesson et al. 2005; Melbourne et al. 2005; Melbourne and Chesson 2006). It also allows one to scale up from experiments conducted locally and for a short time to predict how a phenomenon will play out at larger scales and over longer periods. Scale transition theory is especially applicable in the context of ecomechanics because it makes use of the mechanistic response functions we have explored in the last eighteen chapters.

Having quantified the consequences of variability, we proceed in Chapter 20 to investigate *spectral analysis*, a tool that allows one to specify how variability is distributed among spatial and temporal scales. For example, air temperature varies through time, and these fluctuations can affect the average performance of plants and animals. Spectral analysis allows us to quantify how much of this variation is associated with daily versus seasonal versus decadal temporal scales, helpful knowledge we can use to interpret the environmental factors that drive ecological patterns in time and space.

Lastly, we combine scale transition theory and spectral analysis to explore how the average response of a system varies through time or space. Often, the longer the time or the larger the area over which a process plays out, the greater the variability encountered—a trend known as $1/f$ noise—with concomitant effects on the average. Chapter 21 provides an overview of how average response changes with temporal or spatial scale and how these scaling effects interact with ecology.

That is where we are headed, but it all begins with scale transition theory.

1   SCALE TRANSITION THEORY
A Brief Introduction

At its core, scale transition theory is a recognition of the role of Jensen's inequality in biology. Recall from Chapter 3 that for a nonlinear function $g(x)$,

$$\overline{g(x)} \neq g(\overline{x}). \tag{19.1}$$

That is, the average of the function, $\overline{g(x)}$, does not equal the function of the average, $g(\overline{x})$. In Chapter 13 we noted, for instance, that the average attraction between atoms is much larger than the attraction one would expect given the average dispersion of electrons. The resulting van der Waals forces allow geckos and insects to walk up walls and hang from the ceiling. But, if $\overline{g(x)} \neq g(\overline{x})$, then

$$\overline{g(x)} = g(\overline{x}) + \text{ something.} \tag{19.2}$$

Our task in this chapter is to quantify this something.

A preliminary glimpse of the answer can be had from a consideration of thermal reaction norms, the functions that model how organisms' performance varies with body temperature. There are upper and lower limits to the range at which an organism can sustain the functions necessary to survive ($T_{\max}$ and $T_{\min}$, respectively), and its performance (growth rate, reproductive output, escape velocity, etc.) often peaks somewhere near the middle of this range. In Figure 19.1A. I have plotted the hypothetical reaction norm for a thermal generalist, a species capable of performing across a wide range of temperatures. When the environment is constant at 35°C, the species performs optimally. However, if temperature varies, alternating between 31°C and 39°C, average performance (the midpoint of the line connecting points on the reaction norm at 31°C and 39°C) is less than optimal. If temperature varies more (between 29°C and 41°C, for instance), the decline in performance is greater. These results suggest that the effect of variation—the something in equation 19.2—is proportional to the magnitude of variation.

Variation's effect also depends on a function's shape. Consider, for example, the reaction norms plotted in Figure 19.1B, where the generalist curve is contrasted with the curve for a stenotherm or thermal specialist. In this heuristic example peak performance and optimal temperature are the same as those of the generalist, but the thermal range of the specialist is narrower, making the function more sharply curved. As a result, the same amount of thermal variation has more drastic effects. While alternation between 31°C and 39°C reduces the performance of the generalist by 7%, it reduces the performance of the specialist by 33%.

Physiological performance curves such as these are typically asymmetrical, falling off more rapidly at temperatures above the optimum than below. This asymmetry affects the consequences of variation. For an average temperature below the optimum, a given variation in temperature has less effect than the same amount of variation

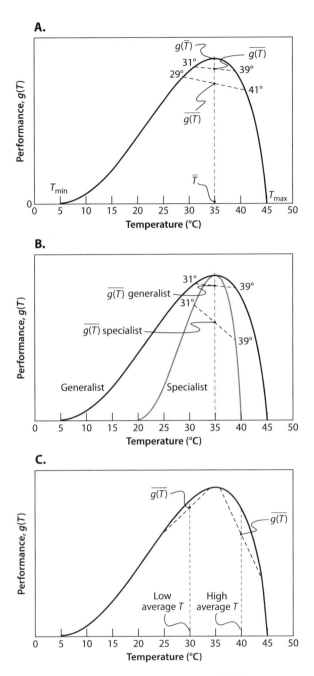

Figure 19.1 A. The average performance of an organism, $\overline{g(T)}$, decreases with increased variation in temperature. **B.** The more steeply curved the performance curve, the greater the sensitivity to thermal variation. (See text.) **C.** For a given variation in temperature, the reduction in average performance is relatively small for temperatures below the optimum and relatively large for temperatures above the optimum.

does for an average at or above the optimum (Figure 19.1C). Martin and Huey (2008) used this fact to explain (in part) why a wide variety of insects and lizards prefer an average temperature a few degrees below the temperature at which they perform best. Clearly, as we investigate the effects of variation, we must take each function's shape into account.

Note that the interaction between temperature variation and performance can have ecological consequences. For example, Deutsch et al. (2008) report that, for insects, the width of the reaction norm ($T_{max} - T_{min}$) increases with increasing latitude. Thus, for a given fluctuation in body temperature, the performance of tropical insects is likely to be more compromised than that of temperate and subpolar species. Similar logic applies when generalists and specialists occur in the same environment. If temperature varies, the performance of the specialist is impacted more that that of the generalist.

In summary, the average thermal response of an organism depends not only on the average temperature it encounters but also on how variable those conditions are and how that variation influences performance, which we can predict based on the shape of the response function. Similar conclusions apply to other response functions and the analysis can be scaled up from the indivudal to populations and communities.

These conclusions have special significance in the context of climate change. As the CO_2 content of the atmosphere increases, average air temperature rises, average pH of the oceans decreases, and a whole host of climatic factors (e.g., average rainfall and wind speed) are altered. It would be a grave mistake, however, to predict biology's response to these changes based solely on these projected *averages*, we also need to account for variation (Dowd et al. 2015).

Global climate models predict, for instance, that average temperature will rise fastest near the poles and slowest near the equator (IPCC 2013). Taken at face value, this would suggest that the effects of temperature on metabolism in plants and cold-blooded animals will be greatest in polar regions. But metabolic rate is a nonlinear function of temperature (Chapter 11). If temperature fluctuates—which it inevitably does—average metabolic rate could be substantially different from the metabolic rate predicted from average temperature. In fact, shifts in average temperature are likely to be accompanied by increased variation through both space and time (IPCC 2013) and, when this variation is taken into account, the prediction reverses. Due to the deleterious effects of variation on stenotherms, the metabolic effects of rising temperature should be greatest near the equator (Dillon et al. 2010).

In short, to predict the consequences of climate change, we need a method to account for the effects on nonlinearities in response functions and how they interact with environmental, physiological, and behavioral variability. This is the job of scale transition theory.

2   BACKGROUND

Before we dive into the theory, it will be helpful to refresh our familiarity with the basic statistical calculations introduced in Chapter 4, starting with the average. If we have n measurements of a particular variable x ($x_1, x_2, x_3, \ldots, x_n$) the average value of x is

$$\bar{x} = \frac{1}{n} \sum_{i=1}^{n} x_i. \tag{19.3}$$

As a statistic, the average serves two main purposes. First, it provides a measure of the central tendency of x, essentially a guess as to what to expect in the future. Suppose we are interested in the daily range of air temperature (daily maximum − daily minimum), and we want to predict what the range will be next June 13. If we have information regarding the temperature range on June 13 from past years, the average of these historical values is our best guess as to what to expect in the future. It is for this reason that the average is also known as the *expectation* (see Denny and Gaines 2000).

The average is also useful in the calculation of cumulative effects. For instance, the sum of calories ingested plays a major role in determining an animal's annual growth: if we know the average daily feeding rate, we can multiply by 365 to calculate the intake accumulated in a year. Other examples abound: The reproductive output of an individual may vary from year to year, but in the context of evolutionary dynamics it is the individual's total lifetime reproductive output—average output times years of reproduction—that determines its fitness. Lift varies from moment to moment as an insect beats its wings, but as long as average lift exceeds the animal's weight, it can stay aloft.

Although useful as a means of prediction and an index of accumulation, the average provides no information about this chapter's central topic—variation. Variation in a set of data is instead quantified by the variance, σ_x^2, the average of squared deviations from the mean:[1]

$$\sigma_x^2 = \frac{1}{n} \sum_{i=1}^{n} (x_i - \bar{x})^2 . \tag{19.4}$$

For our purposes, it is illuminating to expand the squared term in this summation:

$$\sigma_x^2 = \frac{1}{n} \sum_{i=1}^{n} \left(x_i^2 - 2x_i \bar{x} + \bar{x}^2 \right)$$

$$= \underbrace{\frac{1}{n} \sum_{i=1}^{n} x_i^2}_{1} - \underbrace{\frac{2}{n} \sum_{i=1}^{n} x_i \bar{x}}_{2} + \underbrace{\frac{1}{n} \sum_{i=1}^{n} \bar{x}^2}_{3} . \tag{19.5}$$

Term 1 is simply the mean of the squared values of x, that is, $\overline{x^2}$. In term 3, we add $\bar{x}^2$ to itself n times and then divide by n, so this term is just $\bar{x}^2$—the mean, squared.

Term 2 requires a bit more scrutiny. Because $\bar{x}$ is a constant, we can pull it outside the summation. Working through the algebra, we then find that

$$-\frac{2}{n} \sum_{i=1}^{n} x_i \bar{x} = -2\bar{x} \left(\frac{1}{n} \sum_{i=1}^{n} x_i \right)$$

$$= -2\bar{x}(\bar{x})$$

$$= -2\bar{x}^2. \tag{19.6}$$

---

[1] Here we return to the use of the symbol σ we introduced in Chapter 4; in the present context, σ^2 is the variance, not the square of tensile or compressive stress.

With these results in hand, we can add our three terms to calculate the variance:

$$\sigma_x^2 = \overline{x^2} - 2\overline{x}^2 + \overline{x}^2$$

$$= \overline{x^2} - \overline{x}^2. \tag{19.7}$$

In short, the variance of x (an index of the variability of our data) is equal to the mean square of x minus the mean of x, squared.

Equation 19.7 is a pleasingly simple relationship and an interesting quirk of statistics, but why have I foisted it on you here? To see why, let's rearrange terms:

$$\overline{x^2} = \overline{x}^2 + \sigma_x^2. \tag{19.8}$$

When written this way, the connection to Jensen's inequality (equation 19.2) becomes apparent. The mean value of x^2—that is, $\overline{x^2}$—differs from the square of the mean ($\overline{x}^2$), as Jensen's inequality predicts. However, Jensen's inequality tells us only that $\overline{x^2} \neq \overline{x}^2$. The calculation we have just performed does more: it tells us *how much* $\overline{x^2}$ differs from $\overline{x}^2$. As predicted, the magnitude of the difference depends on the variability of x, its variance.

Equation 19.8 thus embodies the essence of scale transition theory. As we will see in Chapter 21, the variance an individual organism encounters—in body temperature for instance—is a function of how far it moves or how long it waits; in other words, on the spatial or temporal scale at which it lives. Similarly, variance in competitive interactions, recruitment, species diversity, and other ecological factors are likely to depend on the spatial extent of a population or community and the period over which it interacts with itself and the environment. Thus, because average response depends on variance, and variance depends on scale, equation 19.8 provides a conceptual recipe for quantifying the effects of scale on organismal, population, and community performance. The term *scale transition* refers to this connection between variance and scale.

Before we get too excited about this result, however, it is necessary to take a step back. Equation 19.8 is a special case; it works only because $g(x) = x^2$ is a quadratic function, that is, a function that fits the general form $g(x) = ax^2 + bx + c$. (In this case, $a = 1$, and $b = c = 0$.) Unfortunately, there is no formula that provides similarly precise answers for other types of functions; $\overline{x^{0.5}} \neq \overline{x}^{0.5} + \sigma_x^2$, for instance. In general, $\overline{g(x)}$ does not precisely equal $g(\overline{x}) + \sigma_x^2$.

But extreme precision isn't necessarily a requirement; in biology, it is often sufficient to have a good approximation. And there *is* a relationship that allows us to calculate an excellent approximation of $\overline{g(x)}$ for a wide variety of response functions (Chesson et al. 2005; for a derivation, see Supplement 19.1):

$$\overline{g(x)} \approx g(\overline{x}) + \frac{1}{2}\frac{d^2 g(\overline{x})}{dx^2}\sigma_x^2. \tag{19.9}$$

Here $d^2 g(\overline{x})/dx^2$ is the second derivative of $g(x)$—its acceleration—evaluated at the mean, $\overline{x}$. To simplify the notation, let's introduce the symbol $\mathcal{S}_T$, the *scale transition coefficient*, as an index of the acceleration (and therefore shape) of $g(x)$:

$$\mathcal{S}_T = \frac{1}{2}\frac{d^2 g(\overline{x})}{dx^2}. \tag{19.10}$$

Thus, in streamlined form,

$$\overline{g(x)} \approx g(\bar{x}) + \mathcal{S}_T \sigma_x^2. \tag{19.11}$$

Whereas equation 19.8 embodies the essence of scale transition theory, equation 19.11 provides the practical recipe.[2] As predicted by our consideration of reaction norms, equation 19.8 tells us that the difference between $\overline{g(x)}$ and $g(\bar{x})$ depends on both the amount of variation in x (i.e., σ_x^2) and the shape of the function, indexed by $\mathcal{S}_T$.

Note also that the sign of $\mathcal{S}_T$ tells us whether the average of the function is greater than or less than the function of the average:

- If $\mathcal{S}_T > 0$, the function is concave upward (it accelerates) and $\overline{g(x)} > g(\bar{x})$.
- If $\mathcal{S}_T < 0$, the function is concave downward (it decelerates) and $\overline{g(x)} < g(\bar{x})$.
- If $\mathcal{S}_T = 0$, the function is linear, and $\overline{g(x)} = g(\bar{x})$.

All this is in accord with what we know about Jensen's inequality. (Scale transition coefficients for some common response functions are derived in Supplement 19.2.)

3 SCALE TRANSITION THEORY IN ACTION
Intermediate Disturbance

Scale transition theory as outlined here is readily employed in theoretical models of population dynamics (for a review, see Chesson et al. (2005)). For example, consider the intermediate disturbance hypothesis, which links species diversity in a community to the interval between disturbance events. Every time an extreme event occurs (fire in a forest, for example, or a tsunami impacting a coral reef), the community is pruned to a low diversity. In the interval between events, diversity initially increases through time as new species recruit, but as succession progresses, diversity can subsequently decline as competition favors a few dominant organisms. Diversity is thus a nonlinear function of t, the time available for succession, that is, the time between disturbance events.

Consider a simple model of the intermediate disturbance hypothesis in which diversity D is parabolically related to t (the gray line in Figure 19.2A):

$$D(t) = kt - \frac{kt^2}{t_{max}}. \tag{19.12}$$

Here k is a constant and t_{max} is the interval length at which competition has run its full course and diversity is minimal. In this context, $D(t)$ is the *local diversity*, the diversity we would observe in one particular small patch of a community.

---

[2] It is easy to show that equation 19.11—the basic scale transition formula—works when $g(x) = x^2$. In this case $\mathcal{S}_T = 1$, and we recover equation 19.8 exactly. More generally, equation 19.11 works well with any response function that can be accurately described by its second-order Taylor expansion (see Supplement 19.1), that is, to functions whose third, fourth, and higher-order derivatives are small compared to their second derivative.

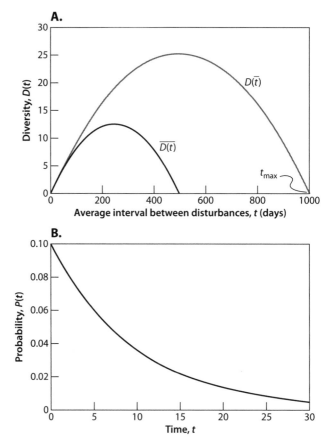

Figure 19.2 A. If stressful events occur randomly with a fixed probability per unit time, the stochastic variation in interevent intervals decreases average diversity below what it would be if interevent intervals were constant. **B.** The Poisson distribution of intervals ($p = 0.1$).

What would average diversity be if we measured it over many different patches, each operating on its own time course? Only if disturbance is synchronous among all patches would equation 19.12 provide an accurate estimate. But, in nature, disturbance often occurs haphazardly in space, and this variability can affect the multipatch diversity we measure.

Let's suppose that in each patch, extreme events occur independently and randomly with constant low probability p per unit time. In this case, $\bar{t}$, the mean interevent interval length for any given patch, is

$$\bar{t} = \frac{1}{p}, \tag{19.13}$$

and the distribution of interval lengths can be accurately described by the Poisson interval distribution (Supplement 19.3):

$$P(t) = pe^{-pt}. \tag{19.14}$$

Here P is the probability that an interval chosen at random has length greater than t. Short intervals are most likely, although some intervals can be quite long

(Figure 19.2B). It can be shown (Supplement 19.3) that, given the Posisson distribution of intervals, the variance in interval length is

$$\sigma_t^2 = \bar{t}^2 \tag{19.15}$$

$$= \frac{1}{p^2}. \tag{19.16}$$

Thus, for a given probability of encountering disturbance events, both the average length of intervals between events (that is, $\bar{t}$, the average time available for succession) and the variance in interval length ($\bar{t}^2$) are set.

We can incorporate these values into the scale-transition formula to calculate the multipatch average diversity:

$$\overline{D}(t) = D(\bar{t}) + \mathcal{S}_T \sigma_t^2. \tag{19.17}$$

For this response function $\mathcal{S}_T$ is

$$\mathcal{S}_T = \frac{1}{2} \frac{d^2 D(\bar{t})}{dt^2}$$

$$= -\frac{k}{t_{max}}, \tag{19.18}$$

and, as we have just seen, $\sigma_t^2 = \bar{t}^2$ (equation 19.15). Thus

$$\overline{D}(t) = \left(k\bar{t} - \frac{k\bar{t}^2}{t_{max}} \right) - \frac{k\bar{t}^2}{t_{max}} \tag{19.19}$$

$$= k\bar{t} - \frac{2k\bar{t}^2}{t_{max}}. \tag{19.20}$$

This relationship is plotted as the black line in Figure 19.2A. $\overline{D}$ is lower than D because, even when average interval time is short, the random nature of disturbance provides occasional long, disturbance-free intervals that allow succession to be completed. Thus, when disturbance events happen purely by chance, diversity—averaged over space—is lower than one would expect from measurements made on a single patch undergoing succession.

Equation 19.20 is expressed as a function of average interval time, but it can be expressed equally well in terms of the probability of disturbance, p:

$$\overline{D}(p) = \frac{k}{p} - \frac{2k}{t_{max}p^2}. \tag{19.21}$$

There is an intermediate probability of disturbance ($4/t_{max}$) that results in maximum average diversity.

Note that the same approach could be used to estimate diversity in a single patch measured over multiple disturbance events.

4   COVARIANCE

To this point we have dealt only with functions of a single variable: performance as a function of temperature and diversity as a function of the probability of disturbance. There are, however, many important biological phenomena that are functions of multiple variables. It is beyond the purview of this chapter to treat in depth the theory of scale transition for multivariable functions, but it will be worth our time to consider two important—but relatively simple—examples with a message that provides an entrée into the more complex aspects of scale transition theory.

A large fraction of marine invertebrates and algae reproduce using external fertilization. Haploid sperm and eggs are extruded into the water column, where their swimming—often augmented by turbulence—causes them to collide. Once a sperm has contacted an egg, fertilization follows, and a new diploid individual is produced. The average rate of contact depends not on the concentration of either gamete alone, but rather on their co-occuring concentration. If s is the concentration of sperm and e the concentration of eggs (number per m^3), the rate of contact is

$$R = k_f s e. \tag{19.22}$$

Here k_f (m$^3 \cdot$ s^{-1}) is a constant proportional to the speed at which gametes move and to the projected area of an egg (that is, to the size of target the egg offers a sperm (Vogel et al. 1982; Kiørboe 2008)). If the concentrations of sperm and eggs are constant, equation 19.22 provides all the information we need to calculate fertilization rate. The situation changes, however, if concentrations vary. In this case,

$$\overline{R} = k_f \overline{se}, \tag{19.23}$$

where $\overline{se}$ is the average product of s and e. Because $\overline{se}$ depends on the product of s and e rather than each factor separately, it introduces complexity that we need to take into account.

The discussion that follows is akin to our analysis of Reynolds stress in Chapter 6, but the logic bears repeating. Consider a hypothetical experiment: in the vicinity of spawning animals, we sample the water at a particular location at random times and measure the concentration of sperm and eggs in each sample. In our first sample we measure s_1 and e_1; in the second sample, s_2, e_2; and so on. After we have taken n samples, we calculate the average concentrations of sperm and eggs—$\overline{s}$ and $\overline{e}$, respectively. Having measured the means, we can then go back and calculate for each sample how far s deviates from $\overline{s}$ and e deviates from $\overline{e}$. In other words, we can calculate

$$\Delta s_i = s_i - \overline{s}, \tag{19.24}$$
$$\Delta e_i = e_i - \overline{e}, \tag{19.25}$$

where $i = 1$ through n. The results are graphed in Figure 19.3A.

There is a wide range for both Δs and Δe, indicating that the concentrations of both sperm and eggs can differ from one sample to the next. But these fluctuations aren't independent. When Δs is positive, Δe tends to be positive. Conversely, when Δs is negative, Δe is negative. Because Δs and Δe tend to have the same sign, the product of deviations, $\Delta s_i \Delta e_i$, tends to be positive. We can quantify this tendency by calculating the average product of co-occurring deviations, the covariance of s and e:

$$\text{Cov}(s, e) = \frac{1}{n} \sum_{i=1}^{n} (\Delta s_i \Delta e_i). \tag{19.26}$$

In this example, $\text{Cov}(s, e) > 0$, indicating that there is a positive correlation between s and e: where there are few sperm, there are few eggs; where there are lots of sperm, there are lots of eggs.

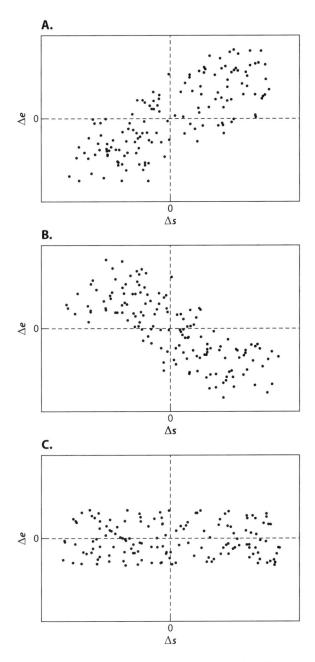

Figure 19.3 Concentrations of sperm and eggs (here expressed as the deviation from mean concentrations) can be positively correlated (**A**), negatively correlated (**B**), or not correlated (**C**).

The correlation need not be positive, however. In the example shown in Figure 19.3B, when Δs is positive, Δe tends to be negative, and vice versa, such that the average product of deviations is negative, so where one finds a lot of sperm, one finds few eggs. And if there is no pattern to the co-occurring concentrations—that is, if s and e fluctuate independently—the covariance is approximately zero (e.g., Figure 19.3C).

Just as $\overline{g(x)}$ is affected by σ_x^2 in the recipe for scale transition (equation 19.11), covariance also affects the mean rate of fertilization. Through a derivation similar to that we performed for the variance (Supplement 19.4), it can be shown that

$$\overline{se} = (\overline{s}\,\overline{e}) + \text{Cov}(s, e). \tag{19.27}$$

In other words, the mean of the product of s and e (which controls the average rate of fertilization, equation 19.22) is equal to the product of the means of those variables *plus the covariance between them.* Thus, if the concentrations of sperm and eggs are positively correlated, the average rate of contact can be high even if average concentrations are low. Conversely, if concentrations are negatively correlated, the rate of fertilization can be low even if average concentrations are high. In short, when dealing with a function of two variables, a second scale transition term—their covariance—needs to be taken into account.

The biological importance of covariance has been elegantly demonstrated by John Crimaldi and his colleagues. Prior to Crimaldi's work, studies of external fertilization used traditional models of turbulent diffusion to predict the average concentrations of sperm and eggs in the sort of flows encountered in nature. Because turblence is highly effective at dispersing gametes (see Chapter 4), these average concentrations are quite low, suggesting that external fertilization is very inefficient (Denny and Shibata 1989). This conclusion led to a paradox: if external fertilization is so inefficient, why has it been retained in so many diverse species?

Using different dyes as stand-ins for sperm and eggs, Crimaldi and his colleagues (2008) directly measured the co-occurring concentration of gametes in laboratory flows. Due to the effects of viscosity, s and e have a high, positive covariance for a short period after they are released. As a result, the average co-occurring concentration, $\overline{se}$, is substantial, even though the mean concentration of both gametes is low. Crimaldi's work thus suggests that—because of spatial covariance—external fertilization can be several orders of magnitude more efficient than average conditions would suggest, helping to explain how this form of reproduction can be successful and why so many species use it.

Similar effects arise in predator-prey interactions. For example, one of the classic models of population dynamics—the Lotka-Volterra equations—involves the co-occurring concentration of predators (e.g., lynx) and their prey (e.g., snowshoe hares). The rate of change in the population of hares, R_h, is governed by two factors: the rate at which hares reproduce and the rate at which they are eaten by lynx. The more hares there are, the more young they can produce in a year, so the rate at which young are born can be modeled as rh, where r is the hares' intrinsic rate of increase (number per year) and h is their population density (number per m^2). As with the contact rate between sperm and egg, the rate at which lynx eat hares is proportional to the product of their population densities. The rate of consumption is $k_p hl$, where l is the population density of lynx (number per m^2) and k_p is a coefficient describing lynx's foraging efficiency (m$^2 \cdot$ s^{-1}). The overall annual rate of hares' population growth is thus the difference between reproduction and lynx consumption:

$$R_h = rh - k_p hl. \tag{19.28}$$

A. No covariance

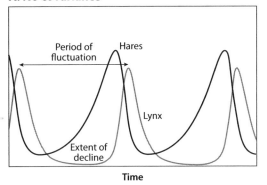

B. Positive covariance

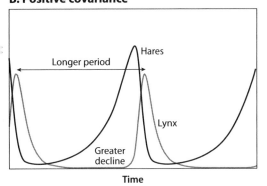

C. Negative covariance

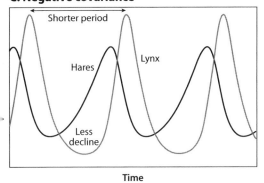

Figure 19.4 Spatial or temporal correlation can affect predator-prey interactions. **A.** No covariance between the densities of hares and lynx. **B.** Positive covariance between species. **C.** Negative covariance between species.

In turn, the rate of change in the lynx population is set by the difference between the rate at which lynx convert prey (captured hares) into baby lynx and the rate at which lynx die:

$$R_l = \epsilon k_p hl - ml. \qquad (19.29)$$

Here ϵ is the efficiency with which captured hares are assimilated, and m is the density-independent mortality rate of lynx. Equations 19.28 and 19.29 are linked—the rate of population change in hares depends on the population density of lynx, and vice versa—and this linkage leads to a stable oscillation of the two populations (Figure 19.4A).

As presented here the Lotka-Volterra equations tacitly assume that h and l do not vary from one location to another. If they vary, the average rates of change depend on the average density of each population ($\bar{h}$ and $\bar{l}$) and the average product of their densities ($\overline{hl}$):

$$\bar{R}_h = r\bar{h} - k_p\overline{hl}, \qquad (19.30)$$
$$\bar{R}_l = \epsilon k_p\overline{hl} - m\bar{l}. \qquad (19.31)$$

It is here that the covariance comes into play. If population densities vary in space, $\overline{hl}$ is affected by $\text{Cov}(h, l)$:

$$\overline{hl} = (\bar{h}\,\bar{l}) + \text{Cov}(h, l). \qquad (19.32)$$

If, for example, lynx are adept at congregating where hares are dense and avoiding areas where hares are scarce, $\text{Cov}(h, l)$ is positive, and $\overline{hl}$ is larger than $\bar{h}\,\bar{l}$. As a result, before their population crashes, lynx can draw the hare population down to lower levels than they otherwise could, and the hare population is, therefore, slower to recover. In this way, postitive covariance lengthens the period of population oscillation for both lynx and hares (Figure 19.4B).

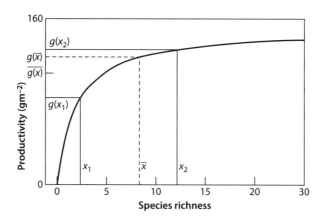

Figure 19.5 Variation in species richness reduces the average productivity of a community (redrawn from Benedetti-Cecchi (2005)).

Conversely, if hares are adept at moving to locations where lynx densities are low and avoiding areas where lynx congregate, $\mathrm{Cov}(h, l)$ is negative, and $\overline{hl}$ is smaller than $\overline{h}\,\overline{l}$. In this case, lynx can't pare hare populations down to very low levels, the period of oscillation is shorter, and (perhaps counterintuitively) the peak densities of lynx are actually higher (Figure 19.4C).

These results must be taken with a large grain of salt. The version of the Lotka-Volterra equations presented here (equations 19.30 and 19.31) is too simple to adequately describe real-world population dynamics. For example, both equations assume that the rate of prey capture increases without limits as prey concentration increases (a Holling type I relationship). It is more likely that capture rate has a limit and can, therefore, be better described by a Holling type II or III relationship (see Chapter 3). Equation 19.30 also assumes that the hare population can grow without limit, but that can't be true, and the rate of population increase would be better described by some form of the logistic equation. Inclusion of these details in the Lotka-Volterra equations changes the math a bit, but the overall message stays the same. If densities vary through space, the rate of population growth for both hares and lynx depends on the covariance of their densities.

In summary, when a response function involves more than one input variable, average response depends not only on the fluctuation in individual variables, but also on how these variables are correlated. Using scale transition theory to predict the average response of a complex system can therefore involve both the variance and covariance of the systems' components.

5   EXAMPLES FROM NATURE

Application of scale transition theory to real-world examples has been less extensive than its application in theory, due at least in part to the practical difficulties involved in collecting sufficient information about variances and covariances. Nonetheless, two examples illustrate scale transition theory's potential.

A wide variety of studies have shown that the productivity of a community (the rate at which it stores energy in organic compounds) is a function of its species richness (Figure 19.5). However, because this relationship decelerates, any spatial variation in species richness within a community reduces the community's average productivity below that expected at average species richness. In a meta-analysis of data from both terrestrial and marine systems, Benedetti-Cecchi (2005) found that this effect can be substantial—for high (but realistic) levels of variance, mean productivity decreases by up to 45%. As humans impose mounting pressure on ecological communities, spatial variation in species richness is likely to increase, with a concomitant (and worrisome) decrease in average productivity beyond what one would expect from land loss alone.

In a detailed study of population dynamics, Melbourne and colleagues (2005) investigated the rate at which biomass of the intertidal porcelain crab *Petrolisthes cinctipes* increases. The rate of growth of crab biomass depends on the competitive interactions among adults, the density of predators, and the rate at which larvae are recruited, each of which could vary from cobble to cobble in the crabs' heterogeneous habitat. Not only do these three factors have direct effects on biomass—for example, the greater the density of predators, the lower the biomass of crabs—but the factors affect each other. Successful recruitment of larval crabs, for instance, is facilitated by the presence of adults, who protect the juveniles from predation. But successful recruitment leads to higher density of adults, which amplifies intraspecific competition, resulting in the production of fewer larvae.

When all factors are taken into account, there are seven scale transition terms, four that depend on the variance of crab or predator density and three that depend on the spatial covariation of crabs and their predators. When all is said and done, the analysis suggests that, at the scale of connected subpopulations (the metapopulation), the effects of variation—the scale-transition terms—approximately negate the rate at which the population would grow under average conditions. Whereas the crab's population could grow nicely if crabs occurred everywhere at their average density, in fact the population grows very little because the dynamics of the crabs' life history results in a highly aggregated distribution. In summary, scale transition theory provides a mechanism for combining information about the behavior and trophic interactions of individuals with information about temporal and spatial variation in the environment, to account for the large-scale pattern of population dynamics.

I anticipate that use of scale transition theory will become more widespread as advances in technology ease the process of field measurements. For example, the advent of miniature, self-contained temperature loggers has vastly simplified efforts to record the spatial and temporal patterns of variation in body and environmental temperatures, making it feasible to accurately measure the required means, variances, and covariances.

6   CONCEPTS, CONCLUSIONS, AND CAVEATS

Scale transition theory provides a means to quantify the effect of variability on the average response of a system:

- Jensen's inequality tells us that $\overline{g(x)} \neq g(\overline{x})$.
- Simple scale transition theory tells us that

$$\overline{g(x)} \approx g(\overline{x}) + \mathcal{S}_T \sigma_x^2. \tag{19.33}$$

The greater the variance or the larger the absolute value of $\mathcal{S}_T$, the greater the effect variability has on the function's average.
- The sign of $\mathcal{S}_T$ tells us whether variation increases or decreases the average response.
- When two or more variables contribute to a system's response, scale transition theory needs to account for covariation in these factors. Positive covariation in gamete concentrations makes external fertilization practical, for example.
- The effects of variation can be sustantial. Many species of insects and lizards have a preference for temperatures several degrees below the temperature at which they perform optimally, a strategy that acknowledges the detrimental effects of temperature variation. Even more striking, spatial variability in species richness may drastically reduce a community's productivity.
- When scale transition theory operates on mechanistic response functions, it forges a useful link between the physics and physiology that operate at individual locations or times and the consequences that accrue when spatial and temporal variation is present.

As Melbourne and colleagues (2005, 2006) have elegantly demonstrated, the concepts introduced in this chapter can be applied to systems with multiple variables. However, the intensity of the mathematics grows quickly with the number of factors involved. If you are interested in getting a feel for the math, Supplement 19.5 provides a taste; beyond that I urge you to consult the excellent, more technical introduction offered by Chesson and colleagues (2005).

As with any statistical tool, scale transition theory can be misinterpreted and abused. The greatest potential for error arises when one forgets that the equations we have used to describe the effects of variance are—almost always—approximations. Only if the response function in question is precisely linear or quadratic are our scale-transition equations exact. Benedetti-Cecchi (2005), for instance, found that the decrease in average productivity estimated using scale transition theory sometimes differed by as much as 18% from more precise estimates made using a model that directly simulated the effects of variation. In short, as useful a tool as it is, scale transition theory must be used with caution. It is a valuable first step toward understanding the effects of variability, but if precise answers are required, additional steps must be taken to verify the theoretical estimates.

In summary, this chapter provides a tool for answering an important question: given a certain amount of variability, how does a system respond? It is now time to turn our attention to variation itself and ask a different, but equally important, question: given a certain organism, habitat, or ecosystem, how much variability is present in nature? This question forms the topic of the next two chapters.

Chapter 20

Spectral Analysis
Quantifying Variation in Time and Space

In our investigation of variation and its effects on organisms in Chapter 19, we used the variance as our primary metric of variability. But when it comes to characterizing nature's fluctuations, variance is a blunt instrument. For example, in a certain habitat we might find that air temperature fluctuates with a variance of 100° squared, but that bare statistic can't tell us whether that fluctuation happens in minutes, hours, or days. Similarly, prey density might vary by 1000 individuals per hectare, squared, but that variance tells us nothing about the spatial distribution of prey. In order to effectively incorporate the consequences of variation into our understanding of ecomechanics, we need to quantify its temporal and spatial patterns.

To that end, this chapter introduces *spectral analysis*, a basic method for exploring the patterns of variation. Although spectral analysis is a standard statistical tool in physics and engineering, it has escaped the notice of most biologists. I hope that in this and the following chapter I can convince you that it provides a valuable perspective on both life and the physical environment.

As a preview of where we are headed, consider the mystery of the ice ages. Beginning in the mid-1700s, scientists began to notice evidence that at multiple times in the past, ice covered much of the globe: morraines of glacial till far from present-day glaciers, ice-worn boulders where in current times there was no ice. As the evidence mounted, so did the perplexity. What could possibly have caused earth's climate to be so much colder in the past, and why did the episodes reoccur?

A variety of explanations were proposed, but none gained traction until Milutin Milancović (a Serbian mathematician, geophysicist, and astronomer) was interned as a prisoner of war in 1914. Finding himself with time on his hands, he laboriously calculated the dynamics of Earth's rotation and its orbit around the sun, leading him to the conclusion that the amount of solar radiation impinging on Earth—and thereby its temperature—should vary with cycles of approximately 23,000, 41,000, and 100,000 y. Milancović proposed that these celestial oscillations caused Earth's recurrent ice ages.

For decades, Milancović's theory was viewed with skepticism. It didn't help that one of Milancović's most ardent supporters was Alfred Wegener, the German meteorologist infamous for proposing the (then) absurd idea that continents drifted. But, beginning in the 1960s, paleoclimatologists were able to piece together a record of earth's temperature extending back five million years (Figure 20.1A). Clearly, temperature has varied episodically, but how could one tell from this complex record whether those variations matched the cycles predicted by Milancović?

The answer was spectral analysis. Applying the techniques we will develop in this chapter, climatologists have been able to carefully measure the periodicities of the ice ages, and they indeed match those proposed by Milancović. From 2.5 to 1 million years ago, ice ages occurred on average every 41,000 y, and from 1 million years ago to the present, every 100,000 y. Debate continues over the mechanism by which celestial factors trigger ice ages, but—due to spectral analysis—there is no doubt that the Milancović cycles are real.

Knowing the pattern of ice ages provides potentially useful information for biologists. In the era when ice ages occurred every 41,000 y, the average rate of climatic change was faster than in recent times when the glaciers arrived every 100,000 y. Through study of the fossilized remains of these different eras, paleontologists can perhaps find clues as to how nature will react to the current rapid rate of climate change.

Ocean waves offer another example, one that we will develop in this chapter. Wind blowing across the ocean surface creates waves that, as they approach shore, cause water to flow back and forth over benthic organisms such as corals. The higher the waves, the faster the flow; the longer the wave period, the longer water flows in each direction. Together, these factors govern the rate at which carbon and nutrients are transported across the benthic boundary layer to corals' symbiotic algae. For a given sea state—which comprises a complex pattern of surface oscillation (Figure 20.1B)—how fast are carbon and nutrients delivered? To answer this question, we need information about how wave height varies as a function of wave period, information that spectral analysis can provide.

Milancović cycles and ocean waves vary through time, but spectral analysis is equally useful for quantifying the pattern of variation through space. Desert lizards eat beetles, for instance, and the density of beetles varies through space (a hypothetical example is shown in Figure 20.1C). If a lizard travels 20 m across the desert floor, how much variation in prey density will it encounter on average? How much if it travels 40 m? Again, these are questions that spectral analysis can answer.

1   SEQUENCE, SIGNAL, AND POWER

In each of these cases (Milancović cycles, ocean waves, desert beetles) it is the sequence of the data that distinguishes them from data more commonly encountered in standard inferential statistics. Consider, for instance, our data regarding the density of desert beetles. If we collect 500 measurements of prey density along a lizard's foraging path $(x_1, x_2, \ldots, x_{500})$ we can calculate the mean and variance of density:

$$\bar{x} = \frac{1}{500} \sum_{i=1}^{500} x_i, \tag{20.1}$$

$$\sigma_x^2 = \frac{1}{500} \sum_{i=1}^{500} (x_i - \bar{x})^2. \tag{20.2}$$

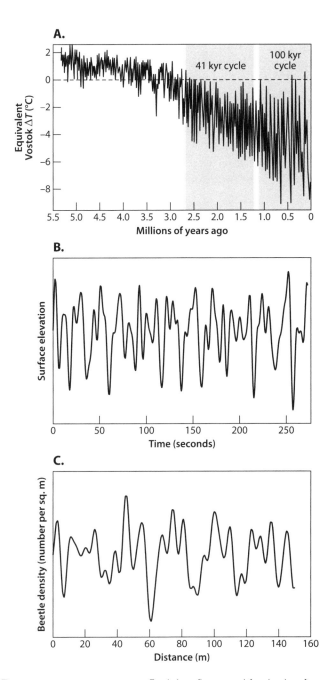

Figure 20.1 The average temperature on Earth has fluctuated for the last five million years (**A**), and the period of oscillation has shifted through time (modified from a figure by R. A. Rohde, Global Warming Art). **B**. A hypothetical temporal record of ocean surface elevation. **C**. A hypothetical record of beetle density as a function of distance along a transect.

The order in which we enter data into these calculations doesn't matter; we could draw the data out of a hat and get the same results. But knowing the data's *sequence* provides information beyond the mean and variance, and sequence is key to the analysis of pattern.

Much of what we know about the analysis of sequential data was first worked out by electrical engineers in the communications industry. As a result, the literature is rife with terms such as *signal* and *power* that have practical meaning when applied to the process of, for instance, getting a telephone conversation from one location to another. However, over the years these terms have been co-opted to apply to situations in which they have little intuitive meaning. Best that we grasp the intent of these terms here at the beginning.

In the context of communications, a *signal* is information conveyed, and quite often the mechanism of conveyance is an electrical voltage. For example, when you talk on the phone, the sound of your voice is converted by a microphone into a varying voltage, and information about this variation is transmitted by wires, optical fibers, or radio waves to another telephone. Upon arrival, the information is transduced back into a voltage signal, which causes a speaker to vibrate around its average position. The resulting sound waves carry your words to the recipient's ear.

Note that it is variation in voltage (rather than voltage itself) that makes the system useful. If the information you transmitted corresponded to a constant voltage, the speaker in the receiving phone would move to a constant position, after which no sound would be produced: no sound, no conveyance of information. Thus, in the context of a telephone conversation, the term signal can be used as an informal synonym for the pattern—the sequence—of deviations from the mean. Signal is often used in this fashion even when the variation being discussed has no association with human communication. For instance, we could easily talk about the wave-height signal of our ocean data or the density signal of our beetles.

If we were to delve deeply into spectral analysis, we would need to differentiate carefully between signal and *noise*. In this context, noise is also used to describe deviations from the mean, but the term is confined to deviations that hinder the conveyance of information. For example, random deviations (which convey no information) can obscure the presence of a signal, and therefore constitute noise. For simplicity, I'll assume in this chapter that all the variation in a data series conveys useful information.

The historical use of the term signal to describe variation in voltage leads naturally to the concept of *power*. Voltage is a measure of the potential energy electrons have for movement, and amperage is a measure of current, the rate at which electrons flow. In an electrical circuit, power—the energy per time it takes to move current—is equal to the voltage V applied to the circuit times the resulting amperage I:

$$\text{electrical power} = VI. \tag{20.3}$$

For example, a typical circuit in your house operates at 120 volts (V) and is capable of safely carrying 20 amperes (A). As a result, you can power a hair dryer that produced 120 V × 20 A = 2400 W of heat.

A second basic relationship (Ohm's law) tells us that the current moving in a circuit is directly proportional to V and inversely proportional to R, the circuit's resistance:

$$I = \frac{V}{R}. \tag{20.4}$$

Inserting this relationship into equation 20.3, we see that electrical power is proportional to the square of voltage:

$$\text{electrical power} = \frac{V^2}{R}.$$ (20.5)

Now consider a voltage that varies through time. Some power is expended in association with the average voltage $\overline{V}$, but there is also electrical power expended by the fluctuations of voltage away from the mean (i.e., the power of the signal's alternating-current (AC) component). By measuring voltage at n points in time and combining equations 20.1 and 20.5, we find that the average power of this fluctuating signal is,

$$\text{average AC power} = \frac{1}{Rn} \sum_{i=1}^{n} \left(V_i - \overline{V} \right)^2.$$ (20.6)

That is, AC power is proportional to the average squared deviation in voltage. But the average squared deviation is, by definition, the signal's variance (see equation 20.2). In other words, when an electrical engineer calculates the variance in voltage, he or she is implicitly calculating one measure of the electrical power of a signal. Thus, for an electrical engineer, power and variance of voltage are intimately related.

As with the term *signal*, the engineering term *power* has been co-opted by the purveyors of spectral analysis. Just as we can talk about the signal of a temporal series of wave heights, we can talk about the series' power, power being an informal substitute for the more appropriate term—variance. Similarly, when exploring the spatial variation in beetle density, the power of the signal has nothing to do with how much energy the insects expend; it is simply the variance of their density.

Note also that the term *power* as used in spectral analysis is different from power as used in inferential statistics. The power of a statistical test refers to its ability to distinguish accurately between the sample means of two populations.

2   FOURIER SERIES

With this terminology in hand, we can proceed to spectral analysis itself. Our first objective is to find a means of describing signals mathematically. Once we have a mathematical model of a given signal, we are then in a position to analyze its pattern. To begin, let's pick back up with our example of ocean waves.

Consider the simple water wave shown in Figure 20.2A. Measured at one point in space, water's surface varies sinusoidally through time, fluctuating up and down around average sea level, $\overline{z}$. For this simple signal,

$$z(t) = \overline{z} + \eta_{max} \cos(2\pi f t).$$ (20.7)

Here η_{max} is the wave's *amplitude* (m), the magnitude of its maximum deviations, and f is the wave's *frequency* (cycles per second, Hz), the number of wave peaks that move past us in unit time. The product $2\pi f t$ is the wave's *phase*, measured in radians. Because the cosine is periodic, the signal starts over again every time ft is an integer. Relative to mean sea level, *elevation η* is

$$\eta(t) = z(t) - \overline{z}$$
$$= \eta_{max} \cos(2\pi f t).$$ (20.8)

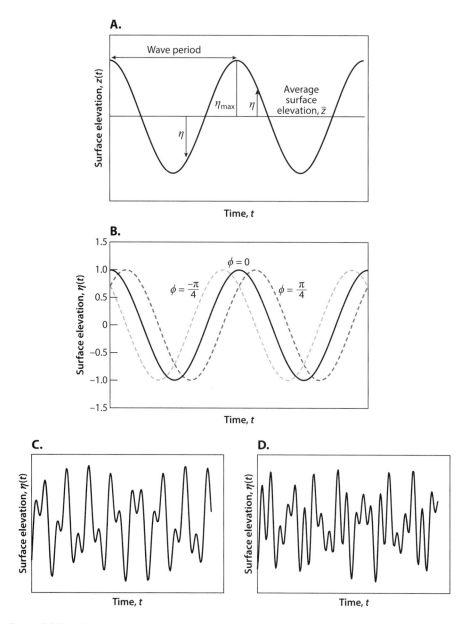

Figure 20.2 A. Nomenclature of an ocean wave. **B.** The phase shift, ø, adjusts the position of the wave in time or space. **C.** The sum of two waves yields a more complex waveform. **D.** The sum of three waves can be even more complex.

This model of the elevation signal assumes that the start of our measurements coincides with the arrival of a peak, that is, that at $t = 0$, $\eta(t) = \eta_{max}$. If that isn't the case, we need to adjust equation 20.8 to account for the difference in time between the initiation of measurements and the arrival of a peak. This is achieved by including a term for the *phase shift*, ϕ, the angle (in radians) corresponding to the temporal shift in the signal:

$$\eta(t) = \eta_{max} \cos(2\pi f t - \phi). \tag{20.9}$$

Now the first peak (η_{max}) occurs at our site when $t = \phi/(2\pi f)$. By varying ϕ between $-\pi$ and π, we can adjust our model to match the temporal location of any wave (Figure 20.2B).

So far, I have described the wave as it varies through time at a particular location. We could describe the wave equally well as a function of location at a particular time. In this case,

$$\eta(x) = \eta_{max} \cos(2\pi f_s x - \phi), \qquad (20.10)$$

where x is distance measured along the path of wave propagation and f_s is *spatial frequency*, the spatial analog of temporal frequency:

$$f_s = \frac{1}{\lambda}. \qquad (20.11)$$

Here λ is wavelength, the distance between successive wave crests.

For simplicity, I will introduce spectral analysis primarily as it relates to a temporal series, but the same principles apply to data measured in space. To model a spatial signal, one simply substitutes x for t and f_s for f; everything else is the same.

To this point, we have conjured up the mathematical description of a single sinusoidal wave, a far cry from the complex signals encountered in nature (e.g., Figure 20.1B, C). Let's see what happens if we combine two of these simple waves with different frequencies and (if we desire) different amplitudes and phases. When combining waves, we assume that they are strictly additive (i.e., each is unaffected by the presence of the other), an assumption known as the *principle of superposition*. In this case,

$$\eta(t) = \eta_{max,1} \cos(2\pi f_1 t - \phi_1) + \eta_{max,2} \cos(2\pi f_2 t - \phi_2). \qquad (20.12)$$

This simple sum can take on a surprising variety of forms (e.g., Figure 20.2C). Adding a third sinusoid,

$$\eta(t) = \eta_{max,1} \cos(2\pi f_1 t - \phi_1) + \eta_{max,2} \cos(2\pi f_2 t - \phi_2) + \eta_{max,3} \cos(2\pi f_3 t - \phi_3). \qquad (20.13)$$

allows us to simulate even more complexity (e.g., Figure 20.2D).

Taken to its logical limit, this concept suggests that with a sufficient number of waveforms, each with an appropriate amplitude and phase, any given pattern of ocean waves can be accurately modeled. Expressed mathematically, we can suppose that

$$\eta(t) = \sum_{i=1}^{\infty} \eta_{max,i} \cos(2\pi f_i t - \phi_i), \qquad (20.14)$$

a suggestion that is, in fact, correct.

The same logic applies not only to ocean waves, but to any signal in nature. In other words, for an arbitrary signal $y(t)$,

$$y(t) = \bar{y} + \sum_{i=1}^{\infty} y_{max,i} \cos(2\pi f_i t - \phi_i). \qquad (20.15)$$

Or, for the signal expressed relative to its mean,

$$y(t) = \sum_{i=1}^{\infty} y_{max,i} \cos(2\pi f_i t - \phi_i). \qquad (20.16)$$

For simplicity, in the rest of this chapter we will assume that signals are measured relative to their means, allowing us to dispense with $\overline{y}$.

Equation 20.16 is simple, important, and (I hope) intuitive, but it leaves open two important questions. First, how does one choose the specific frequencies (f_i) to best model a particular signal? And then, how does one choose the amplitudes ($y_{\max,i}$) and phase shifts (ϕ_i) that correspond to those frequencies? We answer these questions in turn.

3   CHOOSING FREQUENCIES
The Role of Harmonics

The standard method for choosing component frequencies was formulated in the early 1800s by the same Jean Baptiste Joseph Fourier we encountered when exploring the conduction of heat in Chapter 12. The method begins by defining a series' *fundamental frequency*.

Just as for each string on a guitar there is a lowest note that it can play, there is for each time series a lowest frequency of fluctuation the series can accurately model. Consider, for instance, air temperature, which flucutuates both daily with the solar cycle as well as with the longer periods of the weather and seasons. If we measure temperature every minute for 24 h, our record will contain one full daily cycle, allowing us to describe the dial rhythm in detail. But having measured temperature for only a day, we don't have enough information to accurately describe weather-driven and seasonal fluctuations. In other words, just as the length of a guitar string determines its lowest note, the length of our record (in this case, 24 h) sets the lowest frequency of temperature fluctuation—the *fundamental frequency*—that we can reliably model.

We can generalize this conclusion. If we measure a signal for a total period $t_{\max}$, the lowest frequency we can model—the fundamental frequency—is

$$f_f = \frac{1}{t_{\max}}. \tag{20.17}$$

It is important to note that the fundamental frequency is set not by the signal itself, but rather by the way in which we measure it: the longer our record, the lower the fundamental frequency.

Now to the task of choosing the frequency of the sinusoids to use when modeling a particular signal. In a stroke of genius, Fourier proposed that a signal can be modeled most expeditiously using only those frequencies that are integral multiples of the fundamental. That is,

$$y(t) = \sum_{k=1}^{\infty} y_{\max,k} \cos\left(2\pi k f_f t - \phi_k\right). \tag{20.18}$$

Here $k \; (= 1, 2, \ldots, n)$ designates the *harmonic* of each component sinusoid. For example, f_f is its own first harmonic, $2f_f$ is the second harmonic of the fundamental, and so forth. In practice, it is never possible to extend this series to an infinite number of harmonics (a topic we will deal with shortly), but the higher the harmonic included, the finer the detail that can be modeled.

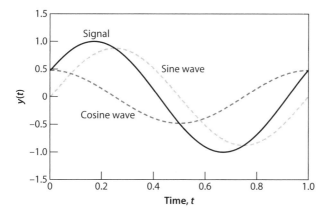

Figure 20.3 The sum of a sine wave and a cosine wave (the signal) is a sinusoidal wave, shifted in phase, with an amplitude that is a function of the amplitudes of the contributing waves.

Equation 20.18 is one form of the *Fourier series*, and it forms the basis of spectral analysis. As noted previously, one can easily adjust the Fourier series to describe a spatial rather than a temporal signal: just substitute x for t and spatial frequency (f_s) for temporal frequency (f).

To summarize: once we have measured a signal (thereby setting t_{max} and f_f), the choice of radian frequencies to use in modeling $y(t)$ is fixed—we use only harmonics of the fundamental frequency: $f_f, 2f_f, 3f_f$, and so on.

4   FOURIER COEFFICIENTS

Although equation 20.18. is a useful model, there are advantages to modifying its form. The basic idea behind this modification is that each component sinusoid in the Fourier series (with a certain amplitude and phase) can itself be modeled as the sum of a cosine wave and a sine wave, both having the same frequency as the component they model. For example, the sinusoid in Figure 20.3 (with $f = 1$, $y_{max} = 1$, and $\phi = 0.5$) can be precisely modeled as the sum of a cosine wave (with $f = 1$ and an amplitude α of 0.878) and a sine wave (again with $f = 1$, but with an amplitude β of 0.479). In general terms, it can be shown (Supplement 20.1) that

$$y_{max,k} \cos \left(2\pi k f_f t - \phi_k\right) = \alpha_k \cos \left(2\pi k f_f t\right) + \beta_k \sin \left(2\pi k f_f t\right), \qquad (20.19)$$

where

$$y_{max,k} = \sqrt{\alpha_k^2 + \beta_k^2} \qquad (20.20)$$

and

$$\phi_k = \arctan \left(\frac{\beta_k}{\alpha_k}\right). \qquad (20.21)$$

The larger α_k and β_k are, the greater the amplitude of the signal is; the larger β_k is relative to α_k, the greater the phase shift is.[1]

Because the sum of a cosine and a sine wave can mimic the effects of y_{max} and ϕ_k, we can rewrite the Fourier series (equation 20.18) as:

$$y(t) = \sum_{k=1}^{\infty} \left[\alpha_k \cos\left(2\pi k f_f t\right) + \beta_k \sin\left(2\pi k f_f t\right) \right].$$ (20.22)

This is the form of the Fourier series used in spectral analysis. The α and β values in this series—the amplitiudes of the component cosine and sine waves, respectively—are known as *Fourier coefficients*. Again, one can easily adjust this series to describe a spatial rather than a temporal signal; just substitute x for t and fundamental spatial frequency ($f_{s,f}$) for fundamental temporal frequency (f_f).

To recap, by choosing the appropriate values for α and β in equation 20.22, and, by using a sufficiently high number of harmonics (k), we can model any signal we desire.[2]

5   THE PERIODOGRAM

Let's postpone for the moment the obvious question of how one would choose appropriate αs and βs. Instead, let's suppose that by some means we have obtained the full set of Fourier coefficients for a signal of interest, providing us with an accurate mathematical model of our signal. We can now use that information to answer our original question—how is the overall variance of a signal divided up among frequencies?

To begin, let's calculate the variance of a generic cosine wave,

$$y(t) = \alpha \cos(\theta).$$ (20.23)

Here, as with our ocean waves, $y(t)$ is the deviation from average, α is the cosine's amplitude, and for simplicity I have replaced $2\pi f t$ with θ. Thus, if we sample the waveform at n equally spaced points in one of its cycles, we can calculate its variance—its mean squared deviation. Recalling that the mean of a cosinusoidal oscillation is zero,

$$\sigma_y^2 = \frac{1}{n} \sum_{i=1}^{n} \left[\alpha^2 \cos^2(\theta_i) - 0 \right]$$

$$= \frac{\alpha^2}{n} \sum_{i=1}^{n} \cos^2(\theta_i).$$ (20.24)

[1] In case you've forgotten your trigonomtery, arctan(x) (the arctangent of x) is the angle whose tangent is x.

[2] There are some discontinuous mathematical functions that cannot be modeled by the Fourier series, but these functions seldom if ever appear in nature.

Now, if we have n points spread evenly across θ from 0 to 2π, the increment between two points is

$$\Delta\theta = \frac{2\pi}{n}. \tag{20.25}$$

Multiplying equation 20.24 by one in the form of $\Delta\theta/\Delta\theta$,

$$\sigma_y^2 = \frac{\alpha^2}{n} \frac{1}{\Delta\theta} \sum_{i=1}^{n} \cos^2(\theta_i)\,\Delta\theta$$

$$= \frac{\alpha^2}{2\pi} \sum_{i=1}^{n} \cos^2(\theta_i)\,\Delta\theta. \tag{20.26}$$

Allowing $\Delta\theta$ to become small, we can take this relationship to its integral form:

$$\sigma_y^2 = \frac{\alpha^2}{2\pi} \int_0^{2\pi} \cos^2(\theta)\,d\theta. \tag{20.27}$$

Unless you are adept at calculus, you would have some difficulty evaluating the integral in this equation, but it is easy to look up the solution:

$$\int_0^{2\pi} \cos^2(\theta)\,d\theta = \pi. \tag{20.28}$$

Thus,

$$\sigma_y^2 = \frac{1}{2}\alpha^2. \tag{20.29}$$

In other words, the variance of a cosine wave is equal to half the square of the wave's amplitude.

The same logic can be applied to a sine wave $[y(t) = \beta\sin(\theta)]$—variance is again half the square of amplitude:

$$\sigma_y^2 = \frac{1}{2}\beta^2 \tag{20.30}$$

We can extend this logic even further. Because the sum of cosine and sine waves is a sinusoid with amplitude $\sqrt{\alpha^2 + \beta^2}$ (equation 20.20) the variance of each combination of harmonic cosine and sine waves in the Fourier series is

$$\sigma_{y,k}^2 = \frac{1}{2}\left(\alpha_k^2 + \beta_k^2\right). \tag{20.31}$$

Let's pause for a moment to take stock of where we have come. The task we assigned ourselves in this chapter was to characterize the pattern of a signal by determining how its overall variance is divided up among component frequencies. Perhaps without your realizing it, equation 20.31 is the answer. If we know the Fourier coefficients for a signal—the αs and βs—equation 20.31 tells us how to calculate the variance for each harmonic component and, therefore, how to specify the variance at each harmonic frequency.

This is the heart and soul of spectral analysis. As such, it deserves a means of visualization. A graph of the harmonic component variances (σ_k^2) as a function of frequency is called a *periodogram*. Figure 20.4A shows hypothetical (but realistic) results for ocean waves. It is clear that the overall variance in elevation cannot be attributed to any single frequency; instead it is distributed among frequencies. In this

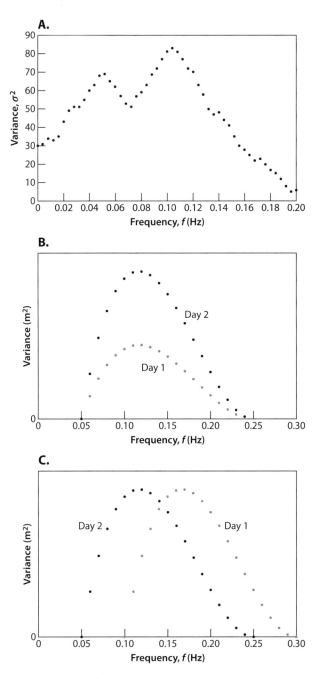

Figure 20.4 A. The periodogram for a hypothetical sea state. Hypothetical periodograms for two sea states: **B.** the same frequency distribution, different variances; **C.** the same variance, different frequency distributions.

case, much of the overall variance is associated with frequencies corresponding to individual waves ($f \approx 0.1$ Hz) and much of the remainder with frequencies typical of groups of waves (a phenomenon known as surf beat, $f \approx 0.05$ Hz).

One last step and we can bring this discussion full circle. Because the harmonic components of a signal act independently of each other (a result of the superposition principle), the overall variance of the entire signal is the sum of its component variances:

$$\sigma_{\text{tot}}^2 = \frac{1}{2} \sum_{k=1}^{\infty} \left(\alpha_k^2 + \beta_k^2 \right) . \tag{20.32}$$

We started this chapter by noting that total variance is a blunt instrument for characterizing pattern. We have now returned to the total variance, but equation 20.32 makes it explicit how the various harmonic components contribute.

To get a feel for the utility of the periodogram, consider the scenario depicted in Figure 20.4B, the periodograms for ocean waves measured on two different days. In this case, waves on the two days have the same frequency characteristics (swell with frequencies centered on 0.12 Hz), but day 2 is wavier than day 1—that is, the overall variance (the sum of individual variances) is greater.

The periodograms shown in Figure 20.4C tell a different story. The sum of values in each of the two periodograms is the same, so we can deduce that the waviness of the ocean (the total variance) was the same on both days. However, on day 1, wave frequencies were centered around 0.18 Hz, whereas on day 2, frequencies were centered around 0.12 Hz. In other words, the days were equally wavy (they had the same vaiance of surface elevation), but on day 1 waves were relatively high-frequency wind chop, while on day 2 they were relatively low-frequency swell.

6   SPECIFYING FOURIER COEFFICIENTS

As we have just seen, if all of a signal's Fourier coefficients are known, we can construct the periodogram. But, given a time series of measurements, how does one choose appropriate coeffcients? To answer this question, consider a short time series in which we have measured some signal y over n (= 4) intervals (Figure 20.5A). The constant interval between measurements is Δt, the *grain* of our measurements. The overall length of our series—its *extent*—is thus $n\Delta t$. What harmonics are required to best model this simple time series, and what values of α and β do they have?

Let's begin with f_f. As we have seen, the overall length of this series ($n\Delta t$) defines its fundamental frequency. Thus, according to equation 20.17;

$$f_f = \frac{1}{n\Delta t}. \tag{20.33}$$

In compliance with Fourier's insight, we therefore model the signal using harmonics of this fundamental frequency.

(Note for future reference that because the Fourier series deals with harmonics of the fundamental, the difference in frequency between successive harmonics (Δf) is the same as f_f itself. For example, the difference in frequency between the second and third harmonics (that is, Δf) is $3f_f - 2f_f = f_f$. This leads to an interesting conclusion: because $f_f = 1/t_{\max}$, the longer the series of data, the smaller f_f is. The smaller f_f is, the smaller the difference in frequency between harmonics (Δf), and the greater the resolution with which we can analyze the frequency dependence of the overall variance. Long data series permit fine frequency resolution.)

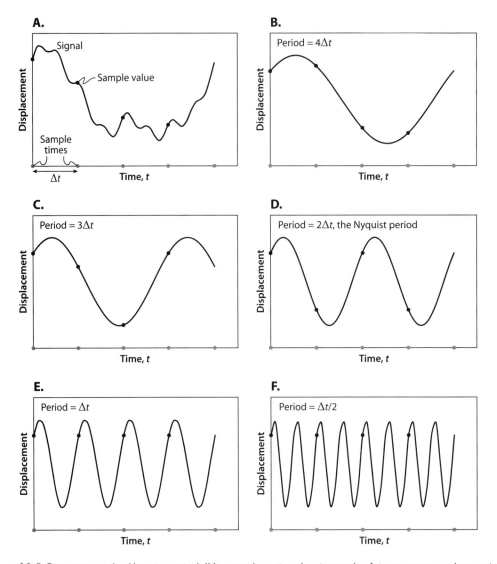

Figure 20.5 Determining the Nyquist period. We sample a signal at intervals of Δt to create a data series with four points (**A**). The ability to discern the components of the signal depends on each component's period. Our sampling regime captures the variation in signals with periods of $4\Delta t$ (**B**), $3\Delta t$ (**C**), and $2\Delta t$ (**D**) but not Δt (**E**) or $\Delta t/2$ (**F**).

Having established a series' fundamental frequency, you might suppose that we could then examine as many harmonics as we like; the more we use, the better we can describe the signal's high-frequency detail. There is a practical limit, however. Let's see what our sampling protocol tells us as we sample signals that represent increasing harmonics of the fundamental frequency. In Figure 20.5B, the signal oscillates with a period of $4\Delta t$; that is, its period of oscillation is equal to the fundamental period set by the length of our short data series. The measurements adequately sample the signal's fluctuations. (Note that if we were to include a fifth point in our sample at a time Δt farther along, it would coincide with the beginning of the second cycle of oscillation and thus would be a redundant measure of the oscillation's behavior. The four points shown are sufficient.) Figure 20.5C shows a new signal with a period of $3\Delta t$ and, thus, a frequency higher than the series' fundamental frequency. Again, our measurements (spaced Δt apart) capture the signal's fluctuation. The same is true for a signal with period $2\Delta t$ (Figure 20.5D). By contrast, when presented with a signal that oscillates with a period of Δt, our measurements sample the signal only once in each period (Figure 20.5E). As a consequence, every measurement yields the same value, and, as far as our measurements can discern, there is no variance at that frequency. The same holds true for signals with periods less than Δt (e.g., Figure 20.5F). In short, we can detect harmonics with periods of $2\Delta t$ or greater but not harmonics with periods of Δt or less. Thus, the highest frequency we can detect (known as the *Nyquist frequency* or *Nyquist limit*[3]) corresponds to the shortest period we can effectively sample, $2\Delta t$:

$$f_{k_{max}} = \frac{1}{2\Delta t}. \qquad (20.34)$$

Recalling that the fundamental frequency is $1/n\Delta t$, we can calculate the harmonic (k_{max}) corresponding to the Nyquist frequency:

$$f_{k_{max}} = \frac{1}{2\Delta t} = k_{max}\frac{1}{n\Delta t}$$

$$k_{max} = \frac{n}{2}. \qquad (20.35)$$

In other words, if we there are n intervals in our time series, only the first $n/2$ harmonics in the Fourier series provide any information about the signal. Thus, the best we can do to approximate the actual signal is the series:

$$y(t) = \sum_{k=1}^{n/2} \left[\alpha_k \cos\left(2\pi k f_f t\right) + \beta_k \sin\left(2\pi k f_f t\right) \right]. \qquad (20.36)$$

This is both good news and bad news. The bad news is that because measured time series always have finite length, our ability to analyze the details of their high-frequency fluctuations is limited to frequencies at or below the Nyquist limit. The good news is that we have to estimate αs and βs for only a limited number of harmonics.

Now to that nagging practical question: how do we find the $n/2$ values for α and β that allow equation 20.36 to best approximate $y(t)$? The algorithms involved are a bit intense, and I refer you to Supplement 20.2 for the details, but an example can convey the calculations' essence—αs and βs are calculated through a series of linear regressions.

Consider a series of 120 daily measurements of water level in an alpine lake (Figure 20.6A). It's clear that water level fluctuates more or less periodically, and our job is to calculate the αs and βs that best describe this behavior. For each harmonic k of the fundamental frequency, we calculate α_k by correlating our water-level data with a cosine wave of amplitude 1 (a unit cosine) at the harmonic frequency. The slope of the regression line is our best estimate of α_k.

---

[3] Named for Harry Nyquist (1889–1976), an electrical engineer working at Bell Laboratories.

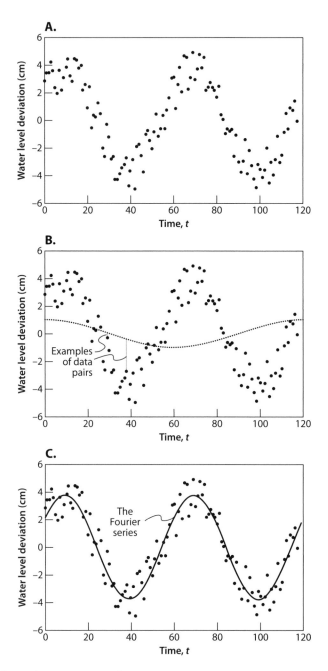

Figure 20.6 Quantifying Fourier coefficients. **A.** The measured record of water level. **B.** Points in the measured record are paired with points of a unit cosine wave of the fundamental frequency. **C.** The calculated fit to the empirical data.

For example, in Figure 20.6B, the measured data are shown along with a unit cosine at the series' fundamental frequency. The water level at time t_1 can be matched with the value of the cosine at time t_1, the level at t_2 with the cosine at t_2, and so on. A plot of all these pairs (one versus the other) is shown in Figure 20.7A. There is a pattern to

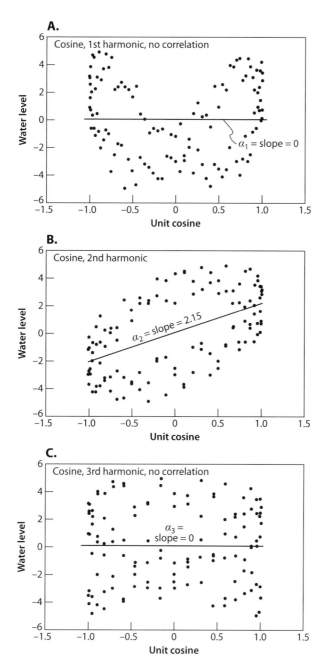

Figure 20.7 Calculating Fourier coefficients: α (A–C) and β (D–F) for the first three harmonics of the water-level signal from Figure 20.6.

this plot but no correlation between the measured data and the cosine—when we fit a line to the graph (a linear regression), we find that it's slope is zero. This tells us that $\alpha_1 = 0$.

I have repeated this procedure for unit cosine waves at the second and third harmonics of the fundamental, with the results shown in Figure 20.7B and C,

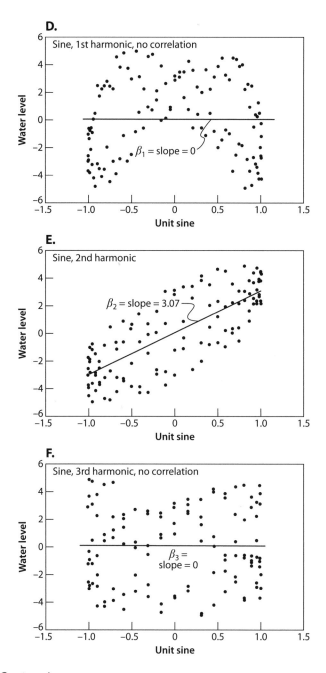

Figure 20.7 Continued.

respectively. For the second harmonic, there is a substantial correlation, and the regression line has a slope of 2.15, so $\alpha_2 = 2.15$. For the third harmonic, there is no correlation and $\alpha_3 = 0$. This procedure could be repeated for higher harmonics to calculate α_4 to $\alpha_{k_{max}}$.

This whole process can then be repeated using unit sine (rather than cosine) waves (Figure 20.7D–F) to calcutate β_1 to β_3. The first and third sine harmonics show no correlation and their regression lines have zero slopes, so β_1 and β_3 are zero. But the data are correlated with the second sine harmonic, and the slope of the regression line is 3.07, so $\beta_2 = 3.07$.

Using the these calculated αs and βs, we can calculate the sum of an abbreviated Fourier series, with results shown in Figure 20.6C. Lo and behold, this sum nicely models the major trend in the measured data, an oscillation at a frequency twice the fundamental of the data series. If we were to extend our efforts to higher harmonics, we would obtain an ever better model of our measurements.

This example is intended to give you insight into the process by which α and β are estimated. It is unlikely, however, to give you a feel for the large number of calculations involved. Estimating α and β is not something one does on a calculator or even in a spreadsheet. Indeed, until 1965 when J. E. Cooley and J. W. Tukey invented the Fast Fourier Transform (FFT), calculating αs and βs for a long series was taxing even for computers. The FFT drastically increases the computational efficiency of calculating αs and βs, and algorithms for the FFT are now included in most programming languages.

In summary, given a series of data, methods are available to estimate α and β for all harmonic components from the fundamental to the Nyquist. With these parameters in hand, one can accurately model the signal, plot the periodogram, and proceed with its interpretation.

7   THE POWER SPECTRUM

There's just one hitch. Although the periodogram admirably portrays how overall variance is divided among frequencies, this is not how the results are traditionally presented. Instead, the frequency-dependent characteristics of a signal are most often graphed as a *power spectrum*, where, as noted earlier, the term power is a stand-in for variance. To understand the switch—and the reason for undertaking it—we begin with a brief digression to statistics.

You may recall from an introductory stats class that the probability of seeing a particular outcome in an experiment can often be calculated using the binomial distribution. A classic example is an experiment in which you toss a coin 10 times and ask what the probability is that you will get a certain number of heads. The answer is provided by the binomial distribution (Figure 20.8A) in which each number of heads—from 0 to 10—is associated with a specific probability. There is, for instance, an 11.72% chance of getting 7 heads in 10 tries (see Denny and Gaines (2000) for an explanation of the binomial distribution).

But calculating binomial probabilities for a large number of trials is laborious. As a practical alternative, statisticians instead rely on the fact that, when dealing with many trials, the binomial distribution approximates a normal (Gaussian) distribution (Figure 20.8B), which is much more easily calculated. In fact, for an infinite number of trials, the two distributions are functionally equivalent, with one important difference. Whereas the binomial distribution specifies a probability for each outcome, the normal distribution provides a *probability density*, the probability per *range* of outcomes. The reason is this: in an infinite number of trials, there is an infinite number of ways that an experiment can play out, so the probability of getting any particular number of heads is effectively zero. A graph of this distribution (an infinite series of zeros) would therefore look like a bare set of axes, a totally uninformative picture.

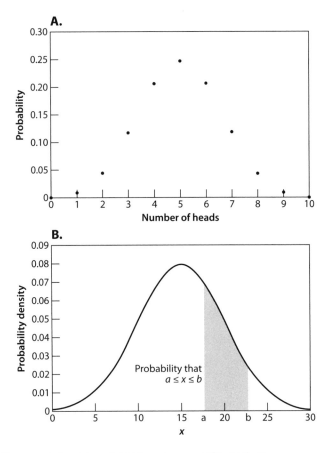

Figure 20.8 Representative binomial (**A**) and normal (**B**) distributions.

When dealing with an infinite number of trials, the best we can do is describe not the probability of each individual outcome, but rather the probability that an outcome chosen at random will fall within a certain range of values.

The normal distribution provides this description. To extract a probability value from a normal curve, one calculates the area under the probability density function corresponding to a given range of outcome values (Figure 20.8B). Because the function covers the probability of all possible outcomes, the total area under the curve is 1.

The power spectrum results from an analogous thought process. From equation 20.31 we know that the variance at each individual harmonic k is

$$\sigma_k^2 = \frac{1}{2} \left(\alpha_k^2 + \beta_k^2 \right). \tag{20.37}$$

Using the same logic as before, we can surmise that the variance associated with individual harmonics is likely to decrease as the number of harmonics in the Fourier series increases, and as we have seen, the number of harmonics we can examine ($n/2$) increases with n, the number of intervals in our series. Thus, as we increase the number of measurements in an attempt to discern more detail in a signal, we simultaneously decrease the magnitude of the variance estimate for each individual frequency.[4] Taken to its logical extreme, this suggests that if we were to measure a signal continuously—that is, at an infinite number of points—our ability to say anything about the distribution of variance would vanish.

To rescue the periodogram from this fate, we need to adjust our expression describing the variance at each harmonic. To counteract the decrease in σ_k^2 as n increases, we multiply σ_k^2 by a term that increases with n. For reasons that will become clear in a moment, we choose $n\Delta t$. Thus our new variable—the *power spectrum, S*—is the product of σ_k^2 and $n\Delta t$. Evaluating S for the kth harmonic,

$$S_k = n\Delta t\sigma_k^2 = \frac{n\Delta t}{2}\left(\alpha_k^2 + \beta_k^2\right). \tag{20.38}$$

We could plot these data as a function of frequency for a spectral equivalent of the periodogram.[5]

Having defined S in this manner, we can immediately turn the definition around. At each individual harmonic,

$$\sigma_k^2 = \frac{S_k}{n\Delta t}. \tag{20.40}$$

But we know from equation 20.33 that

$$\frac{1}{n\Delta t} = f_f. \tag{20.41}$$

Furthermore, as noted previously, because the Fourier series calculates variance at harmonics of the fundamental frequency, Δf, the difference in frequency between harmonics, is equal to f_f. Thus, at each harmonic,

$$\sigma_k^2 = S_k\Delta f, \tag{20.42}$$

and summed across harmonics,

$$\sigma_{\text{tot}}^2 = \sum_{k=1}^{n/2} S_k\Delta f. \tag{20.43}$$

To this point, we have simply come up with another way of stating equation 20.32, the equation that gave us the periodogram. The value of these machinations appears when n becomes very large. Taken to the limit as n approaches infinity, Δf becomes df, and

$$\sigma_{\text{tot}}^2 = \int_0^\infty S\,df. \tag{20.44}$$

In other words, the overall variance of a signal is equal to the area under the curve of its power spectrum. Furthermore, the variance associated with any range of frequencies

---

[4] Only if a harmonic *exactly* matched the perfectly constant frequency of a signal would amplitude stay constant as the number of harmonics increases. For real-world signals, an exact match and perfect constancy are essentially impossible.

[5] There is one exception to this rule. At the Nyquist frequency, $k = n/2$:

$$S_{k_{\text{max}}} = n\Delta t\alpha_{k_{\text{max}}}^2. \tag{20.39}$$

The difference of a factor of 2 in the form of this expression is due to the fact that $\beta_{k_{\text{max}}} = 0$ (see Supplement 20.2).

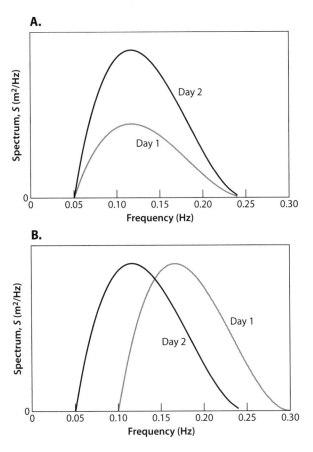

Figure 20.9 Hypothetical spectra for ocean waves. **A.** Two sea states with the same frequency distribution but different variances. **B.** Two sea states with the same variance but different frequency distributions.

is the area under that portion of the overall spectrum. For example, the variance in the range between $f = a$ and $f = b$, is

$$\sigma_{a,b}^2 = \int_a^b S \, df. \tag{20.45}$$

Let's consider some hypothetical spectra for ocean waves (Figure 20.9). In panel A, surface elevations recorded on two days have the same frequency characteristics, but day 2 is wavier than day 1 (the variance in surface elevation is greater) because the area under its spectrum is greater. We draw different conclusions from the spectra shown in Figure 20.9B. In this case, because the area under the two spectra is the same, we can deduce that the waviness of the ocean was the same on both days. However, on day 1, wave frequencies were centered around 0.18 s^{-1}, whereas on day 2, wave frequencies were centered around 0.12 s^{-1}. In other words, the days were equally wavy, but on day 1 the waves were wind chop, while on day 2 they were swell.

These examples should give you an intuitive feel for how to interpret the spectra you might encounter in the literature. But at this point, you might be scratching your head. The spectra in Figure 20.9 correspond exactly to the periodograms in Figure 20.4.

Since the spectrum doesn't contain any more information than the periodogram, why bother with the translation?

The answer is twofold. Recall from the preceding discussion that construction of a spectrum assumes that the signal can be described continuously. This is impossible for real-world measurements (one would need an infinite series of data) but is easy in a theoretical model. For instance, a mathematically inclined physical oceanographer might assume a particular size and shape for the power spectrum of ocean waves. He or she could then express the spectrum as a continuous function of f, and proceed to evaluate its properties according to equation 20.45.

By contrast, spectra calculated from real-world data are, at their core, always periodograms. Because actual measurements can be made only at discrete times and for a finite period, the Fourier series can be calculated for only a finite number of frequencies. In this respect, it would be best if the data were presented as a periodogram rather than translating them into a spectrum with its implied aura of continuity. But that sort of truth in advertising would raise another problem. Consider a study in which theory and measurement are combined, the goal being to use theory to predict the real world. If the theoretical results were presented as a spectrum and the measured results as a periodogram, it would be difficult for the reader to compare them quantitatively. The traditional solution to this problem is to present data (both theoretical and measured) as spectra. As long as you keep in mind that any spectrum resulting from measured data is really a series of individual points, no problems should arise.

8 SPECTRAL ANALYSIS IN BIOLOGY

8.1 Population Cycles in Snowshoe Hares

We will put spectral analysis to use in Chapter 21 when we consider the issue of scaling in ecology, but to provide a preview of the grandeur to come, let's explore two examples of the utility of spectral analysis in ecomechanics.

Populations of arctic mammals often cycle through periods of boom and bust (Krebs et al. 2001), the classic case being the decadal cycle of snowshoe hares. Several factors contribute to the periodic fluctuations in hare populations—the availability of food, predation by lynx and birds, and the physiological response of hares to the stress in their lives—but, remarkably, despite the inherently local nature of these drivers, cycles of population fluctuation are synchronized across vast areas. Hare populations in much of the Canadian Arctic fluctuate in unison across habitats that differ drastically in precipitation, average temperature, terrain, and flora.

It has long been speculated that maintenance of this synchronicity requires cues from some external time giver, some factor—presumably one in the physical environment—that applies equally across all local populations. In an influential paper in 1993, Sinclair and colleagues made the brash suggestion that sunspots provided the necessary cue.

In support of this proposition, Sinclair and his colleagues obtained a lengthy time series of hare population density, relying on indirect evidence from tree rings. When populations are dense, hares are forced to eat the growing tips of spruce trees for

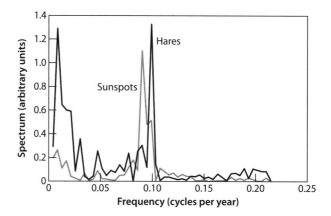

Figure 20.10 The spectra for snowshoe hares and sunspots (drawn from data in Sinclair et al. 1993).

winter forage.[6] A nipped tip leaves a dark mark in the subsequent tree ring, thereby providing a history of hare population cycles back to the eighteenth century.

The periodic fluctuation in sunspots has been recorded by astronomers for the same period, and when Sinclair and colleagues compared the power spectra of the two phenomena, they matched within the resolution of the technique; both have spectral peaks at frequencies (cycles per year) of 0.090 to 0.099 (Figure 20.10). With this suggestive evidence in hand, they proceeded to analyze their data in greater depth, concluding that during times of elevated solar activity, not only do the sunspot and hare cycles have the same period, but they tend to maintain a constant phase relationship. When solar activity is low, the phase of the hare cycle gradually shifts, only to be brought back into synchrony when the sun acts up again.

The intriguing thought that something as distant and subtle as sunspots might act as a synchonizing influence in population biology has sparked considerable interest and controversy. The evidence of Sinclair and colleagues strongly suggests a mechanistic connection between solar activity and hare populations. However, experiments have elegantly demonstrated that the proximal causes of population fluctuations are predation pressure (from lynx, for instance) and its effects on hare reproductive efficiency (Krebs et al. 2001). If the solar cycle cues the system, it seems that it must act indirectly through this predator-prey relationship. Other than some vague ideas about climatic effects, Sinclair and his colleagues have not been able to identify a plausible mechanism, so the question of a connection between solar activity and snowshoe hares remains open.

However, the current lack of a mechanistic tie between hares and sunspots should not cause us to lose sight of the main point here. As with ice ages and celestial mechanics, without spectral analysis we wouldn't know that there was a phenemenon that required investigation.

---

[6] You might wonder how hares get to the tops of trees to eat their growing tips. This is the arctic, with its low temperatures and short growing season. A 50-y-old spruce tree is only about a meter high, within the reach of an adult hare.

8.2   Sea Urchins and Algal Dominance

In the nearshore waters of Japan, benthic seaweeds have a characteristic zonation. From the intertidal zone to a depth of 2 to 3 m, rocky substrata are covered by a dense canopy of kelps and red algae. Deeper, this canopy abruptly disappears, giving way to barren areas in which crustose corallines are the only seaweeds. From a strictly mechanical perspective, this zonation is perplexing. Kelps and red algae (which are susceptible to damage from breaking waves) are found only in the shallows where wave-induced velocities are highest, while crustose corallines (which are virtually immune to wave-induced disturbance) dominate where velocities are benign.

The answer to this paradox involves the interaction of seaweeds with their primary herbivore, the sea urchin *Strongylocentrotus nudus*. Where flow is slow, these urchins are effective grazers, but when average velocity exceeds $0.4 \text{ m} \cdot \text{s}^{-1}$, they retreat to their burrows. Kawamata (1998) suggested that water velocity thus serves as an indirect control on algal abundance. Urchins can't feed in the shallows where water velocities are high, allowing seaweeds to flourish, even though some are damaged. Deeper, urchins are free to graze, forming the barrens. To test this hypothesis, however, he needed to predict how velocity varied with depth at his study site. This is where spectral analysis comes in.

From basic wave theory (Denny 1988), we know that maximum water velocity at the substratum (where urchins live) is a function of wave amplitude η_{max}, wave frequency f, and depth of the water column d:

$$u_{max}(f, d) = \sqrt{\eta_{max}^2 U(f, d)^2}. \tag{20.46}$$

Here the function U describes how velocity varies with depth, depending on wave frequency:

$$U(f, d) = \frac{1}{\sinh\left[\frac{4\pi^2 f^2 d}{g \tanh \sqrt{\frac{4\pi^2 f^2 d}{g}}}\right]}. \tag{20.47}$$

Values of U are shown as a function of frequency and depth in Figure 20.11.[7] Provided we know wave amplitude and frequency, equations 20.46 and 20.47 allow us to calculate the velocity imposed on an urchin living at a certain depth.

Our analysis would be straightforward if waves had a fixed amplitude and frequency, but as we have seen, the ocean surface isn't that simple. Let's suppose that we have recorded surface elevation at a site where urchins live (e.g., Figure 20.2B). How can we use this messy empirical information about surface elevation to predict average velocity as a function of depth?

Of course, spectral analysis is the answer. From the empirical data, we can calculate the periodogram. From the periodogram, we know what η_{max}^2 is for each component

---

[7] The hyperbolic sine and tangent (sinh and tanh) are defined as

$$\sinh(x) = \frac{1}{2}\left(e^x - e^{-x}\right), \tag{20.48}$$

$$\cosh(x) = \frac{1}{2}\left(e^x + e^{-x}\right), \tag{20.49}$$

$$\tanh(x) = \frac{\sinh(x)}{\cosh(x)}. \tag{20.50}$$

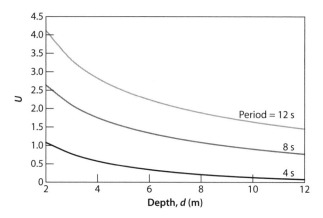

Figure 20.11 The function U used to calculate velocity as a function of wave period and water-column depth, d.

wave frequency f, and from these frequency-specific variances, we can use equations 20.46 and 20.47 to calculate each frequency's contribution to the overall velocity at a given depth. We then sum these frequency-specific velocities to calculate the overall average maximum velocity, u_{max}. In short, in order to predict the velocities imposed on urchins as a function of depth, Kawamata needed to know not only the overall waviness at his site, but also how the variance of surface elevation was spread across frequencies.

To this end, he obtained periodograms for the surface elevation every day for five years. From these data, he followed a procedure similar to that described here to calculate daily average u_{max} as a function of depth, and from these he estimated the fraction of time available for urchins to feed. At 2 m depth, urchins could feed for less then 2.6% of the time, insufficient to keep seaweeds in check. At 10 m, they could feed more than 70% of the time, plenty of opportunity to maintain the algal barrens. Thus, with the help of spectral analysis, Kawamata could explain the distribution of his seaweeds.

9 CONCEPTS, CONCLUSIONS, AND CAVEATS

The mean and variance are insufficient to quantify the pattern of fluctuation in a sequence of data. At the very least, we need to take into account how overall variance is divided up among the component frequencies of the signal. This is a job for spectral analysis, which relies on the ability of the Fourier series to model a signal in time or space. By calculating α and β for the harmonic components of a signal, we can construct the data's periodogram and power spectrum, which allow us to make comparisons among signals (as Sinclair et al. did with the fluctuations in sunspots and arctic hares), to use the distribution of variance among frequencies to make other calculations (as Kawamata did to predict the effect of water velocity on urchin foraging), or to identify periodicities (as for the ice ages). We will put power spectra to use in the next chapter where we describe how the temporal or spatial scale at which an organism interacts with its environment affects the plant or animal's average response.

Be warned that this chapter has introduced spectral analysis only at a conceptual level. There are a whole host of details that need to be taken into account when the method is put to practical use. For example, spectral analysis as described here implicitly assumes that the process from which measurements are taken is *stationary*. Roughly speaking, a procees is stationary if the same statistics (e.g., the mean, variance, and spectrum) are obtained regardless of when or where the data series starts. Many ecological series violate this assumption—air temperature, for instance, is trending upward, so its mean changes through time—and these nonstationary series must be massaged into stationarity before they are analyzed. The brief discussion in Supplement 20.4 will introduce you to this and other issues in the practical use of spectral analysis, allowing you to bridge the gap to introductory texts such as Diggle (1990). Once you are comfortable with the basics, there are several excellent upper-level texts (e.g., Priestley 1981; Bendat and Piersol 1986).

Spectral analysis is only one of several methods available to explore the pattern of sequential data. *Wavelet analysis* is a notable alternative, with some distinct advantages. For example, spectral analysis considers a data series as a whole, correlating the data with cosine and sine waves that have the same extent as the entire data set (e.g., Figure 20.6B). As a result, spectral analysis can accurately quantify the contributions of various frequencies of oscillation to the over all variance, but it can't localize where within a series those oscillations occurred. When applied to the ice-age data in Figure 20.1A, for instance, spectral analysis can tell us that the variance in Earth's temperature has sustantial components at periods of 41,000 and 100,000 y, but it can't tell us that the 41,000-y oscillations ocurred at a diferent time than the 100,000-y oscillations. By contrast, wavelet analysis correlates data not with a continuous cosine or sine wave, but rather with a "wavelet," a particular pattern of oscillation that is localized in time or space. By moving the wavelet through the signal, analyzing the components of frequency as it goes, wavelet analysis can quantify when or where each frequency component contributes. Unfortunately, while there are a host of upper-level texts dealing with wavelets, I know of no truly introductory text.

And finally, I would be remiss if I didn't mention the use of sequential data in making predictions. In this chapter, we have focused on the analysis of measurements already in hand, but the pattern of variation measured in the past can be a potent guide to what will happen in the future. In spatial terms, the pattern of variation in one small part of the world can provide important information about patterns in adjacent areas. The use of sequential data to make predictions falls into the realm of *state-space modeling*, of which *Kalman filtering* is a particularly useful tool. Unfortunately, this type of analysis is well beyond the purview of this text, and I refer you to Priestley (1981) for an introduction.

Chapter 21

Quantifying the Effects of Scale

In this chapter, we return to our exploration of the consequences of Jensen's inequality. Recall from Chapter 19 that the average response of an organism, population, or community depends not only on the average conditions it encounters, but also on the variability of those conditions. The greater the variance, the more the average response $(\overline{g(x)})$ differs from the response one would expect from average conditions alone $(g(\overline{x}))$. This is an intriguing and important conclusion, but it leaves us with an obvious question—how much variation do we find in nature? As with many deceptively simple questions, the answer to this query is rife with contingencies. In particular, for many ecological and environmental systems, variability depends on the *scale* of measurement. For instance, air temperature measured for a minute exhibits some minor variability. When measured for a day, more variability is revealed, and when measured for a year or a decade, even more. Analogously, temperature variability depends on spatial scale: there is little variability among sites separated by short distances and variability increases as more distant sites are included. The same ideas apply to ecological factors: the longer the time frame or the larger the area, the greater the variability in prey density, disturbance, reproductive rates, and so forth. Because a system's response typically depends on the variability encountered and variability changes with time and distance, it is important to understand how the scale of variability presents itself in nature.

In this chapter, we will see that our newfound understanding of power spectra provides a powerful tool for modeling the effect of scale. One model in particular—known as $1/f$ noise—provides a flexible perspective that can be applied in an exceptionally wide variety of circumstances.

1   1/F NOISE

Let's assume that—either by calculation or direct measurement—the overall variation in a system has been determined. Our task is to ascertain how much of this systemwide variation an individual organism actually experiences. As it turns out, we already have the necessary tools, although it takes a bit of explaining to understand how we can use them.

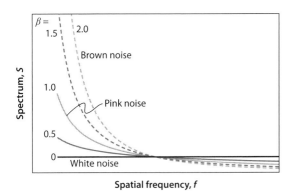

Figure 21.1 Examples of $1/f$-noise spectra with various values of β.

We start with the spectrum. Recall from Chapter 20 that the overall variance of a signal is the integral of its power spectrum, S,

$$\sigma_{\text{tot}}^2 = \int_0^\infty S(f)\,df, \quad (21.1)$$

where frequency f can be expressed in terms of time or space. For time,

$$f = \frac{1}{t}, \quad (21.2)$$

where t is the period over which a component of the signal repeats itself. (In this context, a component is the pair of cosine and sine terms in the Fourier series at a given frequency.) For space,

$$f_s = \frac{1}{\ell}, \quad (21.3)$$

where ℓ is the distance over which a component repeats.

Nature is complex, so one might suppose that spectra differ drastically from one system to another. Indeed, some systems have distinctive spectra: ocean tides, for instance, have spectra characterized by distinct peaks at frequencies determined by celestial mechanics, air temperature has a daily peak, and rainfall an annual peak. However, for many environmental and biological systems, the frequency dependence of variation can be described by a single family of spectra:

$$S(f) = \frac{k_{\text{var}}}{f^\beta}. \quad (21.4)$$

The form of equation 21.4 gives this type of variation its name: $1/f$ noise, (one over f noise) where noise is an informal stand-in for variance. The exponent β, which sets the rate at which S decays with increasing frequency,[1] typically takes on values between 0 and 2 (Figure 21.1), and k_{var} sets the spectrum's overall magnitude. (Note that k_{var} has the same units as the variance multiplied by seconds$^{1-\beta}$ for temporal frequencies or meters$^{1-\beta}$ for spatial frequencies.)

We find that $1/f$ noise is surprisingly widespread in nature. The spectra of fractal objects and fractal time series are $1/f$ noise (Hastings and Sugihara 1993). In biology, everything from nucleotide sequences in DNA to the variation in diversity along transects to extinction rates of species have $1/f$-noise spectra (Halley 1996). And $1/f$ noise is exceptionally common in population biology: in one literature survey of 544 time series, 97% had the characteristics of $1/f$ noise (Inchausti and Halley 2002).

---

[1] β as used here has no connection to the β in the Fourier series. It is unfortunate that the same symbol is traditionally used in both cases, but there just aren't enough symbols to go around.

Given its ubiquity, $1/f$ noise serves as a good general model for exploring the effects of spatial and temporal scale.

2   THE COLOR OF SPECTRAL DATA

Members of the $1/f$-noise family of spectra are often categorized by an analogy to the the spectrum of light. When $\beta = 0$, the overall variance is equally distributed among all frequencies, analogous to white light (Figure 21.1). Consequently, a $1/f$-noise spectrum with $\beta = 0$ is known as *white noise*. For β between 0 and 2, variance at low frequencies (analogous to red light) is large compared to that at high frequencies, and the spectra are said to be reddened—the larger β is, the redder the spectrum. In particular, a spectrum with $\beta = 1$ is referred to as *pink noise*.

Creative departures from these analogies are allowed. For instance, the spatial spectrum of a random walk—Brownian motion—is characterized by $1/f$ noise with $\beta = 2$, and spectra of this sort are thus referred to as *brown noise*. Spectra with $\beta > 2$ are sometimes termed *black noise*.

Attempts have been made to correlate different processes and environments with different characteristic βs. The heat capacity of the ocean discourages high frequency shifts in temperature, so surface temperature and factors that depend on it (air temperature, wind speed, etc.) tend to be reddened in the marine environment. Noting this fact, Steele (1985) proposed that environmental factors have larger βs in marine than in terrestrial systems. His suggestion is supported by Vasseur and Yodzis (2004) who, in a survey of 152 data sets, found that the βs in terrestrial systems typically ranged from 0.1 to 0.75, while those in marine systems typically ranged from 0.75 to 1.5. The terrestrial/marine difference in environmental parameters is apparently not reflected in population biology, however. In their survey of temporal records of population size, Inchausti and Halley (2002) found that βs were higher in terrestrial than in marine populations.

Although it is tempting to search for general conclusions of this sort, at present they must be used with caution. For every system analyzed to date, there is considerable variation in β. For example, in their survey of populations data, Inchausti and Halley (2002) found that the average β was 1.02, supporting the proposition made by Halley (1996) that $\beta = 1$ should be used as the default model of ecological variation. But the βs reported by Inchausti and Halley ranged from less than 0 to greater than 2, which tends to undermine the utility of the mean. It seems that for the foreseeable future, we will need to measure β in each system we investigate.

3   PATTERNS OF 1/F NOISE

In any natural system, there are limits to the range of frequencies an organism, population, or community can experience, and these limits ($f_{\min}$ and $f_{\max}$) determine the variance that will be encountered (Figure 21.2A). For a $1/f$-noise system sampled at frequencies between $f_{\min}$ and $f_{\max}$,

$$\sigma^2 = k_{\text{var}} \int_{f_{\min}}^{f_{\max}} \frac{1}{f^\beta} df. \qquad (21.5)$$

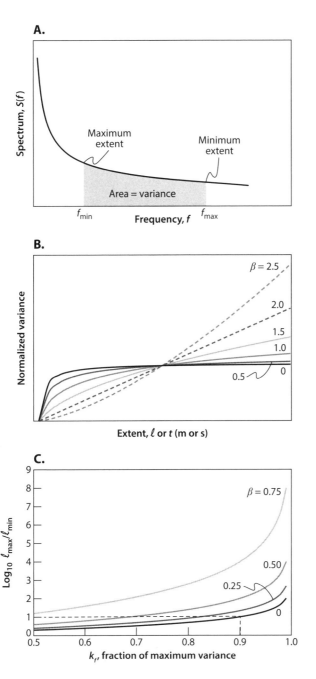

Figure 21.2 A. Total variance encountered by an organism is the area under the spectrum between the frequencies corresponding to minimum and maximum extents. **B.** How total variance scales with extent depends on β. **C.** The extent required to reach fraction k_f of σ_{max}^2.

Solving equation 21.5, we find that for $\beta \neq 1$,

$$\sigma^2 = \frac{k_{var}}{1 - \beta} \left(f_{max}^{1-\beta} - f_{min}^{1-\beta} \right). \tag{21.6}$$

For $\beta = 1$,

$$\sigma^2 = k_{var} \ln \left(\frac{f_{max}}{f_{min}} \right). \tag{21.7}$$

Thus, for $1/f$-noise spectra, if we know k_{var} and β, we can calculate the variance for any range of temporal frequencies an organism might experience. Variance in turn interacts with the system's response functions to determine an organism's, population's, or community's average behavior.

The upper frequency limit, f_{max}, is often determined by the nature of the system. Body temperature may vary through time, for example, but it can never fluctuate faster than the response time set by an organism's thermal inertia (Chapter 12). For a plant or animal that reproduces once per year, annual reproductive output may vary from year to year, but it can't vary within a year. Solar irradiance can change with location, but when it comes to soaking up heat from the sun, variation at spatial scales smaller than the size of an organism is simply averaged out as light is absorbed.

It is important not to confuse f_{max} (the upper frequency limit for organisms, populations, or communities) with $f_{k_{max}}$ (the Nyquist frequency), the highest frequency we can mathematically discern in a series of discrete data (Chapter 20). The Nyquist frequency is set by the interval between data points (Δt or $\Delta \ell$): for temporal series $f_{k_{max}} = 1/(2\Delta t)$, for spatial series $f_{k_{max}} = 1/(2\Delta \ell)$. However, just because a series of discrete data doesn't allow us to discern variation at frequencies above $f_{k_{max}}$, it doesn't imply that variation doesn't exist at higher frequencies. It just means that our sampling interval is too long to detect it. Organisms, populations, and communities sample the world continuously rather than discretely; thus, Δt and $\Delta \ell$ for them are effectively zero, and $f_{k_{max}}$ is effectively infinite. In short, for organisms, populations, and communities, the upper frequency limit f_{max} is set by the nature of the system, as described here.

On the other hand, the lowest frequency an organism, population, or community experiences is open to biological influence. The longer an organism lives, for instance, the more weather-related variation it will face; the farther it forages, the more variation in prey density it will encounter. Thus, in our exploration of the consequences of scale, our primary focus is not on the upper frequency limit—which is fixed—but rather on the effect of the lower limit of frequency variation, f_{min}.

Equations 21.6 and 21.7 are less intuitive than might be desired because they are couched in terms of frequency rather than time or distance. For instance, one has to remember that maximum temporal frequency corresponds to the minimum time over which an organism encounters its environment. Similarly, the minimum spatial frequency is determined by the maximum distance over which organisms interact with each other or the environment. Equations 21.6 and 21.7 can be made more intuitive by restating them in terms of distance or time rather than frequency. For example, substituting the inverse of distance ℓ for spatial frequency f_s and noting that $f_{s,max}$ corresponds to ℓ_{min}, and vice versa, we find that for $\beta \neq 1$,

$$\sigma^2 = \frac{k_{var}}{1 - \beta} \left(\ell_{min}^{\beta-1} - \ell_{max}^{\beta-1} \right). \tag{21.8}$$

For $\beta = 1$,

$$\sigma^2 = k_{var} \ln \left(\frac{\ell_{max}}{\ell_{min}} \right). \tag{21.9}$$

Thus, if we know β and k_{var}—which together characterize the spectrum—we can calculate the variance for any range of distances.

An example shows how this is done. We stretch a tape measure across a forest floor and measure solar irradiance along it, starting at one end and working toward the other. How does the variance in our data change as we increase the overall distance— the *extent*, ℓ_{max}—encompassed by our measurements? According to equations 21.8 and 21.9, if irradiance is a $1/f$-noise process, the longer the section of transect we cover, the greater the variation we record.

Analogous equations apply to time. For $\beta \neq 1$,

$$\sigma_x^2 = \frac{k_{var}}{1 - \beta} \left(t_{min}^{\beta-1} - t_{max}^{\beta-1} \right). \tag{21.10}$$

For $\beta = 1$,

$$\sigma_x^2 = k_{var} \ln \left(\frac{t_{max}}{t_{min}} \right). \tag{21.11}$$

We could, for instance, measure solar irradiance at one spot in the forest. If irradiance through time is a $1/f$-noise process, equations 21.10 and 21.11 allow us to model how the variance in irradiance increases with the increasing extent (t_{max}) of our temporal series.

Let's explore these results for specific values of β (Figure 21.2B). (For simplicity, in each case I have set $k_{var} = 1$.)

- $\beta = 0$. In this case (white noise), variance increases rapidly with an increase in extent, asymptotic to a particular value.
- $0 < \beta < 1$. Variance again approaches an asymptote, but the closer β is to 1, the higher the asymptote and the larger the extent required for variance to plateau.
- $1 \leq \beta < 2$. Variance increases without bound. Any increase in the extent of measurement uncovers new variability, albeit at a steadily decreasing rate.
- $\beta = 2$. This is a special case in which variance increases linearly with an increase in the extent of measurement. Random walks are an important example of a $1/f$-noise process with $\beta = 2$—the variance of location for an organism performing a random walk increases linearly with time (Chapter 4).
- $\beta > 2$. In this case, variance increases at an increasing rate with any increase in the extent of measurement.

In summary, $\beta = 1$ forms the dividing line between spectra that have limited variance at infinite extent ($\beta < 1$) from those with unlimited variance ($\beta \geq 1$). When variance is limited, the larger β is, the slower the system approaches the limit; when variance is unlimited, the larger β is, the more rapidly variance increases with increasing extent.

4   ASYMPTOTES FOR $\beta < 1$

As we have just seen, when β is less than 1, variance approaches an asymptote at large extent, and the magnitude of the asymptote (σ^2_{max}) depends on β. By setting ℓ_{max} or t_{max} to infinity in equations 21.8 and 21.10, we can calculate the asymptote's magnitude:

$$\sigma^2_{max} = \left(\frac{k_{var}}{1-\beta}\right) \ell^{\beta-1}_{min}, \tag{21.12}$$

$$\sigma^2_{max} = \left(\frac{k_{var}}{1-\beta}\right) t^{\beta-1}_{min}. \tag{21.13}$$

For a given k_{var} and ℓ_{min} (or t_{min}), the larger β is (while remaining less than 1), the larger the asymptotic variance.

But in real-world situations (where infinite extents are in short supply), knowledge of the asymptote has little practical value. Instead, it is more important to have a feel for the time or distance at which variance begins to plateau. For example, at what ℓ_{max} does variance reach 50%, or 90%, or 99% of its maximum value? Working through the algebra (Supplement 21.1), we find that the time or distance required to reach a fraction k of σ^2_{max} is

$$t_k = t_{min} (1-k)^{1/(\beta-1)}, \tag{21.14}$$

$$\ell_k = \ell_{min} (1-k)^{1/(\beta-1)}. \tag{21.15}$$

These relationships are graphed in Figure 21.2C for a range of βs. For $\beta = 0$, variance reaches 90% of its maximum value when t (or ℓ) is only 10 times t_{min} (or ℓ_{min}). As β approaches 1, it takes more and more time (or greater and greater distance) to reach a given fraction of the asymptotic variance.

5   ESTIMATING β AND K_{VAR}

So far we have discussed $1/f$ noise in abstract terms. It is now time to talk specifics. How does one determine if the variance in a system is $1/f$ noise? If it is, how does one evaluate β and k_{var}?

The answers lie in a reworking of equation 21.4. Taking the logarithm of each side of the equation, we find that:

$$\log S(f) = \log k_{var} - \beta \log f. \tag{21.16}$$

Thus, if a spectrum is $1/f$ noise, it forms a straight line when the logarithm of S is plotted as a function of the logarithm of f. The slope of the line is $-\beta$, and its y-intercept is $\log k_{var}$ (Figure 21.3).

To apply this approach, spectral estimates are first obtained from empirical data, as discussed in Chapter 20. To recap, one measures the factor in question at a series of equally spaced points in time or space. From that series, one then calculates spectral estimates, $S(f)$, for harmonic frequencies ranging from the fundamental to the Nyquist. One then plots the log of $S(f)$ as a function of the log of f. Fitting a regression line to these log-log data provides a quantitative estimate of β and k_{var}. As with any statistical procedure, care must be exercised when using this approach (see Supplement 21.2); what is important here is that straightforward methods are

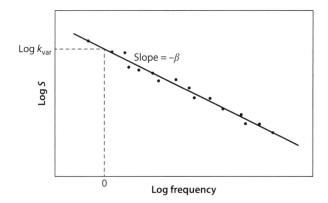

Figure 21.3 When plotted on log-log axes, $1/f$ spectra form a straight line with slope $-\beta$ and a y-intercept of log k_{var}.

available for ascertaining whether a given system can be modeled as $1/f$ noise, and, if it can be, for estimating k_{var} and β.

With β and k_{var} in hand, all we need to do to calcuate σ^2 (equations 21.8–21.11) is to specify ℓ_{min} and ℓ_{max}, or for temporal variance, t_{min} and t_{max}. As noted before, ℓ_{min} is set by the minimum extent at which an organism, population, or community can experience variation. For example, if the parameter of interest is the spatial density of barnacles on an intertidal rock, there is a practical minimum size to the area one could use to measure density—measuring average barnacle density over an area smaller than an individual barnacle would be meaningless. In a case such as this, ℓ_{min} is set by the linear dimension of the standard area (e.g., a quadrat) used to make the density measurements. The minimum interval of time, t_{min}, is similarly set by practical concerns. For example, if one is interested in the variation in maximum daily temperature, one day is the obvious choice for t_{min}. If it is instantaneous temperature that matters, the thermal response time of the organism is an appropriate choice. With ℓ_{min} and t_{min} set, we can then proceed to calculate σ^2 as a function of ℓ_{max} and t_{max}.

Unlike β values, which in typical cases are between 0 and 2, k_{var} can take on almost any value depending on the situation. The spectrum of wave-induced water velocity on rocky shores, for instance, is characterized by a β of approximately 1 across a wide range of shoreline topographies (Denny et al. 2004). However, the k_{var} values of these shores are likely to differ substantially depending on wave exposure: the low velocities of a protected shore will be associated with small k_{var}, the high velocities of an exposed shore with a large k_{var}.

6   SCALE TRANSITION
An Example

Having gained familiarity with $1/f$ noise, we can now apply it in conjunction with scale transition theory to explore a system's average response. To my knowledge, this approach has not been applied to real-world data in a biological context, so we investigate a hypothetical example.

The intertidal zone of rocky shores is home to extensive beds of mussels and barnacles, which form the primary food for predatory snails. Snails are ectotherms, so their body temperature varies with the local environment. Where conditions make for high body temperatures, snails have elevated feeding rates, with concomitant negative impacts on the abundance of local prey.

Let's assume that the shore's physical environment varies such that snail body temperature is a spatial $1/f$-noise process with $\beta = 1$ and $k_{var} = 5°C^2 \cdot$ m. Thus, the longer the stretch of shore occupied by a population, the larger the variation in body temperature individuals encounter.

We now introduce into this system an invasive species of snails. As these predatory invaders spread along the shore from their initial foothold, how does the population's average per-individual rate of mussel consumption vary? To answer this question, we combine scale transition theory with the $1/f$-noise spectrum of body temperature.

The first step is to define the thermal response function R and its scale transition coefficient $\mathcal{S}_T$ (see Chapter 19). We treat predation as a Q_{10} process such that the rate of consumption at temperature T is

$$R(T) = R(T_{ref})Q_{10}^{(T-T_{ref})/10}. \tag{21.17}$$

T_{ref} is a reference temperature, which I arbitrarily take to be 10°C, and let's assume that consumption varies with a Q_{10} of 2. The scale transition coefficient for the Q_{10} relationship is thus (Supplement 19.2)

$$\mathcal{S}_T(T) = \frac{(\ln Q_{10})^2}{200} R(T_{ref})Q_{10}^{(T-T_{std})/10}. \tag{21.18}$$

The next step is to quantify the variance in temperature, which we can easily do because the spatial spectrum of temperature can be modeled as a $1/f$-noise process. As a function of ℓ_{max} (the spatial extent of the population along the shore), σ_T^2 is

$$\sigma_T^2 = k_{var}\ln\left(\frac{\ell_{max}}{\ell_{min}}\right). \tag{21.19}$$

As suggested before, we set ℓ_{min} to 0.01 m, the size of an individual snail.

With R, $\mathcal{S}_T$, and σ_T^2 in hand, we can now formulate the scale transition equation:

$$\overline{R(T)} = R(\overline{T}) + \mathcal{S}_T(\overline{T})\sigma_T^2$$
$$= R(T_{ref})Q_{10}^{(\overline{T}-T_{ref})/10} + \left[\frac{(\ln Q_{10})^2}{200}R(T_{std})Q_{10}^{(\overline{T}-T_{std})/10}\right]k_{var}\ln\left(\frac{\ell_{max}}{\ell_{min}}\right). \tag{21.20}$$

The results are shown in Figure 21.4A. As the snail population spreads—thereby encountering more variability in body temperature—average consumption rate increases substantially. When the invasion spans a kilometer, consumption rate is 24% higher than its initial value, 29% at 10 km, 34% at 100 km. Due to the interaction between the $1/f$-noise environment and the nonlinear nature of metabolic processes, expansion of the invader's range does more than just open new feeding grounds, it also increases the average predation pressure. In turn, the increased rate of removal of mussels and barnacles can have significant consequences at the community scale—fewer mussels and barnacles means less refuge for interstitial species but more open space in which pioneer species can settle.

The same general conclusion applies for different values of β, although the rate of increase varies (Figure 21.4B). (For ease of comparison, I have adjusted k_{var} so that relative consumption rate is the same at the 1-km scale for all βs.) If $\beta = 0.75$, the

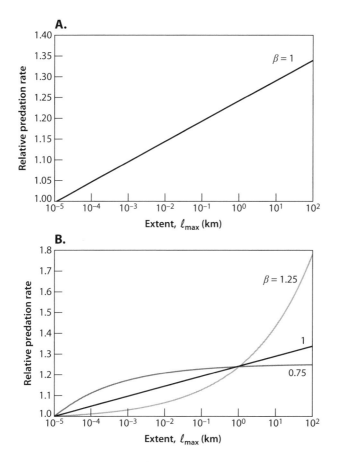

Figure 21.4 A. Average predation rate of a predatory snail as a function of its spatial extent along a shore ($\beta = 1$), expressed relative to the rate at minimum extent. **B.** The pattern of predation rate depends on β.

effects of increasing scale have run their course when the snails have invaded a mere kilometer of shore. By contrast, if $\beta = 1.25$, the consequences of scale become ever more drastic as the invasion proceeds.

Although this hypothetical example is far too simple to accurately model the effects of real snail invasions, I hope that it makes clear the potential for this approach. By combining response functions and spectra (in this case a $1/f$-noise spectrum), we can account for the effects of spatial and temporal variation on the mean response of organisms—effects that can cascade through the ecosystem. Thus, this approach provides a context in which the multitude of response functions we have investigated (drag, boundary-layer thickness, mass transport, etc.) can be applied to real-world situations with all their contingencies.

7   OTHER SPECTRA

The $1/f$-noise spectra are particularly convenient for exploring the effects of scale: they can accurately model a wide variety of real-world fluctuations, and their simple

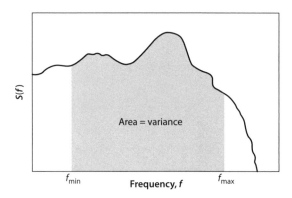

Figure 21.5 Regardless of the shape of the spectrum, total variance encountered by an organism is the area under the spectrum between the frequencies corresponding to minimum and maximum extent.

mathematical form is easily manipulated. But this approach to quantifying variance is by no means limited to systems characterized by $1/f$-noise spectra. Any spectrum can be used to calculate variance as a function of extent (e.g., Figure 21.5), and that variance can then be inserted into scale transition theory. The only difference is that the integration from f_{min} to f_{max} may not be as mathematically straightforward as it is for $1/f$-noise. In a worst-case scenario, the integration can be carried out numerically as described in Supplement 21.3.

8 CONCEPTS, CONCLUSIONS, AND CAVEATS

Jensen's inequality tells us that the average response of a system—an individual, population, or community—can be affected by environmental or biological variability. Scale transition theory allows us to calculate how much a given amount of variance affects the average response, and the spectrum tells us how much variance an organism, population, or community will encounter as it moves through space and time. In particular, the variation in many systems of biological relevance can be described by $1/f$-noise spectra, which allows us to easily model the effects of scale. As our hypothetical example illustrates, the physiological and ecological consequences of variation can be substantial. In short, the ability to tie individual response functions to the vagaries of the real world has the potential to make scale transition theory a valuable tool in ecomechanics.

There are two major caveats regarding the use of $1/f$-noise spectra in this context. First, despite a large effort by mathematicians, physicists, and statisticians, there is currently no well-accepted mechanistic explanation for why so many systems have $1/f$-noise spectra. As a result, there may be aspects of our interpretation of the effects of scale that will need to be revisited if and when the mechanism(s) leading to $1/f$-noise spectra are better understood.

The second caveat applies to $1/f$-noise spectra for which $\beta \geq 1$. In these cases, the equations relating variance to scale suggest that σ^2 can increase without limit as extent tends toward infinity and f_{min} approaches zero (Figure 21.2B). But there are always limits. At very large scales, $1/f$-noise scaling (with $\beta \geq 1$) *must* deviate from reality, and one should be careful to use a $1/f$-noise model only within appropriate bounds.

A.

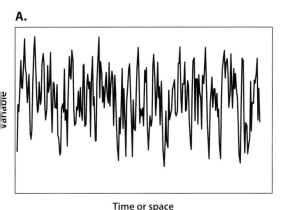

Time or space

B.

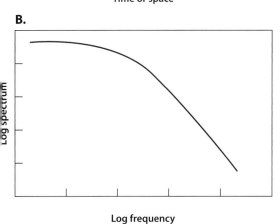

Log frequency

C.

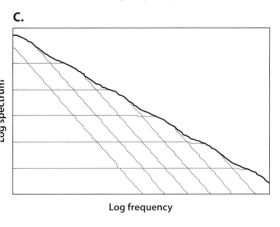

Log frequency

Figure 21.6 A. A first-order autoregressive process. B. The spectrum of the process shown in (A). C. The sum of spectra from multiple individual autoregressive processes can mimic the appearance of a $1/f$ spectrum (redrawn from Inchausti and Halley (2004)).

Indeed, some $1/f$-noise spectra have been shown to deviate from reality at large scales, raising the possibility that there might be a better general descriptor for ecological spectra. One intriguing possibility is the spectrum generated by an autoregressive process. Consider, for instance, a bed of clams in a unidirectional—but turbulent—current. Each clam ingests water, filters some fraction of the food from it, and squirts the remainder back into flow. The amount of food available to a clam in the middle of the bed depends on the unfiltered effluent flowing from its upstream neighbor plus the chance turbulent delivery of food from the mainstream. Described mathematically, the flux $\mathcal{F}$ of food to clam number x in line from the bed's upstream edge is

$$\mathcal{F}(x) = k_{uf}\mathcal{F}(x-1) + \epsilon.$$
$$(21.21)$$

Here $\mathcal{F}(x-1)$ is the amount of food ingested by the clam immediately upstream of clam x, k_{uf} is the fraction left unfiltered by the upstream clam, and ϵ is the random amount of food delivered by turbulence. This equation describes a first-order autoregressive (AR1) process (Priestley 1981). (It is first order because $\mathcal{F}(x)$ depends only on chance and $\mathcal{F}(x-1)$.) Typical results of this process are shown in Figure 21.6A.

The spectrum of an AR1 process is shown in Figure 21.6B. At high frequencies, it resembles a $1/f$-noise spectrum, but the AR1 spectrum flattens out at low frequencies, just as we suppose large-β $1/f$-noise spectra must. Indeed, the distinction between AR1 and $1/f$-noise processes disappears under certain circumstances. When $\alpha = 1$, equation 21.21 describes a random walk, which has a spectrum identical to that for a $1/f$-noise process with $\beta = 2$. When $\alpha = 0$, equation 21.21 describes white noise, a $1/f$-noise process with $\beta = 0$.

It has been proposed that $1/f$-noise spectra in nature are due to the superposition of multiple AR1 spectra, each acting over a different characteristic scale (Figure 21.6C). We will have to wait and see whether this proposition pans out as new data become available.

Chapter 22

Biology of Extremes

In Chapters 3 and 19, we explored the consequences of Jensen's inequality and developed a method—scale transition theory—for estimating the average response to variable conditions. We saw, for example, that drag on a coral averaged over a few minutes increases with any variation in flow speed and that when prey are scarce, their spatial variability can augment the average rate at which they are captured and eaten. By concentrating on the average, however, we have given short shrift to the extremes. In some cases, rather than simply causing shifts in the mean, short-term fluctuations can be drastic enough to kill individuals and disrupt whole populations. In these cases, the average is immaterial. Furthermore, short-term extremes can have long lasting effects: a single successful spawning year can innundate a kelp forest with sea-urchin recruits, and the resulting increase in hebivory can affect the community for years; a single storm can level a forest with consequences that persist for decades; an earthquake can change the path of a river with effects that last for centuries. As a result, extreme events can play critical roles in ecology and evolution (e.g., Gaines and Denny 1993; Katz et al. 2005). They can create bottlenecks in genetic variability (e.g., Gallardo et al. 1995) and can shift communities between alternate stable states (e.g., Barkai and McQuaid 1988; Petraitis et al. 2009). In extraordinary cases, such as the meteor strike that ended the dinosaurs' reign, extreme events can change the entire course of life on Earth.

As important as they are, extreme events have not received due attention from ecologists. This shortfall stems not from a lack of interest or recognition, but rather from the fact that extreme events are by definition rare and, therefore, difficult to document and study. It would take a bold ecologist to study a phenomenon that might—or might not—happen once every few decades, much less one that might happen only once every few centuries.

The study of extreme events also suffers from statistical tradition—application of inferential statistics in biology has focused on the analysis of averages rather than extremes. For example, in two widely used general texts on biological statistics (Sokal and Rohlf 2012; Zar 1999), the term *extreme* isn't even listed in the index.

In this chapter, we explore the nature of extreme ecological events and see how response functions can be used to predict both their likelihood and their consequences. This exploration comes in two parts. First, we delve into the *statistics of extremes*. We will learn how it is possible to extrapolate from a relatively small number of empirical measurements to predict the magnitude and likelihood of extreme events. These extrapolations allow us to accurately quantify events so rare that they may never have been directly observed.

There are systems, however, in which even the small requisite number of empirical measurements are not available. For instance, on rare occasions extreme temperatures can bake seaweeds and animals on rocky shores, causing widespread havoc. But lacking a multidecadal record of the intertidal environment, we don't have sufficient information to directly assess the likelihood of these events. In the second half of this chapter, I introduce a method—the *environmental bootstrap*—that can often solve this type of problem. Some extreme events depend not on a single factor, but rather on the chance alignment of multiple, individually benign factors. On rocky shores, for example, low tides, high air temperatures, and bright sunlight are each common occurrences. However, only on the rare occasion when they happen at the same time can they do any harm. In cases such as this, the environmental bootstrap allows ecologists to combine mechanistic response functions with the laws of probability to maximally leverage the data available and thereby extend their reach into the statistics of extremes.

1   STATISTICS OF EXTREMES

As a prelude to the statistics of extremes, let's review the statistics with which you are likely to be more familiar—statistics that deal with averages and variances.

Suppose we want to quantify the average noontime body temperature in a population of lizards. Ideally, we would measure the temperature of every individual in the population, but that is seldom possible—typical populations consist of far too many individuals to be measured by any practical means. As an alternative, we select a sample of n individuals at random and take their temperatures. From these data, we calculate the *sample average*, $\bar{x}$, an estimate of the population's average.

However, we would be naive to believe that $\bar{x}$ is an exact measure of the population's true mean temperature. If we were to measure another random sample of individuals, we would in all likelihood find a different mean. A third sample would yield yet another estimate. In short, when forced to work from samples, there is always some uncertainty in our estimate of the true population average.

Fortunately, we can precisely characterize this uncertainty, allowing us to make important inferences about the population from which our samples are drawn. If we were to sample the population a large number of times and construct a frequency diagram of the resulting sample means, the histogram would have a bell-shaped profile. As the number of samples increases, the histogram asymptotically approaches the *normal* (or Gaussian) distribution (Figure 22.1A; see Supplement 22.1 for the details).[1] The peak of the distribution occurs at the function's mean—it is our best guess of the population average—and the breadth of the curve is a measure of our uncertainty in this estimate. The curve's breadth—quantified by its standard deviation

---

[1] When n is small, the distribution of sample means is described by Student's t distribution, the basis for the standard t test. The t distribution asymptotes to the normal distribution as n increases.

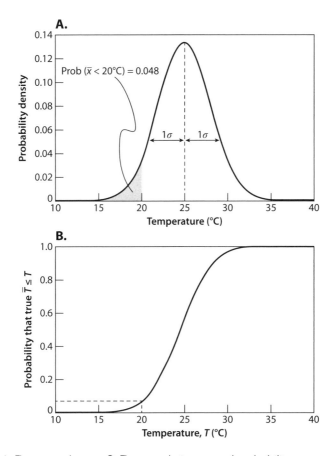

A.

Prob ($\bar{x} < 20°C$) = 0.048

1σ 1σ

Probability density

Temperature (°C)

B.

Probability that true $\bar{T} \leq T$

Temperature, T (°C)

Figure 22.1 A. The normal curve. **B.** The cumulative normal probability curve.

($\sigma_{\bar{x}}$)—is set jointly by n, the number of individuals measured in each sample, and σ_x^2, the variance among these measurements:

$$\sigma_{\bar{x}} = \sqrt{\frac{\sigma_x^2}{n}}. \tag{22.1}$$

The subscript $\bar{x}$ is a reminder that $\sigma_{\bar{x}}$ is a measure of the uncertainty in our estimate of the population mean.[2] The more variable temperatures are within a sample, the wider the normal curve and the greater our uncertainty. Conversely, the more individual measurements we include in each sample, the narrower the distribution of means, and the greater our trust in our estimate of the population mean.

The beauty of the normal curve lies in its generality. Note that in the discussion to this point we have said nothing about the distribution of body temperatures in the lizard population. In fact, within very broad limits, it doesn't matter what underlying

---

[2] The standard deviation of sample means is also known as the *standard error of the mean* or simply the *standard error*. Both terms are misleading, implying as they do that there is something wrong with the mean and its measurement, so I avoid them here.

distribution we sample, the distribution of means will always be asymptotic to the normal. This important fact is known as the *central limit theorem*, and it bears repeating. No matter how strange or disorderly the distribution of values within a population, the averages of random samples drawn from that population behave in an orderly fashion—they approximate a normal distribution.

The asymptotic behavior of means can save a huge amount of work. Because the central limit theorem tells us that sample means are normally distributed, we aren't compelled to demonstrate normality every time we sample a population. Instead, we can take a single sample of n individuals (which allows us to calculate $\bar{x}$ and (via equation 22.1) $\sigma_{\bar{x}}$, and from these values we can estimate the normal distribution to which the data are asymptotic. We can in turn use this asymptotic normal distribution to tell us how confident we are in our description of the population mean.

For example, in a sample taken from our lizard population, we might find that $\bar{x}$ is 25°C and $\sigma_{\bar{x}}$ is 3°C. What is the chance that the true population mean is less than 20°C? To answer this question, we calculate the area under the normal curve to the left of 20°C (Figure 22.1A). In this case, because the variation among sample means is relatively small, there is only a 4.8% chance that the true mean is this far below our estimate.

This procedure can be repeated to calculate the probability associated with any other value of the true mean, with the results shown in Figure 22.1B. This *cumulative normal probability distribution* quantifies the probability that the true mean temperature is less than a chosen value on the x-axis. Although less familiar than the bell-shaped curve from which it is derived, the cumulative distribution is more user friendly. With the cumulative distribution in hand, one can directly quantify probabilities associated with the normal curve. For example, to determine the probability that the true mean is less than or equal to 20°C, no integration is required; one simply draws a line up from 20°C on the abcissa and notes the corresponding probability value (4.8%).

In summary, because the central limit theorem assures us that sample means are asymptotic to a normal distribution, we can make useful, quantitative statements about the characteristics of the population we have sampled.

The discussion to this point should be familiar to anyone who has had an introductory course in inferential statistics. I have included it here both to refresh your memory and to highlight the importance of the central limit theorem. We now make an important connection: *Just as the means of samples have a defined distribution, so do the extremes. And knowing the distribution of extremes allows us to make quantitative predictions about how extremes affect organisms.*

Suppose that we want to know not the average body temperature in our population of lizards, but rather the extreme body temperature. (For simplicity, I explore the case in which we care about maximum temperature, but similar logic could be applied to the minimum.) To determine exactly the highest temperature in the population, we would have to measure every lizard. Instead, we repeat our sampling (selecting n lizards at random), but rather than recording the mean value of each sample, we instead record the highest value. Because we sample at random, it is unlikely that the extreme temperature of one sample is exactly the same as that of the next. Instead, after we have taken multiple samples, we end up with a distribution of extreme values.

As I alluded to before, there is structure to the distribution of these extremes.[3] We skip right over the probability density function of extremes and move right to their cumulative probability distribution. Given a very large number of samples, the distribution of extremes can be described by the *generalized extreme value* (GEV) distribution (Coles, 2001):[4]

$$P_{max}(x) = \exp \left\{ -\left[1 + \xi \frac{x - \zeta}{\psi} \right]^{-1/\xi} \right\}. \tag{22.6}$$

Here $P_{max}(x)$ is the probability that the population extreme, x_{max}, is less than or equal to x. For example, in Figure 22.2A, the probability is 0.88 that maximum temperature in our lizard population is less than or equal to 30°C. The higher the value of x, the more certain we are that the population extreme is less than x.

Unlike the normal distribution, which is determined by two parameters (the mean and standard deviation), the generalized extreme value distribution is governed by three parameters:[5] ζ, the distribution's mode; ψ, which sets the distribution's breadth; and ξ, which sets its shape. Examples of the effects of ζ, ψ, and ξ are shown in Figure 22.2B–D.

Equation 22.6 comes with the following provisos:

$$\text{if } \xi > 0, P_{max}(x) = 0 \text{ for } x \leq \zeta - \frac{\psi}{\xi},$$

$$\text{if } \xi < 0, P_{max}(x) = 1 \text{ for } x \geq \zeta - \frac{\psi}{\xi}. \tag{22.7}$$

This second proviso deserves particular attention. If $\xi < 0$, $P_{max}(x)$ reaches 1 when x is $\left| \frac{\psi}{\xi} \right|$ above the distribution's mode, and, in these cases, $\zeta - \frac{\psi}{\xi}$ is our best estimate of the population maximum. If $\xi > 0$, there is no well-defined limit to the distribution's maximum value.

---

[3] In a few systems, the distribution of values in the population is well-enough known that the distribution of extremes can be calculated directly. The height of ocean waves, for instance, closely approximates a Rayleigh distribution, from which the extreme height can be estimated (Denny and Gaines 2000). Cases such as this are rare in ecology, however.

[4] Gaines and Denny (1993) and Denny and Gaines (2000) use an alternative version of the generalized extreme value equation:

$$P(x) = \exp \left[-\left(\frac{\alpha - \beta x}{\alpha - \beta \epsilon} \right)^{1/\beta} \right]. \tag{22.2}$$

This expression is equal to equation 22.6 with

$$\alpha = \psi - \xi \zeta, \tag{22.3}$$
$$\beta = -\xi, \tag{22.4}$$
$$\epsilon = \zeta. \tag{22.5}$$

[5] For those not thoroughly familiar with the Greeek alphabet, ζ is zeta, ψ is psi, and ξ is xi.

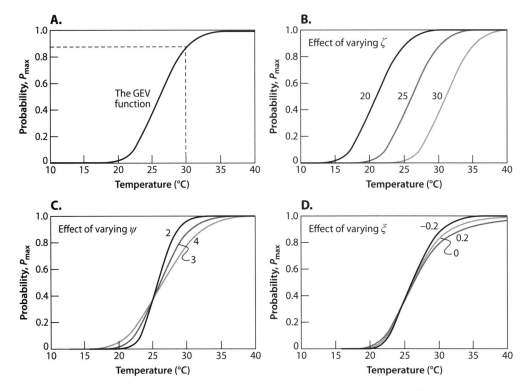

Figure 22.2 A. The cumulative probability curve of the generalized extreme value (GEV) function. The effect of varying ζ, ψ, and ξ are shown in panels **B**, **C**, and **D**, respectively.

Note that equation 22.6 is undefined when $\xi = 0$. However, by allowing ξ to approach zero, it can be shown that when $\xi = 0$,

$$P_{max}(x) = \exp\left\{-\exp\left[-\frac{x - \zeta}{\psi}\right]\right\}. \tag{22.8}$$

In practice, equation 22.8 is used when $|\xi| \leq 0.01$.

It is worth noting a fundamental difference between our estimations of a population's mean and its extreme. Given even a modest sampling, it is likely that the true population mean lies well within the range of values in our samples. By contrast, it is quite likely that the population extreme lies outside the sampled range. In short, finding the true mean is a process of interpolation; finding the true extreme is a process of extrapolation.

As noted earlier, when exploring a population's average, one can gain access to the normal distribution from a single sample, which provides the estimates of $\bar{x}$ and $\sigma_{\bar{x}}$ needed to specify the cumulative probability distribution. Access to the generalized extreme value distribution isn't quite so simple—multiple samples are required to estimate values for ζ, ψ, and ξ.

To calculate the GEV distribution, one begins by collecting N samples (each with n individual measurements) and noting the extreme value (x) of each. These N extremes are then used to construct a rough cumulative probability curve (the dots in Figure 22.3A) using the following procedure. Sample extremes are ranked from the lowest value to the highest, where the rank i of the lowest value is 1 and that of

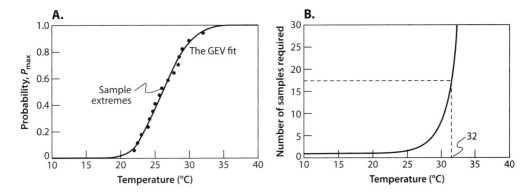

Figure 22.3 The GEV function can be fit to empirical data (A), allowing one to calculate the number of samples required to observe a given extreme (B).

Table 22.1   Hypothetical Temperature Data (see Figure 22.3A). Probability is calculated using equation 22.9.

Rank, i	Probability, P_{max}	Sample Extreme
1	0.059	22.0
2	0.118	22.5
3	0.176	23.0
4	0.235	24.0
5	0.294	24.2
6	0.353	24.7
7	0.412	25.0
8	0.471	25.7
9	0.529	26.0
10	0.588	27.0
11	0.647	27.8
12	0.706	28.2
13	0.765	28.4
14	0.824	29.0
15	0.882	30.0
16	0.941	32.0

the highest is N. The cumulative probability value for each sample extreme is then estimated:

$$P_{max}(x_i) \approx \frac{i}{1+N}. \qquad (22.9)$$

If two or more sample extremes have the same value, they are each assigned the average of their ranks. As an example of how this is done, the data for Figure 22.3A are shown in Table 22.1.

A nonlinear curve-fitting program is then used to calculate the values of ζ, ψ, and ξ that best fit equation 22.6 or 22.8 to this rough probability curve. (Most computer-based statistics packages have nonlinear curve-fitting modules.) The generalized extreme value distribution using these best-fit parameter values is our best estimate of the cumulative probability curve to which our data would asymptote if we were to take a very large number of samples (the solid line in Figure 22.3A). We can then proceed to make inferences from this asymptotic curve. For instance, if ξ is significantly less than 0, we estimate that the absolute maximum value in our population is $\left|\frac{\psi}{\xi}\right|$ above the mode.

Equation 22.6 describes extreme values in terms of their probability of occurrence. However, at times it can be more intuitive to think of extremes in terms of how many samples one would have to take to find an extreme of a given magnitude. For instance, let's assume that in our analysis of lizard body temperature the probability is 0.941 that in a sample of n lizards, the maximum temperature is less than or equal to 32°C; that is, $P_{max}(32) = 0.941$. Conversely, the probability that in a sample this size temperature is greater than 32°C is $P = 1 - P_{max} = 0.058$, or 1/17. With 1 chance in 17 of experiencing this high a temperature in a single sample, it can be shown that (not surprisingly) we would on average have to take 17 samples to encounter a body temperature this extreme (Figure 22.3B, Supplement 22.2). In general, the mean number of samples (each of a given size) needed to obtain a given extreme is (Coles 2001)

$$\overline{N}_r = \frac{1}{P} = \frac{1}{1 - P_{max}}. \tag{22.10}$$

The lower the probability of an event, the larger the number of samples required to observe it (Figure 22.3B). (Note that $\overline{N}_r$ depends on n, the number of individuals in each sample. The larger n is, the more likely it is to find an extreme of a given magnitude in that sample, and the smaller the number of samples needed (on average) to observe that extreme value.)

To this point, I have couched our discussion in terms of a population of individuals at one instant in time. More often, the statistics of extremes is applied to samples taken at repeated intervals through time. For example, if we desire to estimate the hottest air temperature we are likely to encounter, we could measure temperature continuously for a year, from which we could determine the annual maximum temperature. Repeating this process for 15 to 30 y would give us the multiple estimates of annual maximum temperature we need to construct a rough cumulative probability curve, to which we could then fit an asymptotic GEV function.

When dealing with temporal data, it is often more intuitive to translate information about probability into information about time. The logic is the same as that we used for the number of samples. For instance, let's assume that our samples of annual maximum air temperature tell us that the probability is 0.9 that the maximum temperature in a randomly chosen year is less than or equal to 38°C; that is, $P_{max}(38) = 0.9$. The probability that maximum temperature is greater than 38°C is thus $P = 1 - P_{max} = 0.1$. With a 10% chance of experiencing this extreme temperature in any given year, we would have to wait an average of 10 y before encountering an air temperature greater than 38°C. Thus, the *average return time*—the mean interval between imposition of a given extreme event—is

$$\overline{t}_r = \frac{1}{P} = \frac{1}{1 - P_{max}}, \tag{22.11}$$

where $\bar{t}_r$ has the same units as the sampling interval at which individual extreme values are measured; in this case, because we are talking about annual maxima, return time is measured in years. The lower the probability of an event, the longer its return time.

As an example of how the statistics of extremes can be used, consider the analysis of sea-surface temperature (SST) and its extremes. Steve Gaines and I investigated the fluctuations in SST at a site in central California, where temperatures had been recorded daily beginning in 1919, providing a lengthy record ideally suited to this type of analysis (Gaines and Denny 1993).

Our first task was to massage the raw data to make sure that they contained no trends or predictable oscillations that would affect our ability to discern extreme values. For example, average temperature gradually increased over the period we analyzed. As a consequence, an elevated temperature that would be rare in the 1920s (and therefore could qualify as an extreme) might be common in the 1990s (and therefore would not qualify). To circumvent this problem, we fit a regression line to the 73 y of our data, providing a mathematical description of how average temperature increased through time, and then for each measured daily temperature calculated ΔT, its deviation from this trend. By dealing with these temperature deviations (rather than absolute temperatures), we removed the trend's influence. A ΔT of, say, $10°C$ in 1920 could be treated the same as a $10°C$ deviation in 1990.

We handled seasonal fluctuations in a similar way. Ocean water is predictably warmer in summer than in winter, so a ΔT that is common in July (and therefore would not qualify as an extreme) might be rare in January (and therefore could qualify). To cope with this problem, we identified the maximum ΔT in each month (ΔT_{max}) for each of the 73 y in the time series. From these data we calculated the mean maximum temperature deviation, $\overline{\Delta T}_{max}$, for each month. We then recorded how far each individual monthly maximum ΔT deviated from the $\overline{\Delta T}_{max}$ for that month. These seasonally independent maximum temperature deviations ($\Delta T_{max,si}$) then formed the basis for our analysis. In the parlance of statistics, our transformations of the data rendered them stationary, a concept we encountered in Chapter 20. (Alternatively, the problem of trends and cycles can be handled by including these factors as covariates in the analysis of extremes. See Katz and colleagues (2005) or Coles (2001) for details.)

The GEV we generated using the transformed data showed that the probability is 0.90 that $\Delta T_{max,si}$ is less than $1.5°C$ in any given month. In other words, there is only a 10% chance that in a given month maximum temperature will exceed the monthly $\overline{\Delta T}_{max}$ by more than $1.5°C$. By contrast, the probability is 0.99 that $\Delta T_{max,si}$ is less than $2.5°C$, and thus there is only a 1% chance that this value will be exceeded. In short, the larger the deviation in temperature, the less likely it is to be encountered.

To give these deviations ecological meaning, they need to be placed in a physiological context. For example, giant kelps prefer cold temperatures. Let's suppose (hypothetically) that a single excursion to an absolute temperature of $17.5°C$ is sufficient to kill an entire kelp forest. If this were to happen, the effects would cascade through the entire nearshore community. Currently the warmest month in the year at our site is September, with an average maximum temperature of $15°C$, $2.5°$ below kelp's danger level. The results just cited tell us that there is currently only a 1% chance that next September's sea-surface temperature will be sufficiently extreme to decimate the forest. The situation will be more dire, however, if ocean warming raises the average monthly maximum temperature. If in the year 2100 average September maximum temperature is 16°C (1.5° below the danger level) the chance that the extreme September temperature will be deadly increase tenfold to 10%. (This example assumes that the parameters of the GEV distribution will be the same in 2100 as they are today, an assumption we will discuss later in the chapter.)

2   EXTREME CAVEATS

The statistics of extremes provides a valuable alternative perspective to traditional inferential statistics. By offering insight into the probability of extremes, this approach allows one to speculate productively about rare events. However, like any form of statistics, the statistics of extremes can be abused.

2.1   How Extreme Is Extreme?

Equations 22.6 and 22.8 provide reliable estimates of cumulative probability only when applied to extreme values. Applying these equations to run-of-the-mill data will lead to spurious conclusions. But how does one differentiate between extremes and garden variety fluctuations? Although there is no hard and fast line between the two, the basic criterion is that extremes must be rare. For example, every day has a maximum air temperature, but just because it is the maximum of a single day doesn't make it rare. The maximum air temperature measured in a month—the most extreme value of 28–31 daily maxima—might qualify as extreme. Annual maximum temperature—the most extreme of 365 daily values—has even better credentials.

There is a trade-off, however. When provided with a certain number of measurements, the larger the sample or the longer the intervals into which one chooses to divide those data, the fewer extremes one records. In turn, the fewer extremes in the analysis, the less certain one is in the estimates of ζ, ψ, and ξ and, therefore, the less certain one is of the calculated estimates of population extremes or return times. In an analysis, it is often necessary to experiment with different sample or interval sizes to determine which is most reliable (Gaines and Denny 1993).

An alternative to the GEV approach offers a mechanism both to maximize the number of extremes available and to minimize the subjectivity involved in identifying them. In this procedure, data are chosen relative to a set threshold rather than as the extreme in each interval or sample. This distinction is illustrated in Figure 22.4. In panel A, I have applied the sampling technique used so far. The fluctuating signal is divided into intervals, and the maximum value in each interval is designated as an extreme (shown by the asterisks). In the alternative approach (panel B), the same data are shown, but rather than dividing the data into intervals, I have set a particular threshold k_{th} and defined each peak that exceeds that threshold as an extreme.

This second method of identifying extremes is tied to a different cumulative probability model. If k_{th} is sufficiently high and $\xi \neq 0$ (Coles 2001):

$$P_{\max}(x - k_{th}) = 1 - \left\{ 1 + \xi \left[\frac{x - k_{th}}{\psi + \xi \left(k_{th} - \zeta \right)} \right] \right\}^{-1/\xi} \tag{22.12}$$

Equation 22.12 is valid provided that

$$x - k_{th} > 0 \tag{22.13}$$

A.

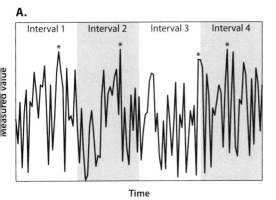

Interval 1   Interval 2   Interval 3   Interval 4

Measured value

Time

B.

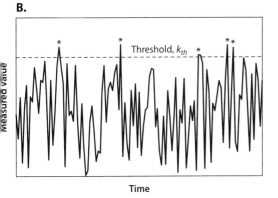

Threshold, k_{th}

Measured value

Time

Figure 22.4 Recording extreme data. **A.** The interval method: each interval has a single extreme. **B.** The points-over-threshold method: once a threshold (k_{th}) is defined, all points above it are treated as extremes.

and

$$1 + \xi \left[\frac{x - k_{th}}{\psi + \xi\,(k_{th} - \zeta)} \right] > 0.$$
(22.14)

This distribution—the *generalized Pareto distribution*—is undefined when $\xi = 0$. In this case, equation 22.12 becomes

$$P_{max}(x - k_{th}) =$$

$$1 - \exp \left[-\frac{x - k_{th}}{\psi + \xi\,(k_{th} - \zeta)} \right]$$
(22.15)

In practice, one sets k_{th} successively higher until the data are well described by equation 22.12 or 22.15. The minimum value of k_{th} that works thus sets a definable criterion for what one means by extreme.

This *peaks over threshold* (POT) approach has the added advantage of maximizing the data available. For example, in Figure 22.4A, two of the highest values fall in the same sampling interval. In the traditional approach to identifying extremes, only one of these values would make it into our data set, a waste of useful information. In the POT approach, both values are included.

2.2   Physical Reality

Another common abuse of the statistics of extremes is intimately tied to its major advantage. Both the GEV and POT distributions allow one to estimate the probability of encountering rare events. This provides a method for extrapolating beyond short-term or small-sample data to make predictions about long-term or population-wide consequences. That's the advantage. The drawback is that, because extreme events are rare, one is seldom in the position to test one's predictions directly. If your short-term data suggest that a given extreme should occur with an average return time of 75 y, you will be long gone before enough 75-y intervals have passed to provide a definitive test.

In the absence of a ready means for verification, it is easy to make rash predictions. For instance, in our investigation of maximum lizard temperatures, one might encounter a GEV distribution for which $\xi > 0$, implying that there is no well-defined upper limit to lizard body temperature. Taking this result at face value, we could conclude that, in a sufficiently large population, body temperature could by chance exceed the temperature needed for spontaneous combustion, and a lizard would occasionally go up in smoke. Although this is a mathematically correct interpretation of the GEV distribution, it is physically improbable. One should always apply common sense to predictions from the statistics of extremes.

2.3   Stationary Distributions

For the statistics of extremes to give reliable answers, the data being analyzed must be stationary. In our example of lizard body temperatures, for example, we assumed that we could repeatedly sample the same population. In practice, although the same individuals might be present from one sample to the next, in the time it takes to obtain the second sample, individuals' temperatures are likely to have changed. Thus, every time we take a sample we are picking from a different population of temperatures. To reliably estimate extreme temperature, we need to know that (even though the population changes from one sample to the next) its statistical characteristics—its mean, variance, and so forth—stay the same. Similarly, in order to explore the extremes of sea-surface temperature, we needed to make the data stationary by removing both a long-term trend and seasonal fluctuations.

The need for stationary distributions can put biological data at odds with the statistics of extremes. Considerable effort must often be expended to ensure that one's data meet these criteria. Methods to this end are outlined in Gaines and Denny (1993), Denny and Gaines (2000), Coles (2001), and Katz and colleagues (2005). For a formal discussion of statistical stationarity, consult Priestley (1981).

2.4   The Meaning of the Average

As we have defined it, the return time of an event is the average time one would have to wait between occurrences. It is important to note the word *average* in this definition. Just because an extreme event occurs with a return time of a century does *not* imply that the event occurs at precise intervals of 100 y. Instead, due to chance there is considerable variation in the interval between events, and this variation can result in interesting consequences.

Consider, for instance, the following scenario. If they are not killed by floods, creekside trees can live for x years before they die of old age. Let's suppose that the return time between floods is t_r years. If $t_r < x$ and floods occur like clockwork every t_r years, every tree dies in a flood. But if floods occur at random, some intervals might exceed the trees' lifetime—that is, some tree will die of old age—even if the average interval is less than x. For a given combination of life span and return time, what is the probability that a tree will die of old age?

If the probability of a flood happening in a year is p, the probability that a tree is *not* flooded in its first year is

$$P_{surv}(1) = (1 - p).\tag{22.16}$$

The probability that the tree isn't washed away in either the first or second year is

$$P_{surv}(2) = (1 - p) \times (1 - p) = (1 - p)^2.\tag{22.17}$$

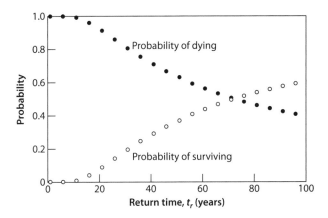

Figure 22.5 The annual probability of dying in a flood decreases as the return time of floods increases.

Extending this logic, we see that the probability that the tree survives x years is

$$P_{surv}(x) = (1 - p)^x. \tag{22.18}$$

Now, the definition of average return time $\bar{t}_r$ (equation 22.11) tells us that p is $1/\bar{t}_r$. Thus,

$$P_{surv}(x) = \left(1 - \frac{1}{\bar{t}_r}\right)^x. \tag{22.19}$$

This probability is graphed in Figure 22.5 as a function of return time for a lifetime (x) of 50 y. Even if the average return time is substantially shorter than a tree's life span, chance ensures that some trees are likely to survive to old age.

Of course this argument can be turned around. Even if trees' lifetime is substantially shorter than the return time of floods, there is a chance they will be washed away before they die of old age. The probability of dying by flood in x years is one minus the probability of surviving those years, so

$$P_{death}(x) = 1 - \left(1 - \frac{1}{\bar{t}_r}\right)^x. \tag{22.20}$$

This function is also shown in Figure 22.5, again for $x = 50$ y. Even when the average time between deluges is considerably longer than its lifetime, a tree stands a good chance of dying in a flood.

In summary, because return time is an average, its variation must be taken into account when considering its biological consequences. We developed this theme in depth in Chapters 19 and 21. In particular, the astute reader will have noted that because the likelihood of an extreme event is low, the probability of having a certain interval between events can be described by the Poisson interval distribution, as we discussed in Chapter 21. Consult Supplement 22.3 to see how P_{surv} and the Poisson interval distribution are connected.

3 AN INTERIM SYNOPSIS

Biology is often shaped by extremes rather than averages. However, because extreme events are inherently rare, special statistical techniques are required for their analysis. The statistics of extremes predicts that extreme values should be asymptotic to a particular cumulative probability distribution, and this assertion—an analogue to the central limit theorem—forms the basis for the estimation of probabilities and return times. Given sufficient data, one can make reliable predictions about the magnitude and probability of extreme events.

However, a quick rereading of that last sentence should clue you in to a potential problem. How does one acquire sufficient information about rare events? Therein lies the topic for the second half of this chapter.

4 THE ENVIRONMENTAL BOOTSTRAP

Some extreme events are *simple* in the statistical sense of the term. That is, for practical purposes, they either happen or they don't. Earthquakes, volcanic eruptions, and tsunamis are examples of environmental extremes that qualify as simple events. They happen, they are important, but they are exceedingly difficult to predict.

In part, this lack of predictability is due to our incomplete understanding of the mechanics of these phenomena. For example, we don't currently know enough about the mechanics of earthquakes to be able to predict when and where they will happen. But our inability is also due to the lack of sufficient historical data. Given a long-enough record of earthquakes on a given fault, we could determine whether they are purely random events (which occur with a certain fixed probability per unit time) or if there is some pattern to their occurrence. In either case, even without detailed knowledge of mechanism we could at least calculate the likelihood of occurrence. But lacking a detailed historical record (which for earthquakes would need to be thousands of years long), we can't reliably predict even the probability of these simple events.

By contrast, many extreme ecological events are due not to one single factor (e.g., the fault slipped), but rather to the simultaneous imposition of multiple, individually benign factors (Paine et al. 1998). For example, in the arid range land of New Mexico, fires and the occasional rainstorm are a normal part of the landscape, and individually neither poses a problem for stream insects. Much of the rainfall from a typical storm is soaked up by vegetation, which buffers the delivery of water to streams, and, because they live underwater, lotic insects are insulated from the effects of fires. But the individually benign effects of fire and rain turn nasty when—by chance—they arrive in quick succession. A range fire can temporarily reduce the cover of vegetation. If a storm arrives before the plants recover, rain isn't absorbed and instead floods the streams. The resulting scour can devastate populations of stream insects, which may take years to recover (Vieira et al. 2004). Thus, the chance alignment of otherwise normal conditions—fire and rain—can form an extreme event.

Rocky shores provide another example. The hydrodynamic forces imposed on intertidal mussels depend on three factors: ocean waviness, wave period, and tidal height. Each of these factors varies independent of the others, and seldom does any one pose a threat. Typically, when waves are high, they have a short period or the tide is out, and the swell loses its oomph before reaching the shore. High tides occur with a predictable pattern, but it is unusual for them to coincide with long-period, high waves. Every once in awhile, though, all three conditions occur by chance at the same time, and mussel beds are decimated.

The shores of central California were hammered by such an event on January 11, 2001. Wave height was within the top 1% of heights observed at this site, and wave period was well above average. What set this event apart, though, was that it coincided with the highest tide of the year. The rare combination of high, long-period waves and an exceptionally high tide wreaked havoc on the intertidal community. Boulders were tossed around like beach balls. (One particular boulder, with a mass of more than 2 tons, was lifted 5 m vertically out of a surge channel and washed 30 m upshore without hitting anything in between.) Large swathes of mussel beds were ripped out. Because mussels are the competitive dominant for space on the rock, the consequences of this compound extreme event cascaded through the midintertidal community and signs of the devastation from that event were still evident a decade later. Paine and his colleagues (1998) document a variety of other compound extreme events in ecology.

Because the components of these compound events (e.g., fire, rain, high waves) are common occurrences, they are open to observation in a way that earthquakes, volcanic eruptions, and tsunamis are not, providing an opportunity to predict ecological extremes. To capitalize on this opportunity, my colleagues and I devised a statistical technique—the environmental bootstrap—to predict the likelihood of compound events (Denny et al. 2009). This approach is best explained by working through our analysis for the thermal risk of intertidal limpets.

The first order of business was to obtain as long a record as possible of the environmental factors of interest. Fortunately, we had recorded air temperature, solar irradiance, and wind speed (the primary determinants of body temperature) every 10 min for 7 y. The data for air temperature are shown in Figure 22.6A.

The next task was to quantify the predictable aspects of these time series. This could be accomplished in several ways; we chose a relatively straightforward procedure. Having recorded each factor for 7 y, we had seven values for each time point within a year: seven values for midnight on January 1 (one value for each year), seven values for 12:10 a.m. on January 1, and so forth. For each of these sets of seven values we could calculate the average, $\bar{x}(t)$, and standard deviation, $\sigma_x(t)$, our best estimates of what to expect at that particular time within a year.[6] An example of these predictable patterns is shown in Figure 22.6B.

The next step was to remove these predictable aspects of the environment from the empirical measurements to provide a time series of the stochastic component, s, of each factor. For each empirical data point, $x(t)$, we substracted the corresponding mean and divided by the corresponding standard deviation:

$$s(t) = \frac{x(t) - \bar{x}(t)}{\sigma_x(t)}. \tag{22.21}$$

The final result was a set of *normalized deviations* that have no predictable daily or seasonal pattern (i.e., they are stationary; see Figure 22.6C).

---

[6] Actually, the procedure for calculating the average and standard deviation was a bit more complicated, but it used the same concept (see Denny et al. 2009).

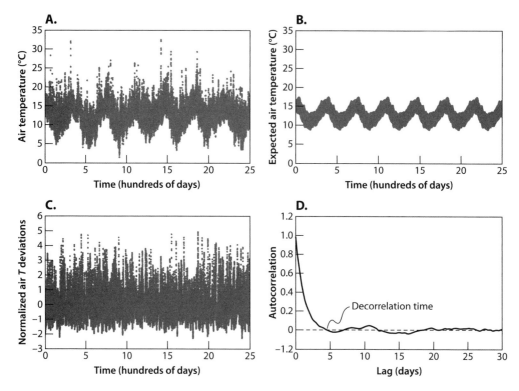

Figure 22.6 The environmental bootstrap takes empirical data (daily maximum air temperature, (**A**)) and separates it into a predictable signal (**B**) and the normalized deviations from that prediction (**C**). **D.** The autocorrelation function for air temperature.

Having thus separated the predictable aspects of the data from the stochastic, the next task was to recombine the two in a random fashion to simulate how one year might have played out if the environmental dice had rolled differently. To do this, we resampled the time series of normalized deviations. For example, we picked at random a short segment (n days) of the time series of normalized deviations of air temperature. The segment started at midnight on the first of the n days and ran through 11:50 p.m. on the last of the n days. We then matched each normalized deviation in our n-day segment with the the corresponding average and standard deviation in our record of predictable values. The first normalized deviation [$s(t_1)$] was matched with the average and standard deviation for midnight on January 1 [$\bar{x}(t_1)$, $\sigma_x(t_1)$], the second s [$s(t_2)$] with the $\bar{x}$ and σ_x for 12:10 a.m. [$\bar{x}(t_2)$, $\sigma_x(t_2)$], and so forth.

Each normalized deviation was then denormalized by multiplying it by the matched standard deviation and then adding the matched average. The net result was an n-day record of fluctuations in air temperature that was statistically similar to the measured record of temperatures but with randomly determined stochastic fluctuations (Figure 22.7A).

This process was then iterated. A second n-day segment of normalized air-temperature deviations was chosen at random, matched up with the average and standard deviations for the second n-day period in the year, and denormalized. A third n-day segment was chosen, and so forth, until an entire hypothetical year of air temperatures had been created. The same steps were then applied to the records of solar irradiance and wind speed (Figure 22.7B). Note that when a certain n-day segment of s was chosen, the data weren't removed. They remained to be potentially chosen again, a procedure known as *sampling with replacement.*

Two aspects of this resampling procedure bear notice. First, the choice of n-day segments is random, but consistent across factors. For example, if, at one iteration in creating our hypothetical record, the segment chosen for normalized air-temperature deviations began at day 312 of the time series, the segments for normalized solar irradiance and wind speed also began at day 312 (Figure 22.7B). Because segments were chosen in this coordinated fashion across factors, any cross correlation among factors was inherently retained in the resulting time series. For example, if, when air temperature was above average, wind speed also tended to be above average, this relationship was retained in the hypothetical year.

How did we pick n for our n-day segments? The choice of n depends on the correlation of normalized deviations not across factors, but rather across time within each factor. Normalized deviations close together in time tend to resemble each other. For instance, if air temperature is above average now, it is likely to be above average ten minutes from now, and it was likely above average 10 min ago. On the other hand, if air temperature is above average now, it would be difficult to to predict what it will be a week from now. In short, the *autocorrelation* between normalized deviations decreases with increasing separation (known as the *lag*) between points (Figure 22.6D). The lag time at which the autocorrelation becomes indistiguishable from zero is known as the *decorrelation time*, and it provided an appropriate benchmark for the choice of n. If n is equal to or greater than the decorrelation time for a given factor, an n-day segment of normalized deviations contains all the statistically predictive information available for that factor. Thus, by setting n to the longest decorrelation time among the relevant factors, we assured ourselves that we had included in our hypothetical time series all the necessary statistical information.

In summary, having separated the predictable aspects from the stochastic aspects of our data, we had a scheme for resampling the data to construct a plausible time series of how a year's worth of air temperatures, wind speeds, and solar irradiances could have occurred if chance had played out differently. Because to the naive observer this procedure seems to be making up new data, it bears some conceptual resemblance to pulling oneself up by one's bootstraps, and for this reason it and other such procedures are known as *bootstrapping*. This environmental bootstrap procedure is one form of a *moving block bootstrap* (Efron and Tibshurani, 1993).

So far, we have seen how to use the environmental bootstrap to calculate one hypothetical year of environmental data. But why stop there? If we could calculate 1 y, we could calculate 1000 or 10,000 y. In other words, we could use the bootstrap to create a large ensemble of realistic environmental scenarios, a resource seldom available in nature. This ensemble could then be used to investigate the probability of compound extreme events.

It is here that mechanistic response functions and the statistics of extremes came together. Using the principles we introduced in Chapters 11 and 12, we developed a heat-budget model in which solar irradiance, wind speed, and air temperature combine to predict limpet body temperature. We could use this model (coupled with a similar model that tells us when a limpet is submerged by the tide) to synthesize the data from an environmental ensemble of 1000 hypothetical years. The process is shown schematically in Figure 22.7C. At each time step in one of our hypothetical

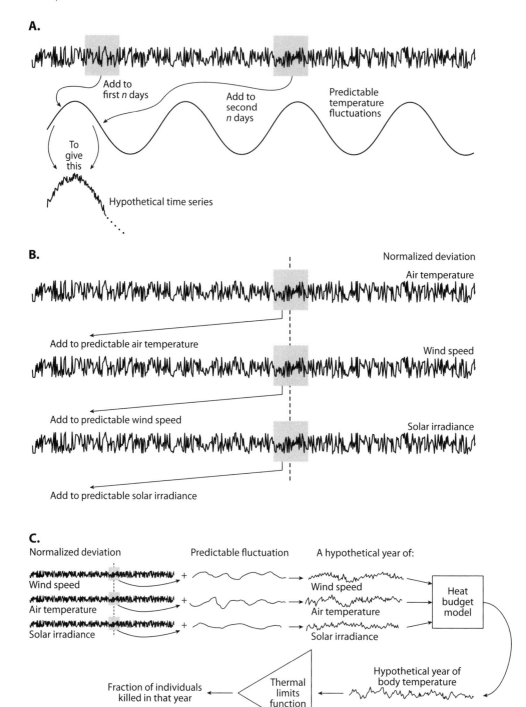

Figure 22.7 The caption appears at the bottom of the next page.

years, values of solar irradiance, wind speed, and air temperature were inserted into the heat-budget model to calculate a value for the body temperature of a limpet on a shore of given orientation (horizontal, or inclined toward the north, west, south or east). The annual maximum of these values was noted, and the procedure was repeated for each of the 1000 y in the ensemble. These 1000 extreme values could then be used to fit a GEV distribution to the hypothetical data, from which we could estimate the return time for any given value of annual maximum body temperature (Figure 22.8A). The probability of encountering such an event (and therefore its return time) varied with the orientation of the substratum to which the animal is attached. An understanding of thermal mechanics thus allowed for a realistic estimate of how often limpets are heated to high temperatures on different aspects of the shore.

Knowledge of temperature was not the final goal, however. To predict the biological consequences of temperature we needed to combine our estimates of maximum body temperature with information about limpets' capacity to cope with thermal stress. From laboratory experiments it was possible to measure the thermal survivorship function for a given species (*Lottia gigantea*, Figure 22.8B), which could then be used to calculate the impact of extreme temperatures.

In a typical year, maximum predicted body temperature was well within limpets' thermal limits, but occasionally low tide coincided with bright sunshine, high air temperatures, and low wind speed, and body temperatures reached lethal levels (Figure 22.8C). On shores facing north, lethal events were predicted to occur only once in several millenia, but, on shores facing south, such events likely occur, on average, approximately once per decade. Although lethal thermal events in *L. gigantea* have not been observed directly at our site, Miller and others (2009) used these calculated return times to hypothesize that *L. gigantea* should be more common high on the shore on pole-facing than equator-facing shores, and these predictions were borne out in the field.

This project illustrates the ability of the environmental bootstrap to simulate realistic, long-term variations in the physical environment, thus providing a valuable ingredient in the study of intertidal ecology. Note again that the utility of the bootstrap relies on our mechanistic understanding of heat transport. Without the heat-budget model, the results of the bootstrap would be just an ensemble of environmental data. The heat-budget model is needed to synthesize those data into body temperature, the biologically relevant factor.

I have explained the environmental bootstrap and its use in the context of extreme temperatures, but the approach can be applied to a wide variety of other phenomena: floods, hydrodynamic forces, species invasions, reproductive success, and so on. Whenever an extreme ecological event is the result of chance alignment of run-of-the-mill fluctuations—and the consequences of these fluctuations can be

---

Figure 22.7 A. Hypothetical time series are created by sampling random blocks of normalized deviations, denormalizing them and adding them to the predictable signal. **B**. Blocks are chosen randomly but are coordinated across factors. **C**. An example of the environmental bootstrap. Hypothetical time series are created for the pertinent variables and a heat-budget model is applied to these data to create a time series of body temperatures. This series is then coupled with a physiological response function to predict the fraction of individuals killed.

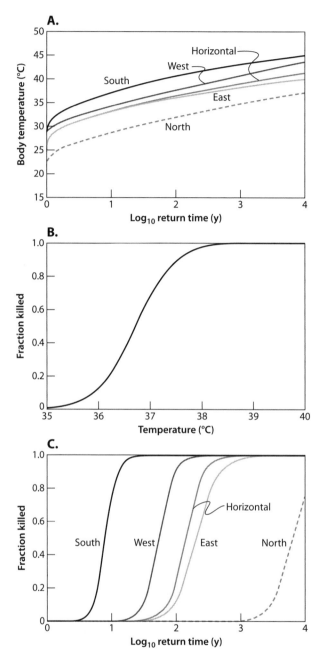

Figure 22.8 The environmental bootstrap applied to the owl limpet at a site in the Northern Hemisphere. **A.** The predicted maximum body temperature associated with each return time. **B.** The thermal survivorship curve for the limpet.**C.** Predicted fraction of limpets killed as a function of return time and shoreline orientation (redrawn from Denny et al. (2009)).

modeled—a procedure similar to that outlined here can provide an estimate of cumulative probability of extreme events.

5   EVOLUTION AND EXTREMES

In addition to its utility in providing data for the statistics of extremes, the ability of the environmental bootstrap to create realistic, long-term environmental scenarios opens the door to a mechanistic investigation of the effects of environmental variability on evolution. For example, Wes Dowd and I coupled bootstrapped simulations of the intertidal environment with the heat budget for *L. gigantea* to explore the evolution of thermal safety factors (Denny and Dowd 2012). This study was spurred by our observation that intertidal limpets are capable of surviving temperatures much higher than any individual is likely to experience in its lifetime. How could they evolve such substantial safety factors?

We began by supposing that each year a limpet acquires a set amount of food energy, which (after meeting basal needs) it can then devote either to produce gametes or to protect itself from temperature stress. In other words, we assume that there is an energetic cost to maintaining the molecular machinery responsible for tolerance and that energy spent on thermal defense subtracts from the energy available for reproduction.

We then supposed that the allocation of energy to thermal defense is under genetic control—each gene involved in thermal defense comes in two forms: a crank-it-up allele and a turn-it-down allele. The more crank-it-up alleles a limpet has, the more energy it spends on defense and the higher its thermal tolerance (Figure 22.9A). At the end of each year, each limpet produces a number of gametes proportional to its remaining energy and contributes those gametes to a common pool. Gametes are then chosen at random to replace individuals that die from old age or thermal stress. The greater the number of gametes produced by an individual, the greater the probability that its genes will make it into the next generation.

We then applied this genetic model to a population of limpets experiencing the environmental conditions specified by a 2000-y-long environmental bootstrap. Benign thermal years (of which there were many) favored the production of gametes by individuals genetically constituted to spend little on thermal defense. During these low-stress intervals the average thermal tolerance of the population gradually declined (Figure 22.9B). However, chance occasionally produced a hot year, killing many of these intolerant limpets and reducing the population to the few animals genetically constituted to spend heavily on thermal defense. These few tolerant individuals supplied the gamete pool for that year. In this fashion, each extreme event led to a population bottleneck that increased the population's average thermal tolerance.

Our simulations suggest that it is these rare extreme events that control a population's thermal tolerance. When averaged over ten runs of the simulation, we found that at the end of 2000 y limpets had evolved such that their mean thermal limit was 5°C to 7°C above the average annual maximum body temperature (Figure 22.9C). In other words, limpets evolved a thermal safety margin, which allowed the species to survive extreme conditions that few individuals in the population would ever experience. The thermal tolerance predicted by our model is very similar to that found in nature, tentative verification of our approach.

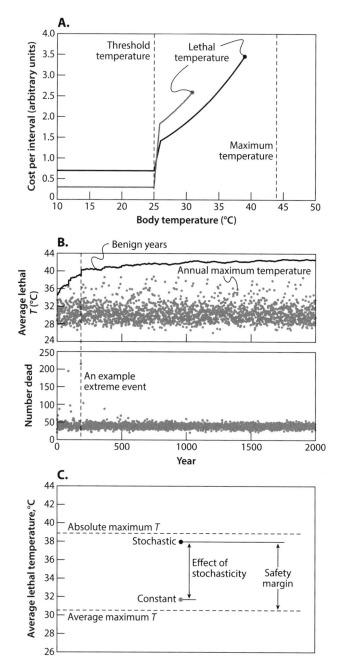

Figure 22.9 The evolution of thermal tolerance in limpets. **A.** A model for the cost of mounting a thermal defense (see the text). **B.** Results from one simulation of the limpet population's interaction with a hypothetical series of annual temperature maxima. **C.** The average thermal limit in a stochastic environment is higher than that in a constant environment with the same average temperature (redrawn from Denny and Dowd (2012)).

To test whether it was indeed the year-to-year variability in temperature that drives the evolution of thermal tolerance, we ran our model without environmental variability—each of the 2000 y in these alternative simulations had a maximum temperature equal to the average maximum in our previous experiment. Absent variability, the evolved safety margin was much smaller (Figure 22.9C).

The ability of the environmental bootstrap to simulate realistic, long-term variations in the physical environment—a mechanistic approach to predicting extremes—thus provides a valuable tool for the study of the evolution of thermal tolerance.

6   BOOTSTRAP CAVEATS

By randomly shuffling segments of normalized deviations, the environmental bootstrap simulates the manner in which chance determines how factors might come into coincidence. There are, however, limits to what the bootstrap can do.

6.1   Absolute Extremes

First, it is unlikely that the environmental bootstrap can simulate the absolute extreme of a given system. The procedure operates only on the data provided, and it is unlikely that any short-term time series contains the long-term extreme value of any factor. Furthermore, because normalized deviations are resampled so as to maintain alignment across factors, random resampling cannot bring the extreme values of different factors into coincidence. Only if those maxima occurred at exactly the same time in the original time series can they ever line up in the bootstrap.

These limitations would be a problem if it were the extreme values of the individual factors that caused a compound extreme event. But the intriguing nature of compound extremes is that they are not the result of an extreme in any single factor but rather the result of the savage alignment of normal events. It is much more likely that moderately deviant values of several factors will come into alignment to produce an extreme event than that an equivalent event can be caused by the extreme deviation in any single factor (see Denny et al. 2009).

6.2   Simulations Aren't Forecasts

Because the environmental bootstrap can calculate multiple years of simulated environmental data, it is tempting to view that capability as a way of forecasting the future. But, as I have described the procedure so far, the hypothetical years of data produced by the bootstrap are not forecasts. In creating an ensemble of 1000 individual years of resampled data, we are simply replaying the same year over and over again, the only difference being the roll of the dice. Because we always sample the same stationary series of normalized deviations, we exclude any change in the statistical nature of stochastic fluctuation. For example, as global climate changes in response to the accumulation of greenhouse gases, both the average and standard deviation of air temperature are expected to increase. These shifts are excluded from the bootstrap as I have described it so far, and thus the results of the bootstrap do not provide accurate forecasts.

That being said, changes in both average and variation could be included in a bootstrap model. To adjust for a temporal change in average conditions, one could add any trend to the predictable portion of the data (Gaines and Denny 1993). Similarly, one could adjust the averaged variances to reflect any predicted temporal changes. With these adjustments, the environmental bootstrap can be used to create replicate time series of a changing environment (see for example, Denny and Dowd 2012). Of course, this ensemble of forecasts is only as accurate as the predictions of the underlying temporal changes and relies on the difficult-to-validate assumption that the parameters of the GEV or POT distribution do not shift through time.

7   CONCEPTS AND CONCLUSIONS

The statistics of extremes allows us to extrapolate from small-sample or short-term data to predict extreme values for whole populations and deep time. Use of the statistics of extremes can be hampered, however, by the time consuming and arduous task of accumulating enough empirical data. For ecologically extreme events caused by the coincidence of normal factors, the task of gathering information can be expedited by the environmental bootstrap. By separating short-term environmental information into its average and stochastic components, and then recombining these components at random, the bootstrap allows one to calculate an ensemble of extreme values for use in the statistics of extremes. When appropriately combined with mechanistic response functions, these bootstrap simulations open new avenues for the exploration of evolution and the dynamics of populations and communities.

Chapter 23

Pattern and Self-Organization

As the last four chapters have documented, spatial and temporal variation in the environment can affect when and where plants and animals can survive, thereby producing patterns in ecological communities. But neither spatial nor temporal variation is necessary for pattern formation. Even in apparently homogeneous environments biological patterns emerge. In uniformly arid scrublands, for instance, plants arrange themselves into striking waves, spirals and labyrinthine clumps (e.g., Klausmeier 1999; Couteron and Lejeune 2001; Reitkerk et al. 2002), and bud moth populations travel in waves through uniform stretches of forest (Björnstad et al. 2002). Similar patterns can be found in nearshore marine habitats. On the blandly uniform mud- and sandflats of the Dutch coast, juvenile mussels aggregate in orderly rows and clumps (van de Koppel et al. 2005; de Jager et al. 2011). In contrast, on the wave-swept shores of the Pacific Northwest, mussel beds form a patchwork of densely packed individuals interspersed with gaps of open space. In this case, the pattern is not in the arrangement of clumps and gaps, but rather in the gaps' size distribution, which can be accurately described by a power law (Guichard et al. 2003).

The surprising presence of pattern in environments that—without biology—would be homogeneous raises several questions. What are the mechanisms by which plants and animals organize themselves in the absence of environmental cues? If we understand these mechanisms, can we predict when and where a given pattern will occur? Is there some biological advantage to pattern formation? The effort to answer these questions has been led by theoretical ecologists drawing on insights from developmental biology and physics (Solé and Bascompte 2006). A thorough treatment of the theory of self-organization is well beyond the scope of this text, but it will be worthwhile to explore two mechanisms of pattern formation that demonstrate how theory plays out in practice. In particular, these examples highlight the potential for mechanistic response functions to inform the prediction of pattern.

1　FACILITATION-INHIBITION

In 1952, Alan Turing[1] outlined a theory of organismal development in which facilitation and inhibition interact at different spatial scales to produce wave- or spot-like patterns. In a nutshell, Turing envisioned a system in which cells in a developing plant or animal respond to two chemical cues—a facilitator and an inhibitor—each of which can diffuse through the organism. When facilitator is present, cells respond by producing more facilitator and by expressing some morphological characteristic—they might turn blue, for instance. Since the production of facilitator leads to the production of more facilitator, facilitation forms a positive feedback loop. However, when cells produce facilitator, they also produce inhibitor, which (if sufficiently concentrated) reverses the effect of the facilitator—inhibitor decreases the production of facilitator and returns cells to their colorless form.

Positive feedback from facilitator and negative feedback from inhibitor prime the organism for pattern formation, but one more element is required—the inhibitor must diffuse faster than the facilitator. When this is the case, the effect of the facilitator remains local, while that of the inhibitor spreads to larger spatial scales, and it is this difference in the scale of influence that results in pattern formation.

Consider, for instance, a uniform, planar array of colorless cells. The action begins when a few cells (chosen at random) begin to produce both facilitator and inhibitor. The facilitator diffuses slower than the inhibitor, so in the vicinity of these random cells the concentration of facilitator is higher than that of inhibitor. Cells respond locally by turning blue, and the additional facilitator they produce influences their neighbors to do likewise. But at a larger scale, the rapid diffusion of inhibitor keeps distant cells from producing facilitator, stopping positive feedback before it can get started. Depending on the relative rates of diffusion for facilitator and inhibitor, this tug-of-war between positive and negative feedback can result in a wide variety of patterns: dots, waves, spirals, and expanding rings.

Although Turing developed his theory in the context of organismal development, the same basic principles apply universally (Rietkerk and van de Koppel 2008; van de Koppel et al. 2015). Any system in which there is local facilitation and distant inhibition can spontaneously produce pattern, and the mussel beds alluded to above provide a convenient model for theory and experimentation.

1.1　Patterns in Mussel Beds

Tidal flats of mud or sand provide a homogeneous, blank slate on which plants and animals can arrange themselves. Aside from the planar substratum, the relevant characteristics of this habitat are its benign tidal flows ($< 1 \text{ m} \cdot \text{s}^{-1}$) and the suspended particles the water carries—the flow of water brings food (primarily phytoplankton) to filter-feeding invertebrates. Despite the habitat's uniformity, young mussels (*Mytilus edulis*) form distinctive rows or clusters (Figure 23.1A).

Van de Koppel et al. (2005, 2008) proposed that, as for cells in Turing's hypothetical models, the large-scale pattern of mussels on tidal flats is due to local facilitation and

---

[1] Turing (1912–54) was a British mathematician and cryptanalyst who made major contributions to computer science.

A.

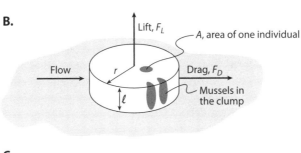

B.

Lift, F_L

A, area of one individual

Flow

r

ℓ

Drag, F_D

Mussels in the clump

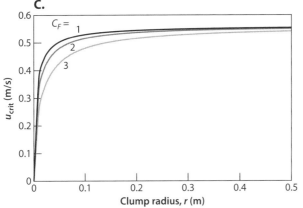

C.

$C_F =$ 1

2

3

u_{crit} (m/s)

0.6

0.5

0.4

0.3

0.2

0.1

0

0   0.1   0.2   0.3   0.4   0.5

Clump radius, r (m)

Figure 23.1 Self organization in a mudflat mussel bed. **A.** Mussels form distinct clumps (photo courtesy of J. Widdows FRPS). **B.** A model for the forces acting on a mussel clump (see the text). **C.** Critical velocity increases with clump radius, with an asymptote of $0.57 \, \mathrm{m \cdot s^{-1}}$.

distant inhibition. In this case, facilitation involves the ability to resist hydrodynamic forces. When mussels aggregate and adhere to each other, the likelihood is reduced that they will be dislodged by drag. Thus, interindividual adhesion acts as a local facilitator—the larger the aggregation of individuals, the more resistant they are to hydrodynamic disturbance. By contrast, the depletion of food acts as a large-scale inhibitor. As water flows over a clump of feeding mussels, phytoplankton are filtered out of the boundary layer. The greater the distance from a clump's leading edge, the less food remains. As a result, it is disadvantageous to be in the middle of too large a clump, all the available food having been eaten by mussels on the periphery. Unlike the advantage of sticking to one's neighbors—which acts at the scale of the individual— the disadvantages accruing from the ingestion of phytoplankton act at the scale of the group.

In response to this system of small-scale facilitation and large-scale inhibition, a pattern is formed as mussels move about. Solitary mussels travel until they find other mussels to attach to, thereby forming clumps. If an aggregation gets too big, hungry mussels jump ship and head off to find a smaller aggregation.

Van de Koppel et al. (2008) demonstrated the efficacy of this explanation through a combination of laboratory experiments, theoretical predictions, and field observations. In the laboratory, they measured the behavioral response of mussels to the size of clump they were in—mussels in small clumps tend to stay put; mussels in large clumps tend to disperse. Based on this behavior, they formulated a computer model to predict what patterns would be found under field conditions, and these predictions closely matched those observed in nature. In the field they then supplied extra food to mussels in the center of clusters and noted that clump size increased, demonstrating that it is food, rather than cluster size itself, that elicits mussel's dispersal behavior.

In addition, van de Koppel and his colleagues moved beyond the mechanics of how patterns are formed in mussel beds to address the evolutionary question of *why* these patterns are formed. They measured the growth rates of individuals in the field and his colleague found that overall productivity of the bed (the increase in mussel biomass per area per time) was higher when mussels clustered than it was for a continuous bed, in which many mussels are likely to be underfed. Indeed, the growth rate of mussels in clumps was no different from that of solitary mussels. The decreased growth rate of mussels in uniform beds provides a potential selective factor for the evolution of pattern-forming behavior.

Taking this analysis even further, de Jager and colleagues (2011) investigated how mussels might have evolved the behavior necessary to form clusters. In the laboratory, they scattered individual mussels on a uniform horizontal surface and videotaped them as they moved. They found that mussels perform Lévy walks, random walks in which step lengths (ℓ) vary. Most steps are short, but occasionally an individual takes one long step in a random direction, leading to the sort of superdiffusion we discussed briefly in Chapter 4. In Lévy walks, the distribution of step lengths varies according to a power law:

$$P(\ell) \propto \ell^{-k_L}. \tag{23.1}$$

Here $P(\ell)$ is the probability of taking a step of length ℓ, and k_L typically varies between 0 and 3. When $k_L = 0$, animals are likely to head off in one random direction and thus may well miss finding another mussel. When $k_L = 3$, the walk approximates a traditional diffusive process, which allows the mussel to thoroughly investigate a small area but reduces the overall area searched. Lévy walks with $k_L \approx 2$ are an effective compromise—thorough enough not to miss much and directed enough to search a large area. The mussels examined by de Jager and her colleagues performed Lévy walks with $k_L = 2$, suggesting that evolution has led them to an optimal behavior for effectively forming spatial pattern.

In their examinations of mussel dynamics, van de Koppel and others (2008) and de Jager and others (2011) took a behavioral approach to pattern formation—they assumed that conditions in the environment were such that interindividual adhesion was sufficient to resist hydrodynamic forces and boundary-layer depletion was sufficient to inhibit growth and asked what behavior led to clustering. Alternatively, one could take a physical approach to the phenomenon. Let's assume that mussels have the behavioral repertoire necessary to form clusters. Under what environmental conditions is pattern formation advantageous? Simpler still, under what conditions is pattern formation possible?

First, let's have a look at the mechanics of facilitation. Consider the hypothetical circular clump of mussels shown in Figure 23.1B. Each mussel is ℓ long and takes up area A in the clump, so in a clump of radius r, there are n mussels:

$$n \approx \frac{\pi r^2}{A}. \tag{23.2}$$

Byssal threads are of little use on soft substrata, so the primary force holding mussels in place against drag is their frictional resistance to being pushed downstream.

Frictional resistance (F_F) depends on the net weight of the clump—the more forcefully the clump is pulled down by gravity, the more force required to dislodge it sideways. If each individual has a weight in water of F_g newtons, the total weight of a circular clump is nF_g. However, any lift (F_L) acting on the clump tends to offset the mussels' weight, so the clump's net weight in flow is

$$F_{net} = nF_g - F_L. \tag{23.3}$$

Multiplying net weight by the coefficient of static friction, C_F, we can calculate the clump's frictional resistance to drag:

$$F_F = C_F \left(nF_g - F_L \right) \tag{23.4}$$

$$= C_F \left(\frac{\pi r^2 F_g}{A} - F_L \right). \tag{23.5}$$

The next step is to model the hydrodynamic forces acting on the clump. Water flowing past the clump exerts a drag on the clump's projected area ($2r\ell$),

$$F_D = \rho_f u^2 C_D r\ell, \tag{23.6}$$

and a lift on the clump's planform area (πr^2):

$$F_L = \frac{1}{2}\rho_f u^2 C_L \pi r^2. \tag{23.7}$$

To maintain its position on the seafloor, the clump must resist these forces. The clump is dislodged when F_D exceeds F_F, that is, when

$$\rho_f u^2 C_D r\ell > C_F \left(\frac{\pi r^2 F_g}{A} - \frac{1}{2}\rho_f u^2 C_L \pi r^2 \right). \tag{23.8}$$

Given this relationship, we can solve for the critical velocity at which a clump is dislodged:

$$u_{\text{crit}} = \sqrt{\frac{\pi C_F F_g r}{\rho_f A \left(C_D \ell + \frac{1}{2} C_F C_L \pi r\right)}}. \tag{23.9}$$

As a clump's radius increases, u_c has an asymptote

$$u_{\text{crit}} = \sqrt{\frac{2 F_g}{\rho_f A C_L}}. \tag{23.10}$$

Note that this asymptote is independent of C_F.

For mussels of the size found in clumps, $\ell = 5$ cm, $A = 3.8 \times 10^{-4}$ m^2, $F_g = 0.055$ N, $C_L = 0.88$ (Denny 1987), and C_D is approximately 1. Using these values, we can calculate u_{crit} as a function of clump radius (Figure 23.1C). Here, u_{crit} increases rapidly with clump radius as required for van de Koppel's model but has an asymptote of 0.57 m · s^{-1}. In other words, clumping is a practical means of avoiding dislodgment but only for benign flows. Indeed, the organized pattern of mussels is disrupted during winter storms (van de Koppel et al. 2005).

In theory, we could use our understanding of boundary-layer flow to make similar predictions in regard to the phytoplankton concentrations at which patterns could form. Presumably, if phytoplankton were sufficiently concentrated, mussels could form a continuous cover and still not be food limited, thereby removing the system's large-scale inhibition. Unfortunately, the calculation of food limitation as a function of cluster size requires more information about the trophic requirements and filtration rates of mussels than is currently available.

1.2   Patterns of Vegetation in Arid Scrub Lands

Similar facilitation/inhibition dynamics can operate in terrestrial ecosystems. Arid and semiarid grazing lands cover nearly 30% of Earth's terrestrial surface. In some places, plants form a more-or-less continuous cover; typical grasslands are an example. In other locations, plants are uniformly rare and the terrain is barren. Under some conditions, however, vegetation forms discrete patterns. In sub-Saharan Africa, for instance, shrubs, trees, and grasses form a pattern known as tiger bush, where bands of vegetation resemble a tiger's stripes (Figure 23.2A).

For many years it was thought that these patterns were due to either spatial variation in soil chemistry or to selective grazing by herbivores. HilleRisLambers and colleagues (2001) proposed, however, that these patterns arise in the context of facilitation/inhibition.

In semiarid regions where tiger bush is found, rain arrives in brief, infrequent downpours. Water is delivered too quickly to be readily absorbed by the ground, and the subsequent redistribution of surface water is the key to pattern formation. As plants grow, they modify the soil beneath them, making it easier for water to infiltrate. By contrast, areas without vegetation often have a surface crust that water has difficulty penetrating. When it rains, surface water builds up on bare soil, resulting in a tendency for it to flow toward vegetated areas where it can soak in. Given that water is the limiting resource for plants in these arid regions, the selective flow of water to areas that already have plant cover provides positive feedback—as water flows to them, plants grow and locally modify the soil, tempting even more water to

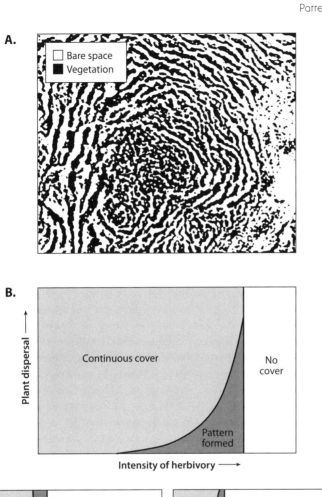

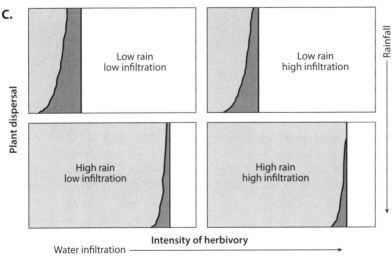

Figure 23.2 A. Tiger bush viewed from the air. **B**. Results of a model for spatial patterning in arid scrublands. **C**. Model results vary with the level of stress imposed by the environment. (See text.) **B** and **C** are redrawn from HilleRisLambers et al. (2001).

come their way in the next rain storm. Because each storm delivers a finite amount of rain, any water that flows to plants is subtracted from water available to bare areas, a reduction that inhibits colonization by plants. Thus rainstorms, water flow, and plants interact to produce the sort of local faciliation/distant inhibition that can lead to pattern formation.

HilleRisLambers and colleagues (2001) explored this idea using a spatially explicit mathematical model. I won't go into the details except to note the components of the model (e.g., the diffusive movement of water in soil and seeds across the landscape) each consisted of a mechanistic response function based on experiment, observation, and physical reasoning. With this spatially explicit physical model in hand, HilleRisLambers and his colleagues solved for the equilibrium pattern of vegetation as a function of rainfall, soil type, herbivory, and dispersal. They found that they could accurately recreate all the variation found in nature: uniform vegetative cover, barren soil, and a spectrum of patterns—everything from discrete clumps of vegetation to tiger stripes.

A typical result for a given soil type and rate of rainfall is shown in Figure 23.2B. Above a critical level of herbivory, there is only bare soil. Below this threshold, patterns can form, but the higher the rate of dispersal, the more intense herbivory must be to maintain the pattern. When the intensity of herbivory is too small, vegetation forms a uniform cover. Note that while the intensity of herbivory affects the capacity to form pattern, the model assumes that intensity is uniform through space and time, so herbivory doesn't create the pattern. This same general picture is maintained across different soil types (soil type affects infiltration) and levels of rainfall (Figure 23.2C).

There are several messages to take home from these results:

- When plants can easily spread and herbivory is mild, cover is uniform. But if dispersal is constrained and the plants are pushed near their limit by herbivory, pattern emerges. In short, pattern is a response to stress.
- Although pattern formation is possible for a wide range of soil types and rainfall, the functional space in which pattern formation can occur (the darkly shaded area on the graph) is larger the more intense the stress imposed. Compare darkly shaded areas between a low-rainfall, low-infiltration (= high-stress) scenario and a high-rainfall, high-infiltration (= low-stress) scenario.
- The conclusion that stress leads to pattern formation suggests that pattern formation will be more common in plants that are not well adapted to drought conditions than in those that are.

The work of HilleRisLambers and his colleagues elegantly demonstrates the potential for mechanistic response functions to account for the formation of pattern in a homogeneous terrestrial environment.

1.3 Summary

Pattern formation resulting from small-scale facilitation and larger-scale inhibition has fascinated ecological modelers ever since Alan Turing proposed the concept half a century ago. Actually demonstrating this form of pattern formation in the field has been a slow process, however, in large part due to the difficulty of accounting for the physical mechanisms involved. It is in this capacity that ecomechanics can contribute. As van de Koppel, HilleRisLambers and their colleagues have shown, the principles outlined in this text can be used to tell us when, where, and how the mechanics of facilitation/inhibition can lead to pattern formation.

2   CRITICALITY

We now turn our attention to a different kind of pattern formation, a process tied to the concept of *criticality*, an idea best introduced through examples.

At the critical temperature of $0°C$, water can abruptly change from liquid to solid. The mechanics of this phase transition is such that, at temperatures near the critical point, small-scale (in this case, molecular) interactions can result in large-scale patterns. For instance, snowflakes—which form near $0°C$—have delicate branching patterns visible to the naked eye even though the scale of these patterns is vastly removed from the molecular interactions that cause them. This emergence of pattern when the system is near a critical state is an example of criticality.

Analogous phenomena occur in ecology. For example, the response of a forest to fire is sensitive to the fraction of area covered by trees. When percent cover is low, a fire started by lightning quickly reaches the edge of the local clump of trees, and burns out. Averaged across the forest, these isolated burnt patches (gaps) in the forest can recover as fast as they are formed, and the fractional coverage of trees is stable through time. There is, however, a critical level of coverage (roughly 40%) at which local clumps of trees begin to overlap, providing a path for fire to consume the entire forest (given this path, the fire can *percolate*). Above this critical coverage, the system is unstable—a single lightning strike can reduce the forest to ashes.

Just as snowflakes form at water's critical temperature, pattern emerges in a forest near the critical level of cover. The pattern is not nearly as obvious as that of a snowflake—which is why it took ecologists so long to notice it—but it is there nonetheless. Near the critical level of cover, the sizes of gaps take on a characteristic distribution that can be described by a power law. If, in a given area of forest, n is the number of gaps with area A:

$$n(A) = a\,A^{-b}. \tag{23.11}$$

Coefficient a and exponent b can differ among forests depending on environmental factors and tree species, but the *pattern* of gap sizes—in this context, the shape of their distribution—remains the same (Figure 23.3A). Note that when plotted on log-log axes, a power-law distribution forms a straight line (Figure 23.3B):

$$\log n(A) = \log a - b \log A. \tag{23.12}$$

The capacity for small-scale interactions (e.g., fire) to produce large-scale patterns near a threshold (e.g., a power-law distribution of gap sizes) is known as criticality, the concept I referred to earlier. Criticality has captured the interest of theoretical ecologists because the patterns formed at critical values are often accompanied by increased resistance to disturbance and increased resilience once disturbed (Pascual and Guichard 2005; Solé and Bascompte 2006). For example, the power-law distribution of gap sizes in a forest near its critical coverage allows this relatively high concentration of trees to avoid catastrophic fires. Fires are contained locally, and the influx of seeds from trees surrounding a gap ensures a rapid recovery.

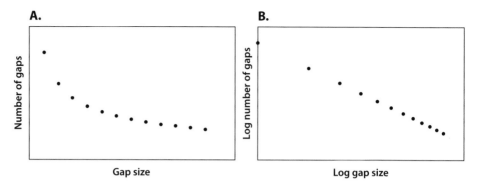

Figure 23.3 A. A hypothetical example of a scale-invariant process: the number of gaps in a forest as a function of gap area. B. The data of A replotted on logarithmic axes.

In contrast to the regular patterns of arid vegetation and sand-flat mussel beds, which are easily discerned (the distinct stripes and clumps easily catch one's eye in photographs), the power-law distribution of gap sizes is more difficult to visualize in a single picture. At any particular scale, a picture of a forest would reveal randomly arranged gaps of various sizes, not what one would normally consider a pattern. The pattern emerges when you analyze photographs taken at different scales: they look the same. In a close up aerial photo, you might find, for instance, that there are half as many gaps with area 20 m^2 as there are with area 10 m^2. In a larger-scale photo, these small gaps wouldn't be visible, but larger gaps would, and they would look similar in that their relative sizes would be the same and they occur at the same frequency—half as many gaps with area 2000 m^2 as those with area 1000 m^2. In other words, when the distribution of gap sizes is a power-law function, the pattern of relative sizes is *scale invariant*.

It is easy to show that a power-law distribution has this characteristic. The number of gaps with size kx relative to the number with size x is

$$\frac{n(kx)}{n(x)} = \frac{a\,(kx)^{-b}}{ax^{-b}} = k^{-b}. \tag{23.13}$$

The ratio of gaps sizes is independent of size, x, and depends only on b.

The phenomenon of criticality has been extensively explored in theory (e.g. Pascual and Guichard 2005), but for practical reasons validation in nature has lagged behind. For example, it would be difficult to convince the authorities to let you burn several replicate forests to test the theory that their gap-size distribution makes them resilient. However, mussel beds again provide a practical alternative system in which to test theory.

On wave-swept rocky shores (as opposed to sandy tidal flats), gaps in mussel beds are produced as hydrodynamic lift dislodges individuals, a process analogous to fires in a forest. Guichard and his colleagues (2003) devised a model to explore the process of gap formation and the characteristics of the resulting gap-size distribution. In short, they proposed that when one mussel is dislodged, adjacent mussels lose a portion of their byssal attachment and are thus weakened. Weak mussels have an increased susceptibility to dislodgment, so, once a gap is initiated, it can grow from its edges if subsequent waves apply sufficient force. If susceptible mussels aren't immediately dislodged, however, they can gradually regain their adhesive strength as they lay down new byssal threads. In this fashion, gaps can form, grow, and potentially stabilize. Additionally, Guichard and his colleagues assumed that gaps can shrink through time as individuals advance inward from a gap's periphery. There is thus a tension between gap formation and gap recovery that governs the pattern of gap sizes within the bed.

Parameterizing their model with empirical estimates for the rate of gap creation, the period of susceptibility, and the rates of edge extension, Guichard and colleagues were able to recreate the gap patterns they observed at three field sites. In particular, they found that their model could mimic the scale-independent distribution of gap sizes observed at one particular site, Boiler Bay.

Building on this tentative validation, Guichard and colleagues then used their model to experiment with the dynamics of gap formation and recovery. They found that, once set up, the system has the tendency to maintain itself near its critical point. A bed with gaps is relatively stable because the emergent pattern of gap size inhibits disturbance from catastrophically percolating through the bed, thereby allowing time for old gaps to close as new gaps are opened. By contrast, if the shore is initially uniformly covered with contiguous mussels, the first gap formed can rapidly propagate through the entire bed as one susceptible mussel after another peels at the gap's edges. In theory, once all mussels are removed, the bed cannot recover because there is no edge from which mussels can extend.

Guichard and his colleagues found that the bed is most stable when the distribution of gap sizes is scale independent. In short, given a few initial gaps, a mussel beds will evolve such that the distribution of gap sizes follows a power law and the bed covers approximately 40% of the rock's surface. This inherent tendency toward stability at the critical point is one example of a particular type of critical phenomenon known as *self-organized criticality* (Pascual and Guichard 2005; Solé and Bascompte 2006).

Under what conditions can Guichard's model operate? The model assumes that water velocities are occasionally sufficient to dislodge mussels. In this case, mussels are dislodged not from sand or mud, but rather from solid rock. The minimum force per area required to dislodge a mussel from rock is approximately $4 \times 10^4 \text{ N} \cdot \text{m}^2$ (Denny et al. 2009). If we rearrange equation 23.7 to calculate lift per area, we can equate it with this minimal adhesive tenacity to calculate the velocity required to dislodge a mussel:

$$4 \times 10^4 = \frac{F_L}{A} \tag{23.14}$$

$$= \frac{1}{2}\rho_f u^2 C_L \tag{23.15}$$

$$= 451 u^2. \tag{23.16}$$

Solving for u, we conclude that a velocity of at least $9.4 \text{ m} \cdot \text{s}^{-1}$ is required to dislodge mussels in a bed attached to rock.

This back-of-the-envelope calculation suggests that the scale-independent pattern predicted by Guichard and his colleagues can manifest itself only when u is greater than approximately $10 \text{ m} \cdot \text{s}^{-1}$. In systems that see only lower velocities, gaps cannot form, and the bed should be continuous. Guichard et al. didn't measure water velocities at their study sites, but at two of their three locations beds were nearly continuous, indicating that waves might not have been forceful enough to elicit pattern formation.

In summary, criticality provides another mechanism through which pattern can form in a homogeneous environment. The mechanisms leading to scale independence near a threshold have not been as well studied as those involved in facilitation/inhibition systems, but, as the pattern of gaps in mussel beds demonstrates, ecological mechanics can be a useful tool in predicting when and where self-organized criticality can arise. This mechanistic approach has proved similarly useful in exploring the patterns of fire-induced gap formation in terrestrial forests; as we discussed in Chapter 12, the interaction of radiative heat transfer and boundary-layer physics helps to determine how fires spread (Finney 1999)

3   CONCEPTS, CONCLUSION, AND CAVEATS

Spatial and temporal variation in the environment can govern when and where plants and animals can survive, but these factors are not necessary for pattern formation. In this chapter we have explored two ways in which pattern can form in systems where, in the absence of biology, the environment is homogeneous:

- The first of these—facilitation/inhibition—is a broadly applicable mechanism that appears to operate in many systems. As long as the scale of inhibition is large relative to the scale of facilitation, patterns can (and often do) form.
- The second mechanism—self-organized criticality—has been extensively explored in theory, although its underlying mechanisms are less well understood and evidence of its presence in nature is less extensive.

Response functions are at the heart of each of these mechanisms of pattern formation, and we have seen how they can be used to predict when and where patterns can form. As our understanding of ecomechanics improves, so will our ability to account for pattern formation.

The study of self-organization is still in its infancy, and one is advised to read the literature with a critical eye. In particular, the concepts of self-organized criticality should be approached with some caution. In many empirical studies, criticality is assumed when some factor (e.g., gap size) is found to have a power-law distribution. Although theory shows that systems at criticality are characterized by power-law distributions, there is scant evidence that the existence of a power-law distribution in nature necessarily implies criticality. Futhermore, care must also be exercised when evaluating empirical data. For example, simply showing that data appear to form a line on a log-log plot is not sufficient to demonstrate that they indeed have a power-law distribution (see Clauset et al. 2009).

This chapter has barely scratched the surface of the complex problem of pattern formation in ecology. For a more thorough introduction, I urge you to consult Solé and Bascompte (2006). I also urge you to keep abreast of the literature. The study of pattern formation is a rapidly expanding field at the cutting edge of ecomechanics.

Chapter 24

Thoughts at the End

W$_e$ have covered a lot of ground in the last twenty-three chapters, everything from basic physics to the emergence of pattern in ecological communities. It is worth taking a moment here at the end to review the major themes of this journey, to highlight some accomplishments of the ecomechanical approach, and briefly to look ahead.

1 TAKE-HOME MESSAGES

In our exploration of the principles of life's physical interactions several general messages have emerged.

1.1 Mechanistic Response Functions Are Key

Mechanistic response functions are arguably the most important tool in science. Because they make explicit, quantitative predictions, they are open to rigorous verification, assuring that, as science explores new ideas, each incremental bit of knowledge builds on a reliable foundation. Furthermore, because they are based on a mechanistic understanding of the world rather than correlations, predictions made using mechanistic functions can be reliably extended beyond the reach of current data. We have seen that physics and engineering are an extraordinarily rich source of biologically relevant response functions, relationships that go far toward explaining how plants and animals operate in the physical environment and how they interact with each other.

1.2 Diffusive Processes Are Everywhere

Random motion at any spatial scale causes objects to disperse, that is, to diffuse. Driven by thermal agitation, turbulence, and locomotion, everything—from molecules to whole organisms—is capable of diffusive transport, and these random walks play a central role in biology. By taking a mechanistic approach to diffusion, we have discovered principles that are common across widely disparate processes, systems, and scales. Boundary layers, for instance, are governed by the principles of diffusion, and these regions of retarded flow have wide-ranging consequences: they influence the forces that organisms experience when they move relative to air or water (forces essential to the study of locomotion); they limit the exchange rates of gases, nutrients, and wastes (factors central to our understanding of physiology); and they control the convective exchange of heat (a key element in determing body temperature). At larger scales, the distance travelled by dispersing propagules and the spread of genes in a population can be predicted using the rules of random walks.

1.3   Fluid-Dynamic Forces Can Be Bad

When air or water moves relative to a plant or animal, forces are imposed that often form impediments to be endured or overcome. Drag imposed by wind and waves has the potential to break trees and corals, limiting where they can survive and for how long, and animals must continually overcome drag as they move in search of food, mates, and shelter. Water is 830-fold denser than air, so fluid-dynamic forces are typically greater in the aquatic realm than in terrestrial habitats, helping to explain why seaweeds are smaller than trees and why fish swim slower than birds can fly.

1.4   Fluid-Dynamic Forces Can Be Good

On the other hand, fluid-dynamic forces are often required for organisms to function effectively. Lift provided by wings allows insects, birds, bats, and some seeds to resist the pull of gravity, thereby augmenting their ability to travel great distances. Thrust generated by drag, lift, and the acceleration reaction provides the force required to push organisms through their fluid media. Ecomechanics' ability to quantify fluid-dynamic forces provides the means to weigh their advantages against their disadvantages, enabling us to predict, for example, whether a particular organism will benefit by migrating to a distant feeding ground.

1.5   Body Temperature Seldom Equals Air Temperature

Body temperature has a profound effect on life—it influences the rates of metabolism, desiccation, growth, reproduction, and locomotion, and extremes of body temperature can be fatal.

Heat-budget models provide an exceptionally powerful means to calculate how body temperature varies in response to the physical environment, allowing us to draw several sweeping conclusions. Because the thermal conductivity of water is so high, all but the largest aquatic organisms have body temperatures very near that of the surrounding water, leveling the thermal playing field for aquatic plants and animals. But the situation is drastically different in air. Because the thermal conductivity of air is low, terrestrial organisms' body temperature is often substantially different from that of the local atmosphere. At times, the difference can be sufficiently extreme to incapacitate or kill organisms, even when air temperature is benign. Since heat-budget models can predict when and where these damaging effects will be imposed, they have the potential to increase the accuracy with which we can predict species range limits. This detailed, mechanistic understanding is likely to have increasing value as ecologists attempt to predict the effects of global climate change.

The low thermal conductivity of air also makes it possible for terrestrial organisms to have body temperatures that differ from one species to the next even when they occupy the same habitat. These species-by-species thermal differences can affect trophic interactions and competition, and thereby community structure. Again, these effects can be predicted by heat-budget models.

1.6   Life Is Shaped By The Properties Of Biological Materials

Through the creative use of ceramics and organic polymers, evolution has provided plants and animals with an astounding variety of structural materials: everything from slimes to rubbers to fibers and crystalline solids. The properties of these materials— their strength, extensibility, stiffness, and resistance to fatigue—can be coupled with beam theory and the theory of adhesion to predict a plant or animal's structural capacity (i.e., its ability to resist the forces it encounters). As with the prediction of body temperature, considerations of structural capacity contribute to our ability to explain when and where organisms can survive, and to predict how they can perform in different locations or changing environments. For example, in coming decades, the decreasing pH of tropical oceans will reduce the strength of coral-reef substratum, thereby increasing the risk that corals will break. The risk differs among species with different shapes, but—with the help of beam theory—we can predict how these environmental changes will affect the species composition of future reefs (see Chapter 1). Conversely, our understanding of the properties of materials provides a valuable perspective on the past. Despite the great diversity of biological materials, the principles of organic chemistry and the physics of fracture set limits to materials' mechanics, helping to elucidate the selective factors that have guided the evolution of morphology. For instance, from our knowldge of the stiffness of a giant kelp's stipe material and the limits to its extensibility, we proposed that it is the strength of the holdfast (limited by the kelp's competition for space on the substratum) that has caused the species to evolve its curious rope-like body plan (Chapter 14).

1.7   Environmental Variation Affects Performance

Jensen's inequality tells us that average performance under variable conditions can differ substantially from performance under average conditions, and these differences can have significant consequences. For instance, climate-change models typically predict how conditions in a particular environment will shift—on average—in coming decades. But Jensen's inequality warns that, when predicting plants' and animals' future performance, we need to take into account not only the shift in average conditions, but also the variation that inevitably accompanies those averages. Again, the importance of response functions comes to the fore. The shape of a response function determines how it interacts with variation.

1.8   Variation Depends on Spatial and Temporal Scale

Because environmental variation can have a profound effect on performance, it is important to consider the effects of variation at the temporal and spatial scales that matter to organisms. Spectral analysis provides a means to this end. Often, variance increases with the temporal and spatial scale at which a plant or animal interacts with its environment. In particular, spectra that have the characteristics of $1/f$ noise provide a convenient model to predict the increasing variation an individual or population will encounter through time and space, and this prediction can then be used in conjunction with scale transition theory to predict performance more accurately. Thus, scale transition theory can have substantial practical importance as ecologists attempt to "scale up" from short-term, local experiments to the long-term trends in whole populations and communities. Yet again, response functions play a key role; in order to apply scale-transition theory appropriately, one must have an accurate (preferably mechanistic) model of how a system operates.

1.9   The Theory of Extremes Opens a Window to Deep Time

Average performance (as predicted by scale transition theory) is a useful metric when conditions are benign, but the average matters little if a single extreme event kills an individual or decimates a population. Often community ecology is governed as much by extreme events as by typical conditions. However, extremes are by definition rare, making them difficult to observe. By vastly expanding the scale of time available for analysis, the combination of mechanistic response functions and the environmental bootstrap provides a way around this stumbling block. Realistic, long-term data series can be created by resampling a short record, enabling biologists to simulate (and the statistics of extremes to quantify) the pattern in which otherwise benign factors come into chance alignment to cause extreme events. These patterns can then be used to explore topics such as the evolution of physiological safety margins and the dynamics of ecological catastrophes.

1.10   The Emergence of Pattern Can Be Predictable

Theoreticians have proposed a variety of mechanisms by which patterns can form in a homogeneous environment (e.g., facilitation/inhibition and self-organized criticality), but pattern formation requires particular conditions. Ecomechanics can be used to specify when and where these conditions are found, and can thereby assist ecologists as they attempt to predict the emergent properties of ecological communities.

2   PLANKTON ECOLOGY

Throughout this text, I have highlighted examples of ecomechanics in action—for example, the ability of heat-budget models and locomotory performance curves to predict the limits to the spread of cane toads in Australia and the ability of fracture mechanics to predict the maximum size of intertidal seaweeds. Let's finish with a final example, one concerning the ecology of marine plankton; it intertwines several of the principles we have explored, and applies them to a system of global importance. We have glimpsed bits and pieces of this story in various chapters, but we are now in a position to bring them together. The tale that follows here is drawn in large part from Thomas Kiørboe's *A Mechanistic Approach to Plankton Ecology* (2008), perhaps the best testament to date of the power and potential of ecomechanics.

Photosynthesis (specifically oxygenic photosynthesis) has profoundly affected life on Earth. Using the energy of light to combine carbon dioxide with water, photosynthesis stores energy in glucose and thereby forms the base of nearly every food chain on the globe. Furthermore, photosynthesis releases oxygen. Over the course of the last two billion years, this beneficial byproduct has accumulated, both dissolved in the ocean and free in the atmosphere. Regardless of the medium, oxygen is required by aerobic respiration to power plants and animals. As an additional bonus, at high altitude atmospheric oxygen is transformed into ozone, where it effectively absorbs most of the harmful ultraviolet light arriving from the sun, allowing plants and animals to abandon seawater's protective shield and invade the land. In short, photosynthesis has been and continues to be one of the most important sets of chemical reactions on Earth, rivaled only by the replication of RNA and DNA.

Approximately half the photosynthesis currently chugging away on Earth is carried out by single-celled phytoplankton in the oceans. As we touched upon in Chapter 4, these cells have a problem. In order to photosynthesize, they must remain near the ocean's surface where sufficient sunlight is available, but the molecular machinery they pack inside themselves makes them denser than the surrounding seawater, causing them to sink. A few species have found ways to adjust their buoyancy to avoid sinking, but these adjustments are energetically expensive, and most species rely on a simpler method to keep their populations afloat.

As I briefly described in Chapter 4, wind-driven tubulence carries phytoplankton on a vertical random walk. While this diffusive transport doesn't affect a population's average sinking rate, its dispersive nature means that, by chance, some cells are wafted near the water's surface. If these few fortunate cells reproduce while high in the water column, the resulting increase in numbers can offset the loss from sinking, and thereby maintain a stable population in the upper ocean. Each individual phytoplankter has no way of predicting whether it will be among the lucky few. So, if the species is to be maintained near the surface, the population as a whole must hedge its bets and divide at short intervals to ensure that a sufficient fraction of those cells that find themselves high in the water column are able to divide, thereby sustaining the population.

It is here that ecomechanics comes into play. Based on a seminal paper published in 1949 by Riley, Stommel, and Bumpus, Kiørboe calculated t_{max}, the maximum reproductive interval phytoplankton can have and still maintain the population's place in the sun. t_{max} is set by the interplay of the turbulent diffusivity ϵ (see Chapter 4), and the cells' sinking speed, w:

$$t_{max} = \frac{16\epsilon}{w^2}. \tag{24.1}$$

The greater the intensity of oceanic turbulence, the larger ϵ is and the more leisurely cells can grow and divide. The faster cells sink, the shorter the reproductive interval must be.

Using results from earlier chapters, we can extend this analysis. In Chapter 4, we concluded that the time (in seconds) it takes a phytoplankton cell to double is

$$t = \frac{r^2 C_N}{3 \mathcal{D} C_\infty}, \tag{24.2}$$

where r is the cell's radius (m), and C_N is the concentration of nitrogen-bearing compounds in the cell (approximately 1500 mol $\cdot$ m^{-3}). $\mathcal{D}$ is the molecular diffusion coefficient of nutrients (nitrate or ammonia) in seawater (1.5 $\times$ 10^{-9} m$^2 \cdot$ s^{-1}), and C_∞ is the ambient concentration of these nutrients, which varies from approximately 10^{-4} mol $\cdot$ m^{-3} in nutrient-poor areas to more than 10^{-2} mol $\cdot$ m^{-3} in nutrient-rich waters. The larger the cell and the lower the ambient concentration of nutrients, the longer it takes for a cell to grow and divide. In Chapter 8, we used low-Re fluid dynamics to calculate the rate at which cells sink. Expressing that relationship (equation 8.8) in terms of cell radius r,

$$w = \frac{2r^2 \rho_e g}{9\mu}. \tag{24.3}$$

Here ρ_e is the cell's effective density (the difference in density between it and seawater, approximately 50 kg $\cdot$ m^{-3}), g is the acceleration of gravity, and μ is seawater's dynamic viscosity (1.1 $\times$ 10^{-3} Pa $\cdot$ s at 20°C). Substituting these relationships into equation 24.1 and solving for r, we arrive at a mechanistic prediction of maximum cell size as a function of a phytoplankter's environment:

$$r_{\max} = 3.15 \left(\frac{\mu^2 \mathcal{D} \epsilon C_\infty}{\rho_e^2 g^2 C_N} \right)^{1/6}. \tag{24.4}$$

Barring any control of cell density, the factors in the denominator of this equation are similar among all phytoplankton, so it is the numerator that is most important. Both viscosity and the diffusivity of nutrients vary with temperature, but viscosity decreases with increasing temperature while the molecular diffusion coefficient increases. As a result, the product $\mu^2 \mathcal{D}$ varies little from one part of the ocean to another. Which leaves us with just two factors, ϵ and C_∞, that exert primary control over phytoplankton size. The more turbulent the ocean's surface layer and the higher the ambient concentration of nutrients, the larger cells can be and still maintain a stable population.

These predictions are borne out in nature. In much of the tropical ocean, winds are gentle (so turbulence is mild) and nutrient concentrations are low. As predicted, the phytoplankton found in these areas are dominated by small species, typically cyanobacteria with radii of only a few micrometers. By contrast, phytoplankton communities where turbulence is intense (e.g., in temperate and polar seas) and nutrients are high (e.g., in areas of upwelling on the eastern edges of oceans[1]) are dominated by diatoms and dinoflagellates with radii of a few tens of μm. In short, equation 24.4 synthesizes a wealth of physical information (viscosity, density, diffusivities, and nutrient concentrations) not only to explain mechanistically why phytoplankton are small (much smaller than terrestrial plants), but also to predict the manner in which their size varies depending on local oceanographic conditions. That's a lot of explanatory value in one simple equation.

Our understanding of the limits to cell size in turn feeds into our understanding of trophic interactions. In general, the larger the organism, the longer it takes to grow and reproduce (a relationship that holds for animals as well as phytoplankton, albeit for different reasons; see Kiørboe (2008)). Small phytoplankton are typically eaten by animals not much larger than themselves (e.g., flagellates). Being roughly the same size, phytoplankton and their grazers reproduce at the same rate. As a result, even if new nutrients are introduced into a system dominated by small cyanobacteria, an increase in the phytoplankton population is quickly matched by an increase in the

---

[1] Along the eastern boundaries of oceans, the combination of alongshore winds and the Coriolis acceleration can draw nutrient-rich water up from the ocean depths, a process known as upwelling (see Denny 2008 for a more thorough explanation).

flagellate population, allowing the grazers to damp out any major fluctuations in phytoplankton population size. By contrast, diatoms and dinoflagellates are too big to be eaten by flagellates; they are instead eaten primarily by copepods, which are easily ten times the size of their prey. Because copepods are large relative to their food, they reproduce much slower. Thus, when abundant light and nutrients are available, diatoms and dinoflagellates can outgrow copepods' control, and the resulting blooms are characteristic of temperate and polar oceans and upwelling areas. In short, our ability to predict phytoplankton cell size from first principles provides valuable insight into the trophic dynamics of the upper ocean, a major component of Earth's life-support system.

Extending a thought introduced in Chapter 8, equations 24.3 and 24.4 can also be used indirectly to understand the role phytoplankton play in regulating the concentration of CO_2 in the atmosphere. As diatoms and dinoflagellates are supplied with nutrients from the ocean depths (by upwelling or wind-driven turbulence), they take in dissolved carbon dioxide that originated in the atmosphere, incorporate it into their cell structure, and then rapidly sink out of the surface layer. This effective *biological pump* is an important regulatory mechanism in Earth's carbon cycle. By contrast, the slow sinking rates of cyanobacteria make for an inefficient pump. Thus, equations 24.3 and 24.4 help to explain when and where the ocean is currently most effective at absorbing the carbon dioxide human society is so assiduously dumping into the atmosphere.

In summary, mechanistic consideration of diffusive processes and low-Re fluid dynamics can provide exceptional insight into fundamental aspects of marine ecology.

Of course, in my interpretation of equation 24.4 I have glossed over a multitude of details:

- Some phytoplankton do regulate their density, allowing them to increase their size without suffering the consequences of an increased sinking rate, perhaps gaining a refuge from grazing by copepods.
- There are limits to how much turbulence is beneficial; if turbulence is too intense, cells can spend an inordinate fraction of their time deep in the water column, hindering their ability to grow.
- As they sink, phytoplankton may contact each other and adhere. Because of its increased size, the resulting two-cell particle sinks faster, which increases its chances of encountering (and adhering to) a third phytoplankter, and so forth. As this aggregative process snowballs, the rate of export of phytoplankton to the deep ocean increases. It is this augmented export—rather than herbivory—that controls the maximum size of algal blooms.
- Grazers are not 100% efficient at assimilating the phytoplankton they eat. The fecal pellets they consequently produce are relatively large and therefore sink faster than individual phytoplankters, providing an alternative mechanism for carbon sequestration.
- In Chapter 5, we briefly delved into the low-Reynolds-number physics that governs the ability of copepods (and other grazers) to find their food and escape from predators—these physical factors affect the ability of grazers to regulate phytoplankton population size.

These and many other important and intriguing details are covered in depth in Kiørboe (2008), but the details should not detract from the main message. By synthesizing the principles of phytoplankton's physical interactions, ecomechanics offers a powerful tool for the study of marine ecology. By extrapolation, when applied to other systems, ecological mechanics can help to realize what Schoener (1986) termed the "mechanistic ecologist's utopia," the ability to ground ecology in basic principles, allowing us to forge a chain of explanation and prediction from physics to physiology to the ecology of individuals and on to the ecology of communities.

3   A LOOK TO THE FUTURE

As much intellectual territory as we have traversed in this text, I am struck here at the end by how much we have left unexplored. As I noted in Chapter 1, huge topics (e.g., comparative physiology, sensory ecology, low-Re fluid dynamics) have received minimal attention, and other topics have received no attention at all. For example, in Chapter 12, I based the discussion of the environmental niche on heat-budget models' ability to predict when and where plants and animals can survive. But survival is only part of the story. A species' ability to persist in a given environment is also affected by how fast and how large individuals grow and when they start to reproduce. We haven't discussed these life-history variables, but several theories (e.g., dynamic-energy-budget and other bioenergetic models (Kooijman 2010, Nisbet and colleagues 2012)) are being developed that incorporate many of the response functions we have discussed to predict optimal life-history strategies. In a few years I expect that Chapter 12 can be expanded to include these models. Similarly, I expect that in the near future Chapter 23 ("Pattern and Self-Organization") can be subsumed into a broader chapter on the emerging theory of complexity.

But, painful as it is to realize how much I have left out, it was never my intent to include everything. Instead, the topics included here are intended to open your eyes to a new way of thinking about biology. The principles we have discussed are applicable far beyond the limited context of this tome; I hope they will empower you to go exploring.

References

Acton, J. R., and P. T. Squire. 1985. *Solving Equations with Physical Understanding*. Adam Hilger Ltd. Bristol UK.

Adlassnig, W., T. Lendl, M. Peroutk, and I. Lang. 2010. Deadly glue—adhesive traps of carnivorous plants. In J. von Byern and I. Grunwald, eds. *Biological Adhesive Systems*, pp. 15–28. Springer, Vienna.

Alexander, R. McN. 1998. When is migration worthwhile for animals that walk, swim or fly? *J. Avian. Biol.* 29:387–94.

———. 2003. *Principles of Animal Locomotion*. Princeton University Press, Princeton.

Alexander, R. McN., and G. Goldspink, eds. 1977. *Mechanics and Energetics of Animal Locomotion*. Chapman and Hall, London.

Almany, G. R. 2004. Does increased habitat complexity reduce predation and competition in coral reef fish assemblages? *Oikos* 106:275–84.

Angilletta, M. J. 2009. *Thermal Adaptation: A Theoretical and Empirical Synthesis*. Oxford University Press, London.

Askew, G. N., and R. L. Marsh. 2002. Muscle designed for maximum short-term power output: Quail flight muscle. *J. Exp. Biol.* 205:2153–60.

Atkins, P. W. 1984. *The Second Law*. Scientific American Library, New York.

Autumn, K. 2006. Properties, principles, and parameters of the gecko adhesive system. In A. M. Smith and J. A. Callow, eds. *Biological Adhesives*, pp. 225–56. Springer-Verlag, Berlin.

Autumn, K., Y. A. Liang, S. T. Hsieh, W. Zesch, W. P Chan, T. W. Kenny, R. Fearing, and R. J. Full. 2000. Adhesive force of a single gecko foot-hair. *Nature* 405:681–85.

Autumn, K., M. Sitti, Y. A. Liang, A. M. Peattie, W. R. Hansen, S. Sponberg, T. W. Kenny, R. Fearing, J. N. Israelachvilli, and R. J. Full. 2002. Evidence for van der Waals adhesion in gecko setae. *Proc. Natl. Acad. Sci.* 99:12252–56.

Barber, A. H., D. Lu, and N. M. Pugno. 2015. Extreme strength observed in limpet teeth. *J. R. Soc. Interface* 12:20141326.

Barkai, A., and C. McQuaid. 1988. Predator-prey reversal in a marine benthic ecosystem. *Science* 242:62–64.

Barlow, D., and M. A. Sleigh. 1993. Water propulsion speeds and power output by comb plates of the ctenophore *Pleurobrachia pileus* under different conditions. *J. Exp. Bio.* 183:149–63.

Barthlott, W., and C. Neinhuis. 1997. Purity of the sacred lotus, or escape from contaminants in biological surfaces. *Planta* 202:1–8.

Batchelor, G. K. 1967. *An Introduction to Fluid Dynamics*. Cambridge University Press, London.

Bell, E. C. 1995. Environmental and morphological influences on thallus temperature and desiccation of the intertidal alga *Mastocarpus papillatus* Kützing. *J. Exp. Mar. Biol. Ecol.* 191:29–53.

Bell, E. C., and J. M. Gosline. 1996. Mechanical design of mussel byssus: Material yield enhances attachment strength. *J. Exp. Biol.* 199:1005–17.

Bellwood, D. R., T. P. Hughes, C. Folke, and M. Nystrom. 2004. Confronting the coral reef crisis. *Nature* 429:827–33.

Bendat, J. S., and A. G. Piersol. 1986. *Random Data: Analysis and Measurement Procedures* (2nd ed.). John Wiley & Sons, New York.

Benedetti-Cecchi L. 2005. Unanticipated impacts of spatial variance of biodiversity on plant productivity. *Ecol. Lett.* 8:791–99.

Berg, H. C. 1983. *Random Walks in Biology*. Princeton University Press, Princeton.

Berman, A., L. Addadi, A. Kvick, L. Leiserowitz, M. Nelson, and S. Weiner. 1990. Intercalation of sea urchin proteins in calcite: Study of a crystalline composite material. *Science* 250:664–67.

Bhushan, B. 1999. *Principles and Applications of Tribology*. John Wiley & Sons, New York.

———. 2007. Adhesion of multi-level hierarchical attachment systems in gecko feet. *J. Adhesion Sci. Technol.* 21: 1213–58.

Biewener, A. A. 2003. *Animal Locomotion*. Oxford University Press, London.

Bjornstad, O. M., M. Peltonen, A. M. Liebhold, and W. Baltensweiler. 2002. Waves of larch budmoth outbreaks in the European Alps. *Science* 298:1020–23.

Blake, J. R., and M. A. Sleigh. 1974. Mechanics of ciliary locomotion. *Biol. Rev.* 49:85–125.

Blanchette, C. A. 1997. Size and survival of intertidal plants in response to wave action: A case study with *Fucus gardneri*. *Ecology* 78:1563–78.

Blasius, H. 1908. Grenzschichten in Flüssigkeiten mit kleiner Reibung. *Z. Math. Physik, Bd.* 56: 1–37. English translation in NACA-TM-1256.

Blickhan, R., and R. J. Full. 1987. Locomotion energetics of the ghost crab: II. Mechanics of the center of mass during walking and running. *J. Exp. Biol.* 130:155–74.

Boulding, E. G., and M. LaBarbera. 1986. Fatigue damage: Repeated loading enables crabs to open larger bivalves. *Biol. Bull.* 171:538–47.

Brackenbury, J. 1997. Caterpillar kinematics. *Nature* 390:453.

Briggs, L. J. 1950. Limiting negative pressure of water. *J. Appl. Phys.* 21:721–22.

Bryson, W. 2004. *A Short History of Nearly Everything*. Black Swan, London.

Bunsell, A. R., and J. Renard. 2005. *Fundamentals of Fibre Reinforced Composite Materials*. Institute of Physics Publishing, London.

Bush, J. W. M., and D. L. Hu. 2006. Walking on water: Biolocomotion at the interface. *Ann. Rev. Fluid Mech.* 38:339–69.

Bush, J.W.M., D. L. Hu, and M. Prakash. 2008. The integument of water-walking arthropods: form and function. *Adv. Insect Physiol.* 34:117–92.

Caldwell, R. L. 1979. A unique form of locomotion in a stomatopod—backward somersaulting. *Nature* 282:71–73.

Campbell, G. S., and J. M. Norman. 1998. *An Introduction to Environmental Biophysics* (2nd ed.). Springer, New York.

Campolo, D., S. D. Jones, and R. S. Fearing. 2003. Fabrication of gecko foot-hair like nano structures and adhesion to random rough surfaces. *IEE Nano* August 12–14, 2003, San Francisco.

Cannell, M. G., and J. Morgan. 1987. Young's modulus of living branches and tree trunks. *Tree Physiol.* 3:355–64.

Carrington, E. 2002. The ecomechanics of mussel attachment: From molecules to ecosystems. *Integr. Comp. Biol.* 42:846–52.

Carrington, E., and J. M. Gosline. 2004. Mechanical design of mussel byssus: Load cycle and strain rate dependence. *Amer. Malac. Bull.* 18:135–42.

Carrington, E., J. H. Waite, G. Sarà, and K. P. Sebens. 2015. Mussels as model system for integrative ecomechanics. *Annu. Rev. Mar. Sci.* 7:443–69.

Carslaw, H. S., and J. C. Jaeger. 1959. *Conduction of Heat in Solids* (2nd ed.). Oxford University Press, London.

Casas, J., and O. Dangles. 2010. Physical ecology of fluid flow sensing in arthropods. *Ann. Rev. Entomol.* 55:505–20

Casas, J., T. Steinmann, and O. Dangles. 2008. The aerodynamic signature of running spiders. *PLoS ONE* 3:e2216.

Casas, J., T. Steinmann, and G. Krijnen. 2010. Why do insects have such a high density of flow-sensing hairs? Insights from the hydrodynamics of biomimetic MEMs sensors. *J. Roy. Soc. Interface.* 7:1487–95.

Cassie, A.B.D., and S. Baxter. 1944. Wettability of porous surfaces. *Trans. Faraday Soc.* 40:546–51.

———. 1945. Large contact angles of plant and animal surfaces. *Nature* 155:21–22.

Cheer, A. Y., and M.A.R. Koehl. 1987. Paddles and rakes: Fluid flow through bristled appendages of small organisms. *J. Theor. Biol.* 129:17–39.

Chesson, P., M. J. Donahue, B. Melbourne, and A. L. Sears. 2005. Scale transition theory for understanding mechanisms in metacommunities. In M. Holyoak, M. A. Leibold, and

R. D. Holt, eds. *Metacommunities: Spatial Dynamics and Ecological Communities*, pp. 279–306. University of Chicago Press, Chicago.

Clauset, A., C. R. Shailz, and M.E.J. Newman. 2009. Power-law distributions in empirical data. *SIAM Rev.* 51:661–703.

Clobert, J., M. Baguette, T. G. Benton, and J. M. Bulock. 2012. *Dispersal Ecology and Evolution*. Oxford University Press, Oxford, UK.

Cohen, J. D., and M. A. Finney. 2010. An examination of fuel particle heating during fire spread. *VI Internatl. Conf. Forest Fire Res.*

Coles, S. 2001. *An Introduction to Statistical Modeling of Extreme Values*. Springer-Verlag, London.

Connell, J. H. 1978. Diversity in tropical rain forests and coral reefs. *Science* 199:1302–10.

Couteron, P., and O. Lejeune. 2001. Periodic spotted patterns in semi-arid vegetation explained by propagation-inhibition model. *J. Ecol.* 89:616–28.

Crimaldi, J. P., J. K. Thompson, J. H. Rosman, R. J. Lowe, and J. R. Koseff. 2002. Hydrodynamics of larval settlement: The influence of turbulent shear stress events at potential recruitment sites. *Limnol. Oceanogr.* 47:1137–51.

Crimaldi, J. P., J. R. Cadwell, and J. B. Weiss. 2008. Reaction enhancement of isolated scalars due to vortex mixing. *Phys. Rev. E74*, 016307.

Crimaldi, J. P. and R. K. Zimmer. 2014. The physics of broadcast spawning in benthic invertebrates. *Ann. Rev. Mar. Sci.* 6:141–65.

Dahlquist, C. A. 1969. Pressure-sensitive adhesives. In R.L. Patrick, ed. *Treatise on Adhesion and Adhesives*, pp. 219–60. Marcel Dekker, New York.

Daniel, I. M., and O. Ishai. 2005. *Engineering Mechanics of Composite Materials* (2nd ed.). Oxford University Press, London.

Davidson, P.A. 2004. *Turbulence*: *An Introduction for Scientists and Engineers*. Oxford University Press, London.

Davis, A. J., L. S. Jenkinson, J. H. Lawton, B. Shorrocks, and S. Wood. 1998. Making mistakes when predicting shifts in species range in response to global warming. *Nature* 391:783–86.

Dayton, P. K. 1975. Experimental evaluation of ecological dominance in a rocky intertidal algal community. *Ecol. Monogr.* 45:137–59.

de Jager, M., F. J. Weissing, P.M.J. Herman, B. A. Nolet, and J. van de Koppel. 2011. Lévy walks evolve through interaction between movements and environmental complexity. *Science* 332:1551–53.

DeMont, M. E. 1990. Tuned oscillations in the swimming scallop *Pecten maximus*. *Can. J. Zool.* 68:786–91

DeMont, M. E., and J. M. Gosline. 1988. Mechanics of jet propulsion in the hydromedusan jellyfish, *Polyorchis penicillatus*: III. A natural resonating bell; the presence and importance of a resonant phenomenon in the locomotor structure. *J. Exp. Biol.* 134:347–61.

De Podesta, M. 2002. *Understanding the Properties of Matter* (2nd ed.).Taylor & Francis, New York.

Denny, M. W. 1976. The physical properties of spider's silk and their role in the design of orb-webs. *J. Exp. Biol.* 65:483–506.

———. 1980. Locomotion: the cost of gastropod crawling. *Science* 208:1288–90.

———. 1981. A quantitative model for the adhesive locomotion of the terrestrial slug, *Ariolomax columbianus*. *J. Exp. Biol.* 91:195–217.

———. 1987. Lift as a mechanism of patch initiation in mussel beds. *J. Exp. Mar. Biol. Ecol.* 113:231–45.

———. 1988. *Biology and the Mechanics of the Wave-Swept Environment*. Princeton University Press, Princeton.

———. 1989. A limpet shell that reduces drag: laboratory demonstration of a hydrodynamic mechanism and an exploration of its effectiveness in nature. *Can. J. Zool.* 67:2098–2106.

———. 1993. *Air and Water: The Biology and Physics of Life's Media*. Princeton University Press, Princeton.

———. 1995. Predicting physical disturbance: Mechanistic approaches to the study of survivorship on wave-swept shores. *Ecol. Mongr.* 65:371–418.

———. 2000. Limits to optimization: fluid dynamics, adhesive strength, and the evolution of shape in limpet shells. *J. Exp. Biol.* 203:2603–22.

———. 2008. *How the Ocean Works: An Introduction to Oceanography.* Princeton University Press, Princeton.

Denny, M. W., and C. A. Blanchette. 2000. Hydrodynamics, shell shape, behavior and survivorship in the owl limpet *Lottia gigantea. J. Exp. Biol.* 203:2623–39.

Denny, M. W., V. Brown, E. Carrington, G. Kraemer, and A. Miller. 1989. Fracture mechanics and the survival of wave-swept macroalgae. *J. Exp. Mar. Biol. Ecol.* 127:211–28.

Denny, M. W., E. Cowen, and B. P. Gaylord. 1997. Flow and flexibility II: The roles of size and shape in determining wave forces on the bull kelp, *Nereocystis luetkeana. J. Exp. Biol.* 200:3165–83.

Denny, M. W., T. Daniel, and M.A.R. Koehl. 1985. Mechanical limits to size in wave-swept organisms. *Ecol. Monogr.* 55:69–102.

Denny, M. W., and W. W. Dowd. 2012. Biophysics, environmental stochasticity, and the evolution of thermal safety margins in intertidal limpets. *J. Exp. Biol.* 215:934–47.

Denny, M. W., and S. Gaines. 2000. *Chance in Biology: Using Probability to Explore Nature.* Princeton University Press, Princeton.

Denny, M. W., B. P. Gaylord, B. Helmuth, and T. Daniel. 1998. The menace of momentum: dynamic forces on flexible organisms. *Limnol. Oceanogr.* 43:955–68.

Denny, M. W., and J. M. Gosline. 1980. The physical properties of the pedal mucus of the terrestrial slug, *Ariolimax columbianus. J. Exp. Biol.* 88:375–93.

Denny, M. W., and C.D.G. Harley. 2006. Hot limpets: Predicting body temperature in a conductance-mediated thermal system. *J. Exp. Biol.* 209:2409–19

Denny, M. W., and B. Helmuth. 2009. Confronting the physiological bottleneck: A challenge from ecomechanics. *Integr. Comp. Biol.* 49:197–201.

Denny, M. W., B. Helmuth, G. H. Leonard, C.D.G. Harley, L.J.H. Hunt, and E. K. Nelson. 2004. Quantifying scale in ecology: lessons from a wave-swept shore. *Ecol. Monogr.* 74:513–32.

Denny, M. W., L.J.H. Hunt, L. P. Miller, and C.D.G. Harley. 2009. On the prediction of extreme ecological events. *Ecol. Monogr.* 79:397–421.

Denny, M. W., K. Mach, S. Tepler, and P. Martone. 2013. Indefatigable: An erect coralline alga is highly resistant to fatigue. *J. Exp. Biol.* 216:3772–80.

Denny, M. W., E. K. Nelson, and K. S. Mead. 2002. Revised estimates of the effects of turbulence on fertilization in the purple sea urchin, *Strongylocentrotus purpuratus. Biol. Bull.* 203:275–77.

Denny, M. W., and M. F. Shibata. 1989. Consequences of surf-zone turbulence for settlement and external fertilization. *Am. Nat.* 134:859–89.

Deutsch, C. A., J. J. Tewsbury, R. B. Huey, K. S. Sheldon, C. K. Ghalambor, D. C. Haak, and P. R. Martin. 2008. Impacts of climate warming on terrestrial ectotherms across latitude. *Proc. Natl. Acad. Sci. (USA)* 105:6668–72.

Diggle, P. J. 1990. *Time Series: A Biostatistical Introduction.* Oxford Science Publications, Oxford UK.

Dillon, M. E., G. Wang, and R. B. Huey. 2010. Global metabolic impacts of recent climate warming. *Nature* 467:704–6.

Done, T. 1983. Coral zonation: Its nature and significance. In D. Barnes, ed. *Perspectives on Coral Reefs*, pp. 107–47. Australian Institute of Marine Science, Townsville, Australia.

Donovan, D., J. Baldwin, and T. Carefoot. 1999. The contribution of anaerobic energy to gastropod crawling and a re-estimation of minimum cost of transport in the abalone, *Haliotis kamtschatkana* (Jonas). *J. Exp. Mar. Biol. Ecol.* 235:273–84.

Donovan, D., and T. Carefoot. 1997. Locomotion in the abalone *Haliotis kamtschatkana*: pedal morphology and cost of transport. *J. Exp. Biol.* 200:1145–53.

Dorgan, K. M., S. R. Arwade, and P. A. Jumars. 2007. Burrowing in marine muds by crack propagation: Kinematics and forces. *J. Exp. Biol.* 210:4198–4212.

Dorgan, K. M., P. A. Jumars, B. D. Johnson, B. P. Boudreau, and E. Landis. 2005. Burrow elongation by crack propagation. *Nature* 433:475.

Dorgan, K. M., S. Lefebvre, J. H. Stillman, and M.A.R. Koehl. 2011. Energetics of burrowing by the cirratulid polychaete *Cirriformia moorei*. *J. Exp. Biol.* 214:2202–14.

Dowd, W. W., F. A. King, and M. W. Denny. 2015. Thermal variation, thermal extremes, and the physiological performance of individuals. *J. Exp. Biol.* 218:1956–67.

Dudley R. 2000. *The Biomechanics of Insect Flight: Form, Function, Evolution.* Princeton University Press, Princeton.

Dudley, R, and C. P. Ellington. 1990. Mechanics of forward flight in bumblebees: I. Kinematics and morphology. *J. Exp. Biol.* 148:19–52.

Dupuy, F., T. Steinmann, D. Pierre, J. P. Christidès, G. Cummins, C. Lazzari, J. Miller, and J. Casas. 2012. Responses of cricket cercal interneurons to realistic stimuli in the field. *J. Exp. Biol.* 215:2382–89.

Dusenberry, D. B. 2009. *Living at Micro Scale.* Harvard University Press, Cambridge.

Eckman, J. E., A.R.M. Nowell, and P. A. Jumars. 1981. Sediment destabilization by animal tubes. *J. Mar. Res.* 39:361–74.

Edwards, J., D. Whitaker, S. Klionsky, and M. J. Laskowski. 2005. A record-breaking pollen catapult. *Nature* 435:164.

Efron, B., and R. J. Tibshirani. 1993. *An Introduction to the Bootstrap.* Chapman & Hall/CRC, Boca Raton.

Einstein, A. 1905. On the motion of small particles suspended in liquids at rest required by the molecular-kinetic theory of heat. Translated and annotated in J. Stachel. 1998. *Einstein's Miraculous Year.* Princeton University Press, Princeton.

Ellington, C. P. 1984. The aerodynamics of hovering insect flight, III: Kinematics. *Phil. Trans. R. Soc. Lond.* 305:41–78.

Elsner, J. B., J. P. Kossin, and T. Jagger. 2008. The increasing intensity of the strongest tropical cyclones. *Nature* 455:92–95.

Etnier, S. A., and S. Vogel. 2000. Reorientation of daffodil (*Narcissus: Amaryllidaceae*) flowers in wind: drag reduction and torsional flexibility. *Am. J. Bot.* 87:29–32.

Evans, M., N. Hastings, and B. Peacock. 2000. *Statistical Distributions* (3rd ed.). Wiley Series in Probability and Statistics, John Wiley & Sons, New York.

Ewoldt, R. H., C. Clasen, A. E. Hosoi, and G. H. McKinley. 2007. Rheological fingerprinting of gastropod pedal mucus and synthetic complex fluids for biomimicking adhesive locomotion. *Soft Mat.* 3:634–43.

Farquhar, T., and Y. Zhao. 2006. Fracture mechanics and its relevance to botanical structures. *Am. J. Bot.* 93:1449–54.

Federle, W. 2006. Why are so many adhesive pads hairy? *J. Exp. Biol.* 209:2611–21.

Ferry, J. D. 1980. *Viscoelastic Properties of Polymers* (3rd ed.). John Wiley & Sons, New York.

Feynman, R. P., R. B. Leighton, and M. Sands. 1963. *The Feynman Lectures on Physics.* Addison-Wesley, Reading, MA.

Finney, M. A. 1999. Mechanistic modeling of landscape fire patterns. In D. J. Mlandenoff and W. L. Baker, eds. *Spatial Modeling of Forest Landscape Change: Approaches and Applications,* pp. 186–209. Cambridge University Press, New York.

Fish, F. E., P. W. Weber, M. M. Murray, and L. E. Howle. 2011. The tubercles on humpback whales' flippers: Application of bio-inspired technology. *Integr. Comp. Biol.* 51:203–13.

Fletcher, N. H. 1992. *Acoustic Systems in Biology.* Oxford University Press, London.

Florioli, R. Y., J. von Langen, and J. H. Waite. 2000. Marine surfaces and the expression of specific byssal adhesive protein variants in *Mytilus*. *Mar. Biotechnol.* 2:352–63.

Friedrichs, M., G. Graf, and B. Springer. 2000. Skimming flow induced over a simulated polychaete tube lawn at low population densities. *Mar. Ecol. Progr. Ser.* 192:219–28.

Full, R. J. 1987. Locomotion energetics of the ghost crab I: Metabolic cost and endurance. *J. Exp. Biol.* 130:137–53.

———. 1997. Invertebrate locomotor systems. In W. Dantzler, ed. *The Handbook of Comparative Physiology,* pp. 853–930. Oxford University Press, London.

Gaines, S. D., and M. W. Denny. 1993. The largest, smallest, highest, lowest, longest, and shortest: Extremes in ecology. *Ecology* 74:1677–92.

Gallardo, M. H., N. Kohle, and C. Araneda. 1995. Bottleneck effects in local populations of fossorial Ctenomys (Rodentia, Ctenomyidae) affected by vulcanism. *Heredity* 74:638–46.

Gates, D. M. 1980. *Biophysical Ecology.* Springer-Verlag, New York.

Gaylord, B. P. 1997. Consequences of wave-induced water motion to nearshore macroalgae. PhD diss., Stanford University.

———. 1999. Detailing agents of physical disturbance: wave-induced velocities and accelerations on a rocky shore. *J. Exp. Mar. Biol. Ecol.* 239:85–124.

Gaylord, B. P., and M. W. Denny. 1997. Flow and flexibility I: Effects of size, shape, and stiffness in determining wave forces on stipitate kelps, *Eisenia arborea* and *Pterygophora californica. J. Exp. Biol.* 200:3141–64.

Gaylord, B. P., B. B. Hale, and M. W. Denny. 2001. Consequences of transient fluid forces for compliant benthic organisms. *J. Exp. Biol.* 204:1347–60.

Gaylord, B. P., D. C. Reed, P. T. Raimondi, L. Washburn, and S. R. McLean. 2002. A physically based model of macroalgal spore dispersal in the wave and current-dominated nearshore. *Ecology* 83:1239–51.

Gerard, V. A. 1986. Photosynthetic characteristics of giant kelp (*Macrocystis pyrifera*) determined *in situ. Mar. Biol.* 90:473–82.

Gere, J. M., and S. P. Timoshenko. 1984. *Mechanics of Materials* (2nd ed.). Brooks/Cole Engineering Division, Monterey, CA.

Gibb, A. C., B. C. Jayne, and G. V. Lauder. 1994. Kinematics of pectoral fin locomotion in bluegill sunfish *Lepomis macrochirus. J. Exp. Biol.* 189:133–61.

Glanville, E. J., and F. Seebacher. 2006. Compensation for environmental change by complementary shifts of thermal sensitivity and thermoregulatory behavior in an ectotherm. *J. Exp. Biol.* 209:4869–77.

Gorb, S., Y. Jian, and M. Scherge. 2000. Ultrastructural architechture and mechanical properties of attachment pads in *Tettigonia viridissima* (Orthoptera Tettigoniidae). *J. Comp. Physol. A* 186:821–31.

Gordon, J. E. 1976. *The New Science of Strong Materials: Or Why We Don't Fall Through the Floor.* Penguin Books, London.

Gosline, J. M. 2016. *Principles of Structural Biomaterials.* University of California Press, Oakland, CA.

Gosline, J. M., P. A. Guerette, C. S. Ortlepp, and K. N. Savage. 1999. The mechanical design of spider silks: From fibroin sequence to mechanical function. *J. Exp. Biol.* 202:3295–3303.

Gosline, J. M., and R. E. Shadwick. 1983. The role of elastic energy storage mechanisms in swimming: an analysis of mantle elasticity in escape jetting in the squid, *Loligo opalescens. Canadian J. Zoology* 61:1421–31.

Gould, S. J. 1985. The median is not the message. *Discover* 6:40–42.

Griffith, A. A. 1921. The phenomena of rupture and flow in solids. *Phil. Trans. Roy, Soc. of London* 221:163–98.

Guasto, J. S., R. Rusconi, and R. Stocker. 2011. Fluid mechanics of planktonic microorganisms. *Annu. Rev. Fluid Mech.* 44:373–400.

Guichard, F., P. M. Halpin, G. W. Allison, J. Lubchenco, and B. A. Menge. 2003. Mussel disturbance dynamics: signatures of oceanographic forcing from local interactions. *Amer. Nat.* 161:889–904.

Gumbel, E. J. 1958. *Statistics of Extremes.* Columbia University Press, New York.

Hale, B. 2001. Macroalgal materials: foiling fracture from fluid forces. PhD diss., Stanford University.

Hall-Craggs, E.B.C. 1965. An analysis of the jump of the lesser galago (*Galago senegalensis*). *J. Zool.* 147:20–29.

Halley, J. 1996. Ecology, evolution, and 1/f noise. *Trends Ecol. Evol.* 11:33–37.

Hanna, G., W. Jon, and W.P.J Barnes. 1991. Adhesion and detachment of the toe pads of tree frogs. *J. Exp. Biol.* 155:103–25.

Happel, J., and H. Brenner. 1973. *Low Reynolds Number Hydrodynamics: With Special Application to Particulate Media* (2nd ed.). Kluwer Academic Publishers, Hingham, MA.

Harley, C.D.G. 2008. Tidal dynamics, topographic orientation, and temperature-mediated mass mortalities on rocky shores. *Mar. Ecol. Progr. Ser.* 371:37–46.

———. 2013. Linking ecomechanics and ecophysiology to interspecific interaction and community dynamics. *Ann. N.Y. Acad. Sci.* 1297:73–82.

Harley, C.D.G., and R. T. Paine. 2009. Contingencies and compounded rare perturbations dictate sudden distributional shifts during periods of gradual climate change. *Proc. Natl. Acad. Sci. (USA)* 106:11172–76.

Harley, C.D.G., M. W. Denny, K. J. Mach, and L. P. Miller. 2009. Thermal stress and morphological adaptations in limpets. *Func. Ecol.* 23:292–301.

Harrington, M. J., and J. H. Waite. 2007. Holdfast heroics: Comparing the molecular and mechanical properties of *Mytilus californianus* byssal threads. *J. Exp. Biol.* 210:4307–18.

Hastings, H. M., and G. Sugihara. 1993. *Fractals: A Users Guide for the Natural Sciences*. Oxford University Press, London.

Heglund, N. C., and C. R. Taylor. 1988. Speed, stride frequency and energy cost per stride: How do they change with body size and gait? *J. Exp. Biol.* 138:301–18.

Heinrichs, B. 1993. *Warm-Blooded Insects: Strategies and Mechanisms of Thermoregulation*. Harvard University Press, Cambridge.

Helmuth, B., and M. W. Denny. 2003. Predicting wave exposure in the rocky intertidal zone: Do bigger waves always lead to larger forces? *Limnol. Oceanogr.* 48:1338–45.

Hijmans, R. J., and C. H. Graham. 2006. The ability of climate envelope models to predict the effect of climate change on species distributions. *Glob. Chang. Biol.* 12:2272–81.

HilleRisLambers, R., M. Rietkerk, F. van den Bosch, H.H.T. Prins, and H. de Kroon. 2001. Vegetation pattern in semi-arid grazing systems. *Ecology* 82:50–61.

Hochachka, P. W., and G. N. Somero. 2002. *Biochemical Adaptation: Mechanism and Process in Physiological Evolution*. Oxford University Press, London.

Hoegh-Guldberg, O., P. J. Mumby, A. J. Hooten, R. S. Steneck, P. Greenfield, E. Gomez, C. D. Harvell, et al. 2007. Coral reefs under rapid climate change and ocean acidification. *Science* 318:1737–42.

Hoerner, S. F. 1965. *Fluid-Dynamic Lift*. Hoerner Fluid Dynamics, Bricktown, NJ.

Holberton, D. V. 1977. Locomotion of protozoa and single cells. In R. McN. Alexander and G. Goldspink, eds. *Mechanics and Energetics of Animal Locomotion*, pp. 279–332. Chapman and Hall, London.

Holbrook, N. M., M. W. Denny, and M.A.R. Koehl. 1991. Intertidal "trees": Consequences of aggregation on the mechanical and photosynthetic properties of sea palms *Postelsia palmaeformis* Ruprecht. *J. Exp. Mar. Biol. Ecol.* 146:39–67.

Holling, C. S. 1959. Some characteristics of simple types of predation and parasitism. *Can. Entomol.* 91:385–98.

Hoppeler, H., and E. R. Weibel. 2005. Scaling functions to body size: Theories and facts. *J. Exp. Biol.* 208:1573–74.

Houlihan, D. F., and A. J. Innes. 1982. Oxygen consumption, crawling speeds, and cost of transport in four Mediterranean intertidal gastropods. *J. Comp. Physiol.* 147:113–21.

Huey, R. B., C. R. Peterson, S. J. Arnold, and W. P. Porter. 1989. Hot rocks and not-so-hot rocks: Retreat-site selection by garter snakes and its thermal consequences. *Ecology* 70:931–44.

Hughes, T. P. 1994. Catastrophes, phase shifts, and large-scale degradation of a Caribbean coral reef. *Science* 265:1547–51.

Humphrey, J.A.C., and F. G. Barth. 2007. Medium flow-sensing hairs: biomechanics and models. *Adv. Insect Physiol.* 34:1–80.

Hurd, C. L. 2000. Water motion, marine macroalgal physiology, and production. *J. Phycol.* 36:453–72.

Hutchinson, G. E. 1957. Concluding remarks. *Cold Spring Harbor Symp. Quant. Biol.* 22:415–27.

Inchausti, P., and J. Halley. 2002. The long-term temporal variability and spectral colour of animal populations. *Evol. Ecol. Res.* 4:1033–48.

————. 2004. The increasing importance of $1/f$-noises as models of ecological variability. *Fluct. Noise Letters* 4:R1–R26.

Incropera, F. P., and D. P. DeWitt. 2002. *Fundamentals of Heat and Mass Transfer* (5th ed.). John Wiley & Sons, New York.

Inglis, C. E. 1913. Stresses in a plate due to the presence of cracks and sharp corners. *Trans. Inst. Naval Architects* 55:219–30.

Innes, A. J. and D. F. Houlihan. 1985. Aerobic capacity and cost of locomotion of a cool temperate gastropod: a comparison with some Mediterranean species. *Comp. Biochem. Physiol.* 80A:487–93.

IPCC. 2013. *Climate Change 2013: The Physical Science Basis. Contribution of working Group 1 to the Fifth Assessment Report of the Intergovernmental Panel on Climate Change.* [Stocker, T. F., D. Qin, G. –K. Plattner, M. Tignor, S. K. Allen, J. Boschung, A. Nauels, Y. Xia, V. Bex, and P. M. Midgely (eds.)]. Cambridge University Press, Cambridge.

Irigoien, X., J. Hulsman, and R. P. Harris. 2004. Global biodiversity patterns of marine phytoplankton and zooplankton. *Nature* 429:863–67.

James, R. S., J. D. Altringham, and D. F. Goldspink. 1995. The mechanical properties of fast and slow skeletal muscles of the mouse in relation to their locomotory function. *J. Exp. Biol.* 198:491–502.

Jameson, W. 1959. *The Wandering Albatross.* William Morrow, New York.

Jensen, J. L. 1906. Sur les fonctions convexes et les inegualites entre les valeurs moyennes. *Actsa Math.* 30:175–93.

Johnsen, S. 2012. *The Optics of Life: A Biologist's Guide to Light in Nature.* Princeton University Press, Princeton.

Jumars, P. A. 1993. *Concepts in Biological Oceanography: An Interdisciplinary Primer.* Oxford University Press, London.

Kaimal, J. C. and J. J. Finnigan. 1994. *Atmospheric Boundary Layers: Their Structure and Measurement.* Oxford University Press, London.

Kamino, K. 2006. Barnacle underwater attachment. In A. M. Smith and J. A. Callow, eds. *Biological Adhesives*, pp. 145–66. Springer-Verlag, Heidelberg.

Kaspari, M., N. A. Clay, J. Luca, S. P. Yanoviak, and A. Kay. 2014. Thermal adaptation generates a diversity of thermal limits in a rainforest ant community. *Global Change Biol.* doi: 10.111/gcb12750.

Katija, K. 2012. Biogenic inputs to ocean mixing. *J. Exp. Biol.* 215:1040–49.

Katz, R. W., G. S . Brush, and M. B. Parlange. 2005. Statistics of extremes: Modeling ecological disturbances. *Ecology* 86:1124–34.

Kawamata S. 1998. Effect of wave-induced oscillatory flow on grazing by a subtidal sea urchin *Strongylocentrotus nudus* (A. Agassiz). *J. Exp. Biol.* 224:31–48.

————. 2012. Experimental evaluation of anti-attachment effect of microalgal mats on grazing activity of the sea urchin *Strongylocentrotus nudus* in oscillating flow. *J. Exp. Biol.* 215:1464–71.

Kearney M., B. L. Phillips, C. R. Tracy, K. A. Christian, G. Betts, and W. P. Porter. 2008. Modelling species distributions without using species distributions: the cane toad in Australia under current and future climates. *Ecography* 31:423–34.

Kearney, M., and W. Porter. 2009. Mechanistic niche modelling: Combining physiological and spatial data to predict species' ranges. *Ecology Letters* 12:334–50.

Kessler, J. O. 1986. The external dynamics of swimming micro-organisms. *Progr. Phycol. Res.* 4:257–307.

Keulegan, G. H., and L. H. Carpenter. 1958. Forces on cylinders and plates in an oscillating fluid. *J. Res. Natl. Bur. Stand.* 60:423–40.

Kier, W. M., and K. K. Smith. 1985. Tongues, tentacles, and trunks: the biomechanics of movement in muscular-hydrostats. *Zool. J. Linn. Soc.* 83:307–24.

King, D., and O. L. Loucks. 1978. The theory of tree bole and branch form. *Rad. Environ. Biophysics.* 15:141–65.

Kiørboe, T. 2008. *A Mechanistic Approach to Plankton Ecology.* Princeton University Press, Princeton.

Klausmeier, C. A. 1999. Regular and irregular patterns in semiarid vegetation. *Science* 284:1826–28.

Kleypas, J. A., R. W. Buddemeier, D. Archer, J-P Gattuso, C. Langdon, and B. N. Opdyke. 1999. Geochemical consequences of increased atmospheric carbon dioxide on coral reefs. *Science* 284:118–20.

Knowlton, N., and J.B.C. Jackson. 2001. The ecology of coral reefs. In M. F. Bertness, S. D. Gaines, and M. E. Hay, eds. *Marine Community Ecology*, pp. 395–422. Sinauer, Sunderland, MA .

Koehl, M. A. R. 1977. Mechanical organization of cantilever-like organisms: sea anemones. *J. Exp. Biol.* 69:127–42.

———. 2011. Hydrodynamics of sniffing by crustaceans. In T. Breithaupt and M. Theil, eds. *Chemical Communication in Crustaceans*, pp. 85–102. Springer-Verlag, New York.

Koehl, M.A.R., and M. G. Hadfield. 2011. Soluble settlement cue in slowly moving water within coral reefs induces larval adhesion to surfaces. *J. Mar. Sys.* 49:75–88.

Kooijman, S.A.L.M. 2010. *Dynamic Energy Budget Theory for Metabolic Organisation*. Cambridge University Press, New York.

Krebs, C. J., R. Boonstra, S. Boutin, and A.R.E. Sinclair. 2001. What drives the 10-year cycle of snowshoe hares? *BioScience* 51:25–35.

Kuhn-Spearing, L. T., H. Kessler, E. Chateau, and R. Ballarini. 1996. Fracture mechanics of *Strombus gigas* conch shell: implications for the design of brittle laminates. *J. Mat. Sci.* 31:6583–94.

Kunze, H. B., S. G. Morgan, and K. M. Lwiza. 2013. Field test of the behavioral regulation of larval transport. *Mar. Ecol. Progr. Ser.* 477:71–87.

Kuo, E.S.L., and E. Sanford. 2009. Geographic variation in the upper thermal limits of an intertidal snail: Implications for climate envelope models. *Mar. Ecol. Progr. Ser.* 388:137–46.

LaBarbera, M. 1983. Why the wheels won't go. *Am. Nat.* 121:395–408.

Lai, J. H., J. C. del Alamo, J. Rodríguez-Rodrígues, and J. C. Lasheras. 2010. The mechanics of adhesive locomotion of terrestrial gastropods. *J. Exp. Biol.* 213:3920–33.

Lamb, H. 1932. *Hydrodynamics* (6th ed.). Dover, New York.

Lambers, H., F. S. Chapin, and T. L. Pons. 2008. *Plant Physiological Ecology*. Springer, New York.

Lauder, G. V., P.G.A. Madden, R. Mittal, H. Dong, and M. Bozkurttas. 2006. Locomotion with flexible propulsors: I. Experimental analysis of pectoral fin swimming in sunfish. *Bioinsp. Biomim.* 1:S25–34

Levin, S. A. 1992. The problem of pattern and scale in ecology: the Robert H. MacArthur Award Lecture. *Ecology* 73:1943–67.

Levin, S. A., T. M. Powell, and J. H. Steele, eds. 1993. *Patch Dynamics*. Springer-Verlag, New York.

Levin, S. A., H. C. Muller-Landau, R. Nathan, and J. Chave. 2003. The ecology and evolution of seed dispersal: a theoretical perspective. *Ann. Rev. Ecol. Evol. Syst.* 34:575–604.

Lima, F. P., N. P. Burnett, B. Helmuth, N. Kish, K. Aveni-DeForge, and D. S. Wethey. 2011. Monitoring the intertidal environment with biomimetic devices. In A. George, ed. *Biomimetic Based Applications*, pp. 499–502. InTech, Rijeka, Croatia

Lin, Q., D. Gourdon, C. Sun, N. Holten-Andersen, T. H. Anderson, J. H. Waite, and J. N. Israelachvili. 2007. Adhesion mechanisms of the mussel foot proteins mfp-1 and mfp-3. *Proc Natl. Acad. Sci. (USA)* 104:3782–86.

Luckhurst, B. E., and K. Luckhurst. 1978. Analysis of the influence of substrate variables on coral reef fish communities. *Mar. Biol.* 49:317–23.

MacArthur, R. H., and R. Levins. 1967. The limiting similarity, convergence, and divergence of coexisting species. *Am. Nat.* 101:377–85.

Mach, K. J. 2009. Mechanical and biological consequences of repetitive loading: Crack initiation and fatigue failure in the red macroalga *Mazzaella*. *J. Exp. Biol.* 212:961–76.

Mach, K. J., B. B. Hale, M. W. Denny, and D. V. Nelson. 2007. Death by small forces: A fracture and fatigue analysis of wave-swept macroalgae. *J. Exp. Biol.* 210:2231–43.

Mach, K. J., D. V. Nelson, and M. W. Denny. 2007. Techniques for predicting the lifetimes of wave-swept macroalgae: A primer on fracture mechanics and crack growth. *J. Exp. Biol.* 210:2213–30.

Mach, K. J., S. K. Tepler, A. V. Staaf, J. C. Bohnhoff, and M. W. Denny. 2011. Failure by fatigue in the field: A model of fatigue breakage for the macroalga *Mazzaella*, with validation. *J. Exp. Biol.* 214:1571–85.

Madin, J. S. 2005. Mechanical limitations of reef corals during hydrodynamic disturbances. *Coral Reefs* 24:630–35.

Madin, J. S., K. P. Black, and S. R. Connolly. 2006. Scaling water motion on coral reefs: From regional to organismal scales. *Coral Reefs* 25:635–44.

Madin, J. S., and S. T. Connolly. 2006. Ecological consequences of major hydrodynamic disturbances on coral reefs. *Nature* 444:477–80.

Madin, J. S., M. J. O'Donnell, and S. R. Connolly. 2008. Climate-mediated changes to post-disturbance coral assemblages. *Biol. Lett.* 4:490–93.

Magariyama Y., S. Sugiyama, N. Muramoto, Y. Maekawa, I. Kawagishi, Y. Imae, and S. Kudo. 1994. Very fast flagellar rotation. *Nature* 371:752.

Maor, E. 1994. *e: The Story of a Number*. Princeton University Press, Princeton.

Marsh, R. L., and J. M. Olson. 1994. Power output of scallop adductor muscle during contractions replicating the *in vivo* mechanical cycle. *J. Exp. Biol.* 193:139–56.

Marshall, D. J., and T. Chua. 2012. Boundary layer convective heating and thermoregulatory behavior during aerial exposure in the rocky eulittoral fringe snail *Echinolittorina malaccana*. *J. Exp. Mar. Biol. Ecol.* 430–31:25–31.

Marshall, D. J., C. D. McQuaid, and G. A. Williams. 2010. Non-climatic thermal adaptation: implications for species responses to climate warming. *Biol. Lett.* 6:669–73.

Martin, R. B. 2003. Fatigue microdamage as an essential element of bone mechanics and biology. *Calcif. Tissue Int.* 73:101–7.

Martin, T. L., and R. B. Huey. 2008. Why "suboptimal" is optimal: Jensen's inequality and ectotherm thermal preference. *Amer. Nat.* 171:E102–E118.

Martinez, M. M. 2001. Running in the surf: Hydrodynamics of the shore crab *Grapsus tenuicrustatus*. *J. Exp. Biol.* 204:3097–3112.

Martone, P. T., and M. W. Denny. 2008. To break a coralline: Wave forces limit the size and survival of articulated fronds. *J. Exp. Biol.* 211:3433–41.

Martone, P. T., M. Boller, I. Burgert, J. Dumais, J. Edwards, K. Mach, N. Rowe, M. Rueggeberg, R. Seidel, and T. Speck. 2010. Mechanics without muscle: Biomechanical inspiration from the plant world. *Integr. Comp. Biol.* 50:888–907.

Martone, P. T., L. Kost, and M. Boller. 2012. Drag reduction in wave-swept macroalgae: Alternative strategies and new predictions. *Am. J. Bot.* 99:806–15.

Massel, S .R. and T. J. Done. 1993. Effects of cyclone waves on massive coral assemblages on the Great Barrier Reef: Meteorology, hydrodynamics and demography. *Coral Reefs* 12:153–66.

Massey, B. S. 1989. *Mechanics of Fluids* (6th ed.). Chapman and Hall, London.

McNair, J. M., J. D. Newbold, and D. D. Hart. 1997. Turbulent transport of suspended particles and dispersing benthic organisms: How long to hit bottom? *J. Theor. Biol.* 188:29–52.

McQuaid, C. D., and P. A. Scherman. 1982. Thermal stress in a high shore intertidal environment: Morphological and behavioural adaptations of the gastropod *Littorina africana*. In G. Chelazzi and M. Vannini, eds. *Behavioral Adaptation to Intertidal Life*, pp. 213–24. Plenum Press, New York.

Meade, K. S., and M.A.R. Koehl. 2000. Stomatopod antennule design: The asymmetry, sampling efficiency and ontogeny of olfactory flicking. *J. Exp. Biol.* 203:3795–3808.

Melbourne, B. A., and P. Chesson. 2006. The scale transition: Scaling up population dynamics with field data. *Ecology* 87:1476–88.

Melbourne, B. A., A. L. Sears, M. J. Donahue, and P. Chesson. 2005. Applying scale transition theory to metacommunities in the field. In M. Holyoak, M.A. Leibold, and R.D. Holt, eds. *Metacommunities: Spatial Dynamics and Ecological Communities*, pp. 307–30. University of Chicago Press, Chicago.

Meyers, M. A., P-Y Chen, A.Y-M. Lin, and Y. Seki. 2008. Biological materials: Structure and mechanical properties. *Progr. Mat. Sci.* 53:1–206.

Middleton, G. V., and J. B. Southard. 1984. *Mechanics of Sediment Movement*. Lecture notes for short course no. 3. *Society of Economic Paleontologists and Mineralogists*. Given in Providence RI. March 13–14.

Miller, L. P. 2008. Life on the edge: morphological and behavioral adaptations for survival on wave-swept shores. PhD diss, Stanford University.

Miller, L. P., C.D.G. Harley, and M. W. Denny. 2009. The role of temperature and desiccation stress in limiting the local-scale distribution of the owl limpet, *Lottia gigantea*. *Func. Ecol.* 23:756–67.

Miller, S. L. 1974. Adaptive design of locomotion and foot form in prosobranch gastropods. *J. Exp. Mar. Biol. Ecol.* 14:99–156.

Mitchell, J. G., L. Pearson, A. Bonazinga, S. Dillon, H. Khuri, and R. Paxinos. 1995. Long lag times and high velocities in the motility of natural assemblages of marine bacteria. *Appl. Environ. Microbiol.* 61:877–82.

Mitchell, S. (translator). 1998. *Tao Te Ching*. Harper Collins, New York.

Monteith, J. L. and M. H. Unsworth. 2008. *Principles of Environmental Physics* (3rd ed.). Academic Press, New York and London.

Moon, F. C. 1992. *Chaotic and Fractal Dynamics: An Introduction for Applied Scientists and Engineers*. Wiley Interscience, New York.

Morison, J. R., M. P. O'Brien, J. W. Johnson, and S. A. Schaaf. 1950. The force exerted by surface waves on piles. *Petroleum Trans. AIME* 189:149–54.

Mumby, P. J. 2006. The impact of exploiting grazers (Scaridae) on the dynamics of Caribbean coral reefs. *Ecol. Appl.* 16:747–69.

Munk, P., and T. Kiørboe. 1985. Feeding behavior and swimming activity of larval herring (*Clupea harengus* L.) in relation to density of copepod nauplii. *Mar. Ecol. Progr. Ser.* I24:15–21.

Muschenheim, D. K. 1987. The dynamics of near-bed seston flux and suspension-feeding benthos. *J. Mar. Res.* 45:473–96.

Muschenheim, D .K., and C. R. Newell. 1992. Utilization of seston flux over a mussel bed. *Mar. Ecol. Prog. Ser.* 85:131–36.

Nachtigall, W. 1974. *Biological Mechanisms of Attachment*. Springer-Verlag, New York.

Nathan, R., W. M. Getz, E. Revilla, M. Holyoak, R. Kadmon, D. Saltz, and P. E. Smouse. 2008. A movement ecology paradigm for unifying organismal movement research. *Proc. Natl. Acad. Sci. (USA)* 105:19052–59.

Niklas, K. J. 1982. Simulated and empiric wind pollination patterns of conifer ovulate cones. *Proc. Natl. Acad. Sci. (USA)* 79:510–14.

Nisbet, R. M., M. Jusup, T. Klanjscek, and L. Pecquerie. 2012. Integrating dynamic energy budget (DEB) theory with traditional bioenergetics models. *J. Exp. Biol.* 215:892–902.

Nobel, P. S. 1991. *Physicochemical and Environmental Plant Physiology*. Academic Press, San Diego.

Norberg, R. A. 1973. Autorotation, self-stability, and structure of single-winged fruits and seeds (samaras) with comparative remarks on animal flight. *Biol. Rev.* 48:561–96.

Norton, T. A. and R. Fetter. 1981. The settlement of *Sargassum muticum* propagules in stationary and flowing water. *J. Mar. Biol. Assoc. UK* 61:929–40.

Nowell, A.R.M., and M. Church. 1979. Turbulent flow in a depth-limited boundary layer. *J. Geophys. Res.* 84:4816–24.

Nowell, A.R.M., and P. A. Jumars. 1984. Flow environments of aquatic benthos. *Annu. Rev. Ecol. Syst.* 15:303–28.

O'Donnell, M. J., M. N. George, and E. Carrington. 2013. Mussel byssus attachment weakened by ocean acidification. *Nature Climate Change* 3:587–90.

Okubo, A., and S. Levin, eds. 2001. *Diffusion and Ecological Problems: Modern Perspectives* (2nd ed.). Springer-Verlag, New York.

O'Riordan, C. A., S. G. Monismith, and J. R. Koseff. 1995. The effect of bivalve ex-current jet dynamics on mass transfer in a benthic boundary layer. *Limnol. Oceanogr.* 40:330–44.

Orr, F. M., L. E. Scriven, and A. P. Rivas. 1975. Pendular rings between solids: Meniscus properties and capillary forces. *J. Fluid Mech.* 67:723–42.

Ortlepp, C., and J. M. Gosline. 2008. The scaling of safety factor in spider draglines. *J. Exp. Biol.* 211:2832–40.

Padilla, D. K. 1993. Rip stop in marine algae: Minimizing the consequences of herbivore damage. *Evol. Ecol.* 7:634–44.

Paine, R. T. 1966. A note on trophic complexity and community stability. *Am. Nat.* 103:91–93.

Paine, R. T. 1974. Intertidal community structure: Experimental studies of the relationship between a dominant competitor and its principal predator. *Oecologia* 15:93–120.

Paine, R. T., and S. A. Levin. 1981. Intertidal landscapes: disturbance and the dynamics of pattern. *Ecol. Monogr.* 51:145–78.

Paine, R. T., M. J. Tegner, and E. A. Johnson. 1998. Compounded perturbations yield ecological surprises. *Ecosystems* 1:535–45.

Parker, A. R., and C. R. Lawrence. 2001. Water capture by a desert beetle. *Nature* 414:33–34.

Pascual, M., and F. Guichard. 2005. Criticality and disturbance in spatial ecological systems. *Trends Ecol. Evol.* 20:88–95.

Patek, S. N., and R. L. Caldwell. 2005. Extreme impact cavitation forces of a biological hammer: Strike forces of the peacock mantis shrimp *Odontodactylus scyllarus*. *J. Exp. Biol.* 208:3655–64

Patek, S. N., W. L. Korff, and R. L. Caldwell. 2004. Deadly strike mechanism of a mantis shrimp. *Nature* 428:819–20.

Patek, S. N., B. N. Nowroozi, J. E. Baio, R. L. Caldwell, and A. P. Summers. 2007. Linkage mechanics and power amplification of the mantis shrimp's raptorial strike. *J. Exp. Biol.* 210:3677–88.

Peck, L. S., M. S. Clark, S. A. Morley, A. Massey, and H. Rossetti. 2009. Animal temperature limits and ecological relevance: Effects of size, activity and rates of change. *Funct. Ecol.* 23:248–56.

Pennycuick, C. J. 1990. Predicting wingbeat frequency and wavelength of birds. *J. Exp. Biol.* 150:171–85.

Petraitis, P. S., E. T. Methratta, E. C. Rhile, N. A. Vidargas, and S. R. Dudgeon. 2009. Experimental confirmation of multiple community states in a marine ecosystem. *Oecologia* 161:139–48.

Pickett, S.T.A., and P. S. White. 1985. *The Ecology of Natural Disturbance and Patch Dynamics.* Academic Press, New York.

Pincebourde, S., E. Sanford, and B. Helmuth. 2008. Body temperature during low tide alters the feeding performance of a top intertidal predator. *Limnol. Oceanogr.* 53:1562–73.

———. 2009. An intertidal sea star adjusts thermal inertia to avoid extreme body temperatures. *Am. Nat.* 174:890–97.

Pocius, A. V. 2002. *Adhesion and Adhesives Technology.* (2nd ed.). Hanser Publishers, Munich.

Podolsky, R. D., and R. B. Emlet. 1993. Separating the effects of temperature and viscosity on swimming and water movement by sand dollar larvae (*Dendraster excentricus*). *J. Exp. Biol.* 176:207–21.

Potin, P., and C. LeBlanc. 2006. Phenolic-based adhesives of marine brown algae. In A. M. Smith and J. A. Callow, eds. *Biological Adhesives*, pp. 105–24. Springer-Verlag, Berlin and Heidelberg.

Potter, K., G. Davidowitz, and H. A. Woods. 2009. Insect eggs protected from high temperatures by limited homeothermy of plant leaves. *J. Exp. Biol.* 212:3448–54.

Power, M. E., D. Tilman, J. A. Estes, B. A. Menge, W. J. Bond, L. S. Mills. G. Daily, J. C. Castilla, J. Lubchenco, and R. T. Paine. 2010. Challenges in the quest for keystones. *BioSci* 46:609–20.

Priestley, M. B. 1981. *Spectral Analysis and Time Series.* Academic Press, New York.

Prosser, C. L., and F. A. Brown. 1961. *Comparative Animal Physiology* (2nd ed.), WB Saunders, Philadelphia.

Quéré, D. 2008. Wetting and roughness. *Annu. Rev. Mater. Res.* 38:71–99.

Quéré, D., and M. Reyssat. 2008. Non-adhesive lotus and other hydrophobic materials. *Phil. Trans. R. Soc. A* 366:1539–56.

Rietkerk, M. and J. van de Koppel. 2008. Regular pattern formation in real ecosystems. *Trends Ecol. Evol,* 23:169–75.

Rietkerk, M., M. C. Boerlijst, F. van Langevelde, R. HilleRisLambers, J. van de Koppel, L. Kumar, H.H.T. Prins, and A. M. de Roos. 2002. Self-organization of vegetation in arid ecosystems. *Amer. Nat.* 160:524–30.

Riisgård, H. U. and P. Larsen. 2010. Particle capture in suspension-feeding invertebrates. *Mar. Ecol. Progr. Ser.* 418:255–93.

Riley, G. A., H. Stommel, and D. F. Bumpus. 1949. Quantitative ecology of the plankton of the western North Atlantic. *Bull. Bingham Oceanographic Collection Yale University.* 12:1–169.

Roark, R. J., and W. C. Young. 1975. *Formulas for Stress and Strain.* McGraw-Hill, New York.

Rominger, J. T. and H. M. Nepf. 2014. Effects of blade flexural rigidity on drag force and transfer rates in model blades. *Limnol. Oceanogr.* 59:2028–41.

Ruel, J. J., and M.,P. Ayres. 1999. Jensen's inequality predicts effects of environmental variation. *Trends Ecol. Evol.* 14:361–66.

Sanford, E. 2002. Water temperature, predation, and the neglected role of physiological rate effects in rocky intertidal communities. *Integr. Comp. Biol.* 42:881–91.

Sarpkaya, T. 1976. In-line and transverse forces on smooth and sand-roughened cylinders in oscillating flow at high Reynolds numbers. Report No. NPS-69SL76062, Naval Postgraduate School, Monterey CA.

Scheffer, M., J. Bascompte, W. A. Brock, V. Brovkin, S. R. Carpenter, V. Dakos, H. Held, E. H. van Nes, M. Rietkerk, and G. Sugihara. 2009. Early warning signals for critical transitions. *Nature* 461:53–59.

Schlichting, H. and K. Gersten. 2000. *Boundary Layer Theory* (8th ed.). Springer-Verlag, New York.

Schmidt-Nielsen, K. 1984. *Scaling: Why Animal Size Is So Important.* Cambridge University Press, Cambridge, MA.

———. 1997. *Animal Physiology: Adaptation and Environment* (5th ed.). Cambridge University Press, Cambridge, MA.

Schoener, T. W. 1986. Mechanistic approaches to community ecology: A new reductionism. *Amer. Zool.* 26:81–106.

Sebens, K. P. 1982. The limits to indeterminate growth: An optimal size model applied to passive suspension feeders. *Ecology* 63:209–22.

Seymour, R. S. and P. Schultze-Motel. 1996. Thermoregulating lotus flowers. *Nature* 383:305.

Sharp, N.C.C. 1997. Timed running speed of a cheetah (*Acinonyx jubatus*). *J. Zool.* 241:493–94.

Shaw, M. T., and W. J. McKnight. 2005. *Introduction to Polymer Viscoelasticity.* Wiley Interscience, New York.

Sherwood, S. C., and M. Huber. 2010. An adaptability limit to climate change due to heat stress. *Proc. Natl. Acad. Sci.* 107:9552–55.

Silvester, N. R., and M. A. Sleigh. 1984. Hydrodynamic aspects of particle capture by *Mytilus*. *J. Mar. Biol. Assoc. UK* 64:859–79.

Sinclair, A.R.E., J. M. Gosline, G. Holdsworth, C. J. Krebs, S. Boutin, J.N.M. Smith, R. Boonstra, and M. Dale. 1993. Can the solar cycle and climate synchronize the snowshoe hare cycle in Canada? Evidence from tree rings and ice cores. *Am. Nat.* 141:173–98.

Sleigh, M. A., and J. R. Blake. 1977. Methods of ciliary propulsion and their size limitations. In T. J. Pedley, ed. *Scale Effects in Animals Locomotion*, pp. 243–55. Academic Press, London.

Smith, A. M. 1991a. Negative pressure generated by octopus suckers: A study of the tensile strength of water in nature. *J. Exp. Biol.* 157:257–71.

———. 1991b. The role of suction in the adhesion of limpets. *J. Exp. Biol.* 161:151–69.

———. 2002. The structure and function of adhesive gels from invertebrates. *Integr. Comp. Biol.* 42:1164–71.

———. 2006. The biochemistray and mechanics of gastropod adhesive gels. In A. M. Smith and J. A. Callow, eds. *Biological Adhesives*, pp. 167–82. Springer-Verlag, Berlin.

Smith, A. M., T. J. Quick, and R. L. St. Peter. 1999. Differences in the composition of adhesive and non-adhesive mucus from the limpet *Lottia limatula*. *Biol. Bull.* 196:34–44.

Sokal, R. R., and F. J. Rohlf. 2012. *Biometry* (4th ed.). W. H. Freeman and Company, New York.

Solé, R. V., and J. Bascompte. 2006. *Self-Organization in Complex Ecosystems*. Princeton University Press, Princeton.

Somero, G. N. 2002. Thermal physiology and vertical zonation of intertidal animals: Optima, limits, and cost of living. *Integr. Comp. Biol.* 42:780–89.

Sousa, W. P. 1979. Experimental investigations of disturbance and ecological succession in a rocky intertidal algal community. *Ecol. Monogr.* 49:227–54.

Steinmann, T., J. Casas, G. Krijnen, and O. Dangles. 2006. Air-flow sensitive hairs: Boundary layers in oscillatory flows around arthropod appendages. *J. Exp. Biol.* 209:4398–4408.

Steele, J. H. 1985. A comparison of terrestrial and marine ecological systems. *Nature* 313:355–58.

Stewart, R. J., J. C. Weaver, D. E. Morse, and J. H. Waite. 2004. The tube cement of *Phragmatopoma californica*: A solid foam. *J. Exp. Biol.* 207:4727–34.

Suter, R. B. 2003. Trichobothrial mediation of an aquatic escape response: Directional jumps by the fishing spider, *Dolomedes triton*, foil frog attacks. *J. Insect. Sci.* 3:1–7.

Taylor, C. R., N. C. Heglund, and G. M. Maloiy. 1982. Energetics and mechanics of terrestrial locomotion: I. Metabolic energy consumption as a function of speed and body size in birds and mammals. *J. Exp. Biol.* 97:1–21.

Tautz, J., and M. Rostàs. 2008. Honeybee buzz attenuates plant damage by caterpillars. *Curr. Biol.* 18:R1125–26.

Taylor, D., J. G. Hazenburg, and T. C. Lee. 2007. Living with cracks: damage and repair in human bone. *Nature Materials* 6:263–68.

Telling, R. H., C. J. Pickard, M. C. Payne, and J. E. Field. 2000. Theoretical strength and cleavage of diamond. *Phys. Rev. Lett.* 84:5160–63.

Taylor, J.R.A., and S. N. Patek. 2010. Ritualized fighting and biological armor: The impact mechanics of the mantis shrimp telson. *J. Exp. Biol.* 213:3496–3504.

Thompson, M. C. 1961. The flight speed of a red-breasted merganser. *Condor* 63:265.

Tobalske, B. W., T. L. Hedrick, K. P. Dial, and A. A. Biewener. 2003. Comparative power curves in bird flight. *Nature* 421:363–66.

Treloar, L.R.G. 1975. *Physics of Rubber Elasticity*. Clarendon Press, Oxford.

Triblehorn, J. D., and D. D. Yager. 2006. Wind generated by an attacking bat: Anemometric measurements and detection by the praying mantis cercal system. *J. Exp. Biol.* 209:1430–40.

Tucker, V. A., T. J. Cade, and A. E. Tucker. 1998. Diving speeds and angles of a gyrfalcon (*Falco rusticolus*). *J. Exp. Biol.* 201:2061–70.

Tunnicliffe, V. 1981. Breakage and propagation of the stony coral *Acropora cervicornis*. *Proc. Natl. Acad. Sci. (USA)* 78:2427–31.

Turing, A. M. 1952. The chemical basis of morphogenesis. *Phil Trans. Roy. Soc. B* 237:37–72.

Tuteja, A., W. Choi, G. H. McKinley, R. E. Cohen, and M. F. Rubner. 2008. Design parameters for superhydrophobicity and superoleophobicity. *MRS Bulletin* 33:752–58.

van de Koppel, J., T. J. Bouma, and P.M.J. Herman. 2012. The influence of local- and landscape-scale processes on spatial self-organization in estuarine systems. *J. Exp. Biol.* 215:962–67.

van de Koppel, J., J. C. Gascoigne, G. Theraulaz, M. Rietkerk, W. M. Mooij, and P.M.J. Herman. 2008. Experimental evidence for spatial self-organization and its emergent effects in mussel bed ecosystems. *Science* 322:739–42.

van de Koppel, J., M. Rietkerk, N. Dankers, and P.M.J. Herman. 2005. Scale-dependent feedback and regular spatial patterns in young mussel beds. *Amer. Nat.* 165:E66–E77.

van de Koppel, J., T. van der Heide, A. Altieri, B. K. Erikson, T. Bouma, H. Olff, and B. Silliman. 2015. Long-distance interactions regulate the structure and resilience of coastal ecosystems. *Annu.Rev. Mar. Sci.* 7:139–58.

Vasseur, D. A., and P. Yodzis. 2004. The color of environmental noise. *Ecology* 85:1146–52.

Verdugo, P., I. Deyrup-Olsen, M. Aitken, M. Villalon, and D. Johnson. 1987. Molecular mechanism of mucus secretion: I. The role of intragranular charge shielding. *J. Dent. Res* 66:506–8.

Videler, J. J. 1993. *Fish Swimming*. Chapman and Hall Fish and Fisheries, vol. 10. Chapman and Hall, London.

Vieira, N.K.M., W. H. Clements, L. S. Guevara, and B. F. Jacobs. 2004. Resistance and resilience of stream insect communities to repeated hydrological disturbances after a wildfire. *Freshwater Biol.* 49:1243–59.

Visser, A. W., and U. H. Thygesen. 2003. Random motility of plankton: Diffusive and aggregative contributions. *J. Plankton Res.* 25:1157–68.

Viswanathan, G. M., M.G.E. da Luz, E. P. Raposo, and H. E. Stanley. 2011. *The Physics of Foraging: An Introduction to Random Searches and Biological Encounters*. Cambridge University Press, New York.

Vogel, H., G. Czihak, P. Chang, and W. Wolf. 1982. Fertilization kinetics of sea urchin eggs. *Math. Biosci.* 58:189–216.

Vogel, S. 1983. How much air flows through a silkmoth's antenna? *J. Insect Physiol.* 29:597–602.

———. 1994. *Life in Moving Fluids* (2nd ed.). Princeton University Press, Princeton.

———. 2001. *Prime Mover: A Natural History of Muscle*. W. W. Norton & Co., New York.

———. 2003. *Comparative Biomechanics: Life's Physical World*. Princeton University Press, Princeton.

———. 2009. *Glimpses of Creatures in Their Physical World*. Princeton University Press, Princeton.

Vogel, S., and C. Loudon. 1985. Fluid mechanics of the thallus of an intertidal red alga, *Halosaccion glandiforme*. *Biol. Bull.* 168:161–74.

Vreeland, V., J. H. Waite, and L. Epstein. 1998. Polyphenols and oxidases in substratum adhesion by marine algae and mussels. *J. Phycol.* 34:1–8.

Wainwright, S. A., W. D. Biggs, J. D. Currey, and J. M. Gosline. 1976. *Mechanical Design in Organisms*. Princeton University Press, Princeton.

Waite, A, A. Fisher, P. A. Thompson, and P. J. Harrison. 1997. Sinking rate versus cell volume relationships illuminate sinking rate control mechanisms in marine diatoms. *Mar. Ecol. Prog. Ser.* 157:97–108.

Waite, J. H., and C. C. Broomell. 2012. Changing environments and structure-property relationships in marine biomaterials. *J. Exp. Biol.* 215:873–83.

Wake, W. C. 1982. *Adhesion and the Formulation of Adhesives*. Applied Science Publishers, London.

Wakeling, J. M., and I. A. Johnston. 1998. Muscle power output limits fast-start performance in fish. *J. Exp. Biol.* 201:1505–26.

Weaver, J. C., G. W. Milliron, A. Miserez, K. Evans-Lutterodt, S. Herrera, I. Gallana, W. J. Mershon, et al. 2012. The stomatopod dactyl club: A formidable damage-tolerant biological hammer. *Science* 336:1275–80.

Webb, P. W. 1975. Hydrodynamics and energetics of fish propulsion. *Bull. Fish. Res. Bd., Canada* 190:1–158.

Weihs, D. 1974. Energetic advantages of burst swimming in fish. *J. Theoret. Biol.* 48:215–29.

Weiner, J. 1994. *The Beak of the Finch*. Vintage, New York.

Weiner, S. W., and L. Addadi. 1997. Design strategies in mineralized biological materials. *J. Mater. Sci.* 7:689–702.

Wenzel, R. N. 1936. Resistance of solid surfaces to wetting by water. *Industrial and Engineering Chemistry* 28:988–94.

Wethey, D. S. 2002. Biogeography, competition, and microclimate: the barnacle *Chthamalus fragilis* in New England. *Integr. Comp. Biol.* 42:872–80.

Williams, T. M. 1999. The evolution of cost-efficient swimming in marine mammals: Limits in energetic optimization. *Phil Trans. Roy Soc. B* 354:193–201.

Wood, C. J. 1972. The flight of albatrosses (a computer simulation). *Ibis* 115:244–56.

Wu, T. 1977. Introduction to the scaling of aquatic animals. In T. J. Pedley, ed. *Scale Effects in Animal Locomotion*, pp. 203–32. Academic Press, NY.

Yao, C. L., and G. N. Somero. 2012. The impact of acute temperature stress on hemocytes of invasive and native mussles (*Mytilis galloprovincialis* and *M. californianus*): DNA damage, membrane integrity, apoptosis and signaling pathways. *J. Exp. Biol.* 215:4267–77.

Young, I. R., S. Zieger, and A. V. Babinin. 2011. Global trends in wind speed and wave height. *Science* 332:451–55.

Yu, J., W. Wei, E. Danner, R. K. Ashley, J. N. Israelachvili, and J. H. Waite. 2011. Mussel protein adhesion depends on interprotein thiol-mediated redox modulation. *Nature Chemical Biology* 7:588–90.

Zack, T .I., T. Claverie, and S. N. Patek. 2009. Elastic energy storage in the mantis shrimp's fast predatory strike. *J. Exp. Biol.* 212:4002–9.

Zar, J. H. 1999. *Biostatistical Analysis* (4th ed.). Prentice Hall, Upper Saddle River, NJ.

Zheng, Y., X. Gao, and L. Jiang. 2007. Directional adhesion of superhyrdrophobic butterfly wings. *Soft Matter* 3:178–82.

Zimmer, R. K., and J. A. Riffell. 2011. Sperm chemotaxis, fluid shear, and the evolution of sexual reproduction. *Proc. Natl. Acad. Sci. (USA)* 108:13200–5.

Zumdahl, S. S., and S. A. Zumdahl. 2003. *Chemistry* (6th ed.). Houghton Mifflin, New York.

Symbol Index

Symbol	Definition	Page Where First Used
$\mathcal{D}_H$,	heat diffusivity/thermal diffusivity	200
DMT	dislodgement mechanical threshold	331
e	the natural number, 2.7183...	41
	concentration of eggs	372
E	elastic modulus	261
E_f	fiber elastic modulus	274
E_L	loss modulus	264
E_m	matrix elastic modulus	274
E_S	storage modulus	263
$E_{\tan}$	tangent elastic modulus	261
$E_{\parallel}$	modulus of a composite with fibers parallel to the load	274
$E_{\perp}$	modulus of a composite with fibers perpendicular	274
$\exp(x)$	e^x, here, for some arbitrary exponent x	100
f	frequency	29
f_d	frequency of damped oscillation	351
f_f	fundamental frequency	386
f_n	natural frequency	347
f_s	frequency in space	385
F	force	21
F_a	adhesion force	302
F_A	acceleration reaction	136
F_{AM}	added mass force	136
F_D	drag force	34
F_{ext}	external force	353
F_g	weight	21
F_{IL}	in-line force	138
F_L	lift force	139
F_M	force required to accelerate a mass	136
F_V	viscous drag	122
$F_{\perp}$	component of force perpendicular to the moment arm	349
$\mathcal{F}$	flux	68
$\mathcal{F}_H$,	heat flux	200
FI	flatness index	149
g	acceleration of gravity, approximately 9.8 m s^{-2}	21
G	shear modulus	263
$G_{\tan}$	tangent shear modulus	263
h	Planck's constant, 6.626×10^{-34} J Hz^{-1}	29
h_c	convective heat transfer coefficient	206
h_m	mass transfer coefficient	214
h_r	relative humidity	217
h	height	156

Symbol	Definition	Page Where First Used
H	rate of heat emitted by infrared radiation	34
	heat budget components with various subscripts:	202
	$\quad H_{sw}$ short-wave radiative heat transfer	
	$\quad H_{lw}$ long-wave radiative heat transfer	
	$\quad H_{cd}$ conductive heat transfer	
	$\quad H_{cv}$ convective heat transfer	
	$\quad H_{ec}$ evaporative heat transfer	
	$\quad H_{met}$ metabolic heating	
I	second moment of area	323
	a circuit's amperage	382
I	solar irradiance	204
$\mathcal{J}$	advance ratio, the ratio of an animal's airspeed to the tangential speed of its wings	181
J	polar second moment of area (the torsional equivalent to I, the second moment of area)	
J_m	rotational moment of inertia	22
k	various coefficients (various subscripts)	
K	population carrying capacity	41
K_c	fracture toughness	289
$\mathcal{K}$	thermal conductivity	201
ℓ	a length	16
ℓ_c	characteristic length (e.g., length along the axis of flow)	122
$\ell_\perp$	perpendicular distance	21
L	length	208
m	mass	21
M	moment	21
	species-specific heat coefficient	212
M_g	weight-imposed moment	154
M_o	overturning moment imposed by lift	154
M_t	total moment	342
$\mathcal{M}$	mass-specific metabolic rate	189
	potential metabolic rate	42
n	total number (of steps, of particles, of links in a polymer chain, etc.)	47
N	population size	41
	total number	59
$\mathcal{N}$	Avagadro's number, approximately 6.02×10^{23}	63
Nu	Nusselt number	207
$\bar{x}$	'overbar' indicates the mean of variable x	17
p	pressure	25
	constant low probability per unit time	370
P	power	162
$P(x)$	probability of a specified event x occurring	370
Q_{10}	the relative change in rate accompanying a $10°C$ increase in temperature	196
r	intrinsic rate of increase	41
	radius of curvature	321

Symbol	Definition	Page Where First Used
R	net rate energy can be acquired	190
	rate of gamete contact	372
	rate of change in a population	374
	a circuit's resistance	382
	thermal response function	414
Re	Reynolds number	77
Re_f	frond Reynolds number	152
Re_*	roughness Reynolds number	94
$\mathcal{R}_{lw}$	infrared radiance	229
$\Re$	universal gas constant, 8.31 J mol^{-1} K^{-1}	63
s	instantaneous speed	18
	concentration of sperm	372
S	sensitivity	128
	power spectrum	399
S_T	scale transition coefficient	368
Sc	Schmidt number	111
t	time	17
	wall thickness of a cylinder	324
t_n	natural period	347
t_r	thermal response time	223
T	temperature	34
T_{wb}	wet-bulb temperature	217
u	translational velocity in the x direction or speed	18
u_t	terminal sinking speed	124
u_*	shear velocity	66
U	velocity as a function of depth	403
v	translational velocity in the y direction	18
V	volume	25
	a circuit's voltage	382
V_f	volume fraction of fiber material	274
V_m	volume fraction of matrix material	274
w	translational velocity in the z direction	18
W	width	208
W	work, energy	26
$\mathcal{W}$	wing loading	178
x	distance along x-axis	14
y	distance along y-axis	14
z	distance along z-axis	14
z_d	damping depth	229
α	absorptivity	204
	component wave amplitude	387
β	component wave amplitude	387
	exponent of decay in a spectral function	407
γ	shear strain	72
	view factor	232
γ_s	surface energy	305
Γ	circulation	139

Symbol	Definition	Page Where First Used
δ	step length in a random walk	56
	boundary-layer thickness	105
δ_c	concentration boundary layer thickness	111
Δx	a finite change, here, in arbitrary variable x	16
ε	nominal strain, also called engineer's strain	258
ε_t	true strain	258
ϵ	eddy or turbulence diffusivity	96
	emissivity	229
	time	357
	random amount of food delivered by turbulence	417
η	wave amplitude	383
θ	an angle	15
θ_a	advancing contact angle	316
θ_r	retreating contact angle	316
θ_{stat}	static contact angle	307
$\Delta\theta$	contact angle hysteresis	316
ϑ	dimensionless variable	114
κ	von Kármán's constant	98
λ	wavelength	30
Λ_{wat}	latent heat of evaporation	214
μ	dynamic viscosity	63
ν	kinematic viscosity	79
	Poisson's ratios	260
π	ratio between a circle's circumference and its radius, approximately 3.1415	63
ρ	density	15
ρ_e	effective density	154
ρ_f	fluid density	25
ρ_w	water density	25
ρ_ε	strain energy density	265
ρ_{frac}	fracture energy density	285
σ	standard deviation, σ^2 is variance	58
	stress	257
σ_a	adhesive tenacity, the force per area an adhesive can resist	302
σ_n	nominal stress, force per nominal area, also called engineer's stress	258
σ_{SB}	Stefan-Boltzmann constant, $5.67 \times 10^{-8}\,\mathrm{W\,m^{-2}\,K^{-4}}$	229
σ_t	true stress, force per instantaneous area	257
τ	characteristic interval	56
	shear stress	76
ϕ	an angle	142
ϕ_f	phase difference between driving force and response	353
φ	frequency of a sinusoidally varying force	353
Φ	force per volume	89
ω	angular velocity	19

Subscript

0	initial or amplitude
b	body or object
bl	boundary layer
brk	breaking
crit	critical
dir	direct
dif	diffuse
df	dynamic friction
e	effective
eq	equilibrium
lw	long-wave
LA	liquid-air interface
max	maximum
min	minimum
net	outcome minus cost
opt	optimum
r	rock
ref	reference
sf	static friction
std	standard
sw	short-wave
SA	solid-air interface
SL	solid-liquid interface
turb	turbulent
var	variance
x	in the x direction
y	in the y direction
z	in the z direction
∞	mainstream

Author Index

Subject Index

Page numbers in *italics* refer to figures.